ADVANCED MECHANICS OF COMPOSITE MATERIALS

ADVANCED MECHANICS OF COMPOSITE MATERIALS

Valery V. Vasiliev

Distinguished Professor
Department of Mechanics and Optimization of
Processes and Structures
Russian State University of Technology, Moscow

Evgeny V. Morozov

Professor of Mechanical Engineering
Division of Engineering Science & Technology
The University of New South Wales Asia, Singapore

ELSEVIER

Amsterdam • Boston • Heidelberg • London • New York • Oxford
Paris • San Diego • San Francisco • Singapore • Sydney • Tokyo

Elsevier
The Boulevard, Langford Lane, Kidlington, Oxford OX5 1GB, UK
Radarweg 29, PO Box 211, 1000 AE Amsterdam, The Netherlands

First edition 2007

British Library Cataloguing in Publication Data
A catalogue record for this book is available from the British Library

Library of Congress Cataloging-in-Publication Data
A catalog record for this book is available from the Library of Congress

ISBN: 978-0-08-045372-9

For information on all Elsevier publications
visit our web site at books.elsevier.com

Printed and bound in Great Britain

07 08 09 10 10 9 8 7 6 5 4 3 2 1

PREFACE TO THE SECOND EDITION

This book is concerned with the topical problems of mechanics of advanced composite materials whose mechanical properties are controlled by high-strength and high-stiffness continuous fibers embedded in polymeric, metal, or ceramic matrix. Although the idea of combining two or more components to produce materials with controlled properties has been known and used from time immemorial, modern composites have been developed only several decades ago and have found by now intensive applications in different fields of engineering, particularly, in aerospace structures for which high strength-to-weight and stiffness-to-weight ratios are required.

Due to wide existing and potential applications, composite technology has been developed very intensively over recent decades, and there exist numerous publications that cover anisotropic elasticity, mechanics of composite materials, design, analysis, fabrication, and application of composite structures. According to the list of books on composites presented in 'Mechanics of Fibrous Composites' by C.T. Herakovich (1998) there were 35 books published in this field before 1995, and this list should be supplemented now with several new books.

In connection with this, the authors were challenged with a natural question as to what caused the necessity to publish another book and what is the difference between this book and the existing ones. Concerning this question, we had at least three motivations supporting us in this work.

First, this book is of a more specific nature than the published ones which usually cover not only mechanics of materials but also include analysis of composite beams, plates and shells, joints, and elements of design of composite structures that, being also important, do not strictly belong to the field of mechanics of composite materials. This situation looked quite natural since composite science and technology, having been under intensive development only over several past decades, required books of a universal type. Nowadays however, implementation of composite materials has reached the level at which special books can be dedicated to each of the aforementioned problems of composite technology and, first of all, to mechanics of composite materials which is discussed in this book in conjunction with analysis of composite materials. As we hope, thus constructed combination of material science and mechanics of solids enabled us to cover such specific features of material behavior as nonlinear elasticity, plasticity, creep, structural nonlinearity and discuss in details the problems of material micro- and macromechanics that are only slightly touched in the existing books, e.g., stress diffusion in a unidirectional material with broken fibers, physical and statistical aspects of fiber strength, coupling effects in anisotropic and laminated materials, etc.

Second, this book, being devoted to materials, is written by designers of composite structures who over the last 35 years were involved in practically all main Soviet and then Russian projects in composite technology. This governs the list of problems covered in the book which can be referred to as material problems challenging designers and determines the third of its specific features – discussion is illustrated with composite parts and structures designed and built within the frameworks of these projects. In connection with this, the authors appreciate the permission of the Russian Composite Center – Central Institute of Special Machinery (CRISM) to use in the book the pictures of structures developed and fabricated at CRISM as part of the joint research and design projects.

The primary aim of the book is the combined coverage of mechanics, technology, and analysis of composite materials at the advanced level. Such an approach enables the engineer to take into account the essential mechanical properties of the material itself and special features of practical implementation, including manufacturing technology, experimental results, and design characteristics.

The book consists of eight chapters progressively covering all structural levels of composite materials from their components through elementary plies and layers to laminates.

Chapter 1 is an introduction in which typical reinforcing and matrix materials as well as typical manufacturing processes used in composite technology are described.

Chapter 2 is also a sort of introduction but dealing with fundamentals of mechanics of solids, i.e., stress, strain, and constitutive theories, governing equations, and principles that are used in the next chapters for analysis of composite materials.

Chapter 3 is devoted to the basic structural element of a composite material – unidirectional composite ply. In addition to conventional description of micromechanical models and experimental results, the physical nature of fiber strength, its statistical characteristics, and interaction of damaged fibers through the matrix are discussed, and an attempt is made to show that fibrous composites comprise a special class of man-made materials utilizing natural potentials of material strength and structure.

Chapter 4 contains a description of typical composite layers made of unidirectional, fabric, and spatially reinforced composite materials. Conventional linear elastic models are supplemented in this chapter with nonlinear elastic and elastic–plastic analysis demonstrating specific types of behavior of composites with metal and thermoplastic matrices.

Chapter 5 is concerned with mechanics of laminates and includes conventional description of the laminate stiffness matrix, coupling effects in typical laminates and procedures of stress calculation for in-plane and interlaminar stresses.

Chapter 6 presents a practical approach to evaluation of laminate strength. Three main types of failure criteria, i.e., structural criteria indicating the modes of failure, approximation polynomial criteria treated as formal approximations of experimental data, and tensor-polynomial criteria are discussed and compared with available experimental results for unidirectional and fabric composites.

Chapter 7 dealing with environmental and special loading effects includes analysis of thermal conductivity, hydrothermal elasticity, material aging, creep, and durability under long-term loading, fatigue, damping, and impact resistance of typical advanced composites. The effect of manufacturing factors on material properties and behavior is demonstrated for filament winding accompanied with nonuniform stress distribution between the fibers and ply waviness and laying-up processing of nonsymmetric laminate exhibiting warping after curing and cooling.

Chapter 8 covers a specific problem of material optimal design for composite materials and presents composite laminates of uniform strength providing high weight efficiency of composite structures demonstrated for filament-wound pressure vessels, spinning disks, and anisogrid lattice structures.

This second edition is a revised, updated, and extended version of the first edition, with new sections on: composites with high fiber fraction (Section 3.6), composites with controlled cracks (Section 4.4.4), symmetric laminates (Section 5.4), engineering stiffness coefficients of orthotropic laminates (Section 5.5), tensor strength criteria (Section 6.1.3), practical recommendations (Section 6.2), allowable stresses for laminates consisting of unidirectional plies (Section 6.4), hygrothermal effects and aging (Section 7.2), application to optimal composite structures (Section 8.3), spinning composite disks (Section 8.3.2), and anisogrid composite lattice structures (Section 8.3.3).

The following sections have been re-written and extended: Section 5.8 Antisymmetric laminates; Section 7.3.3 Cyclic loading; Section 7.3.4 Impact loading; Section 8.3.1 Composite pressure vessels. More than 40 new illustrations and 5 new tables were added.

The new title 'Advanced Mechanics of Composite Materials' has been adopted for the 2nd edition, which provides better reflection of the overall contents and improvements, extensions and revisions introduced in the present version.

The book offers a comprehensive coverage of the topic in full range: from basics and fundamentals to the advanced modeling and analysis including practical design and engineering applications and can be used as an up-to-date introductory text book aimed at senior undergraduates and graduate students. At the same time it includes a detailed and comprehensive coverage of the contemporary theoretical models at the micro- and macro-levels of material structure, practical methods and approaches, experimental results, and optimization of composite material properties and component performance that can be used by researchers and engineers.

The authors would like to thank several people for their time and effort in making the book a reality. Specifically, we would like to thank our Elsevier editors who have encouraged and participated in the preparation of the first and second editions. These include Ian Salusbury (Publishing editor of the first edition), Emma Hurst and David Sleeman (Publishing editors of the second edition), and Derek Coleman (Development editor). Special thanks are due to Prof. Leslie Henshall, for his work on the text improvements and to Dr. Konstantin Morozov for his help in the development of illustrations in the book. The authors are also grateful to the Central Institute of Special Machinery (CRISM) that supplied many illustrations and case studies.

Valery V. Vasiliev *Evgeny V. Morozov*

CONTENTS

Chapter 1

INTRODUCTION

1.1. Structural materials

Materials are the basic elements of all natural and man-made structures. Figuratively speaking, these materialize the structural conception. Technological progress is associated with continuous improvement of existing material properties as well as with the expansion of structural material classes and types. Usually, new materials emerge due to the necessity to improve structural efficiency and performance. In addition, new materials themselves as a rule, in turn provide new opportunities to develop updated structures and technology, while the latter challenges materials science with new problems and tasks. One of the best manifestations of this interrelated process in the development of materials, structures, and technology is associated with composite materials, to which this book is devoted.

Structural materials possess a great number of physical, chemical and other types of properties, but at least two principal characteristics are of primary importance. These characteristics are the stiffness and strength that provide the structure with the ability to maintain its shape and dimensions under loading or any other external action.

High stiffness means that material exhibits low deformation under loading. However, by saying that stiffness is an important property we do not mean that it should be necessarily high. The ability of a structure to have controlled deformation (compliance) can also be important for some applications (e.g., springs; shock absorbers; pressure, force, and displacement gauges).

Lack of material strength causes an uncontrolled compliance, i.e., in failure after which a structure does not exist any more. Usually, we need to have as high strength as possible, but there are some exceptions (e.g., controlled failure of explosive bolts is used to separate rocket stages).

Thus, without controlled stiffness and strength the structure cannot exist. Naturally, both properties depend greatly on the structure's design but are determined by the stiffness and strength of the structural material because a good design is only a proper utilization of material properties.

To evaluate material stiffness and strength, consider the simplest test – a bar with cross-sectional area A loaded with tensile force F as shown in Fig. 1.1. Obviously, the higher the force causing the bar rupture, the higher is the bar's strength. However, this strength does not only depend on the material properties – it is proportional to the cross-sectional area A.

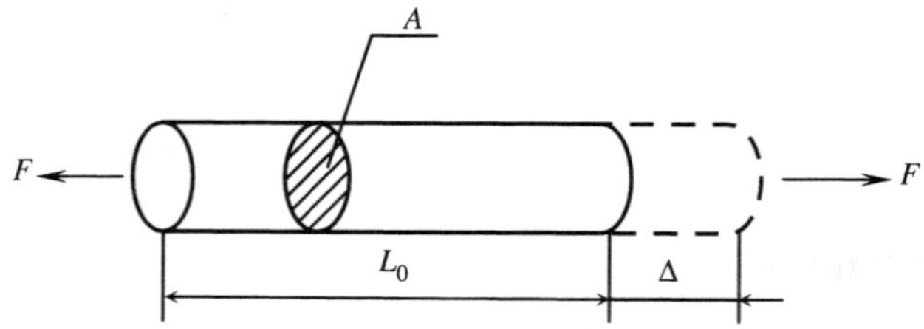

Fig. 1.1. A bar under tension.

Thus, it is natural to characterize material strength by the ultimate stress

$$\overline{\sigma} = \frac{\overline{F}}{A} \tag{1.1}$$

where $\overline{F}$ is the force causing the bar failure (here and subsequently we use the overbar notation to indicate the ultimate characteristics). As follows from Eq. (1.1), stress is measured as force divided by area, i.e., according to international (SI) units, in pascals (Pa) so that 1 Pa = 1 N/m^2. Because the loading of real structures induces relatively high stresses, we also use kilopascals (1 kPa = 10^3 Pa), megapascals (1 MPa = 10^6 Pa), and gigapascals (1 GPa = 10^9 Pa). Conversion of old metric (kilogram per square centimeter) and English (pound per square inch) units to pascals can be done using the following relations: 1 kg/cm^2 = 98 kPa and 1 psi = 6.89 kPa.

For some special (e.g., aerospace or marine) applications, i.e., for which material density, ρ, is also important, a normalized characteristic

$$k_\sigma = \frac{\overline{\sigma}}{\rho} \tag{1.2}$$

is also used to describe the material. This characteristic is called the 'specific strength' of a material. If we use old metric units, i.e., measure force and mass in kilograms and dimensions in meters, substitution of Eq. (1.1) into Eq. (1.2) yields k_σ in meters. This result has a simple physical sense, namely k_σ is the length of the vertically hanging fiber under which the fiber will be broken by its own weight.

The stiffness of the bar shown in Fig. 1.1 can be characterized by an elongation Δ corresponding to the applied force F or acting stress $\sigma = F/A$. However, Δ is proportional to the bar's length L_0. To evaluate material stiffness, we introduce strain

$$\varepsilon = \frac{\Delta}{L_0} \tag{1.3}$$

Since ε is very small for structural materials the ratio in Eq. (1.3) is normally multiplied by 100, and ε is expressed as a percentage.

Naturally, for any material, there should be some interrelation between stress and strain, i.e.,

$$\varepsilon = f(\sigma) \quad \text{or} \quad \sigma = \varphi(\varepsilon) \tag{1.4}$$

These equations specify the so-called constitutive law and are referred to as constitutive equations. They allow us to introduce an important concept of the material model which represents some idealized object possessing only those features of the real material that are essential for the problem under study. The point is that in performing design or analysis we always operate with models rather than with real materials. Particularly, for strength and stiffness analysis, such a model is described by constitutive equations, Eqs. (1.4), and is specified by the form of the function $f(\sigma)$ or $\varphi(\varepsilon)$.

The simplest is the elastic model which implies that $f(0) = 0$, $\varphi(0) = 0$ and that Eqs. (1.4) are the same for the processes of an active loading and an unloading. The corresponding stress–strain diagram (or curve) is presented in Fig. 1.2. The elastic model (or elastic material) is characterized by two important features. First, the corresponding constitutive equations, Eqs. (1.4), do not include time as a parameter. This means that the form of the curve shown in Fig. 1.2 does not depend on the rate of loading (naturally, it should be low enough to neglect inertial and dynamic effects). Second, the active loading and the unloading follow one and the same stress–strain curve as in Fig. 1.2. The work performed by force F in Fig. 1.1 is accumulated in the bar as potential energy, which is also referred to as strain energy or elastic energy. Consider some infinitesimal elongation $\mathrm{d}\Delta$ and calculate the elementary work performed by the force F in Fig. 1.1 as $\mathrm{d}W = F\mathrm{d}\Delta$. Then, work corresponding to point 1 of the curve in Fig. 1.2 is

$$W = \int_0^{\Delta_1} F\mathrm{d}\Delta$$

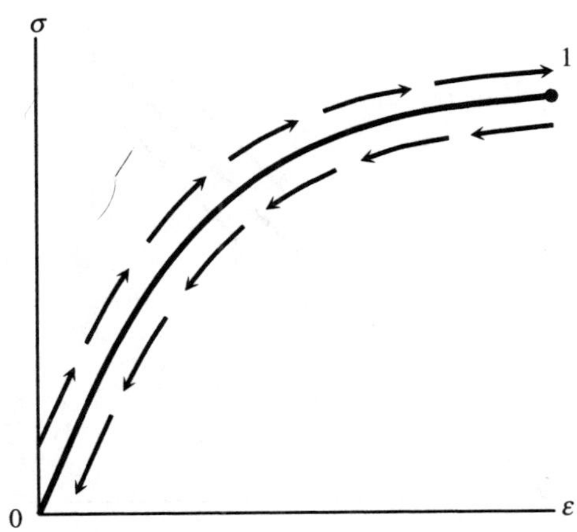

Fig. 1.2. Stress–strain curve for an elastic material.

where Δ_1 is the elongation of the bar corresponding to point 1 of the curve. The work W is equal to elastic energy of the bar which is proportional to the bar's volume and can be presented as

$$E = L_0 A \int_0^{\varepsilon_1} \sigma \, \mathrm{d}\varepsilon$$

where $\sigma = F/A$, $\varepsilon = \Delta/L_0$, and $\varepsilon_1 = \Delta_1/L_0$. Integral

$$U = \int_0^{\varepsilon_1} \sigma \, \mathrm{d}\varepsilon = \int_0^{\varepsilon_1} \varphi(\varepsilon) \mathrm{d}\varepsilon \tag{1.5}$$

is a specific elastic energy (energy accumulated in a unit volume of the bar) that is referred to as an elastic potential. It is important that U does not depend on the history of loading. This means that irrespective of the way we reach point 1 of the curve in Fig. 1.2 (e.g., by means of continuous loading, increasing force F step by step, or using any other loading program), the final value of U will be the same and will depend only on the value of final strain ε_1 for the given material.

A very important particular case of the elastic model is the linear elastic model described by the well-known Hooke's law (see Fig. 1.3)

$$\sigma = E\varepsilon \tag{1.6}$$

Here, E is the modulus of elasticity. It follows from Eqs. (1.3) and (1.6), that $E = \sigma$ if $\varepsilon = 1$, i.e., if $\Delta = L_0$. Thus, the modulus can be interpreted as the stress causing elongation of the bar in Fig. 1.1 to be the same as the initial length. Since the majority of structural materials fail before such a high elongation can occur, the modulus is usually much higher than the ultimate stress $\overline{\sigma}$.

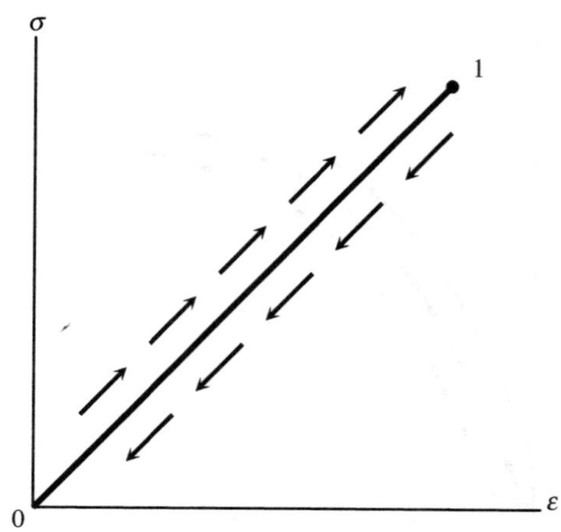

Fig. 1.3. Stress–strain diagram for a linear elastic material.

Similar to specific strength k_σ in Eq. (1.2), we can introduce the corresponding specific modulus

$$k_E = \frac{E}{\rho} \tag{1.7}$$

which describes a material's stiffness with respect to its material density.

Absolute and specific values of mechanical characteristics for typical materials discussed in this book are listed in Table 1.1.

After some generalization, the modulus can be used to describe nonlinear material behavior of the type shown in Fig. 1.4. For this purpose, the so-called secant, E_s, and tangent, E_t, moduli are introduced as

$$E_s = \frac{\sigma}{\varepsilon} = \frac{\sigma}{f(\sigma)} \quad E_t = \frac{d\sigma}{d\varepsilon} = \frac{d\varphi(\varepsilon)}{d\varepsilon} \tag{1.8}$$

While the slope α in Fig. 1.4 determines the conventional modulus E, the slopes β and γ determine E_s and E_t, respectively. As can be seen, E_s and E_t, in contrast to E, depend on the level of loading, i.e., on σ or ε. For a linear elastic material (see Fig. 1.3), $E_s = E_t = E$.

Hooke's law, Eq. (1.6), describes rather well the initial part of stress–strain diagram for the majority of structural materials. However, under a relatively high level of stress or strain, materials exhibit nonlinear behavior.

One of the existing models is the nonlinear elastic material model introduced above (see Fig. 1.2). This model allows us to describe the behavior of highly deformable rubber-type materials.

Another model developed to describe metals is the so-called elastic–plastic material model. The corresponding stress–strain diagram is shown in Fig. 1.5. In contrast to an elastic material (see Fig. 1.2), the processes of active loading and unloading are described with different laws in this case. In addition to elastic strain, ε_e, which disappears after the load is taken off, the residual strain (for the bar shown in Fig. 1.1, it is plastic strain, ε_p) remains in the material. As for an elastic material, the stress–strain curve in Fig. 1.5 does not depend on the rate of loading (or time of loading). However, in contrast to an elastic material, the final strain of an elastic–plastic material can depend on the history of loading, i.e., on the law according to which the final value of stress was reached.

Thus, for elastic or elastic–plastic materials, constitutive equations, Eqs. (1.4), do not include time. However, under relatively high temperature practically all the materials demonstrate time-dependent behavior (some of them do it even under room temperature). If we apply some force F to the bar shown in Fig. 1.1 and keep it constant, we can see that for a time-sensitive material the strain increases under a constant force. This phenomenon is called the creep of the material.

So, the most general material model that is used in this book can be described with a constitutive equation of the following type:

$$\varepsilon = f(\sigma, t, T) \tag{1.9}$$

Table 1.1
Mechanical properties of structural materials and fibers.

Material	Ultimate tensile stress, $\overline{\sigma}$ (MPa)	Modulus, E (GPa)	Specific gravity	Maximum specific strength, $k_\sigma \times 10^3$ (m)	Maximum specific modulus, $k_E \times 10^3$ (m)
Metal alloys					
Steel	400–2200	180–210	7.8–7.85	28.8	2750
Aluminum	140–700	69–72	2.7–2.85	26.5	2670
Titanium	420–1200	110	4.5	26.7	2440
Magnesium	220–320	40	1.8	14.4	2220
Beryllium	620	320	1.85	33.5	17,300
Nickel	400–500	200	8.9	5.6	2250
Metal wires (diameter, μm)					
Steel (20–1500)	1500–4400	180–200	7.8	56.4	2560
Aluminum (150)	290	69	2.7	10.7	2550
Titanium (100–800)	1400–1500	120	4.5	33.3	2670
Beryllium (50–500)	1100–1450	240–310	1.8–1.85	80.5	17,200
Tungsten (20–50)	3300–4000	410	19–19.3	21.1	2160
Molybdenum (25–250)	1800–2200	360	10.2	21.5	3500
Thermoset polymeric resins					
Epoxy	60–90	2.4–4.2	1.2–1.3	7.5	350
Polyester	30–70	2.8–3.8	1.2–1.35	5.8	310
Phenol-formaldehyde	40–70	7–11	1.2–1.3	5.8	910
Organosilicone	25–50	6.8–10	1.35–1.4	3.7	740
Polyimide	55–110	3.2	1.3–1.43	8.5	240
Bismaleimide	80	4.2	1.2	6.7	350
Thermoplastic polymers					
Polyethylene	20–45	6–8.5	0.95	4.7	890
Polystyrene	35–45	30	1.05	4.3	2860
Teflon	15–35	3.5	2.3	1.5	150
Nylon	80	2.8	1.14	7.0	240
Polyester (PC)	60	2.5	1.32	4.5	190
Polysulfone (PSU)	70	2.7	1.24	5.6	220
Polyamide-imide (PAI)	90–190	2.8–4.4	1.42	13.4	360
Polyetheretherketone (PEEK)	90–100	3.1–3.8	1.3	7.7	300
Polyphenylene sulfide (PPS)	80	3.5	1.36	5.9	250
Synthetic fibers					
Capron	680–780	4.4	1.1	70	400
Dacron	390–880	4.9–15.7	1.4	60	1430
Teflon	340–440	2.9	2.3	190	130
Nitron	390–880	4.9–8.8	1.2	70	730
Polypropylene	730–930	4.4	0.9	100	480
Viscose	930	20	1.52	60	1300
Fibers for advanced composites (diameter, μm)					
Glass (3–19)	3100–5000	72–95	2.4–2.6	200	3960
Quarts (10)	6000	74	2.2	270	3360

Table 1.1 (Contd.)

Material	Ultimate tensile stress, $\bar{\sigma}$ (MPa)	Modulus, E (GPa)	Specific gravity	Maximum specific strength, $k_\sigma \times 10^3$ (m)	Maximum specific modulus, $k_E \times 10^3$ (m)
Basalt (9–13)	3000–3500	90	2.7–3.0	130	3300
Aramid (12–15)	3500–5500	140–180	1.4–1.47	390	12,800
Polyethylene (20–40)	2600–3300	120–170	0.97	310	17,500
Carbon (5–11)					
High-strength	7000	300	1.75	400	17,100
High-modulus	2700	850	1.78	150	47,700
Boron (100–200)	2500–3700	390–420	2.5–2.6	150	16,800
Alumina – Al_2O_3 (20–500)	2400–4100	470–530	3.96	100	13,300
Silicon Carbide – SiC (10–15)	2700	185	2.4–2.7	110	7700
Titanium Carbide – TiC (280)	1500	450	4.9	30	9100
Boron Carbide – B_4C (50)	2100–2500	480	2.5	100	10,000
Boron Nitride – BN (7)	1400	90	1.9	70	4700

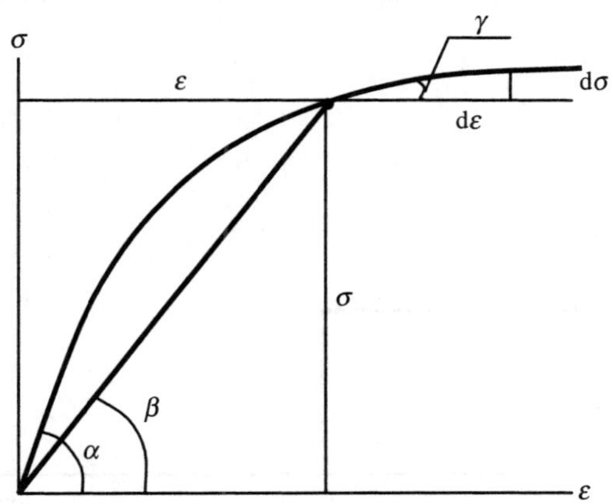

Fig. 1.4. Introduction of secant and tangent moduli.

where t indicates the time moment, whereas σ and T are stress and temperature, corresponding to this moment. In the general case, constitutive equation, Eq. (1.9), specifies strain that can be decomposed into three constituents corresponding to elastic, plastic and creep deformation, i.e.,

$$\varepsilon = \varepsilon_e + \varepsilon_p + \varepsilon_c \tag{1.10}$$

However, in application to particular problems, this model can be usually substantially simplified. To show this, consider the bar in Fig. 1.1 and assume that a force F is applied at the moment $t = 0$ and is taken off at moment $t = t_1$ as shown in Fig. 1.6a. At the

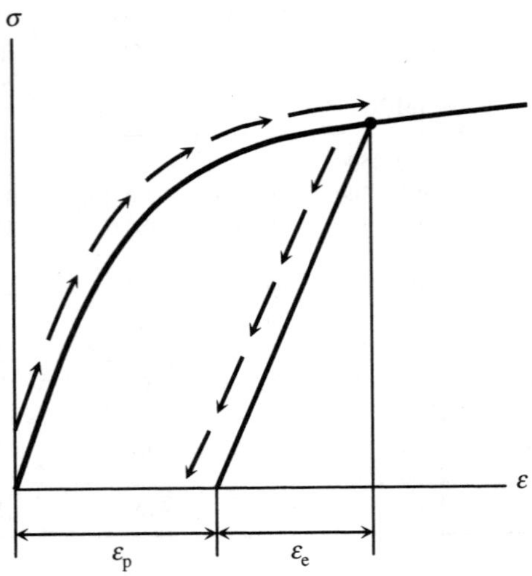

Fig. 1.5. Stress–strain diagram for elastic–plastic material.

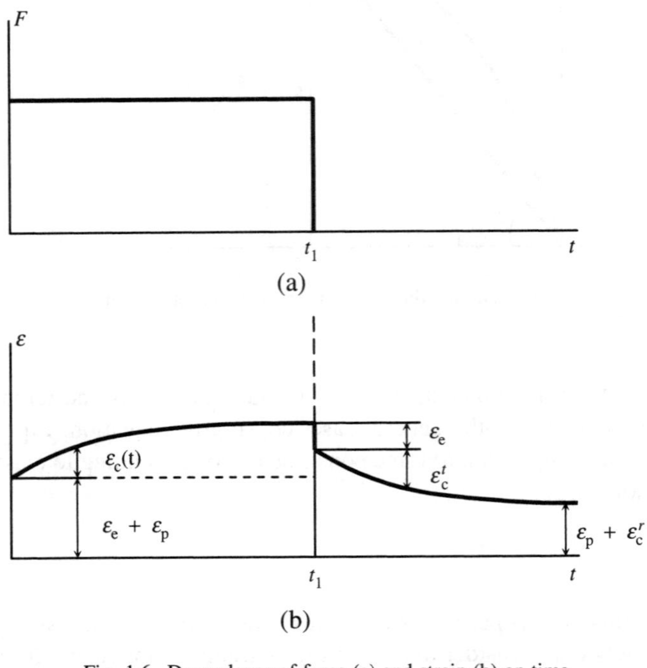

Fig. 1.6. Dependence of force (a) and strain (b) on time.

moment $t = 0$, elastic and plastic strains that do not depend on time appear, and while time is running, the creep strain is developed. At the moment $t = t_1$, the elastic strain disappears, while the reversible part of the creep strain, ε_c^t, disappears with time. Residual strain consists of the plastic strain, ε_p, and residual part of the creep strain, ε_c^r.

Now assume that $\varepsilon_p \ll \varepsilon_e$ which means that either the material is elastic or the applied load does not induce high stress and, hence, plastic strain. Then we can neglect ε_p in Eq. (1.10) and simplify the model. Furthermore, let $\varepsilon_c \ll \varepsilon_e$ which in turn means that either the material is not susceptible to creep or the force acts for a short time (t_1 is close to zero). Thus, we arrive at the simplest elastic model, which is the case for the majority of practical applications. It is important that the proper choice of the material model depends not only on the material nature and properties but also on the operational conditions of the structure. For example, a shell-type structure made of aramid–epoxy composite material, that is susceptible to creep, and designed to withstand the internal gas pressure should be analyzed with due regard to the creep, if this structure is a pressure vessel for long term gas storage. At the same time for a solid propellant rocket motor case working for seconds, the creep strain can be ignored.

A very important feature of material models under consideration is their phenomenological nature. This means that these models ignore the actual material microstructure (e.g., crystalline structure of metals or molecular structure of polymers) and represent the material as some uniform continuum possessing some effective properties that are the same irrespective of how small the material volume is. This allows us, first, to determine material properties testing material samples (as in Fig. 1.1). Second, this formally enables us to apply methods of Mechanics of Solids that deal with equations derived for infinitesimal volumes of material. And third, this allows us to simplify the strength and stiffness evaluation problem and to reduce it to a reasonable practical level not going into analysis of the actual mechanisms of material deformation and fracture.

1.2. Composite materials

This book is devoted to composite materials that emerged in the middle of the 20th century as a promising class of engineering materials providing new prospects for modern technology. Generally speaking any material consisting of two or more components with different properties and distinct boundaries between the components can be referred to as a composite material. Moreover, the idea of combining several components to produce a material with properties that are not attainable with the individual components has been used by man for thousands of years. Correspondingly, the majority of natural materials that have emerged as a result of a prolonged evolution process can be treated as composite materials.

With respect to the problems covered in this book we can classify existing composite materials (composites) into two main groups.

The first group comprises composites that are known as 'filled materials.' The main feature of these materials is the existence of some basic or matrix material whose properties are improved by filling it with some particles. Usually the matrix volume fraction is more than 50% in such materials, and material properties, being naturally modified by the fillers, are governed mainly by the matrix. As a rule, filled materials can be treated as homogeneous and isotropic, i.e., traditional models of mechanics of materials developed for metals and other conventional materials can be used to describe their behavior. This group of composites is not touched on in the book.

The second group of composite materials that is under study here involves composites that are called 'reinforced materials.' The basic components of these materials (sometimes referred to as 'advanced composites') are long and thin fibers possessing high strength and stiffness. The fibers are bound with a matrix material whose volume fraction in a composite is usually less than 50%. The main properties of advanced composites, due to which these materials find a wide application in engineering, are governed by fibers whose types and characteristics are considered below. The following sections provide a concise description of typical matrix materials and fiber-matrix compositions. Two comments should be made with respect to the data presented in these sections. First, only brief information concerning material properties that are essential for the problems covered in this book is presented there, and, second, the given data are of a broad nature and are not expected to be used in design or analysis of particular composite structures. More complete description of composite materials and their components including the history of development and advancement, chemical compositions, physical characteristics, manufacturing, and applications can be found elsewhere (Peters, 1998).

1.2.1. Fibers for advanced composites

Continuous glass fibers (the first type of fibers used in advanced composites) are made by pulling molten glass (at a temperature about 1300°C) through 0.8–3.0 mm diameter dies and further high-speed stretching to a diameter of 3–19 μm. Usually glass fibers have solid circular cross sections. However there exist fibers with rectangular (square or plane), triangular, and hexagonal cross sections, as well as hollow circular fibers. Typical mechanical characteristics and density of glass fibers are listed in Table 1.1, whereas a typical stress–strain diagram is shown in Fig. 1.7.

Important properties of glass fibers as components of advanced composites for engineering applications are their high strength, which is maintained in humid environments but degrades under elevated temperatures (see Fig. 1.8), relatively low stiffness (about 40% of the stiffness of steel), high chemical and biological resistance, and low cost. Being actually elements of monolithic glass, the fibers do not absorb water and do not change their dimensions in water. For the same reason, they are brittle and sensitive to surface damage.

Quartz fibers are similar to glass fibers and are obtained by high-speed stretching of quartz rods made of (under temperature of about 2200°C) fused quartz crystals or sand. The original process developed for manufacturing glass fibers cannot be used because the viscosity of molten quartz is too high to make thin fibers directly. However, this more complicated process results in fibers with higher thermal resistance than glass fibers.

The same process that is used for glass fibers can be employed to manufacture mineral fibers, e.g., basalt fibers made of molten basalt rocks. Having relatively low strength and high density (see Table 1.1) basalt fibers are not used for high-performance, e.g.,

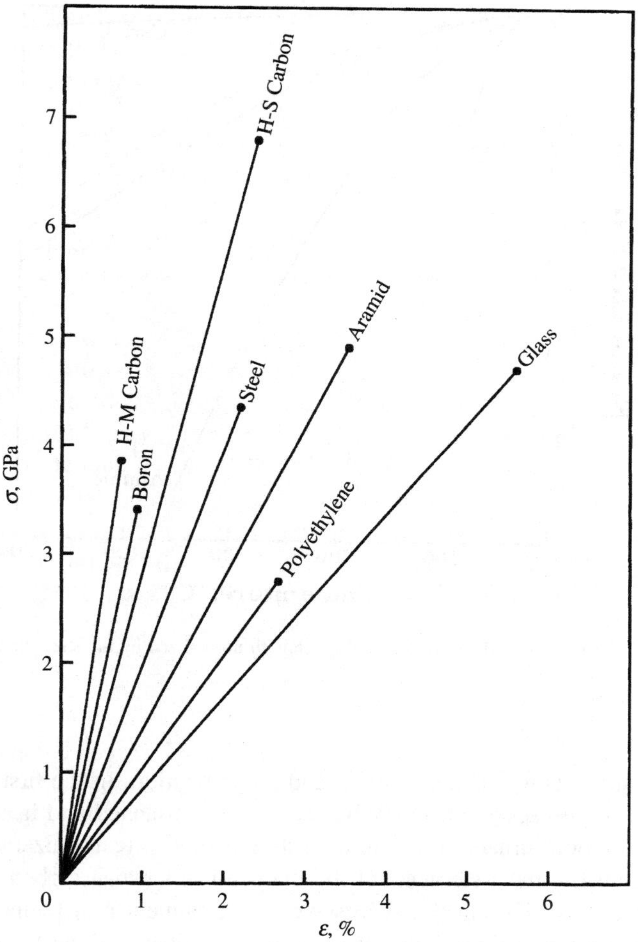

Fig. 1.7. Stress–strain diagrams for typical fibers of advanced composites.

aerospace structures, but are promising reinforcing elements for pre-stressed reinforced concrete structures in civil engineering.

Substantial improvement of a fiber's stiffness in comparison with glass fibers has been achieved with the development of carbon (or graphite) fibers. Modern high-modulus carbon fibers have a modulus that is a factor of about four higher than the modulus of steel, whereas the fiber density is lower by the same factor. Although the first carbon fibers had lower strength than glass fibers, modern high-strength fibers have a 40% higher tensile strength compared to the strength of the best glass fibers, whereas the density of carbon fibers is 30% less than that of glass fibers.

Carbon fibers are made by pyrolysis of organic fibers of which there exist two main types – PAN-based and pitch-based fibers. For PAN-based fibers the process consists of

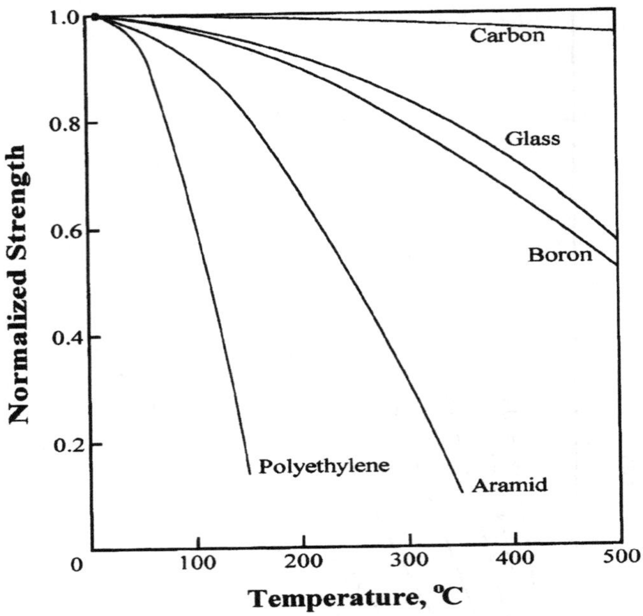

Fig. 1.8. Temperature degradation of fiber strength normalized by the strength at 20°C.

three stages – stabilization, carbonization, and graphitization. In the first step (stabilization), a system of polyacrylonitrile (PAN) filaments is stretched and heated up to about 400°C in an oxidation furnace, while in the subsequent step (carbonization under 900°C in an inert gas media) most elements of the filaments other than carbon are removed or converted into carbon. During the successive heat treatment at a temperature reaching 2800°C (graphitization) a crystalline carbon structure oriented along the fiber's length is formed, resulting in PAN-based carbon fibers. The same process is used for rayon organic filaments (instead of PAN), but results in carbon fibers with lower modulus and strength because rayon contains less carbon than PAN. For pitch-based carbon fibers, the initial organic filaments are made in approximately the same manner as for glass fibers from molten petroleum or coal pitch and pass through carbonization and graphitization processes. Because pyrolysis is accompanied with a loss of material, carbon fibers have a porous structure and their specific gravity (about 1.8) is less than that of graphite (2.26). The properties of carbon fibers are affected by the crystallite size, crystalline orientation, porosity and purity of the carbon structure.

Typical stress–strain diagrams for high-modulus (HM) and high-strength (HS) carbon fibers are plotted in Fig. 1.7. As components of advanced composites for engineering applications, carbon fibers are characterized by very high modulus and strength, high chemical and biological resistance, electric conductivity and very low coefficient of thermal expansion. The strength of carbon fibers practically does not change with temperature up to 1500°C (in an inert media preventing oxidation of the fibers).

The exceptional strength of 7.06 GPa is reached in Toray T-1000 carbon fibers, whereas the highest modulus of 850 GPa is obtained in Carbonic HM-85 fibers. Carbon fibers are anisotropic, very brittle, and sensitive to damage. They do not absorb water and do not change their dimensions in humid environments.

There exist more than 50 types of carbon fibers with a broad spectrum of strength, stiffness and cost, and the process of fiber advancement is not over – one may expect fibers with strength up to 10 GPa and modulus up to 1000 GPa within a few years.

Organic fibers commonly encountered in textile applications can be employed as reinforcing elements of advanced composites. Naturally, only high performance fibers, i.e., fibers possessing high stiffness and strength, can be used for this purpose. The most widely used organic fibers that satisfy these requirements are known as aramid (aromatic polyamide) fibers. They are extruded from a liquid crystalline solution of the corresponding polymer in sulfuric acid with subsequent washing in a cold water bath and stretching under heating. Some properties of typical aramid fibers are listed in Table 1.1, and the corresponding stress–strain diagram is presented in Fig. 1.7. As components of advanced composites for engineering applications, aramid fibers are characterized by low density providing high specific strength and stiffness, low thermal conductivity resulting in high heat insulation, and a negative thermal expansion coefficient allowing us to construct hybrid composite elements that do not change their dimensions under heating. Consisting actually of a system of very thin filaments (fibrils), aramid fibers have very high resistance to damage. Their high strength in the longitudinal direction is accompanied by relatively low strength under tension in the transverse direction. Aramid fibers are characterized with pronounced temperature (see Fig. 1.8) and time dependence for stiffness and strength. Unlike the inorganic fibers discussed above, they absorb water resulting in moisture content up to 7% and degradation of material properties by 15–20%.

The list of organic fibers has been supplemented recently with extended chain polyethylene fibers demonstrating outstanding low density (less than that of water) in conjunction with relatively high stiffness and strength (see Table 1.1 and Fig. 1.7). Polyethylene fibers are extruded from the corresponding polymer melt in a similar manner to glass fibers. They do not absorb water and have high chemical resistance, but demonstrate relatively low temperature and creep resistance (see Fig. 1.8).

Boron fibers were developed to increase the stiffness of composite materials when glass fibers were mainly used to reinforce composites of the day. Being followed by high-modulus carbon fibers with higher stiffness and lower cost, boron fibers have now rather limited application. Boron fibers are manufactured by chemical vapor deposition of boron onto about 12 μm diameter tungsten or carbon fiber (core). Because of this technology, boron fibers have a relatively large diameter, 100–200 μm. They are extremely brittle and sensitive to surface damage. Typical mechanical properties of boron fibers are presented in Table 1.1 and Figs 1.7 and 1.8. Being mainly used in metal matrix composites, boron fibers degrade on contact with aluminum or titanium matrices at the temperature that is necessary for processing (above 500°C). To prevent this degradation, chemical vapor deposition is used to cover the fiber surface with about 5 μm thick layer of silicon carbide, SiC, (such fibers are called Borsic) or boron carbide, B_4C.

There exists a special class of ceramic fibers for high-temperature applications composed of various combinations of silicon, carbon, nitrogen, aluminum, boron, and titanium. The most commonly encountered are silicon carbide (SiC) and alumina (Al_2O_3) fibers.

Silicon carbide is deposited on a tungsten or carbon core-fiber by the reaction of a gas mixture of silanes and hydrogen. Thin (8–15 μm in diameter) SiC fibers can be made by pyrolysis of polymeric (polycarbosilane) fibers at temperatures of about 1400°C in an inert atmosphere. Silicon carbide fibers have high strength and stiffness, moderate density (see Table 1.1) and very high melting temperature (2600°C).

Alumina (Al_2O_3) fibers are fabricated by sintering of fibers extruded from the viscous alumina slurry with rather complicated composition. Alumina fibers, possessing approximately the same mechanical properties as SiC fibers, have relatively large diameter and high density. The melting temperature is about 2000°C.

Silicon carbide and alumina fibers are characterized by relatively low reduction in strength at elevated temperatures (see Fig. 1.9).

Promising ceramic fibers for high-temperature applications are boron carbide (B_4C) fibers that can be obtained either as a result of reaction of a carbon fiber with a mixture of hydrogen and boron chloride at high temperature (around 1800°C) or by pyrolysis of cellulosic fibers soaked with boric acid solution. Possessing high stiffness and strength and

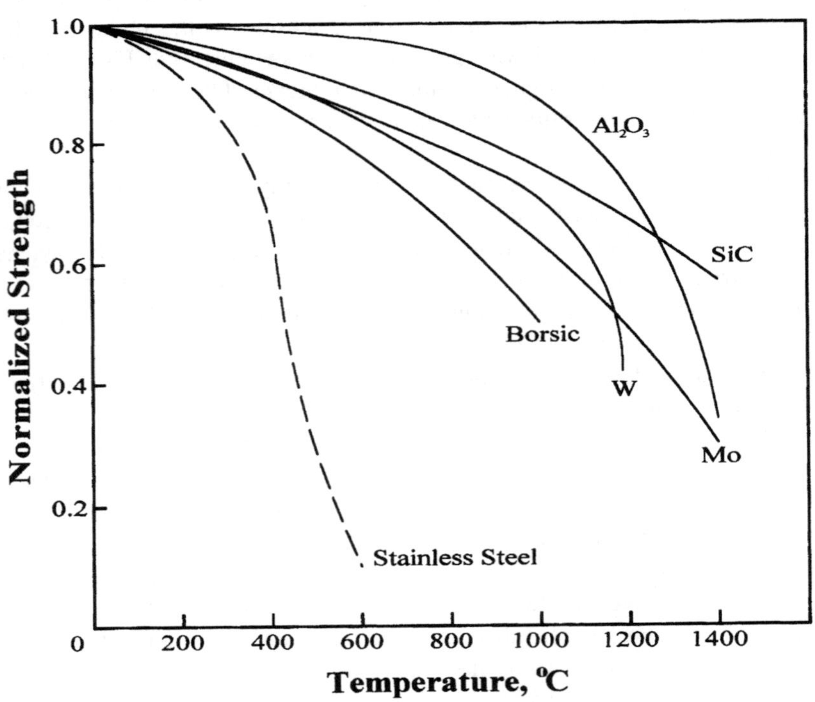

Fig. 1.9. Temperature dependence of high-temperature fibers normalized strength (in comparison with stainless steel).

moderate density (see Table 1.1), boron carbide fibers have very high thermal resistance (up to 2300°C).

Metal fibers (thin wires) made of steel, beryllium, titanium, tungsten, and molybdenum are used for special, e.g., low-temperature and high-temperature applications. Typical characteristics of metal fibers are presented in Table 1.1 and Figs. 1.7 and 1.9.

In advanced composites, fibers provide not only high strength and stiffness but also a possibility to tailor the material so that directional dependence of its mechanical properties matches that of the loading environment. The principle of directional properties can be traced in all natural materials that have emerged as a result of a prolonged evolution and, in contrast to man-made metal alloys, are neither isotropic nor homogeneous. Many natural materials have fibrous structures and utilize high strength and stiffness of natural fibers listed in Table 1.2. As can be seen (Tables 1.1 and 1.2), natural fibers, having lower strength and stiffness than man-made fibers, can compete with modern metals and plastics.

Before being used as reinforcing elements of advanced composites, the fibers are subjected to special finish surface treatments, undertaken to prevent any fiber damage under contact with processing equipment, to provide surface wetting when the fibers are combined with matrix materials, and to improve the interface bond between fibers and matrices. The most commonly encountered surface treatments are chemical sizing performed during the basic fiber formation operation and resulting in a thin layer applied to the surface of the fiber, surface etching by acid, plasma, or corona discharge, and coating of the fiber surface with thin metal or ceramic layers.

With only a few exceptions (e.g., metal fibers), individual fibers, being very thin and sensitive to damage, are not used in composite manufacturing directly, but in the form of tows (rovings), yarns, and fabrics.

A unidirectional tow (roving) is a loose assemblage of parallel fibers consisting usually of thousands of elementary fibers. Two main designations are used to indicate the size of

Table 1.2
Mechanical properties of natural fibers.

Fiber	Diameter (μm)	Ultimate tensile stress, $\overline{\sigma}$ (MPa)	Modulus, E (GPa)	Specific gravity
Wood	15–20	160	23	1.5
Bamboo	15–30	550	36	0.8
Jute	10–50	580	22	1.5
Cotton	15–40	540	28	1.5
Wool	75	170	5.9	1.32
Coir	10–20	250	5.5	1.5
Bagasse	25	180	9	1.25
Rice	5–15	100	6	1.24
Natural silk	15	400	13	1.35
Spider silk	4	1750	12.7	–
Linen	–	270	–	–
Sisal	–	560	–	–
Asbestos	0.2	1700	160	2.5

the tow, namely the K-number that gives the number of fibers in the tow (e.g., $3K$ tow contains 3000 fibers) and the tex-number which is the mass in grams of 1000 m of the tow. The tow tex-number depends not only on the number of fibers but also on the fiber diameter and density. For example, AS4-6K tow consisting of 6000 AS4 carbon fibers has 430 tex.

A yarn is a fine tow (usually it includes hundreds of fibers) slightly twisted (about 40 turns per meter) to provide the integrity of its structure necessary for textile processing. Yarn size is indicated in tex-numbers or in textile denier-numbers (den) such that 1 tex = 9 den. Continuous yarns are used to make fabrics with various weave patterns. There exists a wide variety of glass, carbon, aramid, and hybrid fabrics whose nomenclature, structure, and properties are described elsewhere (Chou and Ko, 1989; Tarnopol'skii et al., 1992; Bogdanovich and Pastore, 1996; Peters, 1998).

An important characteristic of fibers is their processability which can be evaluated as the ratio, $K_p = \overline{\sigma}_s/\overline{\sigma}$, of the strength demonstrated by fibers in the composite structure, $\overline{\sigma}_s$, to the strength of fibers before they were processed, $\overline{\sigma}$. This ratio depends on fibers' ultimate elongation, sensitivity to damage, and manufacturing equipment causing damage to the fibers. The most sensitive to operational damage are boron and high-modulus carbon fibers possessing relatively low ultimate elongation $\overline{\varepsilon}$ (less than 1%, see Fig. 1.7). For example, for filament wound pressure vessels, $K_p = 0.96$ for glass fibers, while for carbon fibers, $K_p = 0.86$.

To evaluate fiber processability under real manufacturing conditions, three simple tests are used – tension of a straight dry tow, tension of tows with loops, and tension of a tow with a knot (see Fig. 1.10). Similar tests are used to determine the strength of individual fibers (Fukuda et al., 1997). For carbon tows, normalized strength obtained in these tests is presented in Table 1.3 (for proper comparison, the tows should be of the same size). As follows from this table, the tow processability depends on the fiber ultimate strain (elongation). The best processability is observed for aramid tows whose fibers have high elongation and low sensitivity to damage (they are not monolithic and consist of thin fibrils).

1.2.2. Matrix materials

To utilize high strength and stiffness of fibers in a monolithic composite material suitable for engineering applications, fibers are bound with a matrix material whose strength and stiffness are, naturally, much lower than those of fibers (otherwise, no fibers would be necessary). Matrix materials provide the final shape of the composite structure and govern the parameters of the manufacturing process. The optimal combination of fiber and matrix properties should satisfy a set of operational and manufacturing requirements that are sometimes of a contradictory nature, and have not been completely met yet in existing composites.

First of all, the stiffness of the matrix should correspond to the stiffness of the fibers and be sufficient to provide uniform loading of fibers. The fibers are usually characterized by relatively high scatter in strength that may be increased due to damage of the fibers caused by the processing equipment. Naturally, fracture of the weakest or damaged fiber should not result in material failure. Instead, the matrix should evenly redistribute the load from

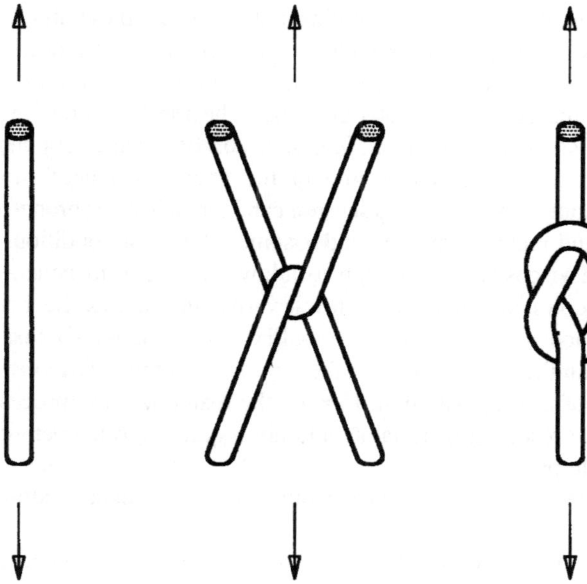

(a) (b) (c)

Fig. 1.10. Testing of a straight tow (a), tows with a loop (b), and tow with a knot (c).

Table 1.3
Normalized strength of carbon tows.

Ultimate strain, $\bar{\varepsilon}$ (%)	Normalized strength		
	Straight tow	Tow with a loop	Tow with a knot
0.75	1	0.25	0.15
1.80	1	0.53	0.18

the broken fiber to the adjacent ones and then load the broken fiber at a distance from the cross section at which it failed. The higher the matrix stiffness, the smaller is this distance, and less is the influence of damaged fibers on material strength and stiffness (which should be the case). Moreover, the matrix should provide the proper stress diffusion (this is the term traditionally used for this phenomenon in the analysis of stiffened structures (Goodey, 1946)) in the material at a given operational temperature. That is why this temperature is limited, as a rule, by the matrix rather than by the fibers. But on the other hand, to provide material integrity up to the failure of the fibers, the matrix material should possess high compliance. Obviously, for a linear elastic material (see Fig. 1.3), a combination of high stiffness and high ultimate strain $\bar{\varepsilon}$ results in high strength which is not the case for modern matrix materials. Thus, close to optimal (with respect to the foregoing requirements) and realistic matrix material should have a nonlinear stress–strain diagram (of the type shown in Fig. 1.5) and possess high initial modulus of elasticity and high ultimate strain.

However, matrix properties, even though being optimal for the corresponding fibers, do not manifest in the composite material if the adhesion (the strength of fiber–matrix interface bonding) is not high enough. High adhesion between fibers and matrices, providing material integrity up to the failure of the fibers, is a necessary condition for high-performance composites. Proper adhesion can be reached for properly selected combinations of fiber and matrix materials under some additional conditions. First, a liquid matrix should have viscosity low enough to allow the matrix to penetrate between the fibers of such dense systems of fibers as tows, yarns, and fabrics. Second, the fiber surface should have good wettability with the matrix. Third, the matrix viscosity should be high enough to retain the liquid matrix in the impregnated tow, yarn, or fabric in the process of fabrication of a composite part. Finally, the manufacturing process providing the proper quality of the resulting material should not require high temperature and pressure to make a composite part.

At present, typical matrices are made from polymeric, metal, carbon, and ceramic materials.

Polymeric matrices are divided into two main types, thermoset and thermoplastic. Thermoset polymers, which are the most widely used matrix materials for advanced composites, include polyester, epoxy, polyimide and other resins (see Table 1.1) cured under elevated or room temperature. A typical stress–strain diagram for a cured epoxy resin is shown in Fig. 1.11. Being cured (polymerized), a thermoset matrix cannot be reset, dissolved, or melted. Heating of a thermoset material results first in degradation of its strength and stiffness, and then in thermal destruction.

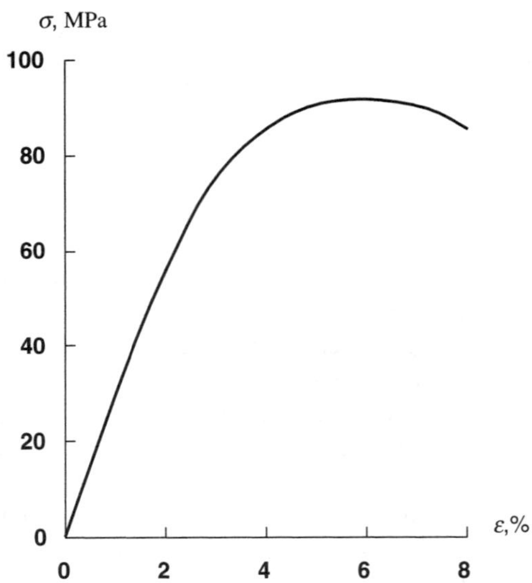

Fig. 1.11. Stress–strain diagram for a typical cured epoxy matrix.

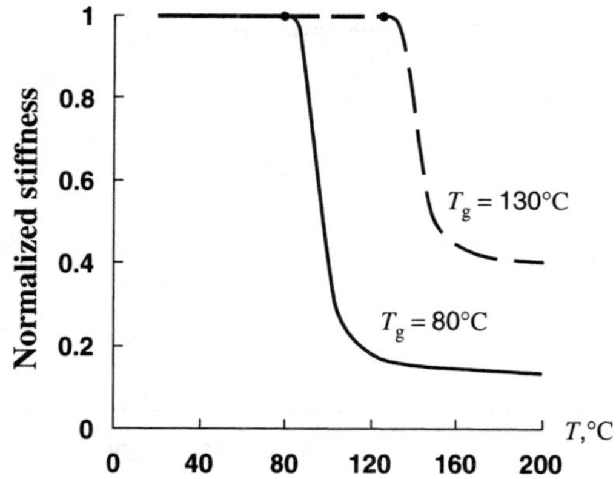

Fig. 1.12. Typical thermo-mechanical diagrams for cured epoxy resins with glass transition temperatures 80°C (———) and 130°C (– – – –).

In contrast to thermoset resins, thermoplastic matrices (PSU, PEEK, PPS and others – see Table 1.1) do not require any curing reaction. They melt under heating and convert to a solid state under cooling. The possibility to re-melt and dissolve thermoplastic matrices allows us to reshape composite parts forming them under heating and simplifies their recycling, which is a problem for thermoset materials.

Polymeric matrices can be combined with glass, carbon, organic, or boron fibers to yield a wide class of polymeric composites with high strength and stiffness, low density, high fatigue resistance, and excellent chemical resistance. The main disadvantage of these materials is their relatively low (in comparison with metals) temperature resistance limited by the matrix. The so-called thermo-mechanical curves are plotted to determine this important (for applications) characteristic of the matrix. These curves, presented for typical epoxy resins in Fig. 1.12, show the dependence of some stiffness parameter on the temperature and allow us to find the so-called glass transition temperature, T_g, which indicates a dramatic reduction in material stiffness. There exist several standard methods to obtain a material's thermo-mechanical diagram. The one used to plot the curves presented in Fig. 1.12 involves compression tests of heated polymeric discs. Naturally, to retain the complete set of properties of polymeric composites, the operating temperature, in general, should not exceed T_g. However, the actual material behavior depends on the type of loading. As follows from Fig. 1.13, heating above the glass transition temperature only slightly influences material properties under tension in the fiber direction and dramatically reduces its strength in longitudinal compression and transverse bending. The glass transition temperature depends on the processing temperature, T_p, at which a material is fabricated, and higher T_p results, as a rule, in higher T_g. Thermoset epoxy matrices cured at a temperature in the range 120–160°C have $T_g = 60-140$°C. There also exist a number of high temperature thermoset matrices (e.g., organosilicone, polyimide, and bismaleimide resins) with $T_g = 250-300$°C and curing temperatures up to 400°C. Thermoplastic matrices are also characterized by a wide range of glass transition temperatures – from 90°C for PPS and 140°C for PEEK to 190°C for PSU and 270°C for PAI

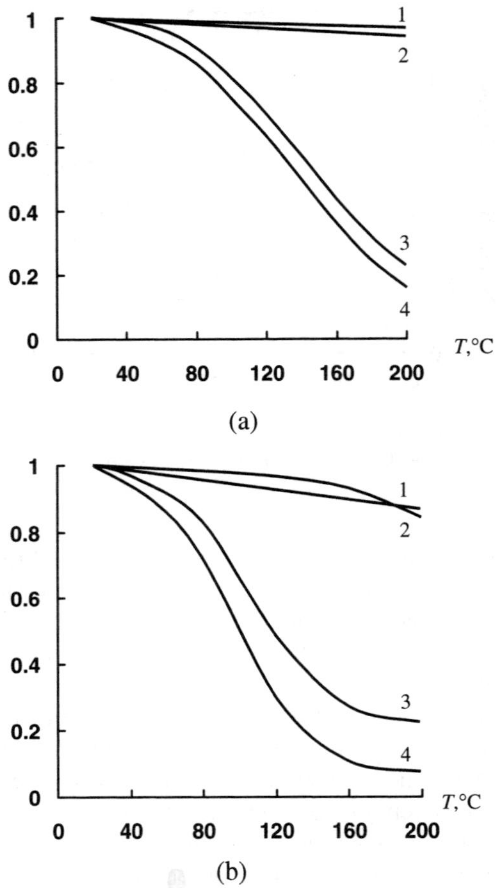

Fig. 1.13. Dependence of normalized longitudinal moduli (1), strength under longitudinal tension (2), bending (3), and compression (4) on temperature for unidirectional carbon composites with epoxy matrices having $T_g = 130°C$ (a) and $T_g = 80°C$ (b).

(see Table 1.1 for abbreviations). The processing temperature for different thermoplastic matrices varies from 300 to 400°C.

Further enhancement in temperature resistance of composite materials is associated with application of metal matrices in combination with high temperature boron, carbon, ceramic fibers, and metal wires. The most widespread metal matrices are aluminum, magnesium, and titanium alloys possessing high plasticity (see Fig. 1.14), whereas for special applications nickel, copper, niobium, cobalt, and lead matrices can be used. Fiber reinforcement essentially improves the mechanical properties of such metals. For example, carbon fibers increase strength and stiffness of such a soft metal as lead by an order of magnitude.

As noted above, metal matrices allow us to increase operational temperatures for composite structures. The dependencies of longitudinal strength and stiffness of

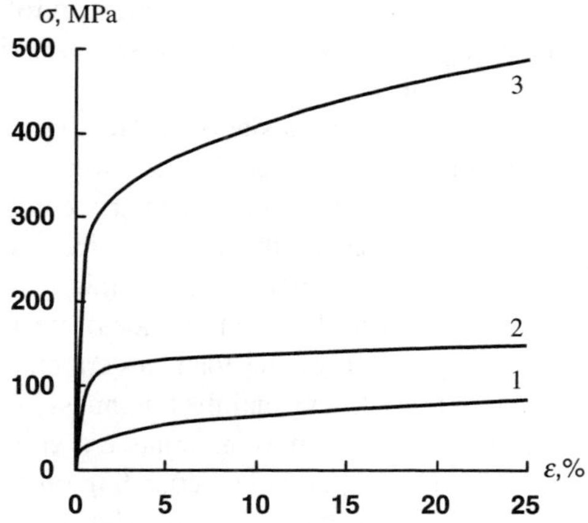

Fig. 1.14. Typical stress–strain curves for aluminum (1), magnesium (2), and titanium (3) matrices.

boron–aluminum unidirectional composite material on temperature, corresponding to the experimental results that can be found in Karpinos (1985) and Vasiliev and Tarnopol'skii (1990), are shown in Fig. 1.15. Naturally, higher temperature resistance requires higher processing temperature, T_p. Indeed, aluminum matrix composite materials are processed at $T_p = 550°C$, whereas for magnesium, titanium, and nickel matrices the appropriate temperature is about 800, 1000, and 1200°C respectively. Some processes also require rather high pressure (up to 150 MPa).

In polymeric composites, the matrix materials play an important but secondary role of holding the fibers in place and providing good load dispersion into the fibers, whereas material strength and stiffness are controlled by the reinforcements. In contrast,

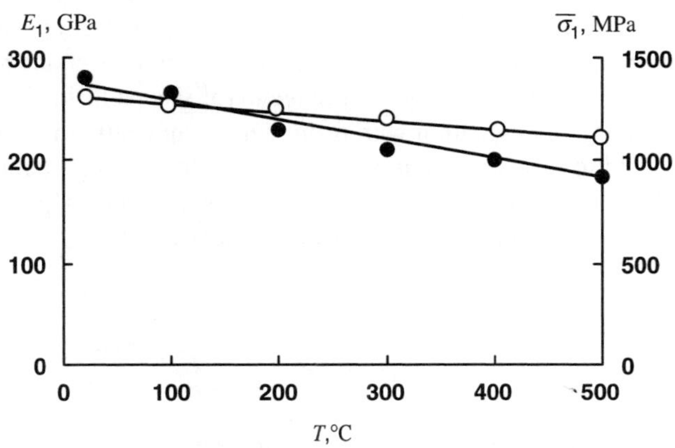

Fig. 1.15. Temperature dependence of tensile strength (•) and stiffness (○) along the fibers for unidirectional boron–aluminum composite.

the mechanical properties of metal matrix composites are controlled by the matrix to a considerably larger extent, though the fibers still provide the major contribution to the strength and stiffness of the material.

The next step in the development of composite materials that can be treated as matrix materials reinforced with fibers rather than fibers bonded with matrix (which is the case for polymeric composites) is associated with ceramic matrix composites possessing very high thermal resistance. The stiffnesses of the fibers which are usually metal (steel, tungsten, molybdenum, niobium), carbon, boron, or ceramic (SiC, Al_2O_3) and ceramic matrices (oxides, carbides, nitrides, borides, and silicides) are not very different, and the fibers do not carry the main fraction of the load in ceramic composites. The function of the fibers is to provide strength and mainly toughness (resistance to cracks) of the composite, because non-reinforced ceramic materials are very brittle. Ceramic composites can operate under very high temperatures depending on the melting temperature of the matrix that varies from 1200 to 3500°C. Naturally, the higher the temperature, the more complicated is the manufacturing process. The main shortcoming of ceramic composites is associated with a low ultimate tensile elongation of the ceramic matrix resulting in cracks appearing in the matrix under relatively low tensile stress applied to the material.

An outstanding combination of high mechanical characteristics and temperature resistance is demonstrated by carbon–carbon composites in which both components – fibers and matrix are made from one and the same material but with different structure. A carbon matrix is formed as a result of carbonization of an organic resin (phenolic and furfural resin or pitch) with which carbon fibers are impregnated, or of chemical vapor deposition of pyrolitic carbon from a hydrocarbon gas. In an inert atmosphere or in a vacuum, carbon–carbon composites can withstand very high temperatures (more than 3000°C). Moreover, their strength increases under heating up to 2200°C while the modulus degrades at temperatures above 1400°C. However in an oxygen atmosphere, they oxidize and sublime at relatively low temperatures (about 600°C). To use carbon–carbon composite parts in an oxidizing atmosphere, they must have protective coatings, made usually from silicon carbide. Manufacturing of carbon–carbon parts is a very energy- and time-consuming process. To convert an initial carbon–phenolic composite into carbon–carbon, it should receive a thermal treatment at 250°C for 150 h, carbonization at about 800°C for about 100 h and several cycles of densification (one-stage pyrolisis results in high porosity of the material) each including impregnation with resin, curing, and carbonization. To refine the material structure and to provide oxidation resistance, a further high-temperature graphitization at 2700°C and coating (at 1650°C) can be required. Vapor deposition of pyrolitic carbon is also a time-consuming process performed at 900–1200°C under a pressure of 150–2000 kPa.

1.2.3. Processing

Composite materials do not exist apart from composite structures and are formed while the structure is fabricated. Being a heterogeneous media, a composite material has two levels of heterogeneity. The first level represents a microheterogeneity induced by at least two phases (fibers and matrix) that form the material microstructure. At the second level the material is characterized by a macroheterogeneity caused by the laminated or more complicated macrostructure of the material which consists usually of a set of layers with different orientations. A number of technologies have been developed by now to manufacture composite structures. All these technologies involve two basic processes during which material microstructure and macrostructure are formed.

The first basic process yielding material microstructure involves the application of a matrix material to the fibers. The simplest way to do it, normally utilized in the manufacturing of composites with thermosetting polymeric matrices, is a direct impregnation of tows, yarns, fabrics, or more complicated fibrous structures with liquid resins. Thermosetting resin has relatively low viscosity (10–100 Pa s), which can be controlled using solvents or heating, and good wetting ability for the majority of fibers. There exist two versions of this process. According to the so-called 'wet' process, impregnated fibrous material (tows, fabrics, etc.) is used to fabricate composite parts directly, without any additional treatment or interruption of the process. In contrast to that, in 'dry' or 'prepreg' processes, impregnated fibrous material is dried (not cured) and thus preimpregnated tapes obtained (prepregs) are stored for further utilization (usually under low temperature to prevent uncontrolled premature polymerization of the resin). An example of a machine for making prepregs is shown in Fig. 1.16. Both processes, having similar advantages and shortcomings, are widely used for composites with thermosetting matrices. For thermoplastic matrices, application of direct impregnation ('wet' processing) is limited by the relatively high viscosity (about 10^{12} Pa s) of thermoplastic polymer solutions or melts. For this reason, 'prepreg' processes with preliminary fabricated tapes or sheets in which fibers are already combined with the thermoplastic matrix are used to manufacture composite parts. There also exist other processes that involve application of heat and pressure to hybrid materials including reinforcing fibers and a thermoplastic polymer in the form of powder, films, or fibers. A promising process (called fibrous technology) utilizes tows, tapes, or fabrics with two types of fibers – reinforcing and thermoplastic. Under heat and pressure, thermoplastic fibers melt and form the matrix of the composite material. Metal and ceramic matrices are applied to fibers by means of casting, diffusion welding, chemical deposition, plasma spraying, processing by compression molding or with the aid of powder metallurgy methods.

The second basic process provides the proper macrostructure of a composite material corresponding to the loading and operational conditions of the composite part that is fabricated. There exist three main types of material macrostructure – linear structure which is appropriate for bars, profiles, and beams, plane laminated structure suitable for thin-walled plates and shells, and spatial structure which is necessary for thick-walled and bulk solid composite parts.

A linear structure is formed by pultrusion, table rolling, or braiding and provides high strength and stiffness in one direction coinciding with the axis of a bar, profile, or a beam. Pultrusion results in a unidirectionally reinforced composite profile made by pulling a bundle of fibers impregnated with resin through a heated die to cure the resin and, to provide the desired shape of the profile cross section. Profiles made by pultrusion and braiding are shown in Fig. 1.17. Table rolling is used to fabricate small diameter tapered tubular bars (e.g., ski poles or fishing rods) by rolling preimpregnated fiber tapes in the form of

Fig. 1.16. Machine making a prepreg from fiberglass fabric and epoxy resin. Courtesy of CRISM.

Fig. 1.17. Composite profiles made by pultrusion and braiding. Courtesy of CRISM.

flags around the metal mandrel which is pulled out of the composite bar after the resin is cured. Fibers in the flags are usually oriented along the bar axis or at an angle to the axis thus providing more complicated reinforcement than the unidirectional one typical of pultrusion. Even more complicated fiber placement with orientation angle varying from 5 to 85° along the bar axis can be achieved using two-dimensional (2D) braiding which results in a textile material structure consisting of two layers of yarns or tows interlaced with each other while they are wound onto the mandrel.

A plane-laminated structure consists of a set of composite layers providing the necessary stiffness and strength in at least two orthogonal directions in the plane of the laminate. Such a plane structure would be formed by hand or machine lay-up, fiber placement, or filament winding.

Lay-up and fiber placement technology provides fabrication of thin-walled composite parts of practically arbitrary shape by hand or automated placing of preimpregnated unidirectional or fabric tapes onto a mold. Layers with different fiber orientations (and even with different fibers) are combined to result in the laminated composite material exhibiting the desired strength and stiffness in given directions. Lay-up processes are usually accompanied by pressure applied to compact the material and to remove entrapped air. Depending on the required quality of the material, as well as on the shape and dimensions of a manufactured composite part, compacting pressure can be provided by rolling or vacuum bags, in autoclaves, or by compression molding. A catamaran yacht (length 9.2 m, width 6.8 m, tonnage 2.2 tons) made from carbon–epoxy composite by hand lay-up is shown in Fig. 1.18.

Filament winding is an efficient automated process of placing impregnated tows or tapes onto a rotating mandrel (Fig. 1.19) that is removed after curing of the composite material. Varying the winding angle, it is possible to control the material strength and stiffness within the layer and through the thickness of the laminate. Winding of a pressure vessel is shown in Fig. 1.20. Preliminary tension applied to the tows in the process of winding induces

Fig. 1.18. Catamaran yacht Ivan-30 made from carbon–epoxy composite by hand lay-up. Courtesy of CRISM.

Fig. 1.19. Manufacturing of a pipe by circumferential winding of preimpregnated fiberglass fabric. Courtesy of CRISM.

Fig. 1.20. Geodesic winding of a pressure vessel.

Fig. 1.21. A body of a small plane made by filament winding. Courtesy of CRISM.

pressure between the layers providing compaction of the material. Filament winding is the most advantageous in manufacturing thin-walled shells of revolution though it can also be used in building composite structures with more complicated shapes (Fig. 1.21).

Spatial macrostructure of the composite material that is specific for thick-walled and solid members requiring fiber reinforcement in at least three directions (not lying in one plane) can be formed by 3D braiding (with three interlaced yarns) or using such textile processes as weaving, knitting, or stitching. Spatial (3D, 4D, etc.) structures used in carbon–carbon technology are assembled from thin carbon composite rods fixed in different directions. Such a structure that is prepared for carbonization and deposition of a carbon matrix is shown in Fig. 1.22.

There are two specific manufacturing procedures that have an inverse sequence of the basic processes described above, i.e., first, the macrostructure of the material is formed and then the matrix is applied to fibers.

The first of these procedures is the aforementioned carbon–carbon technology that involves chemical vapor deposition of a pyrolitic carbon matrix on preliminary assembled and sometimes rather complicated structures made from dry carbon fabric. A carbon–carbon shell made by this method is shown in Fig. 1.23.

The second procedure is the well-known resin transfer molding. Fabrication of a composite part starts with a preform that is assembled in the internal cavity of a mold from dry fabrics, tows, yarns, etc., and forms the macrostructure of a composite part. The shape of this part is governed by the shape of the mold cavity into which liquid resin is transferred under pressure through injection ports.

The basic processes described above are always accompanied by a thermal treatment resulting in the solidification of the matrix. Heating is applied to cure thermosetting resins, cooling is used to transfer thermoplastic, metal, and ceramic matrices to a solid phase,

Fig. 1.22. A 4D spatial structure. Courtesy of CRISM.

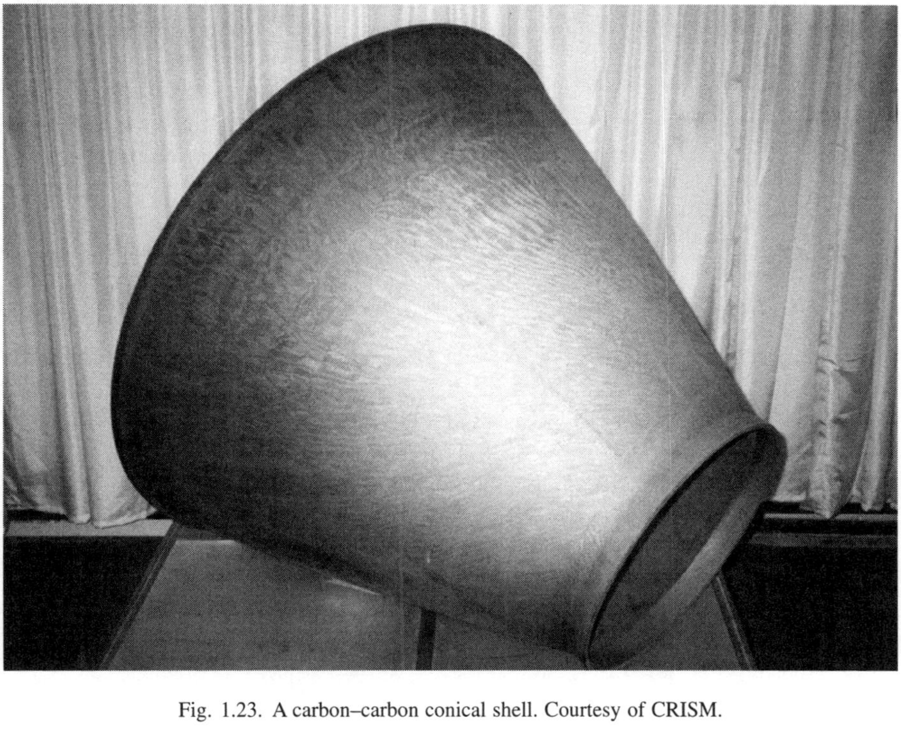

Fig. 1.23. A carbon–carbon conical shell. Courtesy of CRISM.

whereas a carbon matrix is made by pyrolisis. The final stages of the manufacturing procedure involve removal of mandrels, molds, or other tooling and machining of a composite part.

The fabrication processes are described in more detail elsewhere (e.g., Peters, 1998).

1.3. References

Bogdanovich, A.E. and Pastore, C.M. (1996). *Mechanics of Textile and Laminated Composites.* Chapman & Hall, London.

Chou, T.W. and Ko, F.K. (1989). *Textile Structural Composites* (T.W. Chou and F.K. Ko eds.). Elsevier, New York.

Fukuda, H., Yakushiji, M. and Wada, A. (1997). Loop test for the strength of monofilaments. In *Proc. 11th Int. Conf. on Comp. Mat. (ICCM-11), Vol. 5, Textile Composites and Characterization* (M.L. Scott ed.). Woodhead Publishing Ltd., Gold Coast, Australia, pp. 886–892.

Goodey, W.J. (1946). *Stress Diffusion Problems. Aircraft Eng.* June, 195–198; July, 227–234; August, 271–276; September, 313–316; October, 343–346; November, 385–389.

Karpinos, D.M. (1985). *Composite Materials. Handbook* (D.M. Karpinos ed.). Naukova Dumka, Kiev (in Russian).

Peters, S.T. (1998). *Handbook of Composites*, 2nd edn. (S.T. Peters ed.). Chapman & Hall, London.

Tarnopol'skii, Yu.M., Zhigun, I.G. and Polyakov, V.A. (1992). *Spatially Reinforced Composites.* Technomic, Pennsylvania.

Vasiliev, V.V. and Tarnopol'skii, Yu.M. (1990). *Composite Materials. Handbook* (V.V. Vasiliev and Yu.M. Tarnopol'skii eds.). Mashinostroenie, Moscow (in Russian).

Chapter 2

FUNDAMENTALS OF MECHANICS OF SOLIDS

The behavior of composite materials whose micro- and macrostructures are much more complicated than those of traditional structural materials such as metals, concrete, and plastics is nevertheless governed by the same general laws and principles of mechanics whose brief description is given below.

2.1. Stresses

Consider a solid body referred by Cartesian coordinates as in Fig. 2.1. The body is fixed at the part S_u of the surface and loaded with body forces q_v having coordinate components q_x, q_y, and q_z, and with surface tractions p_s specified by coordinate components p_x, p_y, and p_z. Surface tractions act on surface S_σ which is determined by its unit normal n with coordinate components l_x, l_y, and l_z that can be referred to as directional cosines of the normal, i.e.,

$$l_x = \cos(n, x), \quad l_y = \cos(n, y), \quad l_z = \cos(n, z) \tag{2.1}$$

Introduce some arbitrary cross section formally separating the upper part of the body from its lower part. Assume that the interaction of these parts in the vicinity of some point A can be simulated with some internal force per unit area or stress σ distributed over this cross section according to some as yet unknown law. Since the mechanics of solids is a phenomenological theory (see the closure of Section 1.1) we do not care about the physical nature of stress, which is only a parameter of our model of the real material (see Section 1.1) and, in contrast to forces F, has never been observed in physical experiments. Stress is referred to the plane on which it acts and is usually decomposed into three components – normal stress (σ_z in Fig. 2.1) and shear stresses (τ_{zx} and τ_{zy} in Fig. 2.1). The subscript of the normal stress and the first subscript of the shear stress indicate the plane on which the stresses act. For stresses shown in Fig. 2.1, this is the plane whose normal is parallel to the z-axis. The second subscript of the shear stress shows the axis along which the stress acts. If we single out a cubic element in the vicinity of point A (see Fig. 2.1), we should apply stresses to all its planes as in Fig. 2.2 which also shows notations and positive directions of all the stresses acting inside the body referred by Cartesian coordinates.

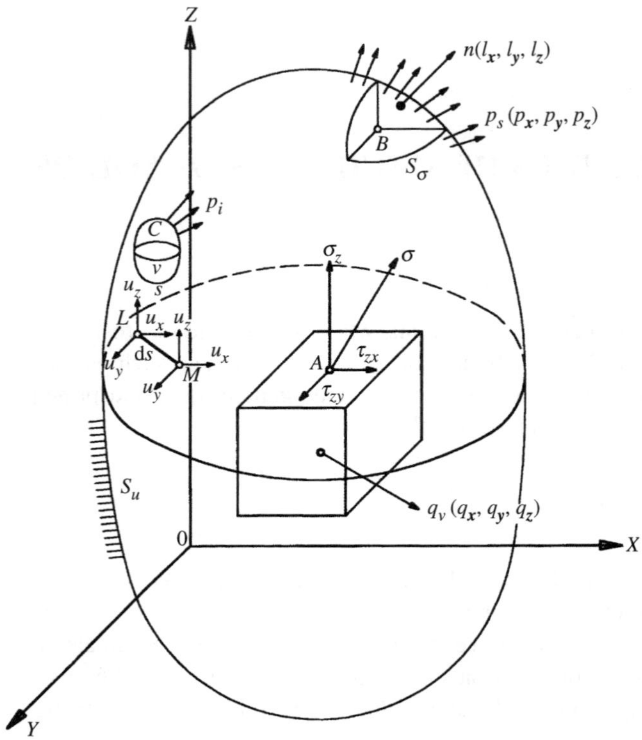

Fig. 2.1. A solid loaded with body and surface forces and referred by Cartesian coordinates.

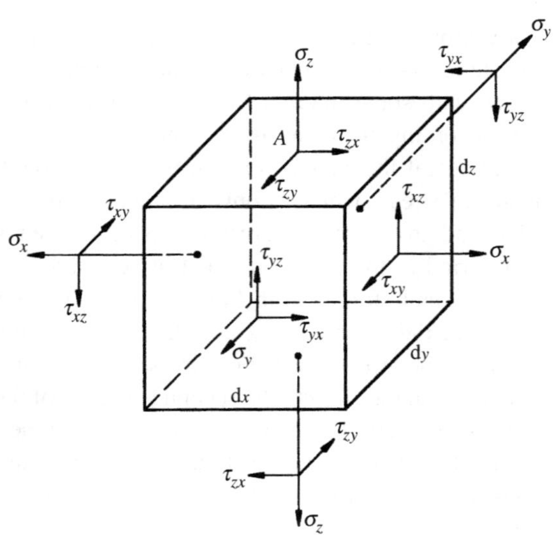

Fig. 2.2. Stress acting on the planes of the infinitely small cubic element.

2.2. Equilibrium equations

Now suppose that the body in Fig. 2.1 is in a state of equilibrium. Then, we can write equilibrium equations for any part of this body. In particular we can do this for an infinitely small tetrahedron singled out in the vicinity of point B (see Fig. 2.1) in such a way that one of its planes coincides with S_σ and the other three planes are coordinate planes of the Cartesian frame. Internal and external forces acting on this tetrahedron are shown in Fig. 2.3. The equilibrium equation corresponding, for example, to the x-axis can be written as

$$-\sigma_x \mathrm{d}S_x - \tau_{yx}\mathrm{d}S_y - \tau_{zx}\mathrm{d}S_z + p_x\mathrm{d}S_\sigma + q_x\mathrm{d}V = 0$$

Here, $\mathrm{d}S_\sigma$ and $\mathrm{d}V$ are the elements of the body surface and volume, whereas $\mathrm{d}S_x = \mathrm{d}S_\sigma l_x$, $\mathrm{d}S_y = \mathrm{d}S_\sigma l_y$, and $\mathrm{d}S_z = \mathrm{d}S_\sigma l_z$. When the tetrahedron is infinitely diminished, the term including $\mathrm{d}V$, which is of the order of the cube of the linear dimensions, can be neglected in comparison with terms containing $\mathrm{d}S$, which is of the order of the square of the linear dimensions. The resulting equation is

$$\sigma_x l_x + \tau_{yx} l_y + \tau_{zx} l_z = p_x \quad (x,\ y,\ z) \tag{2.2}$$

The symbol $(x,\ y,\ z)$, which is widely used in this chapter, denotes permutation with the aid of which we can write two more equations corresponding to the other two axes changing x for y, y for z, and z for x.

Consider now the equilibrium of an arbitrary finite part C of the body (see Fig. 2.1). If we single this part out of the body, we should apply to it body forces q_v and surface tractions p_i whose coordinate components p_x, p_y, and p_z can be expressed, obviously, by Eq. (2.2) in terms of stresses acting inside the volume C. Because the sum of the

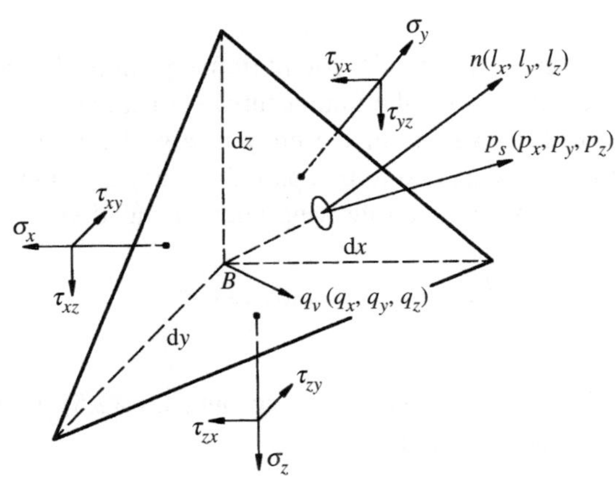

Fig. 2.3. Forces acting on an elementary tetrahedron.

components corresponding, for example, to the x-axis must be equal to zero, we have

$$\iiint_v q_x \mathrm{d}v + \iint_s p_x \mathrm{d}s = 0$$

where v and s are the volume and the surface area of the part of the body under consideration. Substituting p_x from Eq. (2.2) we get

$$\iint_s (\sigma_x l_x + \tau_{yx} l_y + \tau_{zx} l_z)\mathrm{d}s + \iiint_v q_x \mathrm{d}v = 0 \quad (x, y, z) \tag{2.3}$$

Thus, we have three integral equilibrium equations, Eq. (2.3), which are valid for any finite part of the body. To convert them into the corresponding differential equations, we use Green's integral transformation

$$\iint_s (f_x l_x + f_y l_y + f_z l_z)\mathrm{d}s = \iiint_v \left(\frac{\partial f_x}{\partial x} + \frac{\partial f_y}{\partial y} + \frac{\partial f_z}{\partial z}\right)\mathrm{d}v \tag{2.4}$$

which is valid for any three continuous, finite, and single-valued functions $f(x, y, z)$ and allows us to transform a surface integral into a volume one. Taking $f_x = \sigma_x$, $f_y = \tau_{yx}$, and $f_z = \tau_{zx}$ in Eq. (2.4) and using Eq. (2.3), we arrive at

$$\iiint_v \left(\frac{\partial \sigma_x}{\partial x} + \frac{\partial \tau_{yx}}{\partial y} + \frac{\partial \tau_{zx}}{\partial z} + q_x\right)\mathrm{d}v = 0 \quad (x, y, z)$$

Since these equations hold true for whatever the part of the solid may be, provided only that it is within the solid, they yield

$$\frac{\partial \sigma_x}{\partial x} + \frac{\partial \tau_{yx}}{\partial y} + \frac{\partial \tau_{zx}}{\partial z} + q_x = 0 \quad (x, y, z) \tag{2.5}$$

Thus, we have arrived at three differential equilibrium equations that could also be derived from the equilibrium conditions for the infinitesimal element shown in Fig. 2.2.

However, in order to keep part C of the body in Fig. 2.1 in equilibrium the sum of the moments of all the forces applied to this part about any axis must be zero. By taking moments about the z-axis we get the following integral equation

$$\iiint_v (q_x y - q_y x)\mathrm{d}v + \iint_s (p_x y - p_y x)\mathrm{d}s = 0$$

Using again Eqs. (2.2), (2.4), and taking into account Eq. (2.5) we finally arrive at the symmetry conditions for shear stresses, i.e.,

$$\tau_{xy} = \tau_{yx} \quad (x, y, z) \tag{2.6}$$

So, we have three equilibrium equations, Eq. (2.5) which include six unknown stresses $\sigma_x, \sigma_y, \sigma_z$ and $\tau_{xy}, \tau_{xz}, \tau_{yz}$.

Eq. (2.2) can be treated as force boundary conditions for the stressed state of a solid.

2.3. Stress transformation

Consider the transformation of a stress system from one Cartesian coordinate frame to another. Suppose that the elementary tetrahedron shown in Fig. 2.3 is located inside the body and that point B coincides with the origin 0 of Cartesian coordinates x, y, and z in Fig. 2.1. Then, the oblique plane of the tetrahedron can be treated as a coordinate plane $z' = 0$ of a new coordinate frame x', y', z' shown in Fig. 2.4 and such that the normal element to the oblique plane coincides with the z'-axis, whereas axes x' and y' are located in this plane. Component p_x of the surface traction in Eq. (2.2) can be treated now as the projection on the x-axis of stress σ acting on plane $z' = 0$. Then, Eq. (2.2) can be presented in the following explicit form specifying projections of stress σ

$$p_x = \sigma_x l_{z'x} + \tau_{yx} l_{z'y} + \tau_{zx} l_{z'z}$$

$$p_y = \sigma_y l_{z'y} + \tau_{zy} l_{z'z} + \tau_{xy} l_{z'x} \qquad (2.7)$$

$$p_z = \sigma_z l_{z'z} + \tau_{xz} l_{z'x} + \tau_{yz} l_{z'y}$$

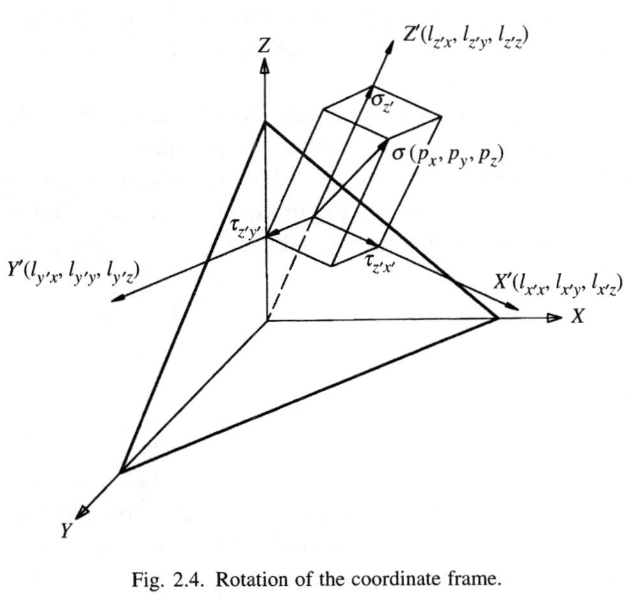

Fig. 2.4. Rotation of the coordinate frame.

Here, l are directional cosines of axis z' with respect to axes x, y, and z (see Fig. 2.4 in which the corresponding cosines of axes x' and y' are also presented). The normal stress $\sigma_{z'}$ can be found now as

$$\begin{aligned}\sigma_{z'} &= p_x l_{z'x} + p_y l_{z'y} + p_z l_{z'z} \\ &= \sigma_x l_{z'x}^2 + \sigma_y l_{z'y}^2 + \sigma_z l_{z'z}^2 + 2\tau_{xy} l_{z'x} l_{z'y} + 2\tau_{xz} l_{z'x} l_{z'z} + 2\tau_{yz} l_{z'y} l_{z'z} \quad (x', y', z')\end{aligned} \tag{2.8}$$

The final result was obtained with the aid of Eqs. (2.6) and (2.7). Changing x' for y', y' for z', and z' for x', i.e., performing the appropriate permutation in Eq. (2.8) we can write similar expressions for $\sigma_{x'}$ and $\sigma_{y'}$.

The shear stress in the new coordinates is

$$\begin{aligned}\tau_{z'x'} &= p_x l_{x'x} + p_y l_{x'y} + p_z l_{x'z} \\ &= \sigma_x l_{x'x} l_{z'x} + \sigma_y l_{x'y} l_{z'y} + \sigma_z l_{x'z} l_{z'z} + \tau_{xy}(l_{x'x} l_{z'y} + l_{x'y} l_{z'x}) \\ &\quad + \tau_{xz}(l_{x'x} l_{z'z} + l_{x'z} l_{z'x}) + \tau_{yz}(l_{x'y} l_{z'z} + l_{x'x} l_{z'y}) \quad (x', y', z')\end{aligned} \tag{2.9}$$

Permutation yields expressions for $\tau_{x'y'}$ and $\tau_{y'z'}$.

2.4. Principal stresses

The foregoing equations, Eqs. (2.8) and (2.9), demonstrate stress transformations under rotation of a coordinate frame. There exists a special position of this frame in which the shear stresses acting on the coordinate planes vanish. Such coordinate axes are called the principal axes, and the normal stresses that act on the corresponding coordinate planes are referred to as the principal stresses.

To determine the principal stresses, assume that coordinates x', y', and z', in Fig. 2.4 are the principal coordinates. Then, according to the aforementioned property of the principal coordinates, we should take $\tau_{z'x'} = \tau_{z'y'} = 0$ and $\sigma_{z'} = \sigma$ for the plane $z' = 0$. This means that $p_x = \sigma l_{z'x}$, $p_y = \sigma l_{z'y}$, and $p_z = \sigma l_{z'z}$ in Eqs. (2.7). Introducing new notations for directional cosines of the principal axis, i.e., taking $l_{z'x} = l_{px}$, $l_{z'y} = l_{py}$, $l_{z'z} = l_{pz}$ we have from Eqs. (2.7)

$$\begin{aligned}&(\sigma_x - \sigma) l_{px} + \tau_{xy} l_{py} + \tau_{xz} l_{pz} = 0 \\ &\tau_{xy} l_{px} + (\sigma_y - \sigma) l_{py} + \tau_{yz} l_{pz} = 0 \\ &\tau_{xz} l_{px} + \tau_{yz} l_{py} + (\sigma_z - \sigma) l_{pz} = 0\end{aligned} \tag{2.10}$$

These equations were transformed with the aid of symmetry conditions for shear stresses, Eq. (2.6). For some specified point of the body in the vicinity of which the principal stresses are determined in terms of stresses referred to some fixed coordinate frame x, y, z

and known, Eqs. (2.10) comprise a homogeneous system of linear algebraic equations. Formally, this system always has the trivial solution, i.e., $l_{px} = l_{py} = l_{pz} = 0$ which we can ignore because directional cosines should satisfy an evident condition following from Eqs. (2.1), i.e.,

$$l_{px}^2 + l_{py}^2 + l_{pz}^2 = 1 \tag{2.11}$$

So, we need to find a nonzero solution of Eqs. (2.10) which can exist if the determinant of the set is zero. This condition yields the following cubic equation for σ

$$\sigma^3 - I_1\sigma^2 - I_2\sigma - I_3 = 0 \tag{2.12}$$

in which

$$\begin{aligned} I_1 &= \sigma_x + \sigma_y + \sigma_z \\ I_2 &= -\sigma_x\sigma_y - \sigma_x\sigma_z - \sigma_y\sigma_z + \tau_{xy}^2 + \tau_{xz}^2 + \tau_{yz}^2 \\ I_3 &= \sigma_x\sigma_y\sigma_z + 2\tau_{xy}\tau_{xz}\tau_{yz} - \sigma_x\tau_{yz}^2 - \sigma_y\tau_{xz}^2 - \sigma_z\tau_{xy}^2 \end{aligned} \tag{2.13}$$

are invariant characteristics (invariants) of the stressed state. This means that if we refer the body to any Cartesian coordinate frame with directional cosines specified by Eqs. (2.1), take the origin of this frame at some arbitrary point and change stresses in Eqs. (2.13) with the aid of Eqs. (2.8) and (2.9), the values of I_1, I_2, I_3 at this point will be the same for all such coordinate frames. Eq. (2.12) has three real roots that specify three principal stresses σ_1, σ_2, and σ_3. There is a convention according to which $\sigma_1 \geq \sigma_2 \geq \sigma_3$, i.e., σ_1 is the maximum principal stress and σ_3 is the minimum one. If, for example, the roots of Eq. (2.12) are 100 MPa, −200 MPa, and 0, then $\sigma_1 = 100$ MPa, $\sigma_2 = 0$, and $\sigma_3 = -200$ MPa.

To demonstrate the procedure, consider a particular state of stress relevant to several applications, namely, pure shear in the *xy*-plane. Let a thin square plate referred to coordinates x, y, z be loaded with shear stresses τ uniformly distributed over the plate thickness and along the edges (see Fig. 2.5).

One principal plane is evident – it is plane $z = 0$, which is free of shear stresses. To find the other two planes, we should take in Eqs. (2.13) $\sigma_x = \sigma_y = \sigma_z = 0$, $\tau_{xz} = \tau_{yz} = 0$, and $\tau_{xy} = \tau$. Then, Eq. (2.12) takes the form

$$\sigma^3 - \tau^2\sigma = 0$$

The first root of this equation gives $\sigma = 0$ and corresponds to plane $z = 0$. The other two roots are $\sigma = \pm\tau$. Thus, we have three principal stresses, i.e., $\sigma_1 = \tau$, $\sigma_2 = 0$, $\sigma_3 = -\tau$. To find the planes corresponding to σ_1 and σ_3 we should put $l_{pz} = 0$, substitute $\sigma = \pm\tau$ into Eqs. (2.10), write them for the state of stress under study, and supplement this set with Eq. (2.11). The final equations allowing us to find l_{px} and l_{py} are

$$\pm\tau l_{px} + \tau l_{py} = 0, \quad l_{px}^2 + l_{py}^2 = 1$$

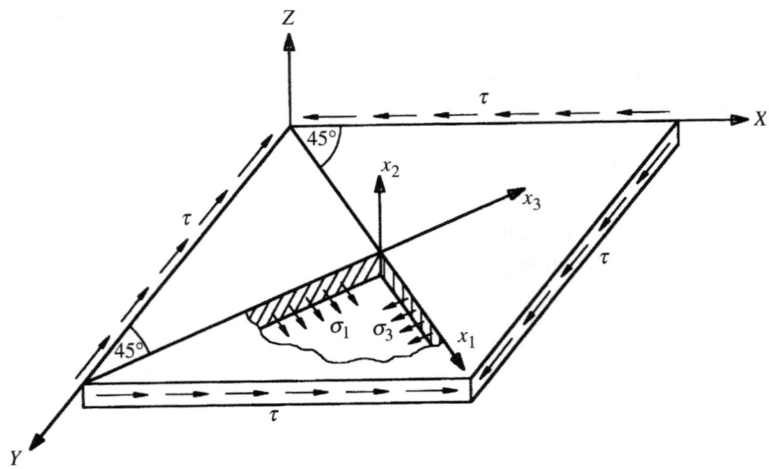

Fig. 2.5. Principal stresses under pure shear.

Solution of these equations yields $l_{px} = \pm 1/\sqrt{2}$ and $l_{py} = \pm 1/\sqrt{2}$, and means that principal planes (or principal axes) make 45° angles with axes x and y. Principal stresses and principal coordinates x_1, x_2, and x_3 are shown in Fig. 2.5.

2.5. Displacements and strains

For any point of a solid (e.g., L or M in Fig. 2.1) coordinate component displacements u_x, u_y, and u_z can be introduced which specify the point displacements in the directions of coordinate axes.

Consider an arbitrary infinitely small element LM characterized with its directional cosines

$$l_x = \frac{\mathrm{d}x}{\mathrm{d}s}, \quad l_y = \frac{\mathrm{d}y}{\mathrm{d}s}, \quad l_z = \frac{\mathrm{d}z}{\mathrm{d}s} \tag{2.14}$$

The positions of this element before and after deformation are shown in Fig. 2.6. Suppose that the displacements of the point L are u_x, u_y, and u_z. Then, the displacements of the point M should be

$$u_x^{(1)} = u_x + \mathrm{d}u_x, \quad u_y^{(1)} = u_y + \mathrm{d}u_y, \quad u_z^{(1)} = u_z + \mathrm{d}u_z \tag{2.15}$$

Since u_x, u_y, and u_z are continuous functions of x, y, z, we get

$$\mathrm{d}u_x = \frac{\partial u_x}{\partial x}\mathrm{d}x + \frac{\partial u_y}{\partial y}\mathrm{d}y + \frac{\partial u_z}{\partial z}\mathrm{d}z \quad (x, y, z) \tag{2.16}$$

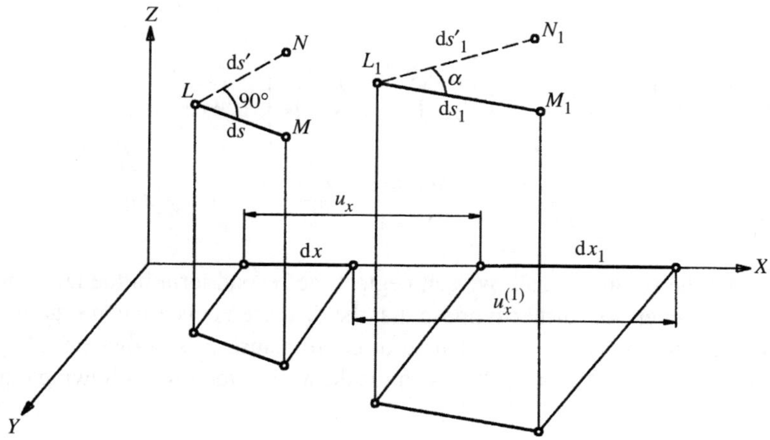

Fig. 2.6. Displacement of an infinitesimal linear element.

It follows from Fig. 2.6 and Eqs. (2.15) and (2.16) that,

$$\mathrm{d}x_1 = \mathrm{d}x + u_x^{(1)} - u_x = \mathrm{d}x + \mathrm{d}u_x = \left(1 + \frac{\partial u_x}{\partial x}\right)\mathrm{d}x + \frac{\partial u_x}{\partial y}\mathrm{d}y + \frac{\partial u_x}{\partial z}\mathrm{d}z \quad (x, y, z) \tag{2.17}$$

Introduce the strain of element *LM* as

$$\varepsilon = \frac{\mathrm{d}s_1 - \mathrm{d}s}{\mathrm{d}s} \tag{2.18}$$

After some rearrangements we arrive at

$$\varepsilon + \frac{1}{2}\varepsilon^2 = \frac{1}{2}\left[\left(\frac{\mathrm{d}s_1}{\mathrm{d}s}\right)^2 - 1\right]$$

where

$$\mathrm{d}s_1^2 = (\mathrm{d}x_1)^2 + (\mathrm{d}y_1)^2 + (\mathrm{d}z_1)^2$$

Substituting for $\mathrm{d}x_1$, $\mathrm{d}y_1$, $\mathrm{d}z_1$ in their expressions from Eq. (2.17) and taking into account Eqs. (2.14), we finally get

$$\varepsilon + \frac{1}{2}\varepsilon^2 = \varepsilon_{xx}l_x^2 + \varepsilon_{yy}l_y^2 + \varepsilon_{zz}l_z^2 + \varepsilon_{xy}l_xl_y + \varepsilon_{xz}l_xl_z + \varepsilon_{yz}l_yl_z \tag{2.19}$$

where

$$\varepsilon_{xx} = \frac{\partial u_x}{\partial x} + \frac{1}{2}\left[\left(\frac{\partial u_x}{\partial x}\right)^2 + \left(\frac{\partial u_y}{\partial x}\right)^2 + \left(\frac{\partial u_z}{\partial x}\right)^2\right] \quad (x, y, z)$$

$$\varepsilon_{xy} = \frac{\partial u_x}{\partial y} + \frac{\partial u_y}{\partial x} + \frac{\partial u_x}{\partial x}\frac{\partial u_x}{\partial y} + \frac{\partial u_y}{\partial x}\frac{\partial u_y}{\partial y} + \frac{\partial u_z}{\partial x}\frac{\partial u_z}{\partial y} \quad (x, y, z) \tag{2.20}$$

Assuming that the strain is small, we can neglect the second term in the left-hand side of Eq. (2.19). Moreover, we further suppose that the displacements are continuous functions that change rather slowly with the change of coordinates. This allows us to neglect the products of derivatives in Eqs. (2.20). As a result, we arrive at the following equation

$$\varepsilon = \varepsilon_x l_x^2 + \varepsilon_y l_y^2 + \varepsilon_z l_z^2 + \gamma_{xy} l_x l_y + \gamma_{xz} l_x l_z + \gamma_{yz} l_y l_z \tag{2.21}$$

in which

$$\varepsilon_x = \frac{\partial u_x}{\partial x}, \quad \varepsilon_y = \frac{\partial u_y}{\partial y}, \quad \varepsilon_z = \frac{\partial u_z}{\partial z}$$

$$\gamma_{xy} = \frac{\partial u_x}{\partial y} + \frac{\partial u_y}{\partial x}, \quad \gamma_{xz} = \frac{\partial u_x}{\partial z} + \frac{\partial u_z}{\partial x}, \quad \gamma_{yz} = \frac{\partial u_y}{\partial z} + \frac{\partial u_z}{\partial y} \tag{2.22}$$

can be treated as linear strain–displacement equations. Taking $l_x = 1, \ l_y = l_z = 0$ in Eqs. (2.22), i.e., directing element *LM* in Fig. 2.6 along the x-axis we can readily see that ε_x is the strain along the same x-axis. Similar reasoning shows that ε_y and ε_z in Eqs. (2.22) are strains in the directions of axes y and z. To find out the physical meaning of strains γ in Eqs. (2.22), consider two orthogonal line elements *LM* and *LN* and find angle α that they make with each other after deformation (see Fig. 2.6), i.e.,

$$\cos\alpha = \frac{\mathrm{d}x_1\mathrm{d}x_1' + \mathrm{d}y_1\mathrm{d}y_1' + \mathrm{d}z_1\mathrm{d}z_1'}{\mathrm{d}s_1\mathrm{d}s_1'} \tag{2.23}$$

Here, $\mathrm{d}x_1$, $\mathrm{d}y_1$, and $\mathrm{d}z_1$ are specified with Eq. (2.17), $\mathrm{d}s_1$ can be found from Eq. (2.18), and

$$\mathrm{d}x_1' = \left(1 + \frac{\partial u_x}{\partial x}\right)\mathrm{d}x' + \frac{\partial u_x}{\partial y}\mathrm{d}y' + \frac{\partial u_x}{\partial z}\mathrm{d}z' \quad (x, y, z)$$

$$\mathrm{d}s_1' = \mathrm{d}s'(1 + \varepsilon') \tag{2.24}$$

Introduce directional cosines of element *LN* as

$$l_x' = \frac{\mathrm{d}x'}{\mathrm{d}s'}, \quad l_y' = \frac{\mathrm{d}y'}{\mathrm{d}s'}, \quad l_z' = \frac{\mathrm{d}z'}{\mathrm{d}s'} \tag{2.25}$$

Since elements *LM* and *LN* are orthogonal, we have

$$l_x l'_x + l_y l'_y + l_z l'_z = 0$$

Using Eqs. (2.14), (2.18), (2.24)–(2.26) and introducing the shear strain γ as the difference between angles $M_1L_1N_1$ and *MLN*, i.e., as

$$\gamma = \frac{\pi}{2} - \alpha$$

we can write Eq. (2.23) in the following form

$$\sin\gamma = \frac{1}{(1+\varepsilon)(1+\varepsilon')}\Big[2(\varepsilon_{xx} l_x l'_x + \varepsilon_{yy} l_y l'_y + \varepsilon_{zz} l_z l'_z) + \varepsilon_{xy}(l_x l'_y + l'_x l_y) + \varepsilon_{xz}(l_x l'_z + l'_x l_z) + \varepsilon_{yz}(l_y l'_z + l'_y l_z)\Big] \quad (2.26)$$

Linear approximation of Eq. (2.26) similar to Eq. (2.21) yields

$$\gamma = 2(\varepsilon_x l_x l'_x + \varepsilon_y l_y l'_y + \varepsilon_z l_z l'_z) + \gamma_{xy}(l_x l'_y + l'_x l_y) + \gamma_{xz}(l_x l'_z + l'_x l_z) + \gamma_{yz}(l_y l'_z + l'_y l_z) \quad (2.27)$$

Here, ε_x, ε_y, ε_z and γ_{xy}, γ_{xz}, γ_{yz} components are determined with Eqs. (2.22). If we now direct element *LM* along the x-axis and element *LN* along the y-axis putting $l_x = 1$, $l_y = l_z = 0$ and $l'_y = 1$, $l'_x = l'_z = 0$, Eq. (2.27) yields $\gamma = \gamma_{xy}$. Thus, γ_{xy}, γ_{xz}, and γ_{yz} are shear strains that are equal to the changes of angles between axes x and y, x and z, y and z, respectively.

2.6. Transformation of small strains

Consider small strains in Eqs. (2.22) and study their transformation under rotation of the coordinate frame. Suppose that x', y', z' in Fig. 2.4 form a new coordinate frame rotated with respect to original frame x, y, z. Since Eqs. (2.22) are valid for any Cartesian coordinate frame, we have

$$\varepsilon_{x'} = \frac{\partial u_{x'}}{\partial x'}, \quad \gamma_{x'y'} = \frac{\partial u_{x'}}{\partial y'} + \frac{\partial u_{y'}}{\partial x'} \quad (x, y, z) \quad (2.28)$$

Here, $u_{x'}$, $u_{y'}$, and $u_{z'}$ are displacements along the axes x', y', z' which can be related to displacements u_x, u_y, and u_z of the same point by the following linear equations

$$u_{x'} = u_x l_{x'x} + u_y l_{x'y} + u_z l_{x'z} \quad (x, y, z) \quad (2.29)$$

Similar relations can be written for the derivatives of displacement with respect to variables x', y', z' and x, y, z, i.e.,

$$\frac{\partial u}{\partial x'} = \frac{\partial u}{\partial x} l_{x'x} + \frac{\partial u}{\partial y} l_{x'y} + \frac{\partial u}{\partial z} l_{x'z} \quad (x, y, z) \tag{2.30}$$

Substituting displacements, Eq. (2.29), into Eqs. (2.28), and transforming to variables x, y, z with the aid of Eqs. (2.30), and taking into account Eqs. (2.22), we arrive at

$$\begin{aligned} \varepsilon_{x'} &= \varepsilon_x l_{x'x}^2 + \varepsilon_y l_{x'y}^2 + \varepsilon_z l_{x'z}^2 + \gamma_{xy} l_{x'x} l_{x'y} + \gamma_{xz} l_{x'x} l_{x'z} + \gamma_{yz} l_{x'y} l_{x'z} \quad (x, y, z) \\ \gamma_{x'y'} &= 2\varepsilon_x l_{x'x} l_{y'x} + 2\varepsilon_y l_{x'y} l_{y'y} + 2\varepsilon_z l_{x'z} l_{y'z} + \gamma_{xy}(l_{x'x} l_{y'y} + l_{x'y} l_{y'x}) \\ &\quad + \gamma_{xz}(l_{x'x} l_{y'z} + l_{x'z} l_{y'x}) + \gamma_{yz}(l_{x'y} l_{y'z} + l_{x'z} l_{y'y}) \quad (x, y, z) \end{aligned} \tag{2.31}$$

These strain transformations are similar to the stress transformations determined by Eqs. (2.8) and (2.9).

2.7. Compatibility equations

Consider strain–displacement equations, Eqs. (2.22), and try to determine displacements u_x, u_y, and u_z in terms of strains ε_x, ε_y, ε_z and γ_{xy}, γ_{xz} γ_{yz}. As can be seen, there are six equations containing only three unknown displacements. In the general case, such a set of equations is not consistent, and some compatibility conditions should be imposed on the strains to provide the existence of a solution. To derive these conditions, decompose derivatives of the displacements as follows

$$\frac{\partial u_x}{\partial x} = \varepsilon_x, \quad \frac{\partial u_x}{\partial y} = \frac{1}{2}\gamma_{xy} - \omega_z, \quad \frac{\partial u_x}{\partial z} = \frac{1}{2}\gamma_{xz} + \omega_y \quad (x, y, z) \tag{2.32}$$

Here

$$\omega_z = \frac{1}{2}\left(\frac{\partial u_y}{\partial x} - \frac{\partial u_x}{\partial y}\right) \quad (x, y, z) \tag{2.33}$$

is the angle of rotation of a body element (such as the cubic element shown in Fig. 2.1) around the z-axis. Three Eqs. (2.32) including one and the same displacement u_x allow us to construct three couples of mixed second-order derivatives of u_x with respect to x and y or y and x, x and z or z and x, y and z or z and y. As long as the sequence of differentiation does not influence the result and since there are two other groups of equations in Eqs. (2.32), we arrive at nine compatibility conditions that can be presented as

$$\frac{\partial \omega_x}{\partial x} = \frac{1}{2}\left(\frac{\partial \gamma_{xz}}{\partial y} - \frac{\partial \gamma_{xy}}{\partial z}\right) \quad (x, y, z)$$

$$\frac{\partial \omega_x}{\partial y} = \frac{1}{2}\frac{\partial \gamma_{yz}}{\partial y} - \frac{\partial \varepsilon_y}{\partial z} \quad (x, y, z), \qquad \frac{\partial \omega_x}{\partial z} = -\frac{1}{2}\frac{\partial \gamma_{yz}}{\partial z} + \frac{\partial \varepsilon_z}{\partial y} \quad (x, y, z) \tag{2.34}$$

These equations are similar to Eqs. (2.32), i.e., they allow us to determine rotation angles only if some compatibility conditions are valid. These conditions compose the set of compatibility equations for strains and have the following final form

$$k_{xy}(\varepsilon, \gamma) = 0, \quad r_x(\varepsilon, \gamma) = 0 \quad (x, y, z) \tag{2.35}$$

where

$$k_{xy}(\varepsilon, \gamma) = \frac{\partial^2 \varepsilon_x}{\partial y^2} + \frac{\partial^2 \varepsilon_y}{\partial x^2} - \frac{\partial^2 \gamma_{xy}}{\partial x \partial y} \quad (x, y, z)$$

$$r_x(\varepsilon, \gamma) = \frac{\partial^2 \varepsilon_x}{\partial y \partial z} - \frac{1}{2}\frac{\partial}{\partial x}\left(\frac{\partial \gamma_{xy}}{\partial z} + \frac{\partial \gamma_{xz}}{\partial y} - \frac{\partial \gamma_{yz}}{\partial x}\right) \quad (x, y, z) \tag{2.36}$$

If strains $\varepsilon_x, \varepsilon_y, \varepsilon_z$ and $\gamma_{xy}, \gamma_{xz}, \gamma_{yz}$ satisfy Eqs. (2.35), we can find rotation angles $\omega_x, \omega_y, \omega_z$ integrating Eqs. (2.34) and then determine displacements u_x, u_y, u_z integrating Eqs. (2.32).

The six compatibility equations, Eqs. (2.35), derived formally as compatibility conditions for Eqs. (2.32), have a simple physical meaning. Suppose that we have a continuous solid as shown in Fig. 2.1 and divide it into a set of pieces that perfectly match each other. Now, apply some strains to each of these pieces. Obviously, for arbitrary strains, the deformed pieces cannot be assembled into a continuous deformed solid. This will happen only under the condition that the strains satisfy Eqs. (2.35). However, even if the strains do not satisfy Eqs. (2.35), we can assume that the solid is continuous but in a more general Riemannian (curved) space rather than in traditional Euclidean space in which the solid existed before the deformation (Vasiliev and Gurdal, 1999). Then, six quantities k and r in Eqs. (2.36), being nonzero, specify curvatures of the Riemannian space caused by small strains ε and γ. The compatibility equations, Eqs. (2.35), require these curvatures to be equal to zero which means that the solid should remain in the Euclidean space under deformation.

2.8. Admissible static and kinematic fields

In solid mechanics, we introduce static field variables which are stresses and kinematic field variables which are displacements and strains.

The static field is said to be statically admissible if the stresses satisfy equilibrium equations, Eq. (2.5), and are in equilibrium with surface tractions on the body surface S_σ where these tractions are given (see Fig. 2.1), i.e., if Eq. (2.2) are satisfied on S_σ.

The kinematic field is referred to as kinematically admissible if displacements and strains are linked by strain–displacement equations, Eqs. (2.22), and displacements satisfy kinematic boundary conditions on the surface S_u where displacements are prescribed (see Fig. 2.1).

Actual stresses and displacements belong, naturally, to the corresponding admissible fields though actual stresses must in addition provide admissible displacements, whereas actual displacements should be associated with admissible stresses. Mutual correspondence between static and kinematic variables is established through the so-called constitutive equations that are considered in the next section.

2.9. Constitutive equations for an elastic solid

Consider a solid loaded with body and surface forces as in Fig. 2.1. These forces induce some stresses, displacements, and strains that compose the fields of actual static and kinematic variables. Introduce some infinitesimal additional displacements $\mathrm{d}u_x$, $\mathrm{d}u_y$, and $\mathrm{d}u_z$ such that they belong to a kinematically admissible field. This means that there exist equations that are similar to Eqs. (2.22), i.e.,

$$\mathrm{d}\varepsilon_x = \frac{\partial}{\partial x}(\mathrm{d}u_x), \quad \mathrm{d}\gamma_{xy} = \frac{\partial}{\partial y}(\mathrm{d}u_x) + \frac{\partial}{\partial x}(\mathrm{d}u_y) \quad (x, y, z) \tag{2.37}$$

and specify additional strains.

Since additional displacements are infinitely small, we can assume that external forces do not change under such variation of the displacements (here we do not consider special cases in which external forces depend on displacements of the points at which these forces are applied). Then we can calculate the work performed by the forces by multiplying forces by the corresponding increments of the displacements and writing the total work of body forces and surface tractions as

$$\mathrm{d}W = \iiint_V (q_x \mathrm{d}u_x + q_y \mathrm{d}u_y + q_z \mathrm{d}u_z)\mathrm{d}V + \iint_S (p_x \mathrm{d}u_x + p_y \mathrm{d}u_y + p_z \mathrm{d}u_z)\mathrm{d}S \tag{2.38}$$

Here, V and S are the body volume and external surface of the body in Fig. 2.1. Actually, we must write the surface integral in Eq. (2.38) only for the surface S_σ on which the forces are given. However, since the increments of the displacements belong to a kinematically admissible field, they are equal to zero on S_u, and the integral can be written for the whole surface of the body. To proceed, we express p_x, p_y, and p_z in terms of stresses with the aid of Eq. (2.2) and transform the surface integral into a volume one using Eq. (2.4). For the sake of brevity, consider only x-components of forces and displacement in Eq. (2.38). We have in several steps

$$\iiint_V q_x \mathrm{d}u_x + \iint_S p_x \mathrm{d}u_x \mathrm{d}s = \iiint_V q_x \mathrm{d}u_x + \iint_S (\sigma_x l_x + \tau_{yx} l_y + \tau_{zx} l_z)\mathrm{d}u_x \mathrm{d}S$$

$$= \iiint_V \left[q_x \mathrm{d}u_x + \frac{\partial}{\partial x}(\sigma_x \mathrm{d}u_x) + \frac{\partial}{\partial y}(\tau_{yx}\mathrm{d}u_x) + \frac{\partial}{\partial z}(\tau_{zx}\mathrm{d}u_x) \right] \mathrm{d}V$$

$$= \iiint_V \left[\left(q_x + \frac{\partial \sigma_x}{\partial x} + \frac{\partial \tau_{yx}}{\partial y} + \frac{\partial \tau_{zx}}{\partial z} \right) \mathrm{d}u_x + \sigma_x \frac{\partial}{\partial x}(\mathrm{d}u_x) \right.$$

$$\left. + \tau_{yx}\frac{\partial}{\partial y}(\mathrm{d}u_x) + \tau_{zx}\frac{\partial}{\partial z}(\mathrm{d}u_x) \right] \mathrm{d}V$$

$$= \iiint_V \left[\sigma_x \mathrm{d}\varepsilon_x + \tau_{xy}\frac{\partial}{\partial y}(\mathrm{d}u_x) + \tau_{xz}\frac{\partial}{\partial z}(\mathrm{d}u_x) \right] \mathrm{d}V$$

The last transformation step has been performed with due regard to Eqs. (2.5), (2.6), and (2.37). Finally, Eq. (2.38) takes the form

$$\mathrm{d}W = \iiint_V (\sigma_x \mathrm{d}\varepsilon_x + \sigma_y \mathrm{d}\varepsilon_y + \sigma_z \mathrm{d}\varepsilon_z + \tau_{xy}\mathrm{d}\gamma_{xy} + \tau_{xz}\mathrm{d}\gamma_{xz} + \tau_{yz}\mathrm{d}\gamma_{yz})\mathrm{d}V \tag{2.39}$$

Since the right-hand side of this equation includes only internal variables, i.e., stresses and strains, we can conclude that the foregoing formal rearrangement actually allows us to transform the work of external forces into the work of internal forces or into potential energy accumulated in the body. For further derivation, let us introduce for the sake of brevity new notations for coordinates and use subscripts 1, 2, 3 instead of x, y, z, respectively. We also use the following notations for stresses and strains

$$\sigma_x = \sigma_{11}, \quad \sigma_y = \sigma_{22}, \quad \sigma_z = \sigma_{33}$$

$$\tau_{xy} = \sigma_{12} = \sigma_{21}, \quad \tau_{xz} = \sigma_{13} = \sigma_{31}, \quad \tau_{yz} = \sigma_{23} = \sigma_{32}$$

$$\varepsilon_x = \varepsilon_{11}, \quad \varepsilon_y = \varepsilon_{22}, \quad \varepsilon_z = \varepsilon_{33}$$

$$\gamma_{xy} = 2\varepsilon_{12} = 2\varepsilon_{21}, \quad \gamma_{xz} = 2\varepsilon_{13} = 2\varepsilon_{31}, \quad \gamma_{yz} = 2\varepsilon_{23} = 2\varepsilon_{32}$$

Then, Eq. (2.39) can be written as

$$\mathrm{d}W = \iiint_V \mathrm{d}U \mathrm{d}V \tag{2.40}$$

where

$$dU = \sigma_{ij} d\varepsilon_{ij} \tag{2.41}$$

This form of equation implies summation over repeated subscripts $i, j = 1, 2, 3$.

It should be emphasized that by now dU is just a symbol, which does not mean that there exists function U and that dU is its differential. This meaning for dU is correct only if we restrict ourselves to the consideration of an elastic material described in Section 1.1. For such a material, the difference between the body potential energy corresponding to some initial state A and the energy corresponding to some other state B does not depend on the way used to transform the body from state A to state B. In other words, the integral

$$\int_A^B \sigma_{ij} d\varepsilon_{ij} = U(B) - U(A)$$

does not depend on the path of integration. This means that the element of integration is a complete differential of function U depending on ε_{ij}, i.e., that

$$dU = \frac{\partial U}{\partial \varepsilon_{ij}} d\varepsilon_{ij}$$

Comparing this result with Eq. (2.41) we arrive at Green's formulas

$$\sigma_{ij} = \frac{\partial U}{\partial \varepsilon_{ij}} \tag{2.42}$$

that are valid for any elastic material. The function $U(\varepsilon_{ij})$ can be referred to as specific strain energy (energy accumulated in the unit of body volume) or elastic potential. The potential U can be expanded into a Taylor series with respect to strains, i.e.,

$$U(\varepsilon_{ij}) = s_0 + s_{ij}\varepsilon_{ij} + \frac{1}{2} s_{ijkl}\varepsilon_{ij}\varepsilon_{kl} + \cdots \tag{2.43}$$

where

$$s_0 = U(\varepsilon_{ij} = 0), \quad s_{ij} = \left.\frac{\partial U}{\partial \varepsilon_{ij}}\right|_{\varepsilon_{ij}=0}, \quad s_{ijkl} = \left.\frac{\partial^2 U}{\partial \varepsilon_{ij} \partial \varepsilon_{kl}}\right|_{\varepsilon_{ij}=0,\, \varepsilon_{kl}=0} \tag{2.44}$$

Assume that for the initial state of the body, corresponding to zero external forces, we have $\varepsilon_{ij} = 0$, $\sigma_{ij} = 0$, $U = 0$. Then, $s_0 = 0$ and $s_{ij} = 0$ according to Eq. (2.42).

For small strains, we can neglect high-order terms in Eq. (2.43) and restrict ourselves to the first system of nonzero terms taking

$$U = \frac{1}{2} s_{ijkl} \varepsilon_{ij} \varepsilon_{kl}$$

Then, Eq. (2.42) yields

$$\sigma_{ij} = s_{ijkl} \varepsilon_{kl} \tag{2.45}$$

These linear equations correspond to a linear elastic model of the material (see Section 1.1) and, in general, include $3^4 = 81$ coefficients of s. However, because $\sigma_{ij} = \sigma_{ji}$ and $\varepsilon_{ij} = \varepsilon_{ji}$, we have the following equations $s_{ijkl} = s_{jikl} = s_{ijlk}$ which reduce the number of independent coefficients to 36. Then, taking into account that the mixed derivative specifying coefficients s_{ijkl} in Eqs. (2.44) does not depend on the sequence of differentiation, we get 15 equations $s_{ijkl} = s_{klij}$ $(ij \neq kl)$. Thus, Eq. (2.45) contains only 21 independent coefficients. Returning to coordinates x, y, z, we can write Eq. (2.45) in the following explicit form

$$\{\sigma\} = [S]\{\varepsilon\} \tag{2.46}$$

where

$$\{\sigma\} = \begin{Bmatrix} \sigma_x \\ \sigma_y \\ \sigma_z \\ \tau_{xy} \\ \tau_{xz} \\ \tau_{yz} \end{Bmatrix} \quad \{\varepsilon\} = \begin{Bmatrix} \varepsilon_x \\ \varepsilon_y \\ \varepsilon_z \\ \gamma_{xy} \\ \gamma_{xz} \\ \gamma_{yz} \end{Bmatrix} \quad [S] = \begin{bmatrix} S_{11} & S_{12} & S_{13} & S_{14} & S_{15} & S_{16} \\ S_{21} & S_{22} & S_{23} & S_{24} & S_{25} & S_{26} \\ S_{31} & S_{32} & S_{33} & S_{34} & S_{35} & S_{36} \\ S_{41} & S_{42} & S_{43} & S_{44} & S_{45} & S_{46} \\ S_{51} & S_{52} & S_{53} & S_{54} & S_{55} & S_{56} \\ S_{61} & S_{62} & S_{63} & S_{64} & S_{65} & S_{66} \end{bmatrix} \tag{2.47}$$

Eq. (2.46) are referred to as constitutive equations. They relate stresses and strains through 21 stiffness coefficients $S_{ij} = S_{ji}$ that specify material mechanical properties within the framework of a linear elastic model of the material. The inverse form of Eq. (2.46) is

$$\{\varepsilon\} = [C]\{\sigma\} \tag{2.48}$$

Strains are expressed in terms of stresses via the matrix of compliance coefficients that can be written as

$$[C]=\begin{bmatrix} \frac{1}{E_x} & -\frac{\nu_{xy}}{E_y} & -\frac{\nu_{xz}}{E_z} & \frac{\eta_{x,\,xy}}{G_{xy}} & \frac{\eta_{x,\,xz}}{G_{xz}} & \frac{\eta_{x,\,yz}}{G_{yz}} \\ -\frac{\nu_{yx}}{E_x} & \frac{1}{E_y} & -\frac{\nu_{yz}}{E_z} & \frac{\eta_{y,\,xy}}{G_{xy}} & \frac{\eta_{y,\,xz}}{G_{xz}} & \frac{\eta_{y,\,yz}}{G_{yz}} \\ -\frac{\nu_{zx}}{E_x} & -\frac{\nu_{zy}}{E_y} & \frac{1}{E_z} & \frac{\eta_{z,\,xy}}{G_{xy}} & \frac{\eta_{z,\,xz}}{G_{xz}} & \frac{\eta_{z,\,yz}}{G_{yz}} \\ \frac{\eta_{xy,\,x}}{E_x} & \frac{\eta_{xy,\,y}}{E_y} & \frac{\eta_{xy,\,z}}{E_z} & \frac{1}{G_{xy}} & \frac{\lambda_{xy,\,xz}}{G_{xz}} & \frac{\lambda_{xy,\,yz}}{G_{yz}} \\ \frac{\eta_{xz,\,x}}{E_x} & \frac{\eta_{xz,\,y}}{E_y} & \frac{\eta_{xz,\,z}}{E_z} & \frac{\lambda_{xz,\,xy}}{G_{xy}} & \frac{1}{G_{xz}} & \frac{\lambda_{xz,\,yz}}{G_{yz}} \\ \frac{\eta_{yz,\,x}}{E_x} & \frac{\eta_{yz,\,y}}{E_y} & \frac{\eta_{yz,\,z}}{E_z} & \frac{\lambda_{yz,\,xy}}{G_{xy}} & \frac{\lambda_{yz,\,xz}}{G_{xz}} & \frac{1}{G_{yz}} \end{bmatrix} \tag{2.49}$$

This matrix is symmetric, and the following 15 symmetry conditions are valid

$$\begin{aligned} &\frac{\nu_{xy}}{E_y}=\frac{\nu_{yx}}{E_x}, \quad \frac{\nu_{xz}}{E_z}=\frac{\nu_{zx}}{E_x}, \quad \frac{\nu_{yz}}{E_z}=\frac{\nu_{zy}}{E_y} \\ &\frac{\eta_{x,\,xy}}{G_{xy}}=\frac{\eta_{xy,\,x}}{E_x}, \quad \frac{\eta_{x,\,xz}}{G_{xz}}=\frac{\eta_{xz,\,x}}{E_x}, \quad \frac{\eta_{x,\,yz}}{G_{yz}}=\frac{\eta_{yz,\,x}}{E_x} \\ &\frac{\eta_{y,\,xy}}{G_{xy}}=\frac{\eta_{xy,\,y}}{E_y}, \quad \frac{\eta_{y,\,xz}}{G_{xz}}=\frac{\eta_{xz,\,y}}{E_y}, \quad \frac{\eta_{y,\,yz}}{G_{yz}}=\frac{\eta_{yz,\,y}}{E_y} \\ &\frac{\eta_{z,\,xy}}{G_{xy}}=\frac{\eta_{xy,\,z}}{E_z}, \quad \frac{\eta_{z,\,xz}}{G_{xz}}=\frac{\eta_{xz,\,z}}{E_y}, \quad \frac{\eta_{z,\,yz}}{G_{yz}}=\frac{\eta_{yz,\,z}}{E_z} \\ &\frac{\lambda_{xy,\,xz}}{G_{xz}}=\frac{\lambda_{xz,\,xy}}{G_{xy}}, \quad \frac{\lambda_{xy,\,yz}}{G_{yz}}=\frac{\lambda_{yz,\,xy}}{G_{xy}}, \quad \frac{\lambda_{xz,\,yz}}{G_{yz}}=\frac{\lambda_{yz,\,xz}}{G_{xz}} \end{aligned} \tag{2.50}$$

The compliance matrix, Eq. (2.49), includes the following engineering constants:

E_x is the modulus of elasticity in the x-direction (x, y, z); ν_{xy} the Poisson's ratio that determines the strain in the x-direction induced by normal stress acting in the orthogonal y-direction (x, y, z); G_{xy} the shear modulus in the xy-plane (x, y, z); $\eta_{x,\,yz}$ the extension–shear coupling coefficient indicating normal strain in the x-direction induced by shear stress acting in the yz-plane (x, y, z); $\eta_{xy,\,z}$ the shear–extension coupling coefficient characterizing shear strain in the xy-plane caused by normal stress acting in the z-direction (x, y, z); and $\lambda_{xy,\,yz}$ the shear–shear coupling coefficient that determines the shear strain taking place in the xy-plane under shear stress acting in the yz-plane (x, y, z).

Having constitutive equations, Eq. (2.46), we can now write the finite expression for elastic potential, U. Substituting stresses into Eq. (2.41) and integrating it with respect to strains, we get the following equation after some transformation with the aid of Eq. (2.46)

$$U = \frac{1}{2}(\sigma_x \varepsilon_x + \sigma_y \varepsilon_y + \sigma_z \varepsilon_z + \tau_{xy}\gamma_{xy} + \tau_{xz}\gamma_{xz} + \tau_{yz}\gamma_{yz}) \tag{2.51}$$

The potential energy of the body can be found as

$$W = \iiint_V U \mathrm{d}V \tag{2.52}$$

The compliance matrix, Eq. (2.49), containing 21 independent elastic constants corresponds to the general case of material anisotropy that practically never occurs in real materials. The most common particular case corresponds to an orthotropic (orthogonally anisotropic) material which has three orthogonal orthotropy (coordinate) axes such that normal stresses acting along these axes do not induce shear strains, whereas shear stresses acting in coordinate planes do not cause normal strains in the direction of these axes. As a result, the stiffness and compliance matrices become uncoupled with respect to normal stresses and strains on one side and shear stresses and strains on the other side. For the case of an orthotropic material, with axes x, y, and z coinciding with the orthotropy axes, Eq. (2.49) takes the form

$$[C] = \begin{bmatrix} \frac{1}{E_x} & -\frac{\nu_{xy}}{E_y} & -\frac{\nu_{xz}}{E_z} & 0 & 0 & 0 \\ -\frac{\nu_{yx}}{E_x} & \frac{1}{E_y} & -\frac{\nu_{yz}}{E_z} & 0 & 0 & 0 \\ -\frac{\nu_{zx}}{E_x} & -\frac{\nu_{zy}}{E_y} & \frac{1}{E_z} & 0 & 0 & 0 \\ 0 & 0 & 0 & \frac{1}{G_{xy}} & 0 & 0 \\ 0 & 0 & 0 & 0 & \frac{1}{G_{xz}} & 0 \\ 0 & 0 & 0 & 0 & 0 & \frac{1}{G_{yz}} \end{bmatrix} \tag{2.53}$$

Symmetry conditions, Eqs. (2.50), reduce to

$$\nu_{xy}E_x = \nu_{yx}E_y, \quad \nu_{xz}E_x = \nu_{zx}E_z, \quad \nu_{yz}E_y = \nu_{zy}E_z$$

These equations have a simple physical meaning. The higher the stiffness, demonstrated by the material in some direction, the less is the strain in this direction under loading in the orthogonal directions. Taking into account the foregoing symmetry conditions, we can conclude that an orthotropic material is characterized with nine independent elastic constants.

The simplest material model corresponds to the isotropic material, whose mechanical properties are the same for any direction or plane of loading. As a result, subscripts indicating coordinate directions and planes in Eq. (2.53) disappear, and it reduces to

$$[C]=\begin{bmatrix} \dfrac{1}{E} & -\dfrac{\nu}{E} & -\dfrac{\nu}{E} & 0 & 0 & 0 \\ -\dfrac{\nu}{E} & \dfrac{1}{E} & -\dfrac{\nu}{E} & 0 & 0 & 0 \\ -\dfrac{\nu}{E} & -\dfrac{\nu}{E} & \dfrac{1}{E} & 0 & 0 & 0 \\ 0 & 0 & 0 & \dfrac{1}{G} & 0 & 0 \\ 0 & 0 & 0 & 0 & \dfrac{1}{G} & 0 \\ 0 & 0 & 0 & 0 & 0 & \dfrac{1}{G} \end{bmatrix} \tag{2.54}$$

The compliance matrix, Eq. (2.54), contains three elastic constants, E, G, and ν. However, only two of them are independent. To show this, consider the case of pure shear for a plate discussed in Section 2.4 (see Fig. 2.5). For this problem, $\sigma_x=\sigma_y=\sigma_z=\tau_{xz}=\tau_{yz}=0$, $\tau_{xy}=\tau$ and Eqs. (2.48) and (2.54) yield

$$\gamma_{xy}=\frac{\tau}{G}$$

The specific strain energy in Eq. (2.51) can be written as

$$U=\frac{1}{2}\tau_{xy}\gamma_{xy}=\frac{1}{2G}\tau^2 \tag{2.55}$$

However, from Section 2.4, pure shear can be reduced to tension and compression in the principal directions (see Fig. 2.5). For these directions, Eqs. (2.48) and (2.54) give

$$\varepsilon_1=\frac{\sigma_1}{E}-\nu\frac{\sigma_3}{E},\quad \varepsilon_3=\frac{\sigma_3}{E}-\nu\frac{\sigma_1}{E}$$

Here $\sigma_1=\tau$, $\sigma_3=-\tau$ and the remaining stresses are equal to zero. The strain energy, Eq. (2.51), can be presented now in the following form

$$U=\frac{1}{2}(\sigma_1\varepsilon_1+\sigma_3\varepsilon_3)=\frac{1+\nu}{E}\tau^2 \tag{2.56}$$

Since Eqs. (2.55) and (2.56) specify one and the same quantity, we get

$$G=\frac{E}{2(1+\nu)} \tag{2.57}$$

Thus, an isotropic material is characterized within the linear elastic model with two independent elastic constants – E and ν.

2.10. Formulations of the problem

The problem of Solid Mechanics is reduced, as follows from the foregoing derivation, to a set of 15 equations, i.e., three equilibrium equations, Eqs. (2.5), six strain–displacement equations, Eqs. (2.22), and six constitutive equations, Eq. (2.46) or (2.48). This set of equations is complete, i.e., it contains 15 unknown functions among which there are six stresses, six strains, and three displacements. Solution of a particular problem should satisfy three boundary conditions that can be written at any point of the body surface. Static or force boundary conditions have the form of Eqs. (2.2), whereas kinematic or displacement boundary conditions are imposed on three displacement functions.

There exist two classical formulations of the problem – displacement formulation and stress formulation.

According to the displacement formulation, we first determine displacements u_x, u_y, and u_z from three equilibrium equations, Eqs. (2.5), written in terms of displacements with the aid of constitutive equations, Eq. (2.46), and strain–displacement equations, Eqs. (2.22). Having found the displacements, we use Eqs. (2.22) and (2.46) to determine strains and stresses.

The stress formulation is much less straightforward than the displacement one. Indeed, we have only three equilibrium equations, Eqs. (2.5), for six stresses which means that the problem of solid mechanics is not, in general, a statically determinate problem. All possible solutions of the equilibrium equations (obviously, there is an infinite number of them because the number of equations is less than the number of unknown stresses) satisfying force boundary conditions (solutions that do not satisfy them, obviously, do not belong to the problem under study) comprise the class of statically admissible stress fields (see Section 2.8). Suppose that we have one of such stress fields. Now, we can readily find strains using constitutive equations, Eq. (2.48), but to determine displacements, we need to integrate a set of six strain–displacement equations, Eqs. (2.22) which having only three unknown displacements are, in general, not compatible. As shown in Section 2.7, this set can be integrated if strains satisfy six compatibility equations, Eqs. (2.35). We can write these equations in terms of stresses using constitutive equations, Eq. (2.48). Thus, the stress formulation of the problem is reduced to a set of nine equations consisting of three equilibrium equations and six compatibility equations in terms of stresses. At first glance it looks like this set is not consistent because it includes only six unknown stresses. However, this is not the case because of the special properties of the compatibility equations. As was noted in Section 2.7, these equations provide the existence of Euclidean space inside the deformed body. But this space automatically exists if strains can be expressed in terms of three continuous displacements as in Eqs. (2.22). Indeed, substituting strains, Eqs. (2.22), into the compatibility equations, Eqs. (2.35), we can readily see that they are identically satisfied for any three functions u_x, u_y, and u_z. This means that the solution of six Eqs. (2.35) including six strains is not unique. The uniqueness is ensured by three equilibrium equations.

2.11. Variational principles

The equations of Solid Mechanics considered in the previous sections can be also derived from variational principles that establish the energy criteria according to which the actual state of the body under loading can be singled out of a system of admissible states (see Section 2.8).

Consider a linear elastic solid and introduce two mutually independent fields of variables: a statically admissible stress field σ'_x, σ'_y, σ'_z, τ'_{xy}, τ'_{xz}, τ'_{yz} and a kinematically admissible field characterized with displacements u''_x, u''_y, u''_z and corresponding strains ε''_x, ε''_y, ε''_z, γ''_{xy}, γ''_{xz}, γ''_{yz}. To construct the energy criteria allowing us to distinguish the actual variables from admissible ones, consider the following integral similar to the energy integral in Eqs. (2.51) and (2.52)

$$I = \iiint_V (\sigma'_x\varepsilon''_x + \sigma'_y\varepsilon''_y + \sigma'_z\varepsilon''_z + \tau'_{xy}\gamma''_{xy} + \tau'_{xz}\gamma''_{xz} + \tau'_{yz}\gamma''_{yz})\mathrm{d}V \tag{2.58}$$

Here, in accordance with the definition of a kinematically admissible field (see Section 2.8),

$$\varepsilon''_x = \frac{\partial u''_x}{\partial x}, \quad \gamma''_{xy} = \frac{\partial u''_x}{\partial y} + \frac{\partial u''_y}{\partial x} \quad (x, y, z) \tag{2.59}$$

Substituting Eqs. (2.59) into Eq. (2.58) and using the following evident relationships between the derivatives

$$\sigma'_x\frac{\partial u''_x}{\partial x} = \frac{\partial}{\partial x}(\sigma'_x u''_x) - u''_x\frac{\partial \sigma'_x}{\partial x}, \quad \tau'_{xy}\frac{\partial u''_x}{\partial y} = \frac{\partial}{\partial y}(\tau'_{xy}u''_x) - u''_x\frac{\partial \tau'_{xy}}{\partial y} \quad \text{etc.,}$$

we arrive at

$$I = \iiint_V \left[\frac{\partial}{\partial x}(\sigma'_x u''_x + \tau'_{xy}u''_y + \tau'_{xz}u''_z) + \frac{\partial}{\partial y}(\tau'_{xy}u''_x + \sigma'_y u''_y + \tau'_{yz}u''_z) \right.$$
$$+ \frac{\partial}{\partial z}(\tau'_{xz}u''_x + \tau'_{yz}u''_y + \sigma'_z u''_z) - \left(\frac{\partial \sigma'_x}{\partial x} + \frac{\partial \tau'_{xy}}{\partial y} + \frac{\partial \tau'_{xz}}{\partial z}\right)u''_x$$
$$\left. - \left(\frac{\partial \sigma'_y}{\partial y} + \frac{\partial \tau'_{xy}}{\partial x} + \frac{\partial \tau'_{yz}}{\partial z}\right)u''_y - \left(\frac{\partial \sigma'_z}{\partial z} + \frac{\partial \tau'_{xz}}{\partial x} + \frac{\partial \tau'_{yz}}{\partial y}\right)u''_z\right]\mathrm{d}V \tag{2.60}$$

Applying Green's integral transformation, Eq. (2.4), to the first three terms under the integral and taking into account that statically admissible stresses should satisfy equilibrium equations, Eqs. (2.5), (2.6), and force boundary conditions, Eq. (2.2), we obtain from Eqs. (2.58) and (2.60)

$$\iiint_V (\sigma'_x\varepsilon''_x + \sigma'_y\varepsilon''_y + \sigma'_z\varepsilon''_z + \tau'_{xy}\gamma''_{xy} + \tau'_{xz}\gamma''_{xz} + \tau'_{yz}\gamma''_{yz})\mathrm{d}V$$

$$= \iint_S (p_x u''_x + p_y u''_y + p_z u''_z)\mathrm{d}S + \iiint_V (q_x u''_x + q_y u''_y + q_z u''_z)\mathrm{d}V \tag{2.61}$$

For actual stresses, strains, and displacements, Eq. (2.61) reduces to the following equation

$$\iiint_V (\sigma_x\varepsilon_x + \sigma_y\varepsilon_y + \sigma_z\varepsilon_z + \tau_{xy}\gamma_{xy} + \tau_{xz}\gamma_{xz} + \tau_{yz}\gamma_{yz})\mathrm{d}V$$

$$= \iint_S (p_x u_x + p_y u_y + p_z u_z)\mathrm{d}S + \iiint_V (q_x u_x + q_y u_y + q_z u_z)\mathrm{d}V \tag{2.62}$$

known as Clapeyron's theorem.

2.11.1. Principle of minimum total potential energy

This principle allows us to distinguish the actual displacement field of the body from kinematically admissible fields. To derive it, assume that the stresses in Eq. (2.61) are actual stresses, i.e., $\sigma' = \sigma,\ \tau' = \tau$, whereas the displacements and the corresponding strains differ from the actual values by small kinematically admissible variations, i.e., $u'' = u + \delta u,\ \varepsilon'' = \varepsilon + \delta\varepsilon,\ \gamma'' = \gamma + \delta\gamma$. Substituting these expressions into Eq. (2.61) and subtracting Eq. (2.62) from the resulting equation, we arrive at

$$\iiint_V (\sigma_x\delta\varepsilon_x + \sigma_y\delta\varepsilon_y + \sigma_z\delta\varepsilon_z + \tau_{xy}\delta\gamma_{xy} + \tau_{xz}\delta\gamma_{xz} + \tau_{yz}\delta\gamma_{yz})\mathrm{d}V$$

$$= \iint_S (p_x\delta u_x + p_y\delta u_y + p_z\delta u_z)\mathrm{d}S + \iiint_V (q_x\delta u_x + q_y\delta u_y + q_z\delta u_z)\mathrm{d}V$$

Assume that under small variation of displacements and strains belonging to the kinematically admissible fields the surface tractions and body forces do not change. Then, we can write the foregoing result in the following form

$$\delta W_\varepsilon - \delta A = 0 \tag{2.63}$$

Here

$$\delta W_\varepsilon = \iiint\limits_V (\sigma_x \delta\varepsilon_x + \sigma_y \delta\varepsilon_y + \sigma_z \delta\varepsilon_z + \tau_{xy}\delta\gamma_{xy} + \tau_{xz}\delta\gamma_{xz} + \tau_{yz}\delta\gamma_{yz})\mathrm{d}V \qquad (2.64)$$

is the variation of the strain energy (internal potential energy of an elastic solid) associated with small kinematically admissible variations of strains and

$$A = \iint\limits_S (p_x u_x + p_y u_y + p_z u_z)\mathrm{d}S + \iiint\limits_V (q_x u_x + q_y u_y + q_z u_z)\mathrm{d}V \qquad (2.65)$$

can be formally treated as work performed by surface tractions and body forces on the actual displacements. Expressing stresses in Eq. (2.64) in terms of strains with the aid of the constitutive equations, Eq. (2.46), and integrating, we can determine W_ε, which is the body strain energy written in terms of strains. The quantity $T = W_\varepsilon - A$ is referred to as the total potential energy of the body. This name historically came from problems in which external forces had a potential function $F = -A$ so that $T = W_\varepsilon + F$ was the sum of internal and external potentials, i.e., the total potential function. Then, the condition in Eq. (2.63) reduces to

$$\delta T = 0 \qquad (2.66)$$

which means that T has a stationary (actually, minimum) value under small admissible variation of displacements in the vicinity of actual displacements. Thus, we arrive at the following variational principle of minimum total potential energy: the actual displacement field, in contrast to all kinematically admissible fields, delivers the minimum value of the body total potential energy. This principle is a variational form of the displacement formulation of the problem discussed in Section 2.10. As can be shown, the variational equations ensuring the minimum value of the total potential energy of the body coincide with the equilibrium equations written in terms of displacements.

2.11.2. Principle of minimum strain energy

This principle is valid for a linear elastic body and establishes the criterion according to which the actual stress field can be singled out of all statically admissible fields. Suppose that displacements and strains in Eq. (2.61) are actual, i.e., $u'' = u$, $\varepsilon'' = \varepsilon$, $\gamma'' = \gamma$, whereas stresses differ from the actual values by small statically admissible variations, i.e., $\sigma' = \sigma + \delta\sigma$, $\tau' = \tau + \delta\tau$. Substituting these expressions in Eq. (2.61) and subtracting Eq. (2.62) for the actual state, we get

$$\delta W_\sigma = 0 \qquad (2.67)$$

where

$$\delta W_\sigma = \iiint\limits_V (\varepsilon_x \delta\sigma_x + \varepsilon_y \delta\sigma_y + \varepsilon_z \delta\sigma_z + \gamma_{xy}\delta\tau_{xy} + \gamma_{xz}\delta\tau_{xz} + \gamma_{yz}\delta\tau_{yz})\mathrm{d}V \qquad (2.68)$$

is the variation of the strain energy associated with the variation of stresses. Expressing strains in terms of stresses with the aid of constitutive equations, Eq. (2.48), and integrating, we can determine W_σ, which is the body strain energy written in terms of stresses. As before, Eq. (2.67) indicates that strain energy, W_σ, has a stationary (in fact, minimum) value under admissible variation of stresses. As a result, we arrive at the following variational principle of minimum strain energy: the actual stress field, in contrast to all statically admissible fields, delivers the minimum value of the body strain energy. This principle is a variational form of the stress formulation of the problem considered in Section 2.10. As can be shown, the variational equations providing the minimum value of the strain energy are compatibility equations written in terms of stresses. It is important that the stress variation in Eq. (2.68) should be performed within the statically admissible field, i.e., within stresses that satisfy equilibrium equations and force boundary conditions.

2.11.3. Mixed variational principles

The two variational principles described above imply variations with respect to either displacements only or stresses only. There exist also the so-called mixed variational principles in which variation is performed with respect to both kinematic and static variables. The first principle from this group follows from the principle of minimum total potential energy considered in Section 2.11.1. Let us expand the class of admissible kinematic variables and introduce displacements that are continuous functions satisfying displacement boundary conditions and strains that are not related to these displacements by strain–displacement equations, Eqs. (2.22). Then we can apply the principle of minimum total potential energy performing a conditional minimization of the total potential energy and introduce Eqs. (2.22) as additional constraints imposed on strains and displacements with the aid of Lagrange's multipliers. Using stresses as these multipliers we can construct the following augmented function

$$T_{\mathrm{L}} = W_\varepsilon - A + \iiint\limits_V \left[\sigma_x\left(\frac{\partial u_x}{\partial x} - \varepsilon_x\right) + \sigma_y\left(\frac{\partial u_y}{\partial y} - \varepsilon_y\right) + \sigma_z\left(\frac{\partial u_z}{\partial z} - \varepsilon_z\right)\right.$$
$$+ \tau_{xy}\left(\frac{\partial u_x}{\partial y} + \frac{\partial u_y}{\partial x} - \gamma_{xy}\right) + \tau_{xz}\left(\frac{\partial u_x}{\partial z} + \frac{\partial u_z}{\partial x} - \gamma_{xz}\right)$$
$$\left. + \tau_{yz}\left(\frac{\partial u_y}{\partial z} + \frac{\partial u_z}{\partial y} - \gamma_{yz}\right)\right]\mathrm{d}V$$

According to the initial principle, Eq. (2.66), $\delta T_{\mathrm{L}} = 0$. Variation of displacements yields, as previously equilibrium equations, variation of stresses results in strain–displacement equations, and variation of strains gives constitutive equations (W_ε should be expressed in terms of strains).

The second form of the mixed variational principle can be derived from the principle of minimum strain energy discussed in Section 2.11.2. Again expand the class of admissible static fields and introduce stresses that satisfy force boundary conditions but do not satisfy equilibrium equations, Eq. (2.5). Then, we can apply the principle of minimum strain energy if we construct an augmented function adding Eqs. (2.5) as additional constraints. Using displacements as Lagrange's multipliers we obtain

$$W_{\mathrm{L}} = W_\sigma + \iiint_V \left[u_x \left(\frac{\partial \sigma_x}{\partial x} + \frac{\partial \tau_{xy}}{\partial y} + \frac{\partial \tau_{xz}}{\partial z} + q_x \right) \right.$$

$$+ u_y \left(\frac{\partial \sigma_y}{\partial y} + \frac{\partial \tau_{xy}}{\partial x} + \frac{\partial \tau_{yz}}{\partial z} + q_y \right)$$

$$\left. + u_z \left(\frac{\partial \sigma_z}{\partial z} + \frac{\partial \tau_{xz}}{\partial x} + \frac{\partial \tau_{yz}}{\partial y} + q_z \right) \right] \mathrm{d}V$$

According to the original principle, Eq. (2.67), $\delta W_{\mathrm{L}} = 0$. The variation with respect to stresses (W_σ should be expressed in terms of stresses) yields constitutive equations in which strains are expressed in terms of displacements via strain–displacement equations, Eqs. (2.22), whereas variation of displacements gives equilibrium equations.

The equations and principles considered in this chapter will be used in the following chapters in the book for the analysis of the mechanics of composite materials.

2.12. Reference

Vasiliev, V.V. and Gurdal, Z. (1999). Optimal structural design. In *Optimal Design* (V.V. Vasiliev and Z. Gurdal eds.). Technomic, Lancaster, pp. 1–29.

Chapter 3

MECHANICS OF A UNIDIRECTIONAL PLY

A ply or lamina is the simplest element of a composite material, an elementary layer of unidirectional fibers in a matrix (see Fig. 3.1), formed when a unidirectional tape impregnated with resin is placed onto the surface of the tool, thus providing the shape of a composite part.

3.1. Ply architecture

As the tape consists of tows (bundles of fibers), the ply thickness (whose minimum value is about 0.1 mm for modern composites) is much higher than the fiber diameter (about 0.01 mm). In an actual ply, the fibers are randomly distributed, as in Fig. 3.2. Since the actual distribution is not known and can hardly be predicted, some typical idealized regular distributions, i.e., square (Fig. 3.3), hexagonal (Fig. 3.4), and layer-wise (Fig. 3.5), are used for the analysis.

A composite ply is generally taken to consist of two constituents: fibers and a matrix whose quantities in the materials are specified by volume, v, and mass, m, fractions

$$v_{\mathrm{f}}=\frac{V_{\mathrm{f}}}{V_{\mathrm{c}}}, \quad v_{\mathrm{m}}=\frac{V_{\mathrm{m}}}{V_{\mathrm{c}}} \tag{3.1}$$

$$m_{\mathrm{f}}=\frac{M_{\mathrm{f}}}{M_{\mathrm{c}}}, \quad m_{\mathrm{m}}=\frac{M_{\mathrm{m}}}{M_{\mathrm{c}}} \tag{3.2}$$

Here, V and M are volume and mass, whereas subscripts f, m, and c correspond to fibers, matrix, and composite material, respectively. Since $V_{\mathrm{c}}=V_{\mathrm{f}}+V_{\mathrm{m}}$ and $M_{\mathrm{c}}=M_{\mathrm{f}}+M_{\mathrm{m}}$, we have

$$v_{\mathrm{f}}+v_{\mathrm{m}}=1, \quad m_{\mathrm{f}}+m_{\mathrm{m}}=1 \tag{3.3}$$

There exist the following relationships between volume and mass fractions

$$v_{\mathrm{f}}=\frac{\rho_{\mathrm{c}}}{\rho_{\mathrm{f}}}m_{\mathrm{f}}, \quad v_{\mathrm{m}}=\frac{\rho_{\mathrm{c}}}{\rho_{\mathrm{m}}}m_{\mathrm{m}} \tag{3.4}$$

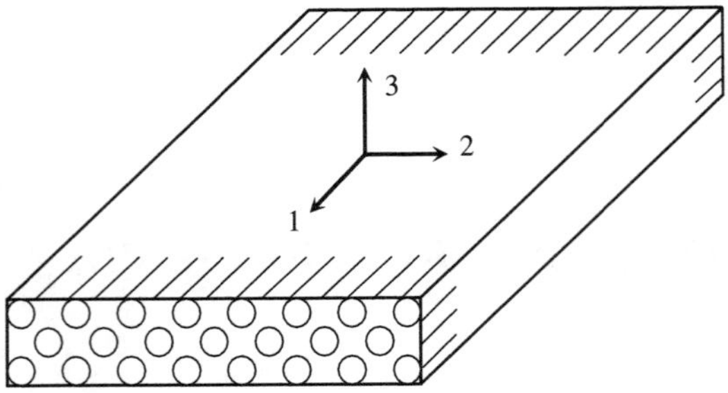

Fig. 3.1. A unidirectional ply.

Fig. 3.2. Actual fiber distribution in the cross-section of a ply ($v_f = 0.65$).

where ρ_f, ρ_m, and ρ_c are the densities of fibers, the matrix, and the composite, respectively. In analysis, volume fractions are used because they enter the stiffness coefficients for a ply, whereas mass fractions are usually measured directly during processing or experimental study of the fabricated material.

Two typical situations usually occur. The first situation implies that we know the mass of fibers used to fabricate a composite part and the mass of the part itself. The mass of fibers can be found if we weigh the spools with fibers before and after they are used or calculate the total length of tows and multiply it by the tow tex-number that is the mass in grams of a 1000-m-long tow. So, we know the values of M_f and M_c and can use the first equations of Eqs. (3.2) and (3.4) to calculate v_f.

Fig. 3.3. Square fiber distribution in the cross-section of a ply ($v_f = 0.65$).

Fig. 3.4. Hexagonal fiber distribution in the cross-section of a ply ($v_f = 0.65$).

Fig. 3.5. Layer-wise fiber distribution in the cross-section of a ply ($\upsilon_f = 0.65$).

The second situation takes place if we have a sample of a composite material and know the densities of the fibers and the matrix used for its fabrication. Then, we can find the experimental value of material density, ρ_c^e, and use the following equation for theoretical density

$$\rho_c = \rho_f \upsilon_f + \rho_m \upsilon_m \tag{3.5}$$

Putting $\rho_c = \rho_c^e$ and taking into account Eqs. (3.3), we obtain

$$\upsilon_f = \frac{\rho_c^e - \rho_m}{\rho_f - \rho_m} \tag{3.6}$$

Consider, for example, a carbon–epoxy composite material with fibers *AS*4 and matrix EPON DPL-862, for which $\rho_f = 1.79\,\mathrm{g/cm^3}$ and $\rho_m = 1.2\,\mathrm{g/cm^3}$. Let $\rho_c^e = 1.56\,\mathrm{g/cm^3}$. Then, Eq. (3.6) yields $\upsilon_f = 0.61$.

This result is approximate because it ignores possible material porosity. To determine the actual fiber fraction, we should remove the resin using matrix destruction, solvent extraction, or burning the resin out in an oven. As a result, we get M_f, and having M_c, can calculate m_f and υ_f with the aid of Eqs. (3.2) and (3.4). Then we find ρ_c using Eq. (3.5) and compare it with ρ_c^e. If $\rho_c > \rho_c^e$, the material includes voids whose volume fraction (porosity) can be calculated using the following equation

$$\upsilon_p = 1 - \frac{\rho_c^e}{\rho_c} \tag{3.7}$$

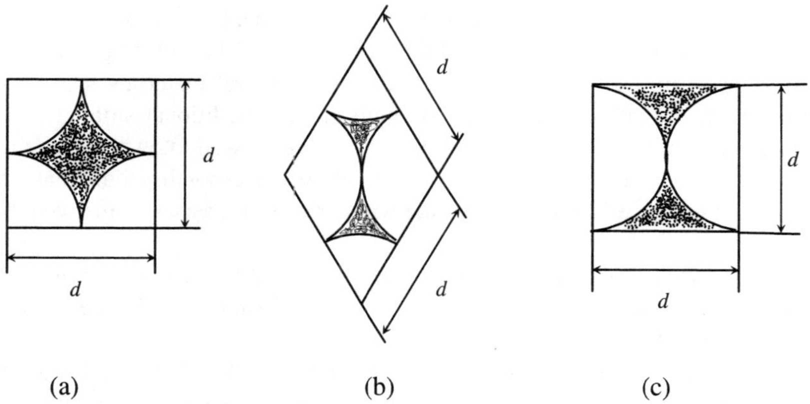

Fig. 3.6. Ultimate fiber arrays for square (a), hexagonal (b), and layer-wise (c) fiber distributions.

For the carbon–epoxy composite material considered above as an example, assume that the foregoing procedure results in $m_f = 0.72$. Then, Eqs. (3.4), (3.5), and (3.7) give $v_f = 0.63$, $\rho_c = 1.58\,\mathrm{g/cm^3}$, and $v_p = 0.013$, respectively.

For real unidirectional composite materials, we normally have $v_f = 0.50 - 0.65$. Lower fiber volume content results in lower ply strength and stiffness under tension along the fibers, whereas higher fiber content, close to the ultimate value, leads to reduction of the ply strength under longitudinal compression and in-plane shear due to poor bonding of the fibers.

Since the fibers usually have uniform circular cross-sections, there exists the ultimate fiber volume fraction, v_f^u, which is less than unity and depends on the fiber arrangement. For typical arrangements shown in Figs. 3.3–3.5, the ultimate arrays are presented in Fig. 3.6, and the corresponding ultimate fiber volume fractions are:

$$\text{Square array} \quad v_f^u = \frac{1}{d^2}\left(\frac{\pi d^2}{4}\right) = \frac{\pi}{4} = 0.785$$

$$\text{Hexagonal array} \quad v_f^u = \frac{2}{d^2\sqrt{3}}\left(\frac{\pi d^2}{4}\right) = \frac{\pi}{2\sqrt{3}} = 0.907$$

$$\text{Layer-wise array} \quad v_f^u = \frac{1}{d^2}\left(\frac{\pi d^2}{4}\right) = \frac{\pi}{4} = 0.785$$

3.2. Fiber–matrix interaction

3.2.1. Theoretical and actual strength

The most important property of advanced composite materials is associated with the very high strength of a unidirectional ply, accompanied with relatively low density.

This advantage of the material is provided mainly by the fibers. Correspondingly, a natural question arises as to how such traditional lightweight materials such as glass or graphite, which were never utilized as primary load-bearing structural materials, can be used to make fibers with strength exceeding the strength of such traditional structural materials as aluminum or steel (see Table 1.1). The general answer is well known: the strength of a thin wire is usually much higher than the strength of the corresponding bulk material. This is demonstrated in Fig. 3.7, showing that the wire strength increases as the wire diameter is reduced.

In connection with this, two questions arise. First, what is the upper limit of strength that can be predicted for an infinitely thin wire or fiber? And second, what is the nature of this phenomenon?

The answer to the first question is given in *The Physics of Solids*. Consider an idealized model of a solid, namely a regular system of atoms located as shown in Fig. 3.8 and find the stress, σ, that destroys this system. The dependence of σ on atomic spacing as given by *The Physics of Solids* is presented in Fig. 3.9. Point O of the curve corresponds to the equilibrium of the unloaded system, whereas point U specifies the ultimate theoretical stress, $\overline{\sigma}_t$. The initial tangent angle, α, characterizes the material's modulus of elasticity, E. To evaluate $\overline{\sigma}_t$, we can use the following sine approximation (Gilman, 1959) for the OU segment of the curve

$$\sigma = \overline{\sigma}_t \sin 2\pi \frac{a - a_0}{a_0}$$

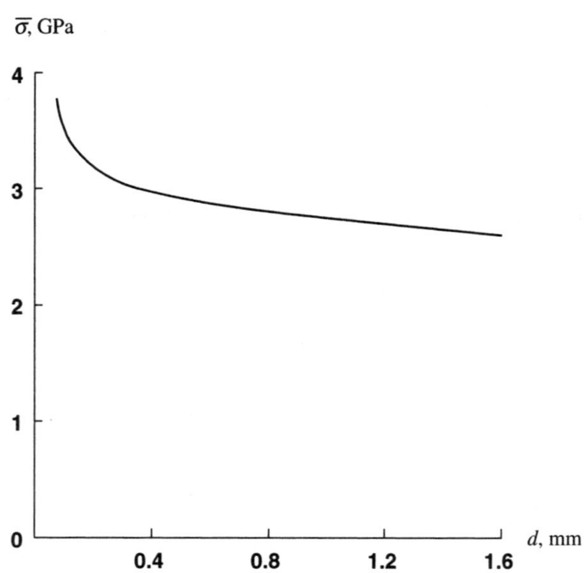

Fig. 3.7. Dependence of high-carbon steel wire strength on the wire diameter.

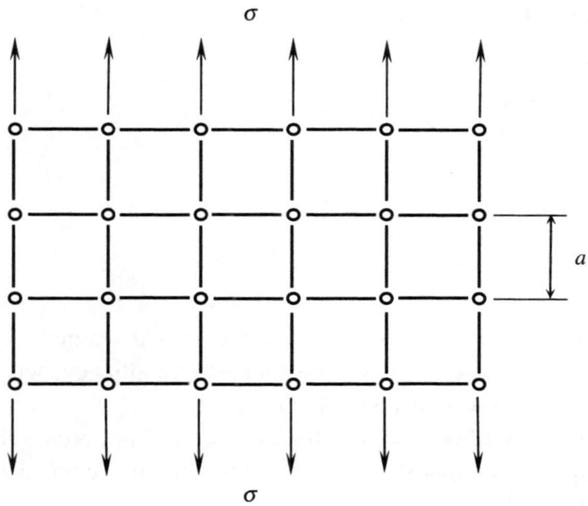

Fig. 3.8. Material model.

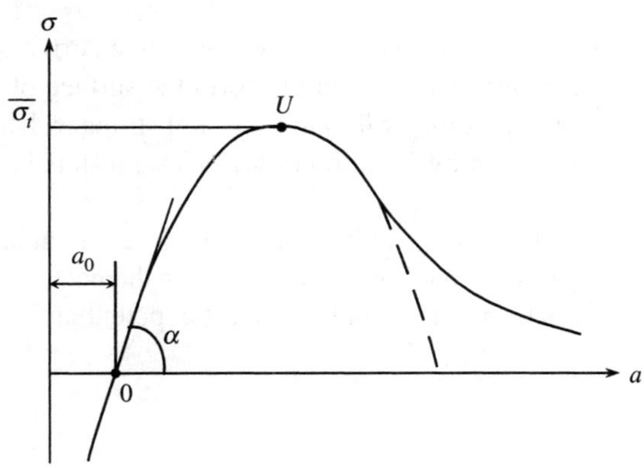

Fig. 3.9. Atoms' interaction curve (——) and its sine approximation (- - - -).

Introducing strain

$$\varepsilon = \frac{a - a_0}{a_0}$$

we arrive at

$$\sigma = \overline{\sigma}_t \sin 2\pi\varepsilon$$

Now, we can calculate the modulus as

$$E = \left(\frac{\mathrm{d}\sigma}{\mathrm{d}\varepsilon}\right)\bigg|_{\varepsilon=0} = 2\pi\overline{\sigma}_{\mathrm{t}}$$

Thus,

$$\overline{\sigma}_{\mathrm{t}} = \frac{E}{2\pi} \tag{3.8}$$

This equation yields a very high value for the theoretical strength. For example, for a steel wire, $\overline{\sigma}_{\mathrm{t}} = 33.4\,\mathrm{GPa}$. Until now, the highest strength reached in 2-μm-diameter monocrystals of iron (whiskers) is about 12 GPa.

The model under study allows us to introduce another important characteristic of the material. The specific energy that should be spent to destroy the material can be presented in accordance with Fig. 3.9 as

$$2\gamma = \int_{a_0}^{\infty} \sigma(a)\mathrm{d}a \tag{3.9}$$

As material fracture results in the formation of two new free surfaces, γ can be referred to as the specific surface energy (energy spent to form the surface of unit area).

The answer to the second question (why the fibers are stronger than the corresponding bulk materials) was in fact given by Griffith (1920), whose results have formed the basis of fracture mechanics.

Consider a fiber loaded in tension and having a thin circumferential crack as shown in Fig. 3.10. The crack length, l, is much less than the fiber diameter, d.

For a linear elastic fiber, $\sigma = E\varepsilon$, and the elastic potential in Eq. (2.51) can be presented as

$$U = \frac{1}{2}\sigma\varepsilon = \frac{\sigma^2}{2E}$$

When the crack appears, the strain energy is released in a material volume adjacent to the crack. Suppose that this volume is comprised of a conical ring whose generating lines are shown in Fig. 3.10 by dashed lines and heights are proportional to the crack length, l. Then, the total released energy, Eq. (2.52), is

$$W = \frac{1}{2}k\pi\frac{\sigma^2}{E}l^2 d \tag{3.10}$$

where k is some constant coefficient of proportionality. On the other hand, the formation of new surfaces consumes the energy

$$S = 2\pi\gamma l d \tag{3.11}$$

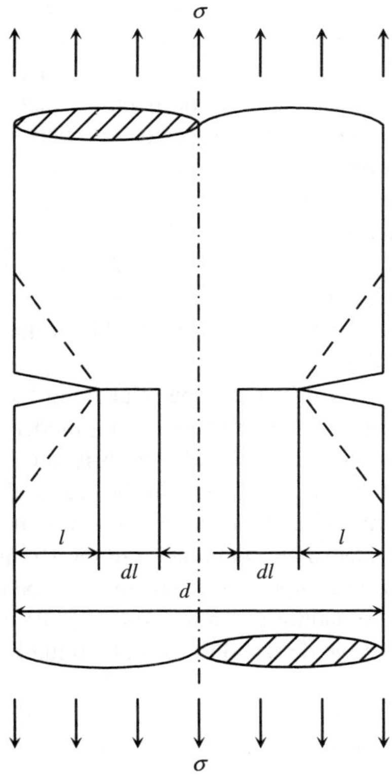

Fig. 3.10. A fiber with a crack.

where γ is the surface energy, Eq. (3.9). Now assume that the crack length is increased by an infinitesimal increment, $\mathrm{d}l$. Then, if for some value of acting stress, σ

$$\frac{\mathrm{d}W}{\mathrm{d}l} > \frac{\mathrm{d}S}{\mathrm{d}l} \tag{3.12}$$

the crack will propagate, and the fiber will fail. Substituting Eqs. (3.10) and (3.11) into inequality (3.12) we arrive at

$$\sigma > \overline{\sigma}_c = \sqrt{\frac{2\gamma E}{kl}} \tag{3.13}$$

The most important result that follows from this condition specifying some critical stress, σ_c, beyond which the fiber with a crack cannot exist is the fact that $\overline{\sigma}_c$ depends on the absolute value of the crack length (not on the ratio l/d). Now, for a continuous fiber, $2l < d$; so, the thinner the fiber, the smaller is the length of the crack that can exist in this fiber and the higher is the critical stress, $\overline{\sigma}_c$. More rigorous analysis shows that, reducing l to a in Fig. 3.8, we arrive at $\overline{\sigma}_c = \overline{\sigma}_t$.

Consider, for example, glass fibers that are widely used as reinforcing elements in composite materials and have been studied experimentally to support the fundamentals of fracture mechanics (Griffith, 1920). The theoretical strength of glass, Eq. (3.8), is about 14 GPa, whereas the actual strength of 1-mm-diameter glass fibers is only about 0.2 GPa, and for 5-mm-diameter fibers, this value is much lower (about 0.05 GPa). The fact that such low actual strength is caused by surface cracks can be readily proved if the fiber surface is smoothed by etching the fiber with acid. Then, the strength of 5-mm-diameter fibers can be increased up to 2 GPa. If the fiber diameter is reduced by heating and stretching the fibers to a diameter of about 0.0025 mm, the strength is increased to 6 GPa. Theoretical extrapolation of the experimental curve, showing the dependence of the fiber strength on the fiber diameter for very small fiber diameters, yields $\overline{\sigma} = 11\,\text{GPa}$, which is close to $\overline{\sigma}_t = 14\,\text{GPa}$.

Thus, we arrive at the following conclusion, clarifying the nature of the high performance of advanced composites and their place among modern structural materials.

The actual strength of advanced structural materials is much lower than their theoretical strength. This difference is caused by defects in the material microstructure (e.g., crystalline structure) or macrocracks inside the material and on its surface. Using thin fibers, we reduce the influence of cracks and thus increase the strength of materials reinforced with these fibers. So, advanced composites comprise a special class of structural materials in which we try to utilize the natural potential properties of the material, rather than the possibilities of technology as we do developing high-strength alloys.

3.2.2. Statistical aspects of fiber strength

Fiber strength, being relatively high, is still less than the corresponding theoretical strength, which means that fibers of advanced composites have microcracks or other defects randomly distributed along the fiber length. This is supported by the fact that fiber strength depends on the length of the tested fiber. The dependence of strength on length for boron fibers (Mikelsons and Gutans, 1984) is shown in Fig. 3.11. The longer the fiber, the higher the probability of a deleterious defect to exist within this length, and the lower the fiber strength. The tensile strengths of fiber segments with the same length but taken from different parts of a long continuous fiber, or from different fibers, also demonstrates the strength deviation. A typical strength distribution for boron fibers is presented in Fig. 3.12.

The first important characteristic of the strength deviation is the strength scatter $\Delta\overline{\sigma} = \overline{\sigma}_{max} - \overline{\sigma}_{min}$. For the case corresponding to Fig. 3.12, $\overline{\sigma}_{max} = 4.2\,\text{GPa}$, $\overline{\sigma}_{min} = 2\,\text{GPa}$, and $\Delta\overline{\sigma} = 2.2\,\text{GPa}$. To plot the diagram presented in Fig. 3.12, $\Delta\overline{\sigma}$ is divided into a set of increments, and a normalized number of fibers $n = N_\sigma/N$ (N_σ is the number of fibers failing at that stress within the increment, and N is the total number of tested fibers) is calculated and shown on the vertical axis. Thus, the so-called frequency histogram can be plotted. This histogram allows us to determine the mean value of the fiber strength as

$$\overline{\sigma}_m = \frac{1}{N}\sum_{i=1}^{N}\overline{\sigma}_i \tag{3.14}$$

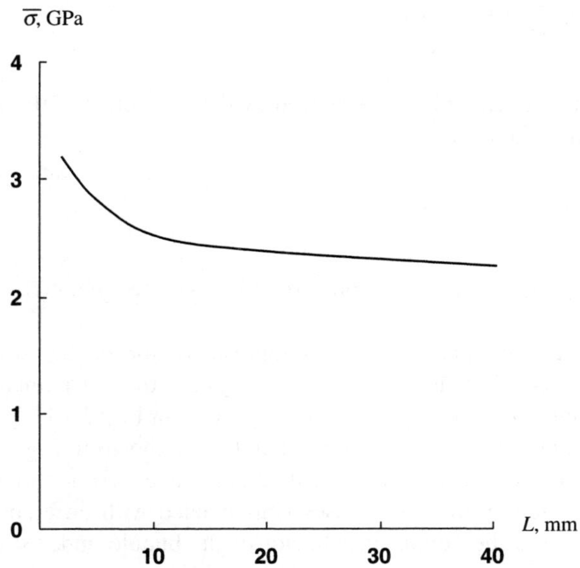

Fig. 3.11. Dependence of strength of boron fibers on the fiber length.

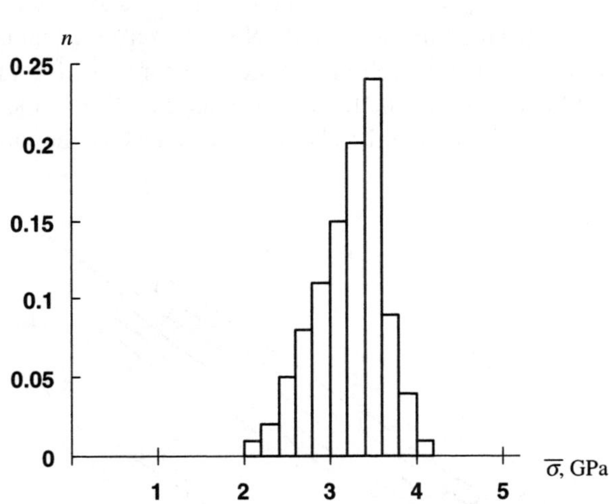

Fig. 3.12. Strength distribution for boron fibers.

and the strength dispersion as

$$d_\sigma = \sqrt{\frac{1}{N-1}\sum_{i=1}^{N}(\overline{\sigma}_\mathrm{m} - \sigma_\mathrm{i})^2} \tag{3.15}$$

The deviation of fiber strength is characterized by the coefficient of the strength variation, which is presented as follows

$$r_\sigma = \frac{d_\sigma}{\overline{\sigma}_\mathrm{m}} 100\% \tag{3.16}$$

For the boron fibers under consideration, Eqs. (3.14)–(3.16) yield $\overline{\sigma}_\mathrm{m} = 3.2\,\mathrm{GPa}$, $d_\sigma = 0.4\,\mathrm{GPa}$, and $r_\sigma = 12.5\%$.

To demonstrate the influence of fiber strength deviation on the strength of a unidirectional ply, consider a bundle of fibers, i.e., a system of approximately parallel fibers with different strength and slightly different lengths, as in Fig. 3.13. Typical stress–strain diagrams for fibers tested under tension in a bundle are shown in Fig. 3.14 (Vasiliev and Tarnopol'skii, 1990). As can be seen, the diagrams have two nonlinear segments. The nonlinearity in the vicinity of zero stresses is associated with different lengths of fibers in the bundles, whereas the nonlinear behavior of the bundle under stresses close to the ultimate values is caused by fracture of the fibers with lower strength.

Useful qualitative results can be obtained if we consider model bundles consisting of five fibers with different strengths. Five such bundles are presented in Table 3.1, showing the normalized strength of each fiber. As can be seen, the deviation of fiber strength is such that the mean strength, $\overline{\sigma}_\mathrm{m} = 1$, is the same for all the bundles, whereas the variation coefficient, r_σ, changes from 31.6% for bundle No. 1 to zero for bundle No. 5. The last row in the table shows the effective (observed) ultimate force, $\overline{F}$, for a bundle. Consider, for example, the first bundle. When the force is increased to $F = 3$, the stresses in all the fibers become $\sigma_j = 0.6$, and fiber No. 1 fails. After this happens, the force $F = 3$ is taken

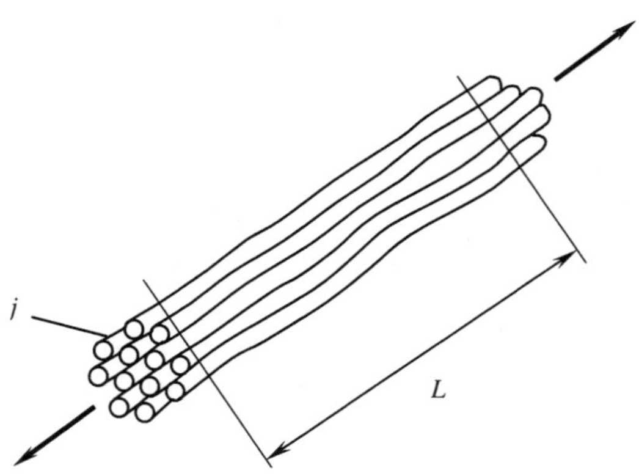

Fig. 3.13. Tension of a bundle of fibers.

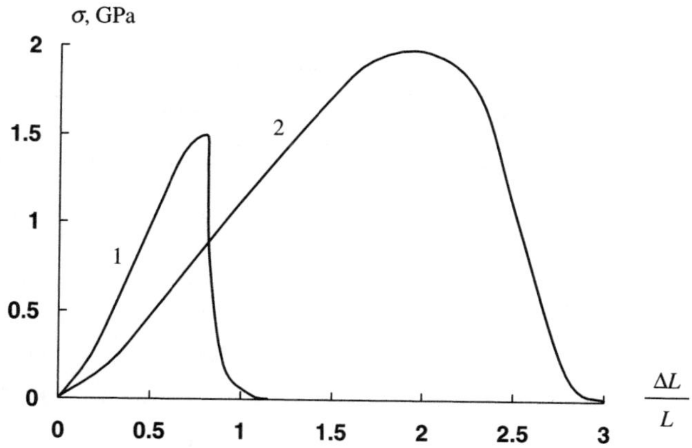

Fig. 3.14. Stress–strain diagrams for bundles of carbon (1) and aramid (2) fibers.

Table 3.1
Strength of bundles consisting of fibers of different strengths.

Fiber number	Bundle number				
	1	2	3	4	5
1	0.6	0.7	0.85	0.9	1.0
2	0.8	0.9	0.9	0.95	1.0
3	1.0	1.0	1.0	1.0	1.0
4	1.2	1.1	1.1	1.05	1.0
5	1.4	1.3	1.15	1.1	1.0
$\overline{\sigma}_m$	1.0	1.0	1.0	1.0	1.0
r_σ (%)	31.6	22.4	12.8	7.8	0
$\overline{F}$	3.2	3.6	4.25	4.5	5.0

by four fibers, and $\sigma_j = 0.75$ ($j = 2, 3, 4, 5$). When the force reaches the value $F = 3.2$, the stresses become $\sigma_j = 0.8$, and fiber No. 2 fails. After that, $\sigma_j = 1.07$ ($j = 3, 4, 5$). This means that fiber No. 3 also fails at force $F = 3.2$. Then, for the two remaining fibers, $\sigma_4 = \sigma_5 = 1.6$, and they also fail. Thus, $\overline{F} = 3.2$ for bundle No. 1. In a similar way, $\overline{F}$ can be calculated for the other bundles in the table. As can be seen, the lower the fiber strength variation, the higher the $\overline{F}$, which reaches its maximum value, $\overline{F} = 5$, for bundle No. 5, consisting of fibers of the same strength.

Table 3.2 demonstrates that strength variation can be more important than the mean strength. In fact, while the mean strength, $\overline{\sigma}_m$, goes down for bundles No. 1–5, the ultimate force, $\overline{F}$, increases. So, it can be better to have fibers with relatively low strength and low strength variation rather than high-strength fibers with high strength variation.

Table 3.2
Strength of bundles consisting of fibers of different strengths.

Fiber number	Bundle number				
	1	2	3	4	5
1	0.6	0.7	0.85	0.9	0.95
2	0.8	0.9	0.9	0.85	0.95
3	1.0	1.2	1.1	1.0	0.95
4	1.6	1.4	1.15	1.05	0.95
5	3.0	1.6	1.4	1.1	0.95
$\overline{\sigma}_m$	1.4	1.16	1.08	1.0	0.95
r_σ (%)	95.0	66.0	22.0	7.8	0
$\overline{F}$	3.2	3.6	4.25	4.5	4.75

3.2.3. Stress diffusion in fibers interacting through the matrix

The foregoing discussion concerned individual fibers or bundles of fibers that are not joined together. This is not the case for composite materials in which the fibers are embedded in the matrix material. Usually, the stiffness of the matrix is much lower than that of fibers (see Table 1.1), and the matrix practically does not take the load applied in the fiber direction. However, the fact that the fibers are bonded with the matrix even having relatively low stiffness changes the mechanism of fiber interaction and considerably increases their effective strength. To show this, the strength of dry fiber bundles can be compared with the strength of the same bundles after they were impregnated with epoxy resin and cured. The results are listed in Table 3.3. As can be seen, composite bundles in which fibers are joined together by the matrix demonstrate significantly higher strength, and the higher the fiber sensitivity to damage, the higher the difference in strength of dry and composite bundles. The influence of a matrix on the variation of strength is even more significant. As follows from Table 3.4, the variation coefficients of composite bundles are lower by an order of magnitude than those of individual fibers.

To clarify the role of a matrix in composite materials, consider the simple model of a unidirectional ply shown in Fig. 3.15 and apply the method of analysis developed for stringer panels (Goodey, 1946).

Table 3.3
Strength of dry bundles and composite bundles.

Fibers	Sensitivity of fibers to damage	Ultimate tensile load $\overline{F}$ (N)		Strength increase (%)
		Dry bundle	Composite bundle	
Carbon	High	14	26	85.7
Glass	Moderate	21	36	71.4
Aramid	Low	66	84	27.3

Table 3.4
Variation coefficients for fibers and unidirectional composites.

Fibers	Variation coefficient r_σ (%)	
	Fibers	Composite
Glass	29	2.0
Carbon	30	4.7
Aramid	24	5.0
Boron	23	3.0

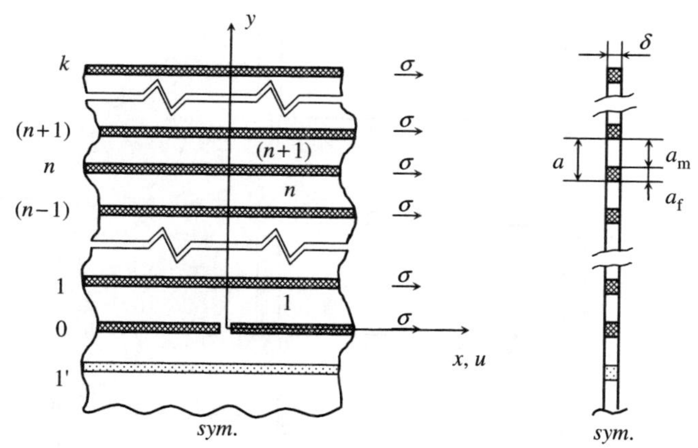

Fig. 3.15. Model of a unidirectional ply with a broken fiber.

Let the ply of thickness δ consist of $2k$ fibers symmetrically distributed on both sides of the central fiber $n = 0$. The fibers are joined with layers of the matrix material, and the fiber volume fraction is

$$v_f = \frac{a_f}{a}, \quad a = a_f + a_m \tag{3.17}$$

Let the central fiber have a crack induced by the fiber damage or by the shortage of this fiber's strength. At a distance from the crack, the fibers are uniformly loaded with stress σ (see Fig. 3.15).

First, derive the set of equations describing the ply under study. Since the stiffness of the matrix is much less than that of fibers, we neglect the stress in the matrix acting in the x direction and assume that the matrix works only in shear. We also assume that there are no displacements in the y direction.

Considering equilibrium of the last ($n = k$) fiber, an arbitrary fiber, and the central ($n = 0$) fiber shown in Fig. 3.16, we arrive at the following equilibrium equations

$$\begin{aligned} a_f\sigma_k' - \tau_k &= 0 \\ a_f\sigma_n' + \tau_{n+1} - \tau_n &= 0 \\ a_f\sigma_0' + 2\tau_1 &= 0 \end{aligned} \tag{3.18}$$

in which $(\,)' = \mathrm{d}(\,)/\mathrm{d}x$.

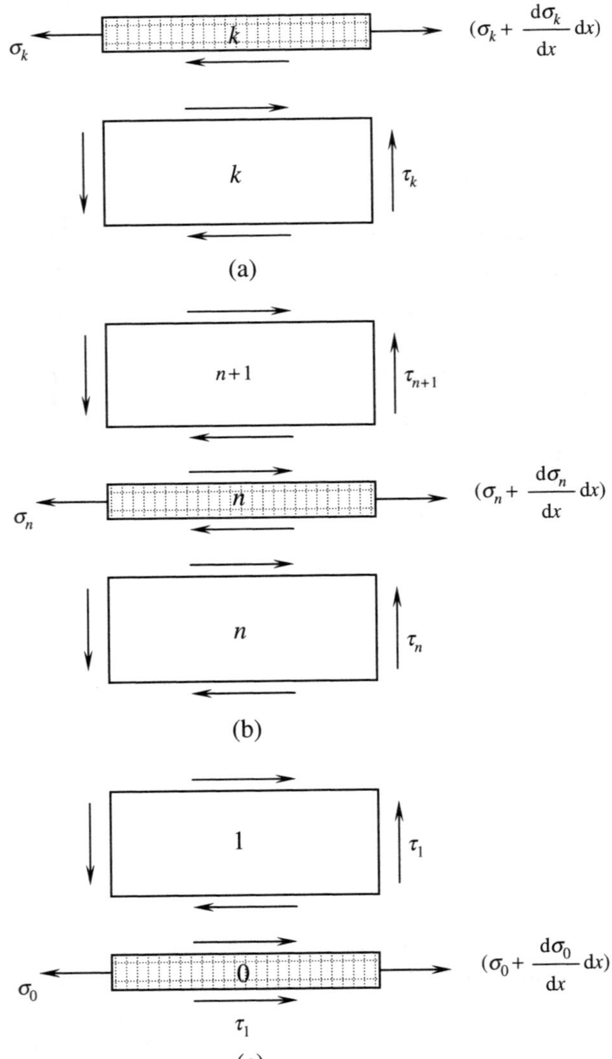

Fig. 3.16. Stresses acting in fibers and matrix layers for the last (a), arbitrary n-th fiber (b), and the central $n = 0$ fiber (c).

Constitutive equations for fibers and the matrix can be written as

$$\sigma_n = E_{\mathrm{f}}\varepsilon_n, \quad \tau_n = G_{\mathrm{m}}\gamma_n \tag{3.19}$$

Here, E_{f} is the fiber elasticity modulus and G_{m} is the matrix shear modulus, whereas

$$\varepsilon_n = u_n' \tag{3.20}$$

is the fiber strain expressed in terms of the displacement in the x direction. The shear strain in the matrix follows from Fig. 3.17, i.e.,

$$\gamma_n = \frac{1}{a_{\mathrm{m}}}(u_n - u_{n-1}) \tag{3.21}$$

This set of equations, Eqs. (3.18)–(3.21), is complete – it includes $10k + 3$ equations and contains the same number of unknown stresses, strains, and displacements.

Consider the boundary conditions. If there is no crack in the central fiber, the solution of the problem is evident and has the form $\sigma_n = \sigma$, $\tau_n = 0$. Assuming that the perturbation of the stressed state induced by the crack vanishes at a distance from the crack, we arrive at

$$\sigma_n(x \to \infty) = \sigma, \quad \tau_n(x \to \infty) = 0 \tag{3.22}$$

As a result of the crack in the central fiber, we have

$$\sigma_0(x = 0) = 0 \tag{3.23}$$

For the remaining fibers, symmetry conditions yield

$$u_n(x = 0) = 0 \quad (n = 1, 2, 3 \ldots k) \tag{3.24}$$

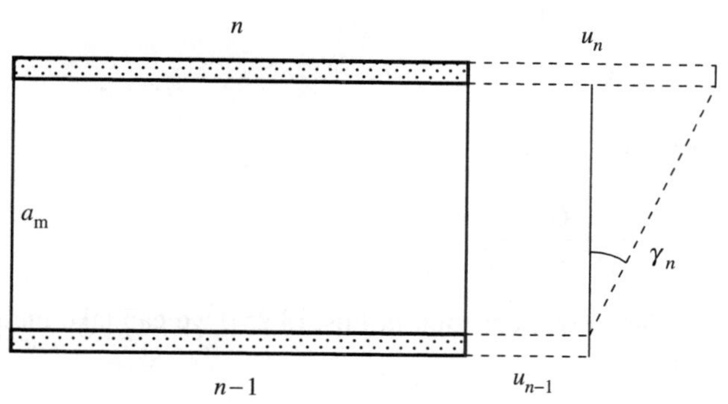

Fig. 3.17. Shear strain in the matrix layer.

To solve the problem, we use the stress formulation and, in accordance with Section 2.10, should consider equilibrium equations in conjunction with compatibility equations expressed in terms of stresses.

First, transform equilibrium equations introducing the stress function, $F(x)$, such that

$$\tau_n = F_n', \quad F_n(x \to \infty) = 0 \tag{3.25}$$

Substituting Eqs. (3.25) into the equilibrium equations, Eqs. (3.18), integrating them from x to ∞, and taking into account Eqs. (3.22) and (3.25), we obtain

$$\begin{aligned} \sigma_k &= \sigma + \frac{1}{a_f} F_k \\ \sigma_n &= \sigma - \frac{1}{a_f}(F_{n+1} - F_n) \\ \sigma_0 &= \sigma - \frac{2}{a_f} F_1 \end{aligned} \tag{3.26}$$

Compatibility equations follow from Eqs. (3.20) and (3.21), i.e.,

$$\gamma_n' = \frac{1}{a_m}(\varepsilon_n - \varepsilon_{n-1})$$

Using constitutive equations, Eqs. (3.19), we can write them in terms of stresses

$$\tau_n' = \frac{G_m}{a_m E_f}(\sigma_n - \sigma_{n-1})$$

Substituting stresses from Eqs. (3.25) and (3.26), and introducing the dimensionless coordinate $\overline{x} = x/a$ (see Fig. 3.15), we finally arrive at the following set of governing equations:

$$\begin{aligned} F_k'' - \mu^2(2F_k - F_{k-1}) &= 0 \\ F_n'' + \mu^2(F_{n+1} - 2F_n + F_{n-1}) &= 0 \\ F_1'' + \mu^2(F_2 - 3F_1) &= 0 \end{aligned} \tag{3.27}$$

in which, in accordance with Eqs. (3.17),

$$\mu^2 = \frac{G_m a^2}{a_f a_m E_f} = \frac{G_m}{v_f(1 - v_f)E_f} \tag{3.28}$$

With due regard to the second equation in Eqs. (3.25), we can take the general solution of Eqs. (3.27) in the form

$$F_n(\overline{x}) = A_n e^{-\lambda \overline{x}} \tag{3.29}$$

Substitution in Eqs. (3.27) yields:

$$A_k\left(2-\frac{\lambda^2}{\mu^2}\right)-A_{k-1}=0 \tag{3.30}$$

$$A_{n+1}-A_n\left(2-\frac{\lambda^2}{\mu^2}\right)+A_{n-1}=0 \tag{3.31}$$

$$A_2-A_1\left(3-\frac{\lambda^2}{\mu^2}\right)=0 \tag{3.32}$$

The finite-difference equation, Eq. (3.31), can be reduced to the following form

$$A_{n+1}-2A_n\cos\theta+A_{n-1}=0 \tag{3.33}$$

where

$$\cos\theta=1-\frac{\lambda^2}{2\mu^2} \tag{3.34}$$

As can be readily checked, the solution for Eq. (3.33) is

$$A_n=B\cos n\theta+C\sin n\theta \tag{3.35}$$

whereas Eq. (3.34) yields, after some transformation,

$$\lambda=2\mu\sin\frac{\theta}{2} \tag{3.36}$$

Substituting the solution, Eq. (3.35), into Eq. (3.30), we obtain, after some transformation,

$$B=-C\tan(k+1)\theta$$

Thus, Eq. (3.35) can be written as

$$A_n=C[\sin n\theta-\cos n\theta\cdot\tan(k+1)\theta] \tag{3.37}$$

Substituting Eq. (3.37) into Eq. (3.32) and performing rather cumbersome trigonometric transformations, we arrive at the following equation for θ

$$\tan k\theta=-\tan\frac{\theta}{2} \tag{3.38}$$

The periodic properties of the tangent function in Eq. (3.38) mean that it has $k+1$ different roots corresponding to intersection points of the curves $z=\tan k\theta$ and $z=-\tan\theta/2$.

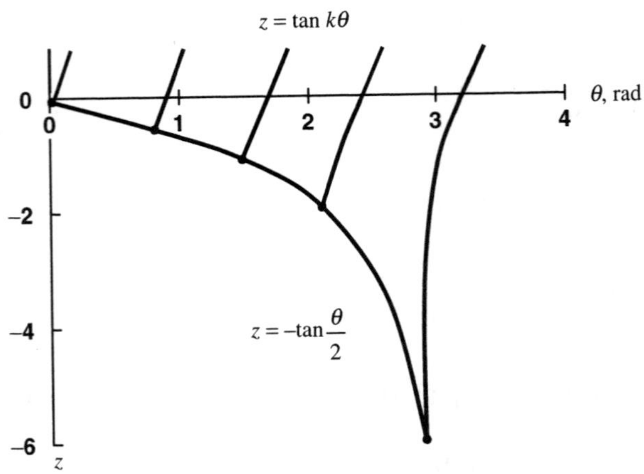

Fig. 3.18. Geometric interpretation of Eq. (3.38) for $k = 4$.

For the case $k = 4$, considered below as an example, these points are shown in Fig. 3.18. Further transformation allows us to reduce Eq. (3.38) to

$$\sin \frac{2k+1}{2}\theta = 0$$

from which it follows that

$$\theta_i = \frac{2\pi i}{2k+1} \qquad (i = 0, 1, 2 \dots k) \tag{3.39}$$

The first root, $\theta_0 = 0$, corresponds to $\lambda = 0$ and $F_n = const$, i.e., to a ply without a crack in the central fiber. So, Eq. (3.39) specifies k roots ($i = 1, 2, 3, \dots, k$) for the ply under study, and the solution in Eqs. (3.29) and (3.37) can be generalized as

$$F_n(\overline{x}) = \sum_{i=1}^{k} C_i[\sin n\theta_i - \cos n\theta_i \cdot \tan(k+1)\theta_i]e^{-\lambda_i \overline{x}} \tag{3.40}$$

where, in accordance with Eq. (3.36),

$$\lambda_i = 2\mu \sin \frac{\theta_i}{2} \tag{3.41}$$

and θ_i are determined by Eq. (3.39).

Using Eq. (3.38), we can transform Eq. (3.40) to the following final form

$$F_n(\overline{x}) = \sum_{i=1}^{k} C_i S_n(\theta_i) e^{-\lambda_i \overline{x}} \tag{3.42}$$

where

$$S_n(\theta_i) = \frac{\sin(2n - 1/2)\theta_i}{\cos(\theta_i/2)} \tag{3.43}$$

Applying Eqs. (3.25) and (3.26), we can find shear and normal stresses, i.e.,

$$\tau_n(\overline{x}) = -\frac{1}{a}\sum_{i=1}^{k} \lambda_i C_i S_n(\theta_i) e^{-\lambda_i \overline{x}} \quad (n = 1, 2, 3 \ldots k) \tag{3.44}$$

$$\sigma_k(\overline{x}) = \sigma + \frac{1}{a_\mathrm{f}}\sum_{i=1}^{k} C_i S_k(\theta_i) e^{-\lambda_i \overline{x}} \tag{3.45}$$

$$\sigma_n(\overline{x}) = \sigma - \frac{1}{a_\mathrm{f}}\sum_{i=1}^{k} C_i [S_{n+1}(\theta_i) - S_n(\theta_i)] e^{-\lambda_i \overline{x}} \quad (n = 1, 2, 3 \ldots k - 1) \tag{3.46}$$

$$\sigma_0(\overline{x}) = \sigma - \frac{2}{a_\mathrm{f}}\sum_{i=1}^{k} C_i S_1(\theta_i) e^{-\lambda_i \overline{x}} \tag{3.47}$$

Displacements can be determined with the aid of Eqs. (3.19), (3.21), and (3.25). Changing x for $\overline{x} = x/a$, we get

$$u_n(\overline{x}) = \frac{a_\mathrm{m}}{aG_\mathrm{m}} F_n'(\overline{x}) + u_{n-1}(\overline{x})$$

For the first fiber ($n = 1$), we have

$$u_1(\overline{x}) = \frac{a_\mathrm{m}}{aG_\mathrm{m}} F_1'(\overline{x}) + u_0(\overline{x})$$

Substituting Eq. (3.42) into these equations, we arrive at

$$u_n(\overline{x}) = -\frac{a_\mathrm{m}}{aG_\mathrm{m}}\sum_{i=1}^{k} C_i \lambda_i S_n(\theta_i) e^{-\lambda_i \overline{x}} + u_{n-1}(\overline{x}) \quad (n = 2, 3, 4 \ldots k) \tag{3.48}$$

$$u_1(\overline{x}) = -\frac{a_\mathrm{m}}{aG_\mathrm{m}}\sum_{i=1}^{k} C_i \lambda_i S_1(\theta_i) e^{-\lambda_i \overline{x}} + u_0(\overline{x}) \tag{3.49}$$

To determine coefficients C_i, we should apply the boundary conditions and write Eqs. (3.23) and (3.24) in the explicit form using Eqs. (3.47)–(3.49). Substituting S_n from

Eq. (3.43) and λ_i from Eq. (3.41), we have

$$\sum_{i=1}^{k} C_i \tan\frac{\theta_i}{2} = \frac{\sigma a_{\mathrm{f}}}{2}$$

$$\sum_{i=1}^{k} C_i \tan\frac{\theta_i}{2} \sin\frac{2n-1}{2}\theta_i = 0 \quad (n = 2, 3, 4 \ldots k)$$

$$\sum_{i=1}^{k} C_i \tan\frac{\theta_i}{2} \sin\frac{\theta_i}{2} = \frac{aG_{\mathrm{m}}}{2\mu a_{\mathrm{m}}} u_0(0)$$

Introducing new coefficients

$$D_i = C_i \tan\frac{\theta_i}{2} \tag{3.50}$$

we arrive at the final form of the boundary conditions, i.e.,

$$\sum_{i=1}^{k} D_i = \frac{\sigma a_{\mathrm{f}}}{2} \tag{3.51}$$

$$\sum_{i=1}^{k} D_i \sin\frac{2n-1}{2}\theta_i = 0 \quad (n = 2, 3, 4 \ldots k) \tag{3.52}$$

$$\sum_{i=1}^{k} D_i \sin\frac{\theta_i}{2} = \frac{aG_{\mathrm{m}}}{2\mu a_{\mathrm{m}}} u_0(0) \tag{3.53}$$

This set contains $k+1$ equations and includes k unknown coefficients D_i and displacement $u_0(0)$.

The foregoing set of equations allows us to obtain the exact analytical solution for any number of fibers, k. To find this solution, some transformations are required. First, multiply Eq. (3.52) by $\sin[(2n-1)\theta_s/2]$ and sum up all the equations from $n = 2$ to $n = k$. Adding Eq. (3.53) for $n = 1$ multiplied by $\sin(\theta_s/2)$, we obtain

$$\sum_{n=1}^{k}\sum_{i=1}^{k} D_i \sin\frac{2n-1}{2}\theta_i \sin\frac{2n-1}{2}\theta_s = \frac{aG_{\mathrm{m}}}{2\mu a_{\mathrm{m}}} u_0(0) \sin\frac{\theta_s}{2}$$

Now, the sequence of summation can be changed, as follows

$$\sum_{i=1}^{k} D_i \sum_{n=1}^{k} \sin\frac{2n-1}{2}\theta_i \sin\frac{2n-1}{2}\theta_s = \frac{aG_{\mathrm{m}}}{2\mu a_{\mathrm{m}}} u_0(0) \sin\frac{\theta_s}{2} \tag{3.54}$$

Using the following known series

$$\sum_{n=1}^{k} \cos(2n-1)\theta = \frac{\sin 2k\theta}{2\sin\theta}$$

we get in several steps

$$R_{is} = \sum_{n=1}^{k} \sin\frac{2n-1}{2}\theta_i \sin\frac{2n-1}{2}\theta_s$$

$$= \frac{1}{2}\sum_{n=1}^{k}\left[\cos\frac{2n-1}{2}(\theta_i-\theta_s) - \cos\frac{2n-1}{2}(\theta_i+\theta_s)\right]$$

$$= \frac{1}{4}\left[\frac{\sin k(\theta_i-\theta_s)}{\sin\frac{1}{2}(\theta_i-\theta_s)} - \frac{\sin k(\theta_i+\theta_s)}{\sin\frac{1}{2}(\theta_i+\theta_s)}\right]$$

$$= \frac{\cos(\theta_i/2)\cos k\theta_i \cos(\theta_s/2)\cos k\theta_s}{\cos\theta_s - \cos\theta_i}\left(\tan k\theta_i \tan\frac{\theta_i}{2} - \tan k\theta_s \tan\frac{\theta_s}{2}\right)$$

Using Eq. (3.38), we can conclude that $R_{is} = 0$ for $i \neq s$. For the case $i = s$, we have

$$R_{ss} = \sum_{n=1}^{k} \sin^2\frac{2n-1}{2}\theta_s = \frac{1}{2}\sum_{n=1}^{k}[1-\cos(2n-1)\theta_s] = \frac{1}{2}\left(k - \frac{\sin 2k\theta_s}{2\sin\theta_s}\right)$$

As a result, Eq. (3.54) yields

$$D_s = \frac{2aG_{\mathrm{m}}u_0(0)\sin(\theta_s/2)\sin\theta_s}{\mu a_{\mathrm{m}}(2k\sin\theta_s - \sin 2k\theta_s)} \quad (s = 1, 2, 3 \ldots k) \tag{3.55}$$

Substituting these coefficients into Eq. (3.51), we can find $u_0(0)$, i.e.,

$$u_0(0) = \frac{\sigma\mu a_{\mathrm{f}} a_{\mathrm{m}}}{4aG_{\mathrm{m}}}\left(\sum_{i=1}^{k}\frac{\sin(\theta_i/2)\sin\theta_i}{2k\sin\theta_i - \sin 2k\theta_i}\right)^{-1} \tag{3.56}$$

Thus, the solution for the problem under study is specified by Eqs. (3.44)–(3.50), (3.55), and (3.56).

For example, consider a carbon–epoxy ply with the following parameters: $E_{\mathrm{f}} = 250\,\mathrm{GPa}$, $G_{\mathrm{m}} = 1\,\mathrm{GPa}$, $v_{\mathrm{f}} = 0.6$, and $k = 4$. The distribution of the normalized stresses in the fibers along the ply is shown in Fig. 3.19, whereas the same distribution of shear stresses in the matrix is presented in Fig. 3.20. As can be seen, in the vicinity of the crack in the central fiber, the load carried by this fiber is transmitted by shear through the matrix to adjacent fibers. At a distance from the end of the fiber, greater than $\bar{l}_i$, the stress in

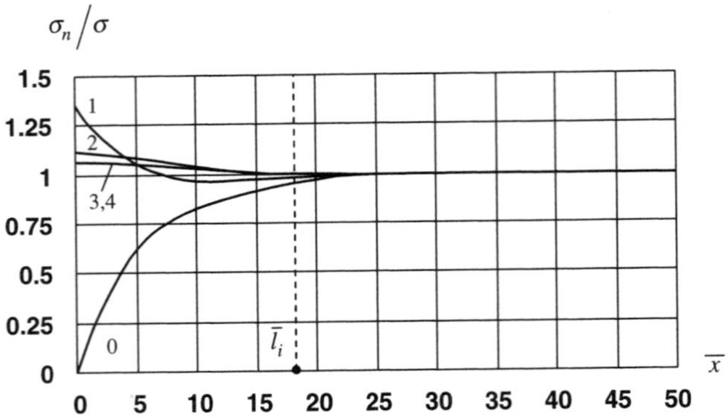

Fig. 3.19. Distribution of normal stresses along the fibers $n = 0, 1, 2, 3, 4$ for $k = 4$, $E_f = 250\,\text{GPa}$, and $G_m = 1\,\text{GPa}$.

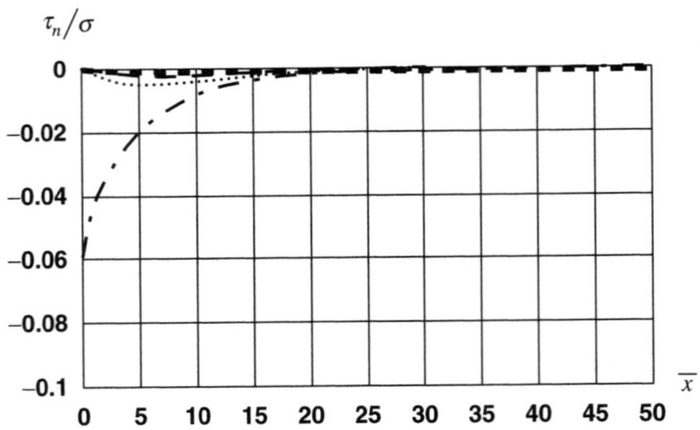

Fig. 3.20. Distribution of shear stresses along the fibers for $k = 4$, $E_f = 250\,\text{GPa}$, and $G_m = 1\,\text{GPa}$. Numbers of the matrix layers: – · – · – $n = 1$; ·········· $n = 2$; – – – – $n = 3$; ▪▪▪▪▪▪ $n = 4$.

the broken fiber becomes close to σ, and for $\bar{x} > \bar{l}_i$, the fiber behaves as if there is no crack. A portion of the broken fiber corresponding to $0 \le \bar{x} \le \bar{l}_i$ is not fully effective in resisting the applied load, and $l_i = \bar{l}_i a$ is referred to as the fiber ineffective length. Since the fiber defects are randomly distributed along its length, their influence on the strength of the ply is minimal if there are no other defects in the central and its adjacent fibers within distance l_i from the crack. To minimize the probability of such defects, we should minimize l_i, which depends on fiber and matrix stiffnesses and material microstructure.

To evaluate l_i, consider Eq. (3.47) and assume that $\sigma_0(\bar{x})$ becomes close to σ if

$$e^{-\lambda_i \bar{l}_i} = k$$

where k is some small parameter indicating how close $\sigma_0(\bar{x})$ should be to σ to neglect the difference between them (as a matter of fact, this difference vanishes only for $\bar{x} \to \infty$). Taking approximately $\lambda_i = 2\mu$ in accordance with Eq. (3.41) and using Eq. (3.28) specifying μ, we arrive at

$$\bar{l}_i = -\frac{1}{2} \ln k \cdot \sqrt{v_f(1 - v_f)\frac{E_f}{G_m}}$$

For $k = 0.01$, we get

$$l_i = 2.3a \cdot \sqrt{v_f(1 - v_f)\frac{E_f}{G_m}} \tag{3.57}$$

For a typical carbon–epoxy ply (see Fig. 3.19) with $a = 0.016$ mm and $v_f = 0.6$, Eq. (3.57) yields 0.29 mm.

Thus, for real composites, the length l_i is very small, and this explains why a unidirectional composite demonstrates much higher strength in longitudinal tension than a dry bundle of fibers (see Table 3.3). Reducing G_m, i.e., the matrix stiffness, we increase the fiber ineffective length, which becomes infinitely large for $G_m \to 0$. This effect is demonstrated in Fig. 3.21, which corresponds to a material whose matrix shear stiffness is much lower than that in the foregoing example (see Fig. 3.19). For this case, $\bar{l}_i = 50$, and Eq. (3.57) yields $l_i = 0.8$ mm. The distribution of shear stresses in this material is shown in Fig. 3.22. Experiments with unidirectional glass–epoxy composites ($E_f = 86.8$ GPa, $v_f = 0.68$, and $a = 0.015$) have shown that reduction of the matrix shear modulus from 1.08 GPa ($l_i = 0.14$ mm) to 0.037 GPa ($l_i = 0.78$ mm) results in reduction of longitudinal tensile strength from 2010 MPa to 1290 MPa, i.e., by 35.8% (Chiao, 1979).

The ineffective length of a fiber in a matrix can be found experimentally by using the single-fiber fragmentation test. For this test, a fiber is embedded in a matrix, and tensile

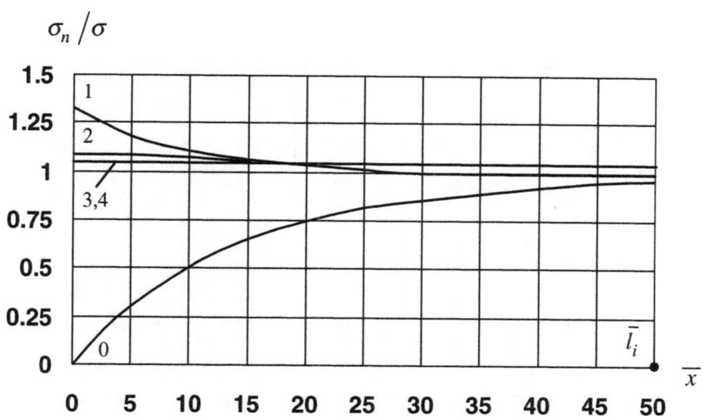

Fig. 3.21. Distribution of normal stresses along the fibres $n = 0, 1, 2, 3, 4$ for $k = 4$, $E_f = 250$ GPa, and $G_m = 0.125$ GPa.

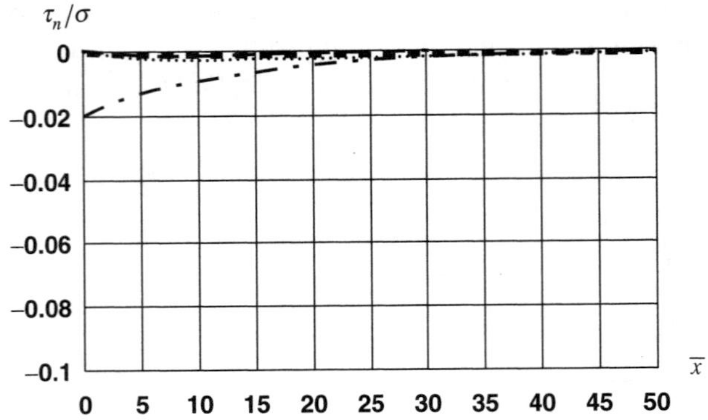

Fig. 3.22. Distribution of shear stresses along the fibers for $k = 4$, $E_f = 250$ GPa, and $G_m = 0.125$ GPa. Numbers of the matrix layers: – · – · – $n = 1$; ·········· $n = 2$; – – – – $n = 3$; ▪▪▪▪▪▪ $n = 4$.

load is applied to the fiber through the matrix until the fiber breaks. Further loading results in fiber fragmentation, and the length of the fiber fragment that no longer breaks is the fiber ineffective length. For a carbon fiber in epoxy matrix, $l_i = 0.3$ mm (Fukuda et al., 1993).

According to the foregoing reasoning, it looks as though the stiffness of the matrix should be as high as possible. However, there exists an upper limit of this stiffness. Comparing Figs. 3.20 and 3.22, we can see that the higher the value of G_m, the higher is the shear stress concentration in the matrix in the vicinity of the crack. If the maximum shear stress, τ_m, acting in the matrix reaches its ultimate value, $\bar{\tau}_m$, delamination will occur between the matrix layer and the fiber, and the matrix will not transfer the load from the broken fiber to the adjacent ones. This maximum shear stress depends on the fiber stiffness – the lower the fiber modulus, the higher the value of τ_m. This is shown in Figs. 3.23 and 3.24, where shear stress distributions are presented for aramid fibers ($E_f = 150$ GPa) and glass fibers ($E_f = 90$ GPa), respectively.

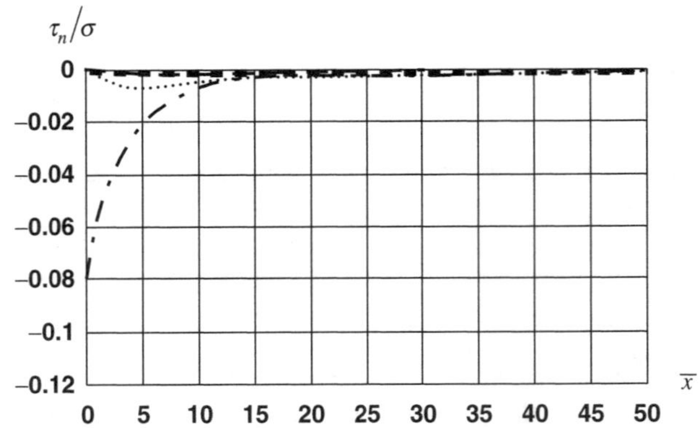

Fig. 3.23. Distribution of shear stresses along the fibers for $k = 4$, $E_f = 150$ GPa, and $G_m = 1$ GPa. Numbers of the matrix layers: – · – · – $n = 1$; ·········· $n = 2$; – – – – $n = 3$; ▪▪▪▪▪▪ $n = 4$.

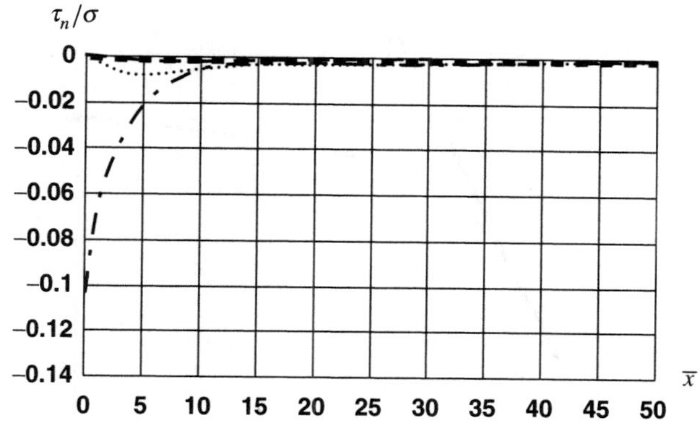

Fig. 3.24. Distribution of shear stresses along the fibers for $k = 4$, $E_f = 90$ GPa, and $G_m = 1$ GPa. Numbers of the matrix layers: –·–·– $n = 1$; ········ $n = 2$; – – – – $n = 3$; ▪▪▪▪▪▪ $n = 4$.

Finally, it should be emphasized that the plane model of a ply, considered in this section (see Fig. 3.15), provides only qualitative results concerning fibers and matrix interaction. In real materials, a broken fiber is surrounded by more than two fibers (at least 5 or 6, as can be seen in Fig. 3.2), so the stress concentration in these fibers and in the matrix is much lower than would be predicted by the foregoing analysis. For a hexagonal fiber distribution (see Fig. 3.4), the stress concentration factor for the fibers does not exceed 1.105 (Tikhomirov and Yushanov, 1980). The effect of fiber breakage on tensile strength of unidirectional composites has been studied by Abu-Farsakh et al. (2000).

3.2.4. *Fracture toughness*

Fracture toughness is a very important characteristic of a structural material indicating resistance of a material to cracks and governed by the work needed to destroy a material (work of fracture). It is well known that there exist brittle and ductile metal alloys, whose typical stress–strain diagrams are shown in Fig. 3.25. Comparing alloys with one and the same basic metal (e.g., steel alloys) we can see that while brittle alloys have higher strength, $\bar{\sigma}$, ductile alloys have higher ultimate elongation, $\bar{\varepsilon}$, and, as a result, higher work of fracture that is proportional to the area under the stress–strain diagram. Though brittle materials have, in general, higher strength, they are sensitive to cracks that, by propagating, can cause material failure for a stress that is much lower than the static strength. That is why design engineers usually prefer ductile materials with lower strength but higher fracture toughness. A typical dependence of fracture toughness on static strength for metals is shown in Fig. 3.26 (line 1). For composites, this dependence is entirely different (line 2) – a higher static strength corresponds usually to higher fracture toughness (Mileiko, 1982). This phenomenon is demonstrated for a unidirectional boron–aluminum composite in Fig. 3.27 (Mileiko, 1982). As can be seen, an increase in fiber volume fraction, v_f, results not only in higher static strength along the fibers (line 1), which is quite natural; it is also

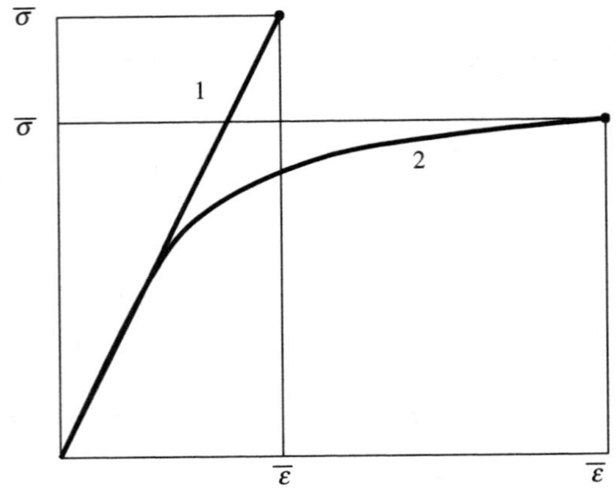

Fig. 3.25. Typical stress–strain diagrams of brittle (1) and ductile (2) metal alloys.

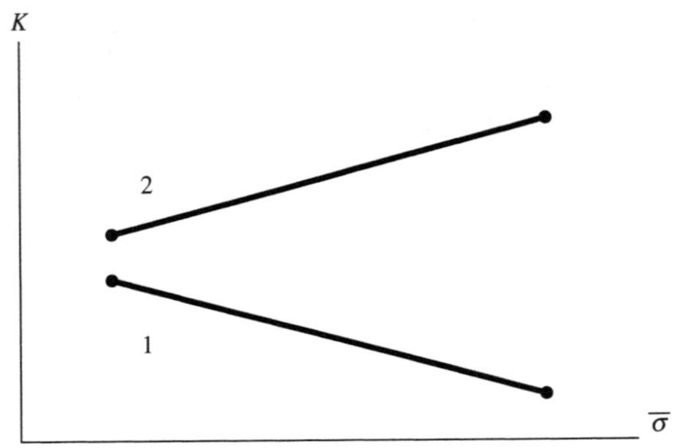

Fig. 3.26. Typical relations between fracture toughness (K) and strength ($\overline{\sigma}$) for metals (1) and composites (2).

accompanied by an increase in the work of fracture (curve 2) and, consequently, in an increase in the material fatigue strength (bending under 10^6 cycles, line 3), which shows a material's sensitivity to cracks.

The reason for such a specific behavior in composite materials is associated with their inhomogeneous microstructure, particularly, with fiber–matrix interfaces that restrain free propagation of a crack (see Fig. 3.28). Of some importance are also fiber defects, local delaminations and fiber strength deviation, which reduce the static strength but increase the fracture toughness. As a result, by combining brittle fibers and brittle matrix, we usually arrive at a composite material whose fracture toughness is higher than that of its components.

Thus, we can conclude that composites comprise a new class of structural materials that are entirely different from traditional man-made materials for several reasons. Firstly, using thin fibers we make an attempt to utilize the high strength capacity that is naturally inherent

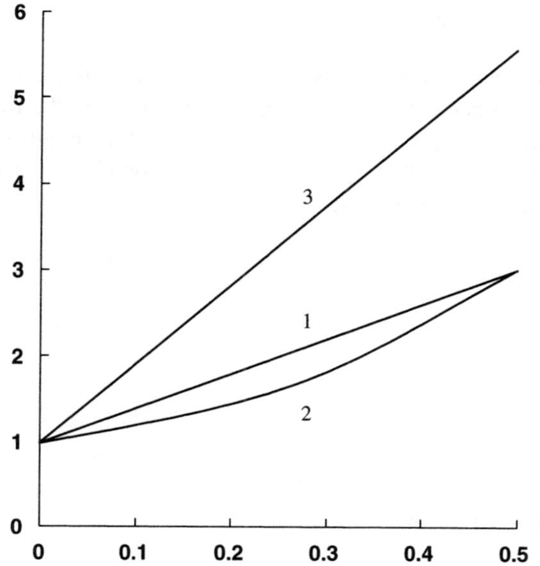

Fig. 3.27. Dependence of static strength (1), work of fracture (2), and fatigue strength (3) on fiber volume fraction for a boron–aluminum composite material.

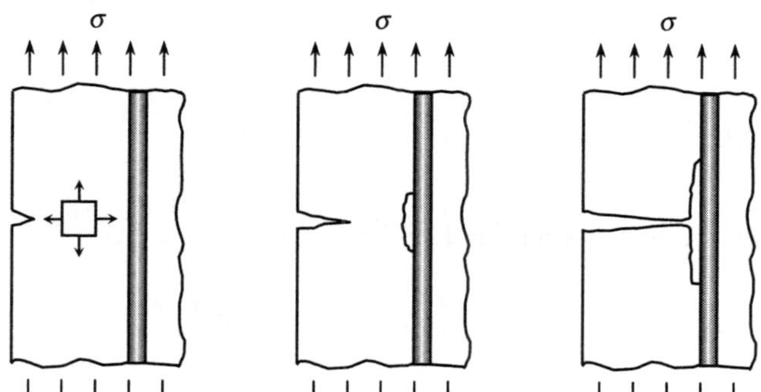

Fig. 3.28. Mechanism of the crack stopping at the fiber–matrix interface.

in all the materials. Secondly, this utilization is provided by the matrix material, which increases the fiber performance and makes it possible to manufacture composite structures. Thirdly, combination of fibers and matrices can result in new qualities of composite materials that are not inherent either in individual fibers or in the matrices, and are not described by the laws of mechanical mixtures. For example, as noted above, brittle fiber and matrix materials, both having low fracture toughness, can provide a heterogeneous composite material with high fracture toughness.

3.3. Micromechanics of a ply

Consider a unidirectional composite ply under the action of in-plane normal and shear stresses as in Fig. 3.29. Since the normal stresses do not change the right angle between axes 1 and 2, and shear stresses do not cause elongations in the longitudinal and transverse directions 1 and 2, the ply is orthotropic, and the corresponding constitutive equations, Eqs. (2.48) and (2.53), yield for the case under study

$$\varepsilon_1 = \frac{\sigma_1}{E_1} - \nu_{12}\frac{\sigma_2}{E_2}$$

$$\varepsilon_2 = \frac{\sigma_2}{E_2} - \nu_{21}\frac{\sigma_1}{E_1} \qquad (3.58)$$

$$\gamma_{12} = \frac{1}{G_{12}}\tau_{12}$$

The inverse form of these equations is

$$\sigma_1 = \bar{E}_1(\varepsilon_1 + \nu_{12}\varepsilon_2)$$

$$\sigma_2 = \bar{E}_2(\varepsilon_2 + \nu_{21}\varepsilon_1) \qquad (3.59)$$

$$\tau_{12} = G_{12}\gamma_{12}$$

where

$$\bar{E}_{1,2} = \frac{E_{1,2}}{1 - \nu_{12}\nu_{21}}$$

and the following symmetry condition is valid

$$E_1\nu_{12} = E_2\nu_{21} \qquad (3.60)$$

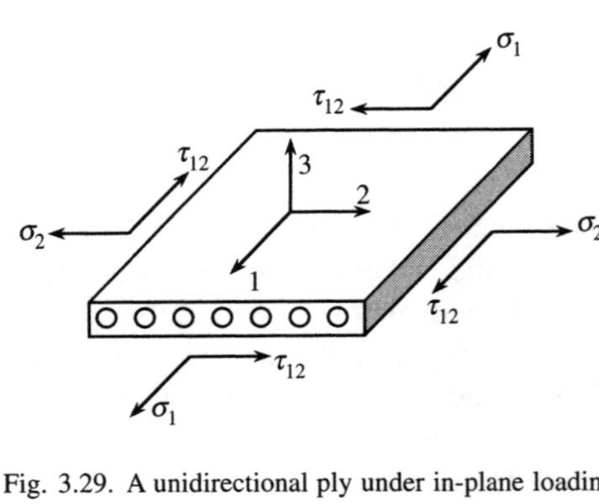

Fig. 3.29. A unidirectional ply under in-plane loading.

The constitutive equations, Eqs. (3.58) and (3.59), include effective or apparent longitudinal, E_1, transverse, E_2, and shear, G_{12}, moduli of a ply and Poisson's ratios ν_{12} and ν_{21}, only one of which is independent, since the second one can be found from Eq. (3.60).

The elastic constants, E_1, E_2, G_{12}, and ν_{12} or ν_{21}, are governed by fibers and matrix properties and the ply microstructure, i.e., the shape and size of the fibers' cross-sections, fiber volume fraction, distribution of fibers in the ply, etc. The task of micromechanics is to derive the corresponding governing relationships, i.e., to establish the relation between the properties of a unidirectional ply and those of its constituents.

To achieve this, we should first know the mechanical characteristics of the fibers and the matrix material of the ply. To determine the matrix modulus, E_{m}, its Poisson's ratio, ν_{m}, and strength, $\overline{\sigma}_{\mathrm{m}}$, conventional material testing specimens and testing procedures can be used (see Figs. 3.30 and 3.31). The shear modulus, G_{m}, can be calculated with the aid of Eq. (2.57). To find the fibers' properties is a more complicated problem. There exist several methods to test elementary fibers by bending or stretching 10–30-mm-long

Fig. 3.30. Specimens of matrix material.

Fig. 3.31. Testing of the matrix specimen.

fiber segments. All of them are rather specific because of the small (about 0.01 mm) fiber diameter, and, what is more important, the fiber properties in a composite material can be different from those of individual fibers (see Section 3.2.3) with the preassigned lengths provided by these methods.

It is worth knowing a fiber's actual modulus and strength, not only for micromechanics but also to check the fiber's quality before they are used to fabricate a composite part. For this purpose, a simple and reliable method has been developed to test the fibers in simulated actual conditions. According to this method, a fine tow or an assembly of fibers is carefully impregnated with resin, slightly stretched to avoid fiber waviness and cured to provide a specimen of the so-called microcomposite material. The microcomposite strand is wrapped over two discs as in Fig. 3.32, or fixed in special friction grips as in Fig. 3.33, and tested under tension to determine the ultimate tensile force $\overline{F}$ and strain ε corresponding to some force $F < \overline{F}$. Then, the resin is removed by burning it out, and the mass of fibers being divided by the strand length and fiber density yields the cross-sectional area of fibers in the strand, A_f. Fiber strength and modulus can be calculated as

$$\overline{\sigma}_f = \frac{\overline{F}}{A_f}, \qquad E_f = \frac{F}{A_f \varepsilon}$$

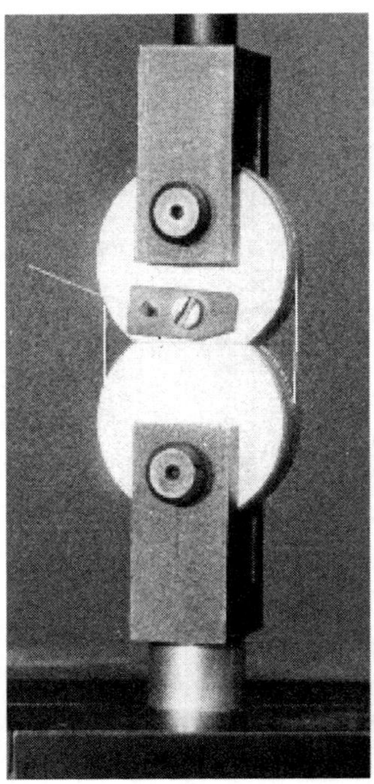

Fig. 3.32. Testing of a microcomposite specimen wrapped over discs.

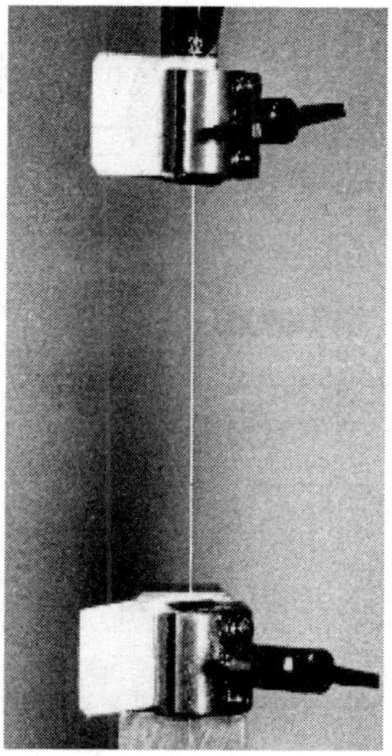

Fig. 3.33. Testing of a microcomposite specimen gripped at the ends.

In addition to fiber and matrix mechanical properties, micromechanical analysis requires information about the ply microstructure. Depending on the level of this information, there exist micromechanical models of different levels of complexity.

The simplest or zero-order model of a ply is a monotropic model ignoring the strength and stiffness of the matrix and assuming that the ply works only in the fiber direction. Taking $E_2 = 0$ and $G_{12} = 0$ in Eqs. (3.59) and putting $\nu_{12} = 0$ in accordance with Eq. (3.60), we arrive at the following equations describing this model

$$\sigma_1 = E_1\varepsilon_1, \quad \sigma_2 = 0, \quad \tau_{12} = 0 \tag{3.61}$$

in which $E_1 = E_f v_f$. Being very simple and too approximate to be used in the stress–strain analysis of composite structures, Eqs. (3.61) are extremely efficient for the design of optimal composite structures in which the loads are carried mainly by the fibers (see Chapter 8).

First-order models allow for the matrix stiffness but require only one structural parameter to be specified – fiber volume fraction, v_f. Since the fiber distribution in the ply is not important for these models, the ply can be presented as a system of strips shown in Fig. 3.34 and simulating fibers (shadowed areas) and matrix (light areas). The structural parameters of the model can be expressed in terms of fiber and matrix volume fractions only, i.e.,

$$\frac{a_f}{a} = v_f, \quad \frac{a_m}{a} = v_m, \quad v_f + v_m = 1 \tag{3.62}$$

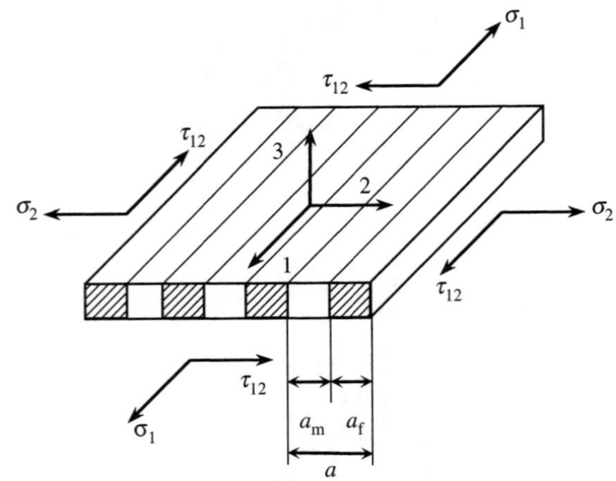

Fig. 3.34. First-order model of a unidirectional ply.

Suppose that the model ply is under in-plane loading with some effective stresses σ_1, σ_2, and τ_{12} as in Fig. 3.34, and find the corresponding effective elastic constants E_1, E_2, G_{12}, ν_{12}, and ν_{21} entering Eqs. (3.58). Constitutive equations for isotropic fiber and matrix strips can be written as

$$
\begin{aligned}
\varepsilon_1^{\mathrm{f,m}} &= \frac{1}{E_{\mathrm{f,m}}}\left(\sigma_1^{\mathrm{f,m}} - \nu_{\mathrm{f,m}}\sigma_2^{\mathrm{f,m}}\right) \\
\varepsilon_2^{\mathrm{f,m}} &= \frac{1}{E_{\mathrm{f,m}}}\left(\sigma_2^{\mathrm{f,m}} - \nu_{\mathrm{f,m}}\sigma_1^{\mathrm{f,m}}\right) \\
\gamma_{12}^{\mathrm{f,m}} &= \frac{1}{G_{\mathrm{f,m}}}\tau_{12}^{\mathrm{f,m}}
\end{aligned}
\tag{3.63}
$$

Here, f and m indices correspond, as stated earlier, to fibers and matrices, respectively.

Let us make some assumptions concerning the model behavior. First, it is natural to assume that effective stress resultant $\sigma_1 a$ is distributed between fiber and matrix strips and that the longitudinal strains of these strips are the same as the effective (apparent) longitudinal strain of the ply, ε_1, i.e.,

$$\sigma_1 a = \sigma_1^{\mathrm{f}} a_{\mathrm{f}} + \sigma_1^{\mathrm{m}} a_{\mathrm{m}} \tag{3.64}$$

$$\varepsilon_1^{\mathrm{f}} = \varepsilon_1^{\mathrm{m}} = \varepsilon_1 \tag{3.65}$$

Second, as can be seen in Fig. 3.34, under transverse tension, the stresses in the strips are the same and equal to the effective stress σ_2, whereas the ply elongation in the transverse direction is the sum of the fiber and matrix strips' elongations, i.e.,

$$\sigma_2^{\mathrm{f}} = \sigma_2^{\mathrm{m}} = \sigma_2 \tag{3.66}$$

$$\Delta a = \Delta a_{\mathrm{f}} + \Delta a_{\mathrm{m}} \tag{3.67}$$

Introducing transverse strains

$$\varepsilon_2 = \frac{\Delta a}{a}, \quad \varepsilon_2^{\mathrm{f}} = \frac{\Delta a_{\mathrm{f}}}{a_{\mathrm{f}}}, \quad \varepsilon_2^{\mathrm{m}} = \frac{\Delta a_{\mathrm{m}}}{a_{\mathrm{m}}}$$

we can write Eq. (3.67) in the following form

$$\varepsilon_2 a = \varepsilon_2^{\mathrm{f}} a_{\mathrm{f}} + \varepsilon_2^{\mathrm{m}} a_{\mathrm{m}} \tag{3.68}$$

The same assumptions can be made for shear stresses and strains, so that

$$\tau_{12}^{\mathrm{f}} = \tau_{12}^{\mathrm{m}} = \tau_{12} \tag{3.69}$$

$$\gamma_{12} a = \gamma_{12}^{\mathrm{f}} a_{\mathrm{f}} + \gamma_{12}^{\mathrm{m}} a_{\mathrm{m}} \tag{3.70}$$

With due regard to Eqs. (3.65), (3.66), and (3.69), constitutive equations, Eqs. (3.63) can be reduced to

$$\varepsilon_1 = \frac{1}{E_{\mathrm{f}}}\left(\sigma_1^{\mathrm{f}} - \nu_{\mathrm{f}}\sigma_2\right), \quad \varepsilon_1 = \frac{1}{E_{\mathrm{m}}}\left(\sigma_1^{\mathrm{m}} - \nu_{\mathrm{m}}\sigma_2\right) \tag{3.71}$$

$$\varepsilon_2^{\mathrm{f}} = \frac{1}{E_{\mathrm{f}}}\left(\sigma_2 - \nu_{\mathrm{f}}\sigma_1^{\mathrm{f}}\right), \quad \varepsilon_2^{\mathrm{m}} = \frac{1}{E_{\mathrm{m}}}\left(\sigma_2 - \nu_{\mathrm{m}}\sigma_1^{\mathrm{m}}\right) \tag{3.72}$$

$$\gamma_{12}^{\mathrm{f}} = \frac{1}{G_{\mathrm{f}}}\tau_{12}, \quad \gamma_{12}^{\mathrm{m}} = \frac{1}{G_{\mathrm{m}}}\tau_{12} \tag{3.73}$$

The first two equations, Eqs. (3.71), allow us to find longitudinal stresses, i.e.,

$$\sigma_1^{\mathrm{f}} = E_{\mathrm{f}}\varepsilon_1 + \nu_{\mathrm{f}}\sigma_2, \quad \sigma_1^{\mathrm{m}} = E_{\mathrm{m}}\varepsilon_1 + \nu_{\mathrm{m}}\sigma_2 \tag{3.74}$$

Equilibrium equation, Eq. (3.64), can be rearranged with the aid of Eqs. (3.62) to the form

$$\sigma_1 = \sigma_1^{\mathrm{f}} v_{\mathrm{f}} + \sigma_1^{\mathrm{m}} v_{\mathrm{m}} \tag{3.75}$$

Substituting Eqs. (3.74) into this equation, we can express ε_1 in terms of σ_1 and σ_2. Combining this result with the first constitutive equation in Eqs. (3.58), we arrive at

$$E_1 = E_{\mathrm{f}} v_{\mathrm{f}} + E_{\mathrm{m}} v_{\mathrm{m}} \tag{3.76}$$

$$\frac{\nu_{12}}{E_2} = \frac{\nu_{\mathrm{f}} v_{\mathrm{f}} + \nu_{\mathrm{m}} v_{\mathrm{m}}}{E_{\mathrm{f}} v_{\mathrm{f}} + E_{\mathrm{m}} v_{\mathrm{m}}} \tag{3.77}$$

The first of these equations specifies the apparent longitudinal modulus of the ply and corresponds to the so-called rule of mixtures, according to which the property of a composite can be calculated as the sum of its constituent material properties, multiplied by the corresponding volume fractions.

Now consider Eq. (3.67), which can be written as

$$\varepsilon_2 = \varepsilon_2^{\mathrm{f}} v_{\mathrm{f}} + \varepsilon_2^{\mathrm{m}} v_{\mathrm{m}}$$

Substituting strains $\varepsilon_2^{\mathrm{f}}$ and $\varepsilon_2^{\mathrm{m}}$ from Eqs. (3.72), stresses σ_1^{f} and σ_1^{m} from Eqs. (3.74), and ε_1 from Eqs. (3.58) with due regard to Eqs. (3.76) and (3.77), we can express ε_2 in terms of σ_1 and σ_2. Comparing this expression with the second constitutive equation in Eqs. (3.58), we get

$$\frac{1}{E_2} = \frac{v_{\mathrm{f}}}{E_{\mathrm{f}}} + \frac{v_{\mathrm{m}}}{E_{\mathrm{m}}} - \frac{v_{\mathrm{f}} v_{\mathrm{m}} (E_{\mathrm{f}} \nu_{\mathrm{m}} - E_{\mathrm{m}} \nu_{\mathrm{f}})^2}{E_{\mathrm{f}} E_{\mathrm{m}} (E_{\mathrm{f}} v_{\mathrm{f}} + E_{\mathrm{m}} v_{\mathrm{m}})} \tag{3.78}$$

$$\frac{\nu_{21}}{E_1} = \frac{\nu_{\mathrm{f}} v_{\mathrm{f}} + \nu_{\mathrm{m}} v_{\mathrm{m}}}{E_{\mathrm{f}} v_{\mathrm{f}} + E_{\mathrm{m}} v_{\mathrm{m}}} \tag{3.79}$$

Using Eqs. (3.76) and (3.79), we have

$$\nu_{21} = \nu_{\mathrm{f}} v_{\mathrm{f}} + \nu_{\mathrm{m}} v_{\mathrm{m}} \tag{3.80}$$

This result corresponds to the rule of mixtures. The second Poisson's ratio can be found from Eqs. (3.77) and (3.78). Finally, Eqs. (3.58), (3.70), and (3.73) yield the apparent shear modulus

$$\frac{1}{G_{12}} = \frac{v_{\mathrm{f}}}{G_{\mathrm{f}}} + \frac{v_{\mathrm{m}}}{G_{\mathrm{m}}} \tag{3.81}$$

This expression can be derived from the rule of mixtures if we use compliance coefficients instead of stiffnesses, as in Eq. (3.76).

Since the fiber modulus is typically many times greater than the matrix modulus, Eqs. (3.76), (3.78), and (3.81) can be simplified, neglecting small terms, and presented in the following approximate form

$$E_1 = E_{\mathrm{f}} v_{\mathrm{f}}, \quad E_2 = \frac{E_{\mathrm{m}}}{v_{\mathrm{m}}\left(1 - \nu_{\mathrm{m}}^2\right)}, \quad G_{12} = \frac{G_{\mathrm{m}}}{v_{\mathrm{m}}}$$

Only two of the foregoing expressions, namely Eq. (3.76) for E_1 and Eq. (3.80) for ν_{21}, both following from the rule of mixtures, demonstrate good agreement with experimental results. Moreover, expressions analogous to Eqs. (3.76) and (3.80) follow practically from the numerous studies based on different micromechanical models. Comparison of predicted and experimental results is presented in Figs. 3.35–3.37, where theoretical dependencies of normalized moduli on the fiber volume fraction are shown with lines. The dots correspond to the test data for epoxy composites reinforced with different fibers

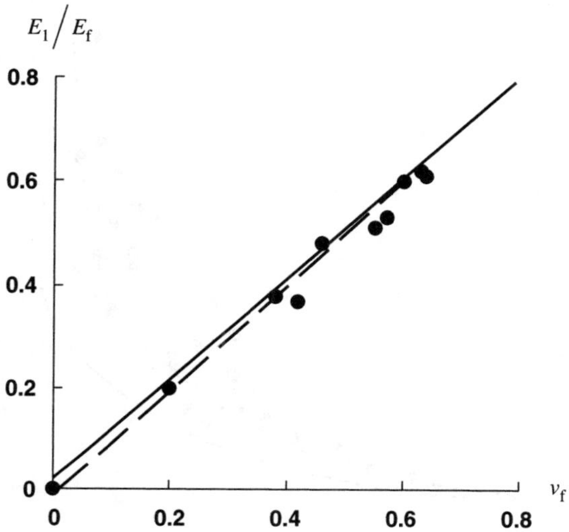

Fig. 3.35. Dependence of the normalized longitudinal modulus on fiber volume fraction. – – – zero-order model, Eqs. (3.61); —— first-order model, Eqs. (3.76); ● experimental data.

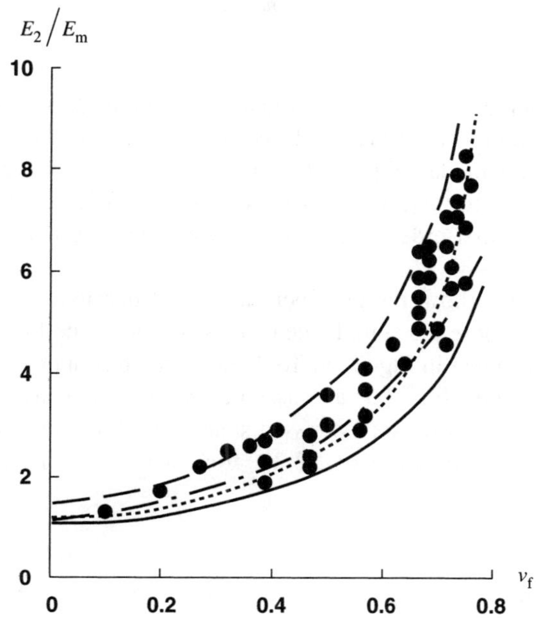

Fig. 3.36. Dependence of the normalized transverse modulus on fiber volume fraction. —— first-order model, Eq. (3.78); .... second-order model, Eq. (3.89); — · — higher-order model (elasticity solution) (Van Fo Fy, 1966); – – – the upper bound; ● experimental data.

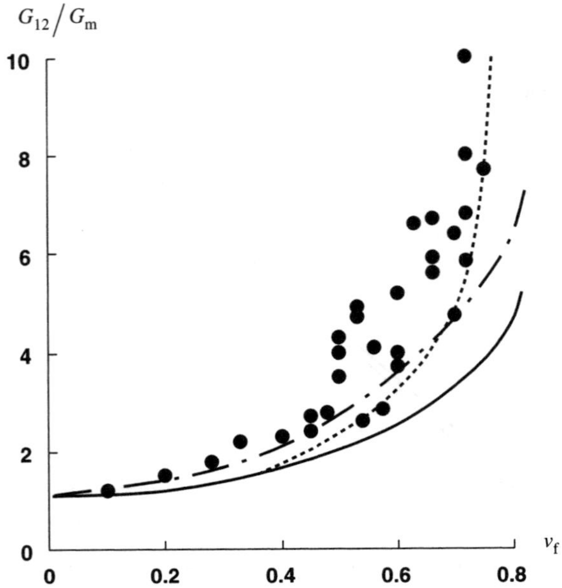

Fig. 3.37. Dependence of the normalized in-plane shear modulus on fiber volume fraction. —— first-order model, Eq. (3.81); .... second-order model, Eq. (3.90); —·— higher-order model (elasticity solution) (Van Fo Fy, 1966); ● experimental data.

that have been measured by the authors or taken from publications of Tarnopol'skii and Roze (1969), Kondo and Aoki (1982), and Lee et al. (1995). As can be seen in Fig. 3.35, not only the first-order model, Eq. (3.76), but also the zero-order model, Eqs. (3.61), provide fair predictions for E_1, whereas Figs. 3.36 and 3.37 for E_2 and G_{12} call for an improvement to the first-order model (the corresponding results are shown with solid lines).

Second-order models allow for the fiber shape and distribution, but, in contrast to higher-order models, ignore the complicated stressed state in the fibers and matrix under loading of the ply as shown in Fig. 3.29. To demonstrate this approach, consider a layer-wise fiber distribution (see Fig. 3.5) and assume that the fibers are absolutely rigid and the matrix is in the simplest uniaxial stressed state under transverse tension. The typical element of this model is shown in Fig. 3.38, from which we can obtain the following equation

$$v_f = \frac{\pi R^2}{2Ra} = \frac{\pi R}{2a} \qquad (3.82)$$

Since $2R < a$, $v_f < \pi/4 = 0.785$. The equilibrium condition yields

$$2R\sigma_2 = \int_{-R}^{R} \sigma_m \mathrm{d}x_3 \qquad (3.83)$$

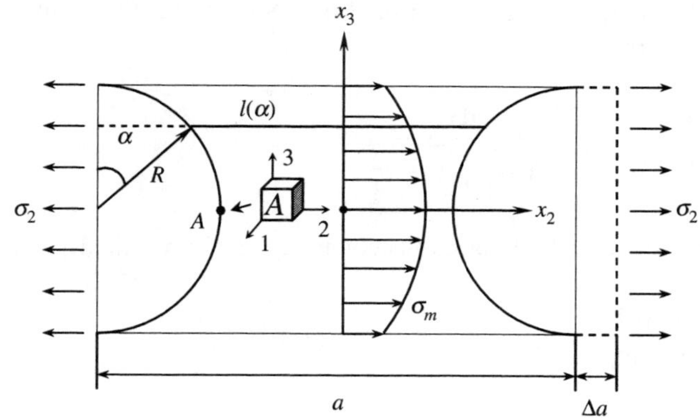

Fig. 3.38. Microstructural model of the second order.

where $x_3 = R\cos\alpha$ and σ_2 is some average transverse stress that induces average strain

$$\varepsilon_2 = \frac{\Delta a}{a} \tag{3.84}$$

such that the effective (apparent) transverse modulus is calculated as

$$E_2 = \frac{\sigma_2}{\varepsilon_2} \tag{3.85}$$

The strain in the matrix can be determined with the aid of Fig. 3.38 and Eq. (3.84), i.e.,

$$\varepsilon_m = \frac{\Delta a}{l(\alpha)} = \frac{\Delta a}{a - 2R\sin\alpha} = \frac{\varepsilon_2}{1 - \lambda\sqrt{1 - (x_3/R)^2}} \tag{3.86}$$

where, in accordance with Eq. (3.82),

$$\lambda = \frac{2R}{a} = \frac{4v_f}{\pi} \tag{3.87}$$

Assuming that there is no strain in the matrix in the fiber direction and there is no stress in the matrix in the x_3 direction, we have

$$\sigma_m = \frac{E_m\varepsilon_m}{1 - \nu_m^2} \tag{3.88}$$

Substituting σ_2 from Eq. (3.85) and σ_m, from Eq. (3.88) into Eq. (3.83) and using Eq. (3.86) to express ε_m, we arrive at

$$E_2 = \frac{E_m}{2R(1-\nu_m^2)} \int_{-R}^{R} \frac{dx_3}{1-\lambda\sqrt{1-x_3^2}}$$

Calculating the integral and taking into account Eq. (3.87), we finally get

$$E_2 = \frac{\pi E_m r(\lambda)}{2v_f\left(1-\nu_m^2\right)} \tag{3.89}$$

where

$$r(\lambda) = \frac{1}{\sqrt{1-\lambda^2}} \tan^{-1}\sqrt{\frac{1+\lambda}{1-\lambda}} - \frac{\pi}{4}$$

Similar derivation for an in-plane shear yields

$$G_{12} = \frac{\pi G_m}{2v_f} r(\lambda) \tag{3.90}$$

The dependencies of E_2 and G_{12} on the fiber volume fraction corresponding to Eqs. (3.89) and (3.90) are shown in Figs. 3.36 and 3.37 (dotted lines). As can be seen, the second-order model of a ply provides better agreement with the experimental results than the first-order model. This agreement can be further improved if we take a more realistic microstructure of the material. Consider the actual microstructure shown in Fig. 3.2 and single out a typical square element with size a as in Fig. 3.39. The dimension a should provide the same fiber volume fraction for the element as for the material under study. To calculate E_2, we divide the element into a system of thin ($h \ll a$) strips parallel to

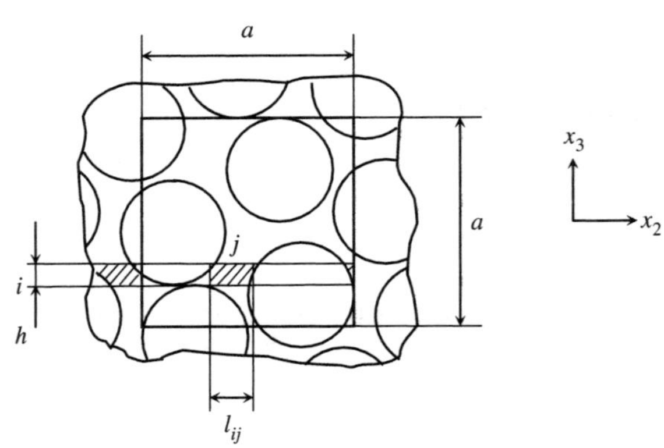

Fig. 3.39. Typical structural element.

axis x_2. The ith strip is shown in Fig. 3.39. For each strip, we measure the lengths, l_{ij}, of the matrix elements, the jth of which is shown in Fig. 3.39. Then, equations analogous to Eqs. (3.83), (3.88), and (3.86) take the form

$$\sigma_2 a = h\sum_i \sigma_{\mathrm{m}}^{(i)}, \quad \sigma_{\mathrm{m}}^{(i)} = \frac{E_{\mathrm{m}}}{1-\nu_{\mathrm{m}}^2}\varepsilon_{\mathrm{m}}^{(i)}, \quad \varepsilon_{\mathrm{m}}^{(i)} = \frac{\varepsilon_2 a}{\sum_j l_{ij}}$$

and the final result is

$$E_2 = \frac{E_{\mathrm{m}}\bar{h}}{1-\nu_{\mathrm{m}}^2}\sum_i \left(\sum_j \bar{l}_{ij}\right)^{-1}$$

where $\bar{h} = h/a$, $\bar{l}_{ij} = l_{ij}/a$. The second-order models considered above can be readily generalized to account for the fiber transverse stiffness and matrix nonlinearity.

Numerous higher-order microstructural models and descriptive approaches have been proposed, including

- analytical solutions in the problems of elasticity for an isotropic matrix having regular inclusions – fibers or periodically spaced groups of fibers,
- numerical (finite element, finite difference methods) stress analysis of the matrix in the vicinity of fibers,
- averaging of stress and strain fields for a media filled in with regularly or randomly distributed fibers,
- asymptotic solutions of elasticity equations for inhomogeneous solids characterized by a small microstructural parameter (fiber diameter),
- photoelasticity methods.

Exact elasticity solution for a periodical system of fibers embedded in an isotropic matrix (Van Fo Fy (Vanin), 1966) is shown in Figs. 3.36 and 3.37. As can be seen, due to the high scatter in experimental data, the higher-order model does not demonstrate significant advantages with respect to elementary models.

Moreover, all the micromechanical models can hardly be used for practical analysis of composite materials and structures. The reason for this is that irrespective of how rigorous the micromechanical model is, it cannot describe sufficiently adequately real material microstructure governed by a particular manufacturing process, taking into account voids, microcracks, randomly damaged or misaligned fibers, and many other effects that cannot be formally reflected in a mathematical model. As a result of this, micromechanical models are mostly used for qualitative analysis, providing us with the understanding of how material microstructural parameters affect the mechanical properties rather than with quantitative information about these properties. Particularly, the foregoing analysis should result in two main conclusions. First, the ply stiffness along the fibers is governed by the fibers and linearly depends on the fiber volume fraction. Second, the ply stiffness across the fibers and in shear is determined not only by the matrix (which is natural), but by the fibers as well. Although the fibers do not take directly the load applied in the transverse direction, they significantly increase the ply transverse stiffness (in comparison with the stiffness of a pure matrix) acting as rigid inclusions in the matrix. Indeed, as can be seen in Fig. 3.34, the higher the fiber fraction, a_f, the lower the matrix fraction, a_m, for the same a, and the higher stress σ_2 should be applied to the ply to cause the same transverse strain ε_2 because only matrix strips are deformable in the transverse direction.

Due to the aforementioned limitations of micromechanics, only the basic models were considered above. Historical overview of micromechanical approaches and more detailed description of the corresponding results can be found elsewhere (Bogdanovich and Pastore, 1996; Jones, 1999).

To analyze the foregoing micromechanical models, we used the traditional approach based on direct derivation and solution of the system of equilibrium, constitutive, and strain–displacement equations. Alternatively, the same problems can be solved with the aid of variational principles discussed in Section 2.11. In their application to micromechanics, these principles allow us not only to determine the apparent stiffnesses of the ply, but also to establish the upper and the lower bounds on them.

Consider, for example, the problem of transverse tension of a ply under the action of some average stress σ_2 (see Fig. 3.29) and apply the principle of minimum strain energy (see Section 2.11.2). According to this principle, the actual stress field provides the value of the body strain energy, which is equal to or less than that of any statically admissible stress field. Equality takes place only if the admissible stress state coincides with the actual one. Excluding this case, i.e., assuming that the class of admissible fields under study does not contain the actual field, we can write the following strict inequality

$$W_\sigma^{\mathrm{adm}} > W_\sigma^{\mathrm{act}} \tag{3.91}$$

For the problem of transverse tension, the fibers can be treated as absolutely rigid, and only the matrix strain energy needs to be taken into account. We can also neglect the energy of shear strain and consider the energy corresponding to normal strains only. With due regard to these assumptions, we use Eqs. (2.51) and (2.52) to get

$$W = \iiint\limits_{V_m} U dV_m \tag{3.92}$$

where V_m is the volume of the matrix, and

$$U = \frac{1}{2}\left(\sigma_1^m \varepsilon_1^m + \sigma_2^m \varepsilon_2^m + \sigma_3^m \varepsilon_3^m\right) \tag{3.93}$$

To find energy W_σ entering inequality (3.91), we should express strains in terms of stresses with the aid of constitutive equations, i.e.,

$$\begin{aligned}
\varepsilon_1^m &= \frac{1}{E_m}\left(\sigma_1^m - \nu_m \sigma_2^m - \nu_m \sigma_3^m\right) \\
\varepsilon_2^m &= \frac{1}{E_m}\left(\sigma_2^m - \nu_m \sigma_1^m - \nu_m \sigma_3^m\right) \\
\varepsilon_3^m &= \frac{1}{E_m}\left(\sigma_3^m - \nu_m \sigma_1^m - \nu_m \sigma_2^m\right)
\end{aligned} \tag{3.94}$$

Consider first the actual stress state. Let the ply in Fig. 3.29 be loaded with stress σ_2 inducing apparent strain ε_2 such that

$$\varepsilon_2 = \frac{\sigma_2}{E_2^{\text{act}}} \tag{3.95}$$

Here, E_2^{act} is the actual apparent modulus, which is not known. With due regard to Eqs. (3.92) and (3.93) we get

$$W = \frac{1}{2}\sigma_2\varepsilon_2 V, \quad W_\sigma^{\text{act}} = \frac{\sigma_2^2}{2E_2^{\text{act}}} V \tag{3.96}$$

where V is the volume of the material. As an admissible field, we can take any state of stress that satisfies the equilibrium equations and force boundary conditions. Using the simplest first-order model shown in Fig. 3.34, we assume that

$$\sigma_1^{\text{m}} = \sigma_3^{\text{m}} = 0, \quad \sigma_2^{\text{m}} = \sigma_2$$

Then, Eqs. (3.92)–(3.94) yield

$$W_\sigma^{\text{adm}} = \frac{\sigma_2^2}{2E_{\text{m}}} V_{\text{m}} \tag{3.97}$$

Substituting Eqs. (3.96) and (3.97) into the inequality (3.91), we arrive at

$$E_2^{\text{act}} > E_2^l$$

where, in accordance with Eqs. (3.62) and Fig. 3.34,

$$E_2^l = \frac{E_{\text{m}} V}{V_{\text{m}}} = \frac{E_{\text{m}}}{v_{\text{m}}}$$

This result, specifying the lower bound on the apparent transverse modulus, follows from Eq. (3.78) if we put $E_{\text{f}} \to \infty$. Thus, the lower (solid) line in Fig. 3.36 represents actually the lower bound on E_2.

To derive the expression for the upper bound, we should use the principle of minimum total potential energy (see Section 2.11.1), according to which (we again assume that the admissible field does not include the actual state)

$$T_{\text{adm}} > T_{\text{act}} \tag{3.98}$$

where $T = W_\varepsilon - A$. Here, W_ε is determined with Eq. (3.92), in which stresses are expressed in terms of strains with the aid of Eqs. (3.94), and A, for the problem under study, is the product of the force acting on the ply and the ply extension induced by this force. Since the force is the resultant of stress σ_2 (see Fig. 3.29), which induces strain ε_2, and same for actual and admissible states, A is also the same for both states, and we can present inequality (3.98) as

$$W_\varepsilon^{\text{adm}} > W_\varepsilon^{\text{act}} \tag{3.99}$$

For the actual state, we can write equations similar to Eqs. (3.96), i.e.,

$$W = \frac{1}{2}\sigma_2\varepsilon_2 V, \quad W_\varepsilon^{\text{act}} = \frac{1}{2}E_2^{\text{act}}\varepsilon_2^2 V \tag{3.100}$$

in which $V = 2Ra$ in accordance with Fig. 3.38. For the admissible state, we use the second-order model (see Fig. 3.38) and assume that

$$\varepsilon_1^{\text{m}} = 0, \quad \varepsilon_2^{\text{m}} = \varepsilon_{\text{m}}, \quad \varepsilon_3^{\text{m}} = 0$$

where ε_{m} is the matrix strain specified by Eq. (3.86). Then, Eqs. (3.94) yield

$$\sigma_1^{\text{m}} = \mu_{\text{m}}\sigma_2^{\text{m}}, \quad \sigma_3^{\text{m}} = \mu_{\text{m}}\sigma_2^{\text{m}}, \quad \sigma_2^{\text{m}} = \frac{E_{\text{m}}\varepsilon_{\text{m}}}{1 - 2\nu_{\text{m}}\mu_{\text{m}}} \tag{3.101}$$

where

$$\mu_{\text{m}} = \frac{\nu_{\text{m}}(1 + \nu_{\text{m}})}{1 - \nu_{\text{m}}^2}$$

Substituting Eqs. (3.101) into Eq. (3.93) and performing integration in accordance with Eq. (3.92), we have

$$W_\varepsilon^{\text{adm}} = \frac{E_{\text{m}}\varepsilon_2^2}{1 - 2\nu_{\text{m}}\mu_{\text{m}}} \cdot \int_{-R}^{R} \mathrm{d}x_3 \int_0^{\frac{a}{2}y} \frac{\mathrm{d}x_2}{y^2} = \frac{\pi R a E_{\text{m}}\varepsilon_2^2 r(\lambda)}{2\upsilon_{\text{f}}(1 - 2\nu_{\text{m}}\mu_{\text{m}})} \tag{3.102}$$

Here,

$$y = 1 - \lambda\sqrt{1 - \left(\frac{x_3}{R}\right)^2}$$

and $r(\lambda)$ is given above; see also Eq. (3.89). Applying Eqs. (3.100) and (3.102) in conjunction with inequality (3.99), we arrive at

$$E_2^{\text{act}} < E_2^u$$

where

$$E_2^u = \frac{\pi E_{\text{m}}}{2\upsilon_{\text{f}}(1 - 2\nu_{\text{m}}\mu_{\text{m}})}$$

is the upper bound on E_2 shown in Fig. 3.36 with a dashed curve.

Taking statically and kinematically admissible stress and strain fields that are closer to the actual states of stress and strain, one can increase E_2^l and decrease E_2^u, making the difference between the bounds smaller (Hashin and Rosen, 1964).

It should be emphasized that the bounds established thus are not the bounds imposed on the modulus of a real composite material but on the result of calculation corresponding to the accepted material model. Indeed, we can return to the first-order model shown in Fig. 3.34 and consider in-plane shear with stress τ_{12}. As can be readily proved, the actual stress–strain state of the matrix in this case is characterized with the following stresses and strains

$$\begin{aligned} &\sigma_1^{\mathrm{m}} = \sigma_2^{\mathrm{m}} = \sigma_3^{\mathrm{m}} = 0, \quad \tau_{12}^{\mathrm{m}} = \tau_{12}, \quad \tau_{13}^{\mathrm{m}} = \tau_{23}^{\mathrm{m}} = 0, \\ &\varepsilon_1^{\mathrm{m}} = \varepsilon_2^{\mathrm{m}} = \varepsilon_3^{\mathrm{m}} = 0, \quad \gamma_{12}^{\mathrm{m}} = \gamma_{12}, \quad \gamma_{13}^{\mathrm{m}} = \gamma_{23}^{\mathrm{m}} = 0 \end{aligned} \tag{3.103}$$

Assuming that the fibers are absolutely rigid and considering stresses and strains in Eqs. (3.103) as statically and kinematically admissible, we can readily find that

$$G_{12}^{\mathrm{act}} = G_{12}^l = G_{12}^u = \frac{G_{\mathrm{m}}}{v_{\mathrm{m}}}$$

Thus, we have found the exact solution, but its agreement with experimental data is rather poor (see Fig. 3.37) because the material model is not sufficiently adequate.

As follows from the foregoing discussion, micromechanical analysis provides only qualitative prediction of the ply stiffness. The same is true for ply strength. Although the micromechanical approach, in principle, can be used for strength analysis (Skudra et al., 1989), it provides mainly better understanding of the failure mechanism rather than the values of the ultimate stresses for typical loading cases. For practical applications, these stresses are determined by experimental methods described in the next section.

3.4. Mechanical properties of a ply under tension, shear, and compression

As is shown in Fig. 3.29, a ply can experience five types of elementary loading, i.e.,

- tension along the fibers,
- tension across the fibers,
- in-plane shear,
- compression along the fibers,
- compression across the fibers.

Actual mechanical properties of a ply under these loading cases are determined experimentally by testing specially fabricated specimens. Since the thickness of an elementary ply is very small (0.1–0.02 mm), the specimen usually consists of tens of plies having the same fiber orientations.

Mechanical properties of composite materials depend on the processing method and parameters. So, to obtain the adequate material characteristics that can be used for analysis of structural elements, the specimens should be fabricated by the same processes that are

Table 3.5
Typical properties of unidirectional composites.

Property	Glass–epoxy	Carbon–epoxy	Carbon–PEEK	Aramid–epoxy	Boron–epoxy	Boron–Al	Carbon–Carbon	Al_2O_3–Al
Fiber volume fraction, v_f	0.65	0.62	0.61	0.6	0.5	0.5	0.6	0.6
Density, ρ (g/cm^3)	2.1	1.55	1.6	1.32	2.1	2.65	1.75	3.45
Longitudinal modulus, E_1 (GPa)	60	140	140	95	210	260	170	260
Transverse modulus, E_2 (GPa)	13	11	10	5.1	19	140	19	150
Shear modulus, G_{12} (GPa)	3.4	5.5	5.1	1.8	4.8	60	9	60
Poisson's ratio, ν_{21}	0.3	0.27	0.3	0.34	0.21	0.3	0.3	0.24
Longitudinal tensile strength, $\overline{\sigma}_1^+$ (MPa)	1800	2000	2100	2500	1300	1300	340	700
Longitudinal compressive strength, $\overline{\sigma}_1^-$ (MPa)	650	1200	1200	300	2000	2000	180	3400
Transverse tensile strength, $\overline{\sigma}_2^+$ (MPa)	40	50	75	30	70	140	7	190
Transverse compressive strength, $\overline{\sigma}_2^-$ (MPa)	90	170	250	130	300	300	50	400
In-plane shear strength, $\overline{\tau}_{12}$ (MPa)	50	70	160	30	80	90	30	120

used to manufacture the structural elements. In connection with this, there exist two standard types of specimens – flat ones that are used to test materials made by hand or machine lay-up and cylindrical (tubular or ring) specimens that represent materials made by winding.

Typical mechanical properties of unidirectional advanced composites are presented in Table 3.5 and in Figs. 3.40–3.43. More data relevant to the various types of particular composite materials could be found in Peters (1998).

We now consider typical loading cases.

3.4.1. Longitudinal tension

Stiffness and strength of unidirectional composites under longitudinal tension are determined by the fibers. As follows from Fig. 3.35, material stiffness linearly increases with increase in the fiber volume fraction. The same law following from Eq. (3.75) is valid for the material strength. If the fiber's ultimate elongation, $\overline{\varepsilon}_f$, is less than that of the matrix (which is normally the case), the longitudinal tensile strength is determined as

$$\overline{\sigma}_1^+ = (E_f v_f + E_m v_m)\overline{\varepsilon}_f \tag{3.104}$$

However, in contrast to Eq. (3.76) for E_1, this equation is not valid for very small and very high fiber volume fractions. The dependence of $\overline{\sigma}_1^+$ on v_f is shown in Fig. 3.44. For very

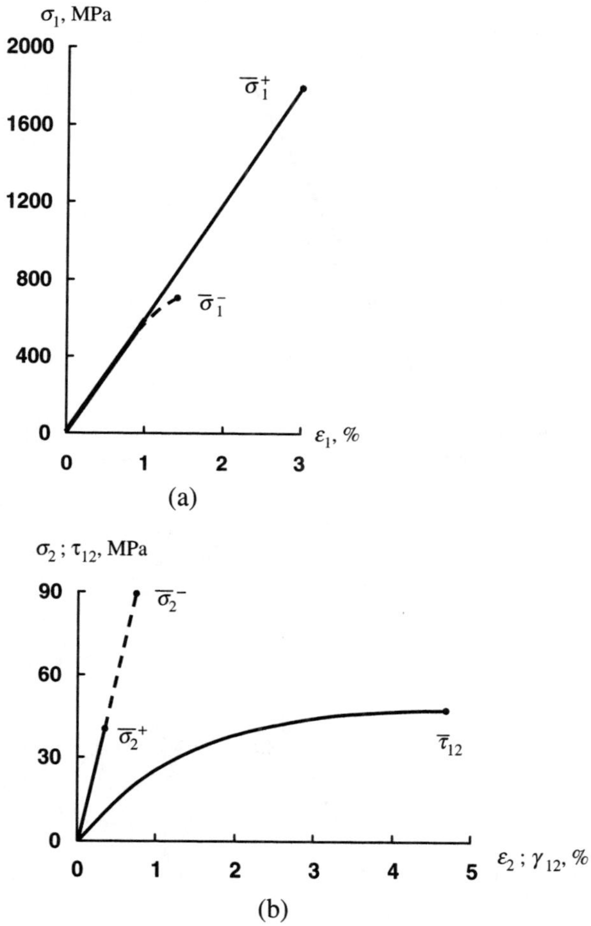

Fig. 3.40. Stress–strain curves for unidirectional glass–epoxy composite material under longitudinal tension and compression (a), transverse tension and compression (b), and in-plane shear (b).

low v_f, the fibers do not restrain the matrix deformation. Being stretched by the matrix, the fibers fail because their ultimate elongation is less than that of the matrix and the induced stress concentration in the matrix can reduce material strength below the strength of the matrix (point B). Line BC in Fig. 3.44 corresponds to Eq. (3.104). At point C, the amount of the matrix reduces below the level necessary for a monolithic material, and the material strength at point D approximately corresponds to the strength of a dry bundle of fibers, which is less than the strength of a composite bundle of fibers bound with the matrix (see Table 3.3).

Strength and stiffness under longitudinal tension are determined using unidirectional strips or rings. The strips are cut out of unidirectionally reinforced plates, and their ends are made thicker (usually glass–epoxy tabs are bonded onto the ends) to avoid specimen

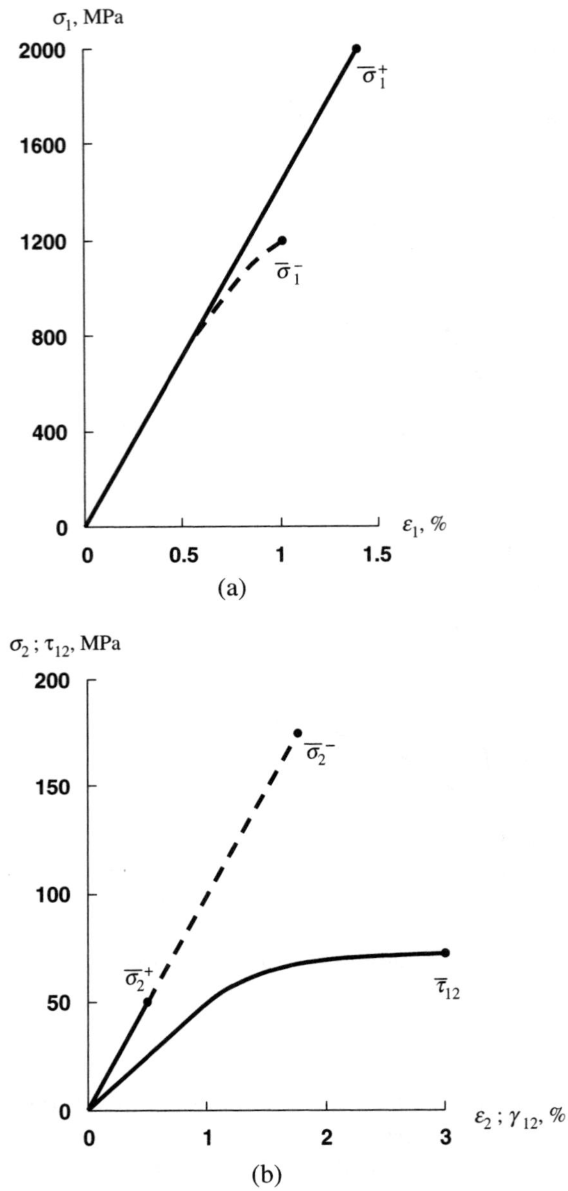

Fig. 3.41. Stress–strain curves for unidirectional carbon–epoxy composite material under longitudinal tension and compression (a), transverse tension and compression (b), and in-plane shear (b).

failure in the grips of the testing machine (Jones, 1999; Lagace, 1985). Rings are cut out of a circumferentially wound cylinder or wound individually on a special mandrel, as shown in Fig. 3.45. The strips are tested using traditional approaches, whereas the rings should be loaded with internal pressure. There exist several methods to apply the pressure

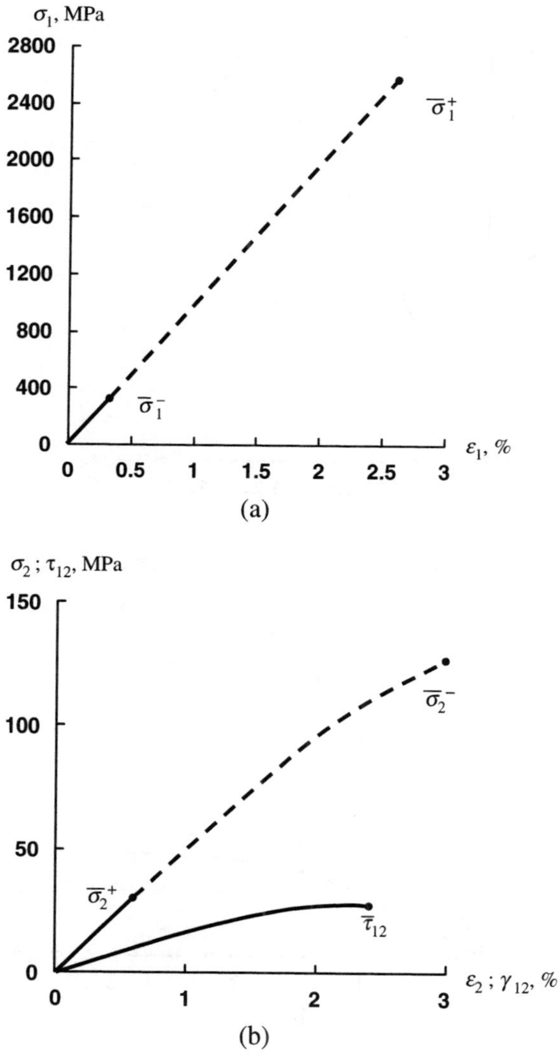

Fig. 3.42. Stress–strain curves for unidirectional aramid–epoxy composite material under longitudinal tension and compression (a), transverse tension and compression (b), and in-plane shear (b).

(Tarnopol'skii and Kincis, 1985), the simplest of which involves the use of mechanical fixtures with various numbers of sectors as in Figs. 3.46 and 3.47. The failure mode is shown in Fig. 3.48. Longitudinal tension yields the following mechanical properties of the material

- longitudinal modulus, E_1,
- longitudinal tensile strength, $\overline{\sigma}_1^+$,
- Poisson's ratio, ν_{21}.

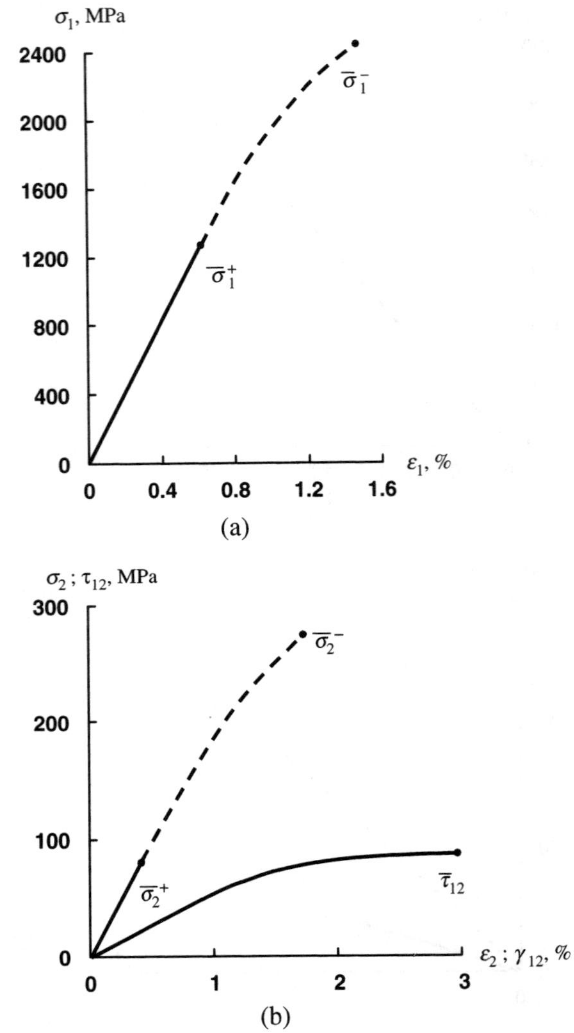

Fig. 3.43. Stress–strain curves for unidirectional boron–epoxy composite material under longitudinal tension and compression (a), transverse tension and compression (b), and in-plane shear (b).

Typical values of these characteristics for composites with various fibers and matrices are listed in Table 3.5. It follows from Figs. 3.40–3.43, that the stress–strain diagrams are linear practically up to failure.

3.4.2. Transverse tension

There are three possible modes of material failure under transverse tension with stress σ_2 shown in Fig. 3.49 – failure of the fiber–matrix interface (adhesion failure), failure

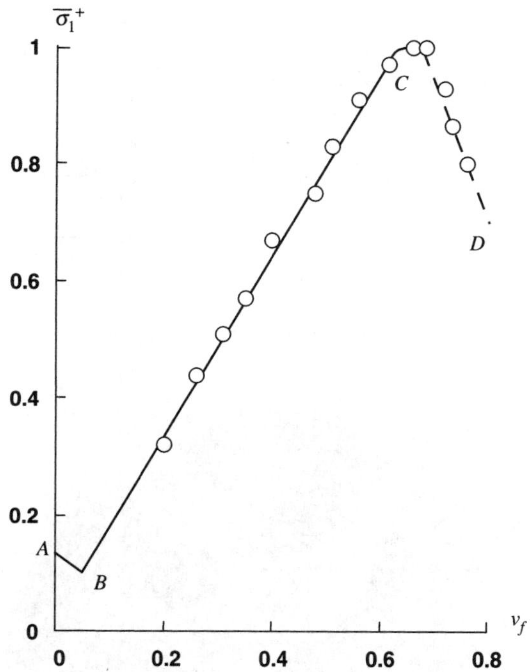

Fig. 3.44. Dependence of normalized longitudinal strength on fiber volume fraction (○ – experimental results).

Fig. 3.45. A mandrel for test rings.

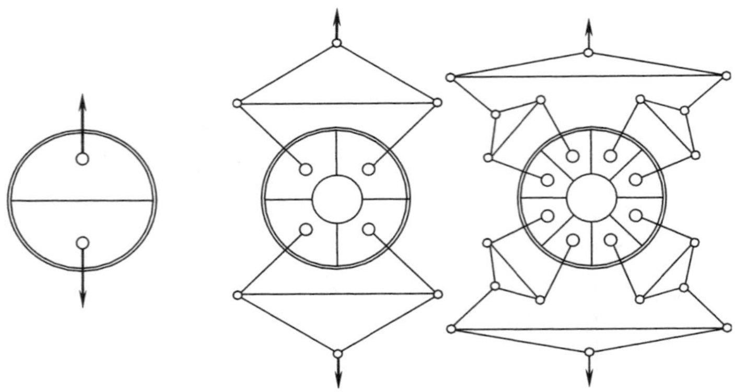

Fig. 3.46. Two-, four-, and eight-sector test fixtures for composite rings.

Fig. 3.47. A composite ring on a eight-sector test fixture.

Fig. 3.48. Failure modes of unidirectional rings.

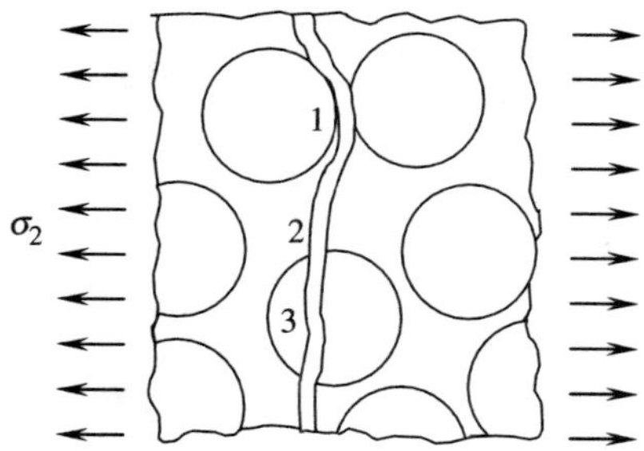

Fig. 3.49. Modes of failure under transverse tension: 1 – adhesion failure; 2 – cohesion failure; 3 – fiber failure.

of the matrix (cohesion failure), and fiber failure. The last failure mode is specific for composites with aramid fibers, which consist of thin filaments (fibrils) and have low transverse strength. As follows from the micromechanical analysis (Section 3.3), material stiffness under tension across the fibers is higher than that of a pure matrix (see Fig. 3.36).

For qualitative analysis of transverse strength, consider again the second-order model in Fig. 3.38. As can be seen, the stress distribution $\sigma_m(x_3)$ is not uniform, and the maximum stress in the matrix corresponds to $\alpha = 90°$. Using Eqs. (3.85), (3.86), and (3.88), we obtain

$$\sigma_m^{max} = \frac{E_m\sigma_2}{(1-\nu_m^2)E_2(1-\lambda)}$$

Taking $\sigma_m^{max} = \overline{\sigma}_m$ and $\sigma_2 = \overline{\sigma}_2^+$, where $\overline{\sigma}_m$ and $\overline{\sigma}_2^+$ are the ultimate stresses for the matrix and composite material, respectively, and substituting for λ and E_2 their expressions in accordance with Eqs. (3.87) and (3.89), we arrive at

$$\overline{\sigma}_2^+ = \overline{\sigma}_m \frac{r(\lambda)}{2v_f}(\pi - 4v_f) \tag{3.105}$$

The variation of the ratio $\overline{\sigma}_2^+/\overline{\sigma}_m$ for epoxy composites is shown in Fig. 3.50. As can be seen, the transverse strength of a unidirectional material is considerably lower than the strength of the matrix. It should be noted that for the first-order model, which ignores the shape of the fiber cross sections (see Fig. 3.34), $\overline{\sigma}_2^+$ is equal to $\overline{\sigma}_m$. Thus, the reduction of $\overline{\sigma}_2^+$ is caused by stress concentration in the matrix induced by cylindrical fibers.

However, both polymeric and metal matrices exhibit, as follows from Figs. 1.11 and 1.14, elastic–plastic behavior, and it is known that plastic deformation reduces the effect of stress concentration. Nevertheless, the stress–strain diagrams $\overline{\sigma}_2^+$–ε_2, shown in Figs. 3.40–3.43, are linear up to the failure point. To explain this phenomenon, consider element A of the matrix located in the vicinity of a fiber as in Fig. 3.38. Assuming that the fiber is absolutely rigid, we can conclude that the matrix strains in directions 1 and 3 are close to zero. Taking $\varepsilon_1^m = \varepsilon_3^m = 0$ in Eqs. (3.94), we arrive at Eqs. (3.101) for stresses, according to which $\sigma_1^m = \sigma_3^m = \mu_m\sigma_2^m$. The dependence of parameter μ_m on the matrix Poisson's

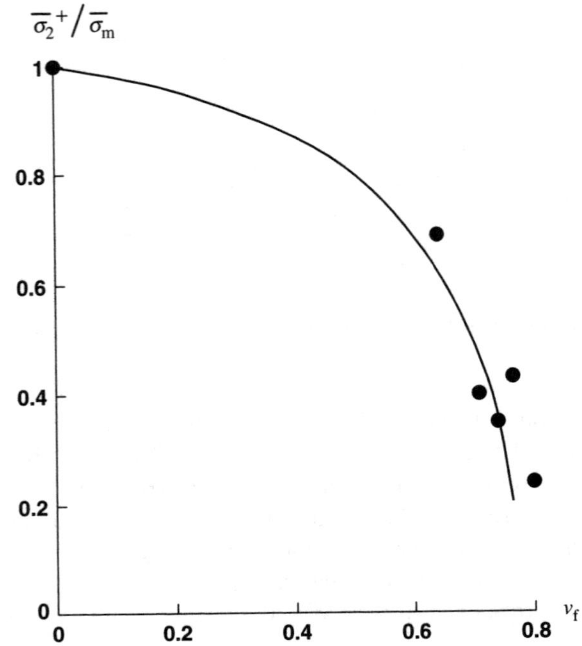

Fig. 3.50. Dependence of material strength under transverse tension on fiber volume fraction: (——) Eq. (3.105); (●) experimental data.

ratio is presented in Fig. 3.51. As follows from this figure, in the limiting case $\nu_m = 0.5$, we have $\mu_m = 1$ and $\sigma_1^m = \sigma_2^m = \sigma_3^m$, i.e., the state of stress under which all the materials behave as absolutely brittle. For epoxy resin, $\nu_m = 0.35$ and $\mu_m = 0.54$, which, as can be supposed, does not allow the resin to demonstrate its rather limited (see Fig. 1.11) plastic properties.

Strength and stiffness under transverse tension are experimentally determined using flat strips (see Fig. 3.52) or tubular specimens (see Fig. 3.53). These tests allow us to determine

- transverse modulus, E_2,
- transverse tensile strength, $\bar{\sigma}_2^+$.

For typical composite materials, these properties are given in Table 3.5.

3.4.3. *In-plane shear*

The failure modes in unidirectional composites under in-plane pure shear with stress τ_{12} shown in Fig. 3.29 are practically the same as those for the case of transverse tension (see Fig. 3.49). However, there is a significant difference in material behavior. As follows from Figs. 3.40–3.43, the stress–strain curves $\tau_{12}-\gamma_{12}$ are not linear, and $\bar{\tau}_{12}$ exceeds $\bar{\sigma}_2^+$.

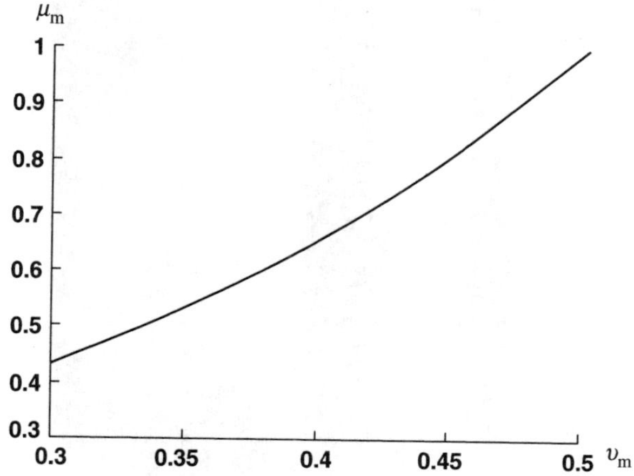

Fig. 3.51. Dependence of parameter μ_m on the matrix Poisson's ratio.

Fig. 3.52. Test fixture for transverse tension and compression of unidirectional strips.

This means that the fibers do not restrict the free shear deformation of the matrix, and the stress concentration in the vicinity of the fibers does not significantly influence material strength because of matrix plastic deformation.

Strength and stiffness under in-plane shear are determined experimentally by testing plates and thin-walled cylinders. A plate is reinforced at 45° to the loading direction and

Fig. 3.53. Test fixture for transverse tension or compression of unidirectional tubular specimens.

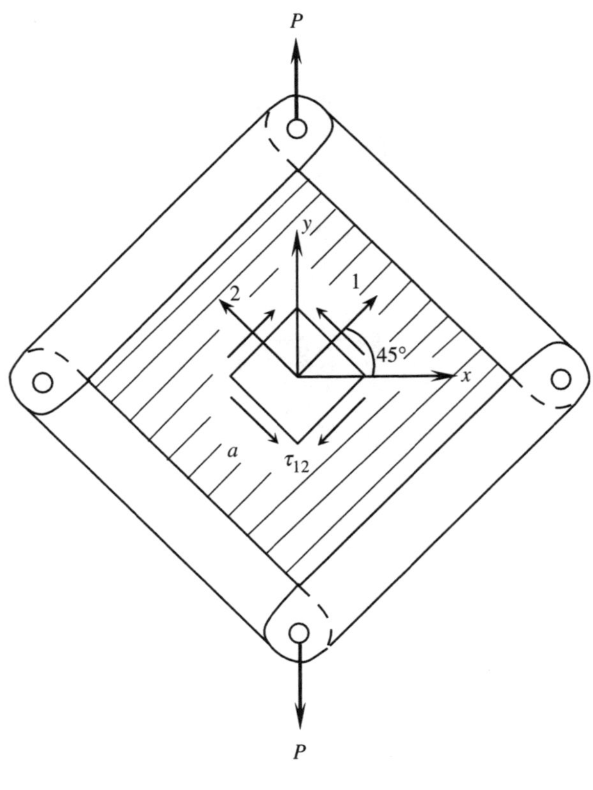

Fig. 3.54. Simulation of pure shear in a square frame.

Fig. 3.55. A tubular specimen for shear test.

is fixed in a square frame consisting of four hinged members, as shown in Fig. 3.54. Simple equilibrium consideration and geometric analysis with the aid of Eq. (2.27) yield the following equations

$$\tau_{12} = \frac{P}{\sqrt{2}ah}, \quad \gamma_{12} = \varepsilon_y - \varepsilon_x, \quad G_{12} = \frac{\tau_{12}}{\gamma_{12}}$$

in which h is the plate thickness. Thus, knowing P and measuring strains in the x and y directions, we can determine $\overline{\tau}_{12}$ and G_{12}. More accurate and reliable results can be obtained if we induce pure shear in a twisted tubular specimen reinforced in the circumferential direction (Fig. 3.55). Again, using simple equilibrium and geometric analysis, we get

$$\tau_{12} = \frac{M}{2\pi R^2 h}, \quad \gamma_{12} = \frac{\varphi R}{l}, \quad G_{12} = \frac{\tau_{12}}{\gamma_{12}}$$

Here, M is the torque, R and h are the cylinder radius and thickness, and φ is the twist angle between two cross-sections located at some distance l from each other. Thus, knowing M and measuring φ, we can find $\overline{\tau}_{12}$ and G_{12}.

3.4.4. Longitudinal compression

Failure under compression along the fibers can occur in different modes, depending on the material microstructural parameters, and can hardly be predicted by micromechanical analysis because of the rather complicated interaction of these modes. Nevertheless, useful qualitative results allowing us to understand material behavior and, hence, to improve properties, can be obtained with microstructural models.

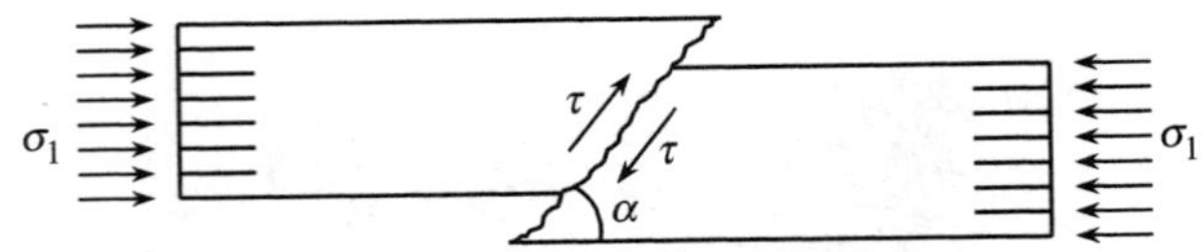

Fig. 3.56. Shear failure under compression.

Consider typical compression failure modes. The usual failure mode under compression is associated with shear in some oblique plane as in Fig. 3.56. The shear stress can be calculated using Eq. (2.9), i.e.,

$$\tau = \sigma_1 \sin\alpha \cos\alpha$$

and reaches its maximum value at $\alpha = 45°$. Shear failure under compression is usually typical for unidirectional composites that demonstrate the highest strength under longitudinal compression. On the other hand, materials showing the lowest strength under compression exhibit a transverse extension failure mode typical of wood compressed along the fibers, and is shown in Fig. 3.57. This failure is caused by tensile transverse strain, whose absolute value is

$$\varepsilon_2 = \nu_{21}\varepsilon_1 \tag{3.106}$$

where ν_{21} is Poisson's ratio and $\varepsilon_1 = \sigma_1/E_1$ is the longitudinal strain. Consider Table 3.6, showing some data taken from Table 3.5 and the results of calculations for epoxy composites. The fourth column displays the experimental ultimate transverse strains $\overline{\varepsilon}_2^+ = \overline{\sigma}_2^+/E_2$

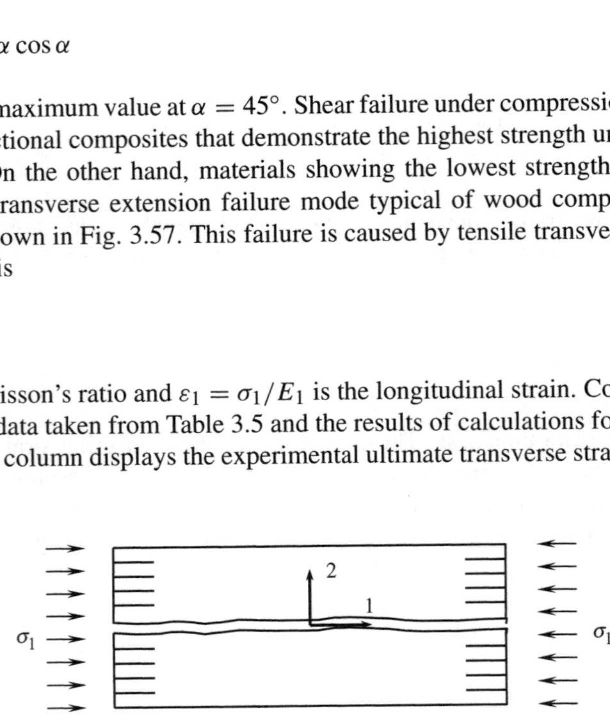

Fig. 3.57. Transverse extension failure mode under longitudinal compression.

Table 3.6
Characteristics of epoxy composites.

Material	Characteristic				
	$\overline{\sigma}_1^-$ (MPa)	$\overline{\varepsilon}_1^-$ (%)	ν_{21}	$\overline{\varepsilon}_2^+$ (%)	$\overline{\varepsilon}_2 = \nu_{21}\overline{\varepsilon}_1^-$
Glass–epoxy	600	1.00	0.30	0.31	0.30
Carbon–epoxy	1200	0.86	0.27	0.45	0.23
Aramid–epoxy	300	0.31	0.34	0.59	0.11
Boron–epoxy	2000	0.95	0.21	0.37	0.20

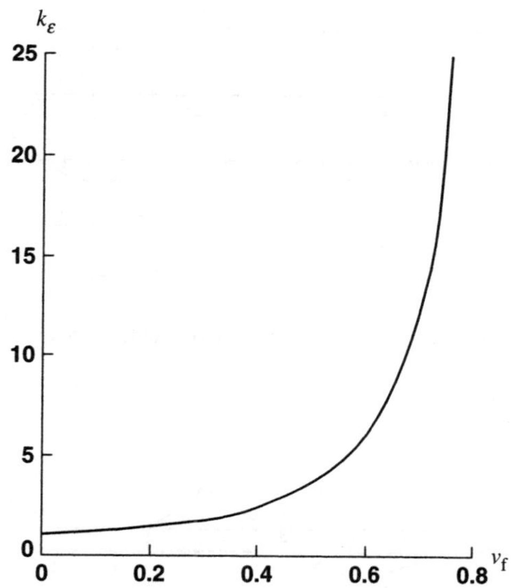

Fig. 3.58. Dependence of strain concentration factor on the fiber volume fraction.

calculated with the aid of data presented in Table 3.5, whereas the last column shows the results following from Eq. (3.106). As can be seen, the failure mode associated with transverse tension under longitudinal compression is not dangerous for the composites under consideration because $\bar{\varepsilon}_2^+ > \bar{\varepsilon}_2$. However, this is true only for fiber volume fractions $v_f = 0.50-0.65$, to which the data presented in Table 3.6 correspond. To see what happens for higher fiber volume fractions, let us use the second-order micromechanical model and the corresponding results in Figs. 3.36 and 3.50. We can plot the strain concentration factor k_ε (which is the ratio of the ultimate matrix elongation, $\bar{\varepsilon}_m$, to $\bar{\varepsilon}_2^+$ for the composite material) versus the fiber volume fraction. As can be seen in Fig. 3.58, this factor, being about 6 for $v_f = 0.6$, becomes as high as 25 for $v_f = 0.75$. This means that $\bar{\varepsilon}_2^+$ dramatically decreases for higher v_f, and the fracture mode shown in Fig. 3.57 becomes quite usual for composites with high fiber volume fractions.

Both fracture modes shown in Figs. 3.56 and 3.57 are accompanied with fibers bending induced by local buckling of fibers. According to N.F. Dow and B.W. Rosen (Jones, 1999), there can exist two modes of fiber buckling, as shown in Fig. 3.59 – a shear mode and a transverse extension mode. To study the fiber's local buckling (or microbuckling, which means that the material specimen is straight, whereas the fibers inside the material are curved), consider a plane model of a unidirectional ply, shown in Figs. 3.15 and 3.60, and take $a_m = a$ and $a_f = \delta = d$, where d is the fiber diameter. Then, Eqs. (3.17) yield

$$v_f = \frac{\bar{d}}{1+\bar{d}}, \quad \bar{d} = \frac{d}{a} \tag{3.107}$$

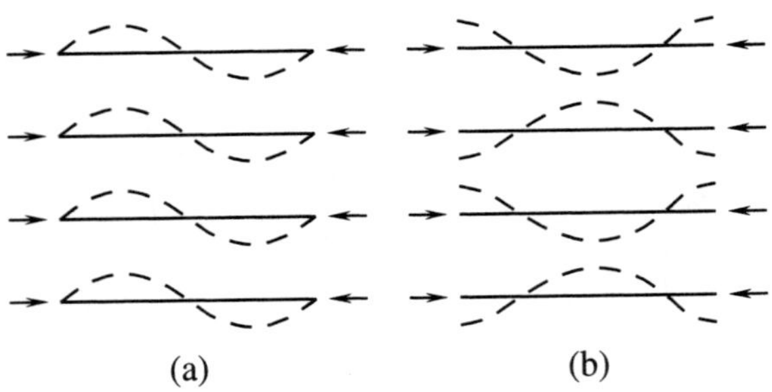

Fig. 3.59. Shear (a) and transverse extension (b) modes of fiber local buckling.

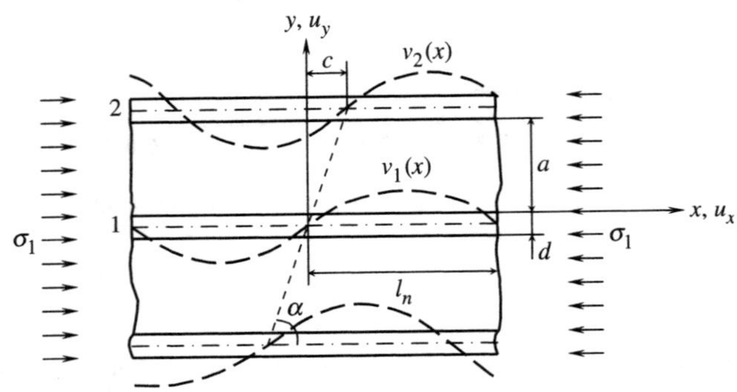

Fig. 3.60. Local buckling of fibers in unidirectional ply.

Because of the symmetry conditions, consider two fibers 1 and 2 in Fig. 3.60 and the matrix between these fibers. The buckling displacement, v, of the fibers can be represented with a sine function as

$$v_1(x) = V \sin \lambda_n x, \quad v_2(x) = V \sin \lambda_n (x - c) \tag{3.108}$$

where V is an unknown amplitude value, the same for all the fibers, $\lambda_n = \pi / l_n$, l_n is a half of a fiber wavelength (see Fig. 3.60), and $c = (a + d) \cot \alpha$ is a phase shift. Taking $c = 0$, we can describe the shear mode of buckling (Fig. 3.59(a)), whereas $c = l_n$ corresponds to the extension mode (Fig. 3.59(b)). To find the critical value of stress σ_1, we use the Timoshenko energy method (Timoshenko and Gere, 1961), yielding the following buckling condition

$$A = W \tag{3.109}$$

Here, A is the work of external forces, and W is the strain energy accumulated in the material while the fibers undergo buckling. Work A and energy W are calculated for a typical ply element consisting of two halves of fibers 1 and 2 and the matrix between

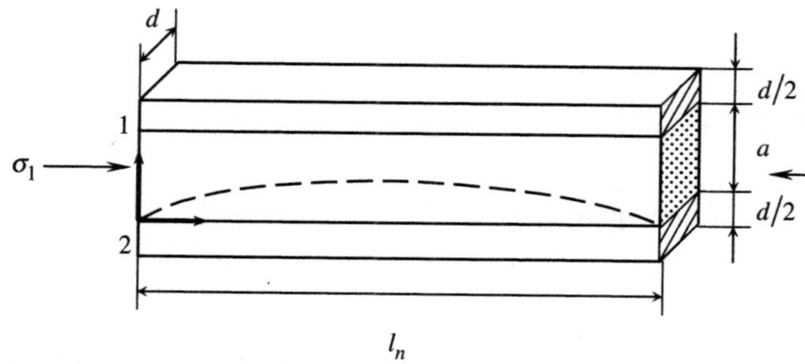

Fig. 3.61. A typical ply element.

them (see Fig. 3.61). The work, A, can be calculated as

$$A = \sigma_1(a+d)d \cdot \delta \tag{3.110}$$

with displacement δ following from Fig. 3.62, i.e.,

$$\delta = l_n - \int_0^{l_n-\delta} \sqrt{1+\left(\frac{\mathrm{d}v_1}{\mathrm{d}x}\right)^2}\,\mathrm{d}x$$

Using conventional assumptions, i.e., taking $(\mathrm{d}v_1/\mathrm{d}x) \ll 1$ and $\delta \ll l$ and substituting v_1 from Eqs. (3.108), we arrive at

$$\delta = \frac{1}{2}\int_0^{l_n}\left(\frac{\mathrm{d}v_1}{\mathrm{d}x}\right)^2 \mathrm{d}x = \frac{1}{4}V^2\lambda_n^2 l_n$$

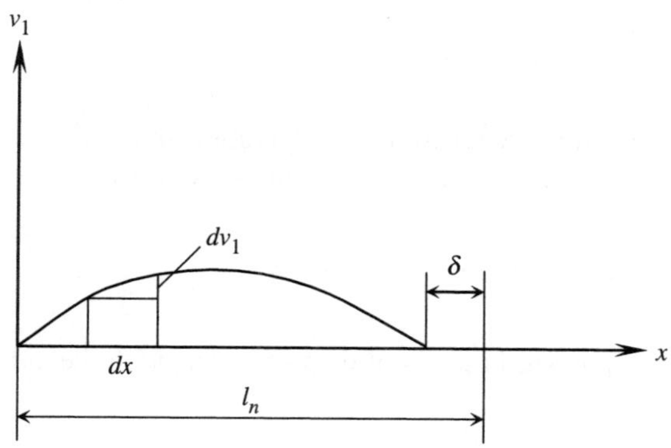

Fig. 3.62. Deformation of a fiber.

Thus, Eq. (3.110) yields

$$A = \frac{\pi^2}{4l_n}\sigma_1 V^2 ad(1+\bar{d}) \tag{3.111}$$

Strain energy consists of three parts, i.e.,

$$W = W_f + W_m^s + W_m^e \tag{3.112}$$

where W_f is the energy of buckled fibers, whereas W_m^s and W_m^e correspond to shear strain and transverse extension of the matrix that supports the fibers. The strain energy of fibers deformed in accordance with Eqs. (3.108) and shown in Fig. 3.61 has the form

$$W_f = \frac{1}{4}D_f \int_0^{l_n} \left[\left(\frac{d^2 v_1}{dx^2} \right)^2 + \left(\frac{d^2 v_2}{dx^2} \right)^2 \right] dx$$

where D_f is the fiber bending stiffness. Substituting Eqs. (3.108) and calculating the integrals, we get

$$W_f = \frac{\pi^4}{4l_n^3} D_f V^2 \tag{3.113}$$

To determine the strain energy of the matrix, we assume that the matrix element shown in Fig. 3.61 is in a state of plane stress (nonzero stresses are σ_x, σ_y, and τ_{xy}), and the equilibrium equations, Eqs. (2.5), can be written as

$$\frac{\partial \sigma_x}{\partial x} + \frac{\partial \tau_{xy}}{\partial y} = 0, \quad \frac{\partial \sigma_y}{\partial y} + \frac{\partial \tau_{xy}}{\partial x} = 0 \tag{3.114}$$

To simplify the solution, we assume that the longitudinal stress, σ_x, acting in the matrix can be neglected in comparison with the corresponding stress acting in the fibers. Thus, we can set $\sigma_x = 0$. Then, Eqs. (3.114) can be integrated and yield

$$\tau_{xy} = \tau(x), \quad \sigma_y = \sigma(x) - \tau'(x)y \tag{3.115}$$

Here, $\tau(x)$ and $\sigma(x)$ are arbitrary functions of integration and $(\ )' = d(\ )/dx$. Neglecting also the Poisson effects, we can express the strains as follows

$$\gamma_{xy} = \frac{\tau(x)}{G_m}, \quad \varepsilon_y = \frac{1}{E_m}[\sigma(x) - \tau'(x)y] \tag{3.116}$$

which can in turn be expressed in terms of displacements with the aid of Eqs. (2.22), i.e.,

$$\gamma_{xy} = \frac{\partial u_x}{\partial y} + \frac{\partial u_y}{\partial x}, \quad \varepsilon_y = \frac{\partial u_y}{\partial y} \tag{3.117}$$

Substituting Eqs. (3.116) into Eqs. (3.117) and integrating, we can determine the displacements as

$$u_x = u(x) + \left[\frac{\tau(x)}{G_m} - v'(x)\right]y - \frac{1}{2E_m}\left[\sigma'(x)y^2 - \frac{1}{3}\tau''(x)y^3\right]$$

$$u_y = v(x) + \frac{1}{E_m}\left[\sigma(x)y - \frac{1}{2}\tau'(x)y^2\right]$$

Here, $u(x)$ and $v(x)$ are functions of integration that, in addition to the functions $\tau(x)$ and $\sigma(x)$, should be found using compatibility conditions at fiber–matrix interfaces. Using Fig. 3.63, we can write these conditions in the following form:

$$u_x(y=0) = -\frac{d}{2}v_1'(x), \quad u_x(y=a) = \frac{d}{2}v_2'(x)$$

$$u_y(y=0) = v_1(x), \quad u_y(y=a) = v_2(x)$$

Satisfying them, we can find $u(x)$ and $v(x)$ directly as

$$u(x) = -\frac{d}{2}V\lambda_n \cos\lambda_n x, \quad v(x) = V\sin\lambda_n x$$

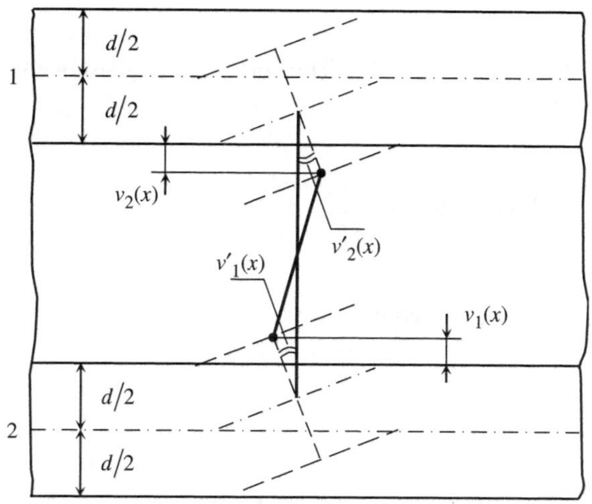

Fig. 3.63. Compatible fiber–matrix deformation.

and derive the following equations for $\sigma(x)$ and $\tau(x)$

$$\sigma(x) = \frac{E_{\rm m}}{a} V[\sin\lambda_n(x-c) - \sin\lambda_n x] + \frac{1}{2}\tau'(x)a \tag{3.118}$$

$$\frac{a^2}{6E_{\rm m}}\tau''(x) - \frac{2}{G_{\rm m}}\tau(x) = -V\lambda_n(1+\overline{d})[\cos\lambda_n(x-c) + \cos\lambda_n x] \tag{3.119}$$

We need a periodic solution of Eq. (3.119) and can find it in the following form

$$\tau(x) = C[\cos\lambda_n(x-c) + \cos\lambda_n x] \tag{3.120}$$

Substituting into Eq. (3.119) and taking into account that $\lambda_n = \pi/l_n$, we have

$$C = V\frac{\pi G_{\rm m}(1+\overline{d})}{2l_n(1+\beta_n)}, \quad \beta_n = \frac{\pi^2 a^2 G_{\rm m}}{12 l_n^2 E_{\rm m}} \tag{3.121}$$

Now, using Eqs. (3.115), (3.118), and (3.120), we can write the final expressions for the stresses acting in the matrix

$$\tau_{xy} = C[\cos\lambda_n(x-c) + \cos\lambda_n x]$$

$$\sigma_y = -\left[C\lambda_n\left(\frac{a}{2} - y\right) - \frac{E_{\rm m}}{a}V\right]\sin\lambda_n(x-c) - \left[C\lambda_n\left(\frac{a}{2} - y\right) + \frac{E_m}{a}V\right]\sin\lambda_n x \tag{3.122}$$

in which C is specified with Eqs. (3.121). The corresponding strain energies of the typical element in Fig. 3.61 are

$$W_{\rm m}^{\rm s} = \frac{ad}{2G_{\rm m}}\int_0^{l_n}\tau_{xy}^2 {\rm d}x, \quad W_{\rm m}^{\rm e} = \frac{ad}{2E_{\rm m}}\int_0^{l_n}\sigma_y^2 {\rm d}x$$

Substituting Eqs. (3.122) and integrating, we arrive at

$$W_{\rm m}^{s} = \frac{adl_n}{2G_{\rm m}}C^2(1+\cos\lambda_n c)$$

$$W_{\rm m}^{\rm e} = \frac{adl_n}{2E_{\rm m}}\left[\frac{\pi^2 a^2}{12 l_n^2}C^2(1+\cos\lambda_n c) + \frac{E_{\rm m}^2}{a^2}V^2(1-\cos\lambda_n c)\right]$$

In conjunction with these results, Eqs. (3.109), (3.111)–(3.113), and (3.121) allow us to determine σ_1, which takes the following final form

$$\sigma_1 = \frac{\pi^2 D_f}{\bar{l}_n^2 \bar{d}\left(1+\bar{d}\right) a^4} + \frac{G_m\left(1+\bar{d}\right)}{2\left(1+\left(\pi^2 G_m / \left(12\bar{l}_n^2 E_m\right)\right)\right)}\left(1+\cos\frac{\pi\bar{c}}{\bar{l}_n}\right) + \frac{2E_m \bar{l}_n^2}{\pi^2\left(1+\bar{d}\right)}\left(1-\cos\frac{\pi\bar{c}}{\bar{l}_n}\right) \quad (3.123)$$

where $\bar{d} = d/a$, $\bar{l}_n = l_n/a$, and $\bar{c} = c/a$. The critical value of σ_1 can be found by minimization of the right-hand part of Eq. (3.123) with respect to $\bar{l}_n$ and $\bar{c}$. However, having in mind only qualitative analysis, we can omit this cumbersome procedure and use Eq. (3.123) for qualitative assessments and estimates.

As follows from this equation, the strength of a unidirectional composite under longitudinal compression should increase with an increase in the fiber bending stiffness. This prediction is definitely supported with experimental data presented in Table 3.6. The highest strength is demonstrated by composites reinforced with boron fibers that have relatively high diameter and high modulus, providing very high fiber bending stiffness. Carbon fibers, also having high modulus but smaller diameter than boron fibers, provide compressive strength that is 40% lower than that of boron composites, but is twice the strength of a composite reinforced with glass fibers having the same diameter as that of carbon fibers, but lower modulus. The lowest strength in compression is demonstrated by composites with aramid fibers. As was already noted, these fibers, although having high tensile stiffness, consist of a system of poorly bonded thin filaments and possess low bending stiffness. As can be seen in Eq. (3.123), compressive strength also increases with an increase in the matrix stiffness. Available experimental results (Woolstencroft et al., 1982; Crasto and Kim, 1993) show that the strength of carbon composites in compression increases linearly, while the matrix shear modulus rises up to $G_m = 1500\,\text{MPa}$, which is the value typical for epoxy resins. For higher values of G_m, the compressive strength does not change, and we can expect that there exists some maximum value of G_m, beyond which the matrix does not allow fibers to buckle, and the material strength is controlled by the fiber strength in compression. Results listed in Table 3.5 support this conclusion. As can be seen, changing an epoxy matrix for an aluminum one with higher stiffness, we do not increase the compressive strength of boron fiber composites. Moreover, by increasing the matrix stiffness, we usually reduce its ultimate elongation. As a result, the material can fail under relatively low stress because of delamination (see Fig. 3.57). An example of such a material can also be found in Table 3.5. Carbon–carbon unidirectional composites with brittle carbon matrix possessing very high stiffness demonstrate very low strength under longitudinal compression.

Fracture of actual unidirectional composites occurs usually as a result of interaction of fracture modes discussed above. Such a fracture is shown in Fig. 3.64. The ultimate stress depends on material structural and manufacturing parameters, has considerable scatter, and can hardly be predicted theoretically. For example, the compressive strength of composites

Fig. 3.64. Failure mode of a unidirectional carbon–epoxy composite under longitudinal compression.

with the same fibers and matrices having the same stiffness but different nature (thermoset or thermoplastic) can be different (Crasto and Kim, 1993).

The strength of composites under longitudinal compression is determined experimentally using ring or flat specimens and special methods to prevent the specimen buckling (Tarnopol'skii and Kincis, 1985). The most accurate results are provided by compression of sandwich specimens with composite facings made from the material under study (Crasto and Kim, 1993).

3.4.5. *Transverse compression*

Under compression across the fibers, unidirectional composites exhibit conventional shear mode of fracture of the type shown in Fig. 3.65. The transverse compression strength is higher than in-plane shear strength (see Table 3.5) due to two main reasons. Firstly, the area of the oblique failure plane is larger than the area of the orthogonal longitudinal ply cross-section in which the ply fails under in-plane shear and, secondly, additional compression across the oblique failure plane (see Fig. 3.65) increases the shear strength. Strength under transverse compression is measured using flat or tubular specimens shown in Figs. 3.52 and 3.53.

Fig. 3.65. Failure under transverse compression.

3.5. Hybrid composites

The foregoing sections of this chapter are concerned with the properties of unidirectional plies reinforced with fibers of a certain type – glass, carbon, aramid, etc. In hybrid composites, the plies can include fibers of two or may be more types, e.g., carbon and glass, glass and aramid, and so on. Hybrid composites provide wider opportunities to control material stiffness, strength, and cost. A promising application of these materials is associated with the so-called thermostable structures, which do not change their dimensions under heating or cooling. For some composites, e.g., with glass or boron fibers, the longitudinal coefficient of thermal expansion is positive, whereas for other materials, e.g., with carbon or aramid fibers, it is negative (see Table 7.1 and Section 7.1.2 of Chapter 7). So, the appropriate combination of fibers with positive and negative coefficients can result in material with zero thermal expansion.

Consider the problem of micromechanics for a unidirectional ply reinforced with two types of fibers. Naturally, the stiffness of these fibers should be different, and we assume that $E_f^{(1)} > E_f^{(2)}$. The first-order model of the ply that generalizes the model in Fig. 3.34 is presented in Fig. 3.66. For tension in the fiber direction, the apparent stress and strain, σ_1 and ε_1, are linked by Hooke's law

$$\sigma_1 = E_1 \varepsilon_1 \tag{3.124}$$

in which the effective modulus is specified by the following equation, generalizing Eq. (3.76)

$$E_1 = E_f^{(1)} v_f^{(1)} + E_f^{(2)} v_f^{(2)} + E_m v_m \tag{3.125}$$

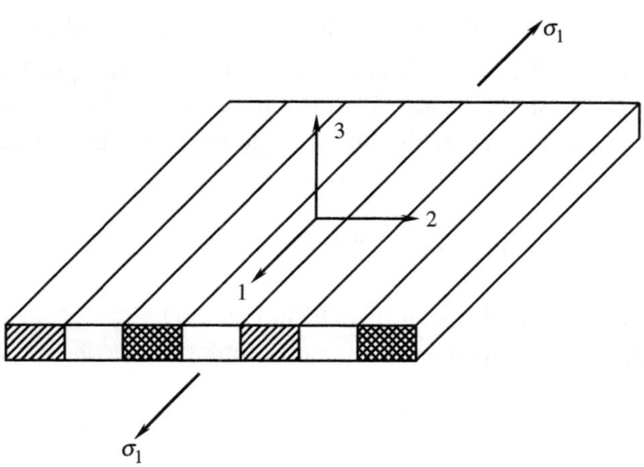

Fig. 3.66. First-order microstructural model of a hybrid unidirectional ply.

Here, $v_f^{(1)}$ and $v_f^{(2)}$ are volume fractions of the fibers of the first and second type, and v_m is the matrix volume fraction, so that

$$v_f^{(1)} + v_f^{(2)} + v_m = 1$$

We also introduce the total volume fraction of the fibers

$$v_f = v_f^{(1)} + v_f^{(2)}$$

and normalized volume fractions of fibers as

$$w_f^{(1)} = \frac{v_f^{(1)}}{v_f}, \qquad w_f^{(2)} = \frac{v_f^{(2)}}{v_f}$$

Obviously,

$$w_f^{(1)} + w_f^{(2)} = 1$$

Then, Eq. (3.125) can be written in the form

$$E_1 = v_f\left[E_f^{(1)} w_f^{(1)} + E_f^{(2)}\left(1 - w_f^{(1)}\right)\right] + E_m(1 - v_f) \tag{3.126}$$

The linear dependence of E_1 on $w_f^{(1)}$ predicted by Eq. (3.126) is in good correlation with the experimental data reported by Zabolotskii and Varshavskii (1984) and is presented in Fig. 3.67.

Since the fibers of hybrid composites have different stiffness, they are characterized, as a rule, with different ultimate elongations. As follows from Fig. 3.68, plotted with the data listed in Table 3.5, there exists an inverse linear dependence between the ply longitudinal modulus and the ultimate elongation $\bar{\varepsilon}_1$. So, assuming $E_f^{(1)} > E_f^{(2)}$, we should take into account that $\bar{\varepsilon}_f^{(1)} < \bar{\varepsilon}_f^{(2)}$. This means that Eq. (3.124) is valid for $\varepsilon_1 \le \bar{\varepsilon}_f^{(1)}$. Strain $\varepsilon_1 = \bar{\varepsilon}_f^{(1)}$ is accompanied with the failure of fibers of the first type. The corresponding part of a possible stress–strain diagram is shown in Fig. 3.69 with the line *OA*. The stress at point *A* is $\sigma_1^{(1)} = E_1\bar{\varepsilon}_f^{(1)}$. After the fibers of the first type fail, the material modulus reduces to

$$E_1^* = E_f^{(2)} v_f\left(1 - w_f^{(1)}\right) + E_m(1 - v_f)$$

This modulus determines the slope of line *OC* in Fig. 3.69.

Since $E_1^* < E_1$, the ply experiences a jump in strain under constant stress $\sigma_1 = \sigma_1^{(1)}$. As follows from Fig. 3.69, the final strain is

$$\varepsilon_1^* = \frac{\sigma_1^{(1)}}{E_1^*}$$

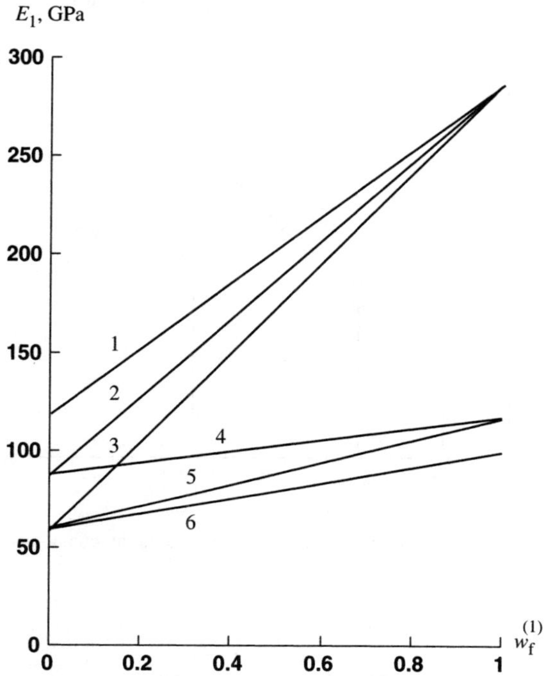

Fig. 3.67. Experimental dependencies of longitudinal modulus on the volume fraction of the higher modulus fibers in hybrid unidirectional composites: 1 – boron–carbon, 2 – boron–aramid, 3 – boron–glass, 4 – carbon–aramid, 5 – carbon–glass, 6 – aramid–glass.

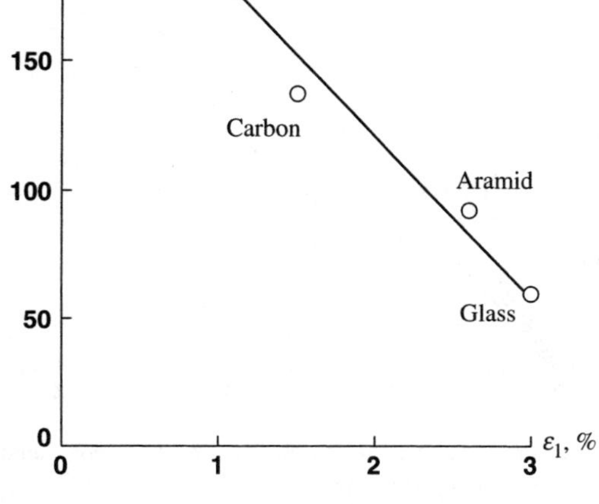

Fig. 3.68. Longitudinal modulus versus ultimate tensile strain for advanced epoxy unidirectional composites.

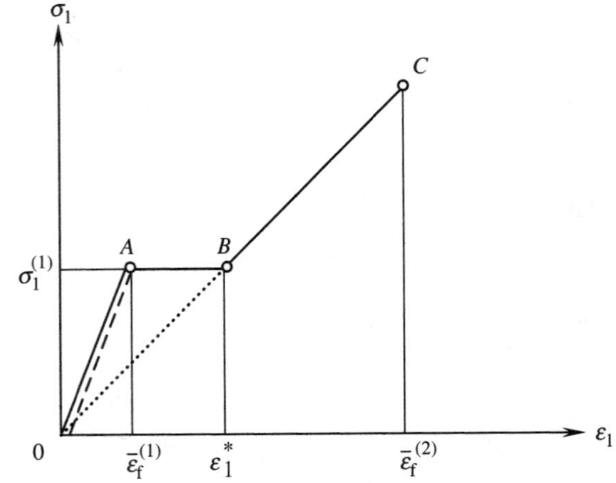

Fig. 3.69. Typical stress–strain diagrams for hybrid unidirectional composites.

There are two possible scenarios of the further material behavior, depending on the relation between strain ε_1^* and the ultimate strain of the fibers of the second type, $\bar{\varepsilon}_f^{(2)}$. If $\varepsilon_1^* \geq \bar{\varepsilon}_f^{(2)}$, these fibers will also fail under stress $\sigma_1^{(1)}$, and the material stress–strain diagram corresponds to the dashed line *OA* in Fig. 3.69. If $\bar{\varepsilon}_f^{(2)} > \varepsilon_1^*$, the material would work up to point *C* in this figure. Experimental diagrams supporting this prediction are shown in Fig. 3.70 (Gunyaev, 1981).

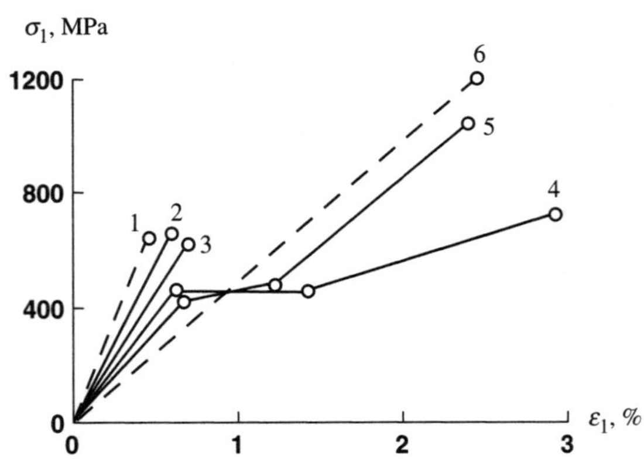

Fig. 3.70. Experimental stress–strain diagrams for hybrid carbon–glass epoxy unidirectional composite with various volume fraction of glass fibers v_g and carbon fibers v_c: 1 – $v_g = 0$; 2 – $v_g = 0.07$; 3 – $v_g = 0.14$; 4 – $v_g = 0.25$; 5 – $v_g = 0.5$; 6 – $v_c = 0$.

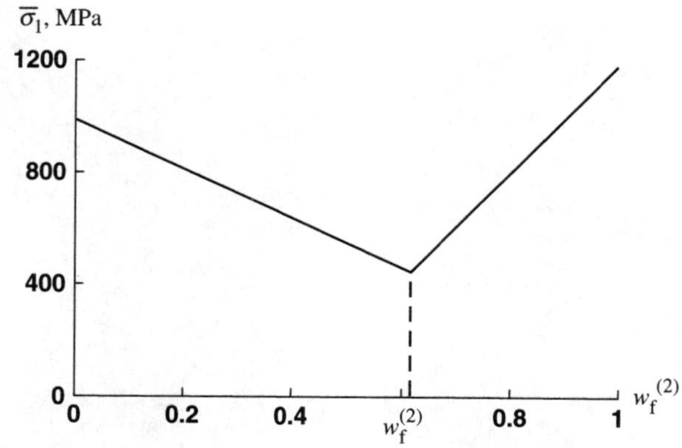

Fig. 3.71. Dependence of the longitudinal strength of unidirectional carbon–glass epoxy composite on the volume fraction of glass fibers.

The threshold value of $w_f^{(2)}$ indicating the minimum amount of the second-type fibers that is sufficient to withstand the load after the failure of the first-type fibers can be found from the condition $\overline{\varepsilon}_1^* = \overline{\varepsilon}_f^{(2)}$ (Skudra et al., 1989). The final result is as follows

$$\overline{w}_f^{(2)} = \frac{E_f^{(1)} v_f \overline{\varepsilon}_f^{(1)} - (1 - v_f) E_m \left(\overline{\varepsilon}_f^{(2)} - \overline{\varepsilon}_f^{(1)}\right)}{v_f \left[E_f^{(1)} \overline{\varepsilon}_f^{(1)} + E_f^{(2)} \left(\overline{\varepsilon}_f^{(2)} - \overline{\varepsilon}_f^{(1)}\right)\right]}$$

For $w_f^{(2)} < \overline{w}_f^{(2)}$, material strength can be calculated as $\overline{\sigma}_1 = E_1 \overline{\varepsilon}_f^{(1)}$ whereas for $w_f^{(2)} > \overline{w}_f^{(2)}$, $\overline{\sigma}_1 = E_1^* \overline{\varepsilon}_f^{(2)}$. The corresponding theoretical prediction of the dependence of material strength on $w_f^{(2)}$ is shown in Fig. 3.71 (Skudra et al., 1989).

3.6. Composites with high fiber fraction

We now return to Fig. 3.44, which shows the dependence of the tensile longitudinal strength of unidirectional composites on the fiber volume fraction v_f. As follows from this figure, the strength increases up to v_f, which is close to 0.7 and becomes lower for higher fiber volume fractions. This is a typical feature of unidirectional fibrous composites (Andreevskaiya, 1966). However, there are some experimental results (e.g., Roginskii and Egorov, 1966) showing that material strength can increase up to $v_f = 0.88$, which corresponds to the maximum theoretical fiber volume fraction discussed in Section 3.1. The reason that the material strength usually starts to decrease at higher fiber volume fractions is associated with material porosity, which becomes significant for materials with a shortage of resin. By reducing the material porosity, we can increase material tensile strength for high fiber volume fractions.

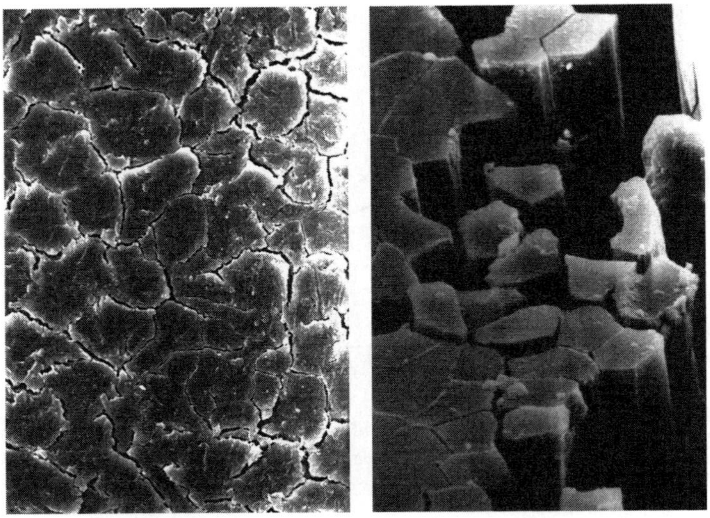

(a) (b)

Fig. 3.72. Cross-section of aramid–epoxy composite with high fiber fraction: (a) initial structure; (b) structure with delaminated fibers.

Moreover, applying the correct combination of compacting pressure and temperature to composites with organic (aramid or polyethylene) fibers, we can deform the fiber cross-sections and reach a value of υ_f that would be close to unity. Such composite materials studied by Golovkin (1985), Kharchenko (1999), and other researchers are referred to as composites with high fiber fraction (CHFF). The cross-section of a typical CHFF is shown in Fig. 3.72.

Table 3.7
Properties of aramid–epoxy composites with high fiber fraction.

Property	Fiber volume fraction, υ_f		
	0.65	0.92	0.96
Density, ρ (g/cm^3)	1.33	1.38	1.41
Longitudinal modulus, E_1 (GPa)	85	118	127
Transverse modulus, E_2 (GPa)	3.3	2.1	4.5
Shear modulus, G_{12} (GPa)	1.6	1.7	—
Longitudinal tensile strength, $\overline{\sigma}_1^+$ (MPa)	2200	2800	2800
Longitudinal compressive strength, $\overline{\sigma}_1^-$ (MPa)	293	295	310
Transverse tensile strength, $\overline{\sigma}_2^+$ (MPa)	22	12	—
Transverse compressive strength, $\overline{\sigma}_2^-$ (MPa)	118	48	—
In-plane shear strength, $\overline{\tau}_{12}$ (MPa)	41	28	18

The properties of aramid–epoxy CHFF are listed in Table 3.7 (Kharchenko, 1999). Comparing traditional composites ($v_f = 0.65$) with CHFF, we can conclude that CHFF have significantly higher longitudinal modulus (up to 50%) and longitudinal tensile strength (up to 30%), whereas the density is only 6% higher. However, the transverse and shear strengths of CHFF are lower than those of traditional composites. Because of this, composites with high fiber fraction can be efficient in composite structures whose loading induces high tensile stresses acting mainly along the fibers, e.g., in cables, pressure vessels, etc.

3.7. Phenomenological homogeneous model of a ply

It follows from the foregoing discussion that micromechanical analysis provides very approximate predictions for the ply stiffness and only qualitative information concerning the ply strength. However, the design and analysis of composite structures require quite accurate and reliable information about the properties of the ply as the basic element of composite structures. This information is provided by experimental methods as discussed above. As a result, the ply is presented as an orthotropic homogeneous material possessing some apparent (effective) mechanical characteristics determined experimentally. This means that, on the ply level, we use a phenomenological model of a composite material (see Section 1.1) that ignores its actual microstructure.

It should be emphasized that this model, being quite natural and realistic for the majority of applications, sometimes does not allow us to predict actual material behavior. To demonstrate this, consider a problem of biaxial compression of a unidirectional composite in the 23-plane as in Fig. 3.73. Testing a glass–epoxy composite material described by Koltunov et al. (1977) shows a surprising result – its strength is about $\overline{\sigma} = 1200\,\text{MPa}$, which is quite close to the level of material strength under longitudinal tension, and material failure is accompanied by fiber breakage typical for longitudinal tension.

The phenomenological model fails to predict this mode of failure. Indeed, the average stress in the longitudinal direction specified by Eq. (3.75) is equal to zero under loading shown in Fig. 3.73, i.e.,

$$\sigma_1 = \sigma_1^f v_f + \sigma_1^m v_m = 0 \tag{3.127}$$

To apply the first-order micromechanical model considered in Section 3.3, we generalize constitutive equations, Eqs. (3.63), for the three-dimensional stress state of the fibers and the matrix as

$$\varepsilon_1^{f,m} = \frac{1}{E_{f,m}}\left[\sigma_1^{f,m} - \nu_{f,m}\left(\sigma_2^{f,m} + \sigma_3^{f,m}\right)\right] \quad (1,2,3) \tag{3.128}$$

Changing 1 for 2, 2 for 3, and 3 for 1, we can write the corresponding equations for ε_2 and ε_3.

Suppose that the stresses acting in the fibers and in the matrix in the plane of loading are the same, i.e.,

$$\sigma_2^{\mathrm{f}} = \sigma_3^{\mathrm{f}} = \sigma_2^{\mathrm{m}} = \sigma_3^{\mathrm{m}} = -\sigma \tag{3.129}$$

and that $\varepsilon_1^{\mathrm{f}} = \varepsilon_1^{\mathrm{m}}$. Substituting $\varepsilon_1^{\mathrm{f}}$ and $\varepsilon_1^{\mathrm{m}}$ from Eqs. (3.128), we get with due regard to Eqs. (3.129)

$$\frac{1}{E_{\mathrm{f}}}\left(\sigma_1^{\mathrm{f}} + 2\nu_{\mathrm{f}}\sigma\right) = \frac{1}{E_{\mathrm{m}}}\left(\sigma_1^{\mathrm{m}} + 2\nu_{\mathrm{m}}\sigma\right)$$

In conjunction with Eq. (3.127), this equation allows us to find σ_1^{f}, which has the form

$$\sigma_1^{\mathrm{f}} = \frac{2\sigma(E_{\mathrm{f}}\nu_{\mathrm{m}} - E_{\mathrm{m}}\nu_{\mathrm{f}})v_{\mathrm{m}}}{E_{\mathrm{f}}v_{\mathrm{f}} + E_{\mathrm{m}}v_{\mathrm{m}}}$$

Simplifying this result for the situation $E_{\mathrm{f}} \gg E_{\mathrm{m}}$, we arrive at

$$\sigma_1^{\mathrm{f}} = 2\sigma\frac{\nu_{\mathrm{m}}v_{\mathrm{m}}}{v_{\mathrm{f}}}$$

Thus, the loading shown in Fig. 3.73 indeed induces tension in the fibers as can be revealed using the micromechanical model. The ultimate stress can be expressed in terms of the fibers' strength $\overline{\sigma}_{\mathrm{f}}$ as

$$\overline{\sigma} = \frac{1}{2}\overline{\sigma}_{\mathrm{f}}\frac{v_{\mathrm{f}}}{\nu_{\mathrm{m}}v_{\mathrm{m}}}$$

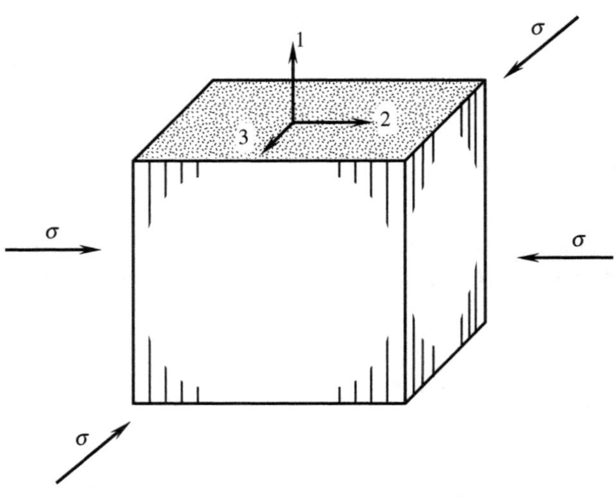

Fig. 3.73. Biaxial compression of a unidirectional composite.

The actual material strength is not as high as follows from this equation, which is derived under the condition that the adhesive strength between the fibers and the matrix is infinitely high. Tension of fibers is induced by the matrix that expands in the 1-direction (see Fig. 3.73) due to Poisson's effect and interacts with fibers through shear stresses whose maximum value is limited by the fiber–matrix adhesion strength. Under high shear stress, debonding of the fibers can occur, reducing the material strength, which is, nevertheless, very high. This effect is utilized in composite shells with radial reinforcement designed to withstand an external pressure of high intensity (Koltunov et al., 1977).

3.8. References

Abu-Farsakh, G.A., Abdel-Jawad, Y.A. and Abu-Laila, Kh.M. (2000). Micromechanical characterization of tensile strength of fiber composite materials. *Mechanics of Composite Materials Structures*, **7**(1), 105–122.

Andreevskaya, G.D. (1966). *High-strength Oriented Fiberglass Plastics*. Nauka, Moscow (in Russian).

Bogdanovich, A.E. and Pastore, C.M. (1996). *Mechanics of Textile and Laminated Composites*. Chapman & Hall, London.

Chiao, T.T. (1979). Some interesting mechanical behaviors of fiber composite materials. In *Proc. of 1st USA-USSR Symposium on Fracture of Composite Materials*, Riga, USSR, 4–7 September, 1978 (G.C. Sih and V.P. Tamuzh eds.). Sijthoff and Noordhoff, Alphen aan den Rijn., pp. 385–392.

Crasto, A.S. and Kim, R.Y. (1993). An improved test specimen to determine composite compression strength. In *Proc. 9th Int. Conf. on Composite Materials (ICCM/9)*, Madrid, 12–16 July 1993, *Vol. 6, Composite Properties and Applications*. Woodhead Publishing Ltd., pp. 621–630.

Fukuda, H., Miyazawa, T. and Tomatsu, H. (1993). Strength distribution of monofilaments used for advanced composites. In *Proc. 9th Int. Conf. on Composite Materials (ICCM/9)*, Madrid, 12–16 July 1993, *Vol. 6, Composite Properties and Applications*. Woodhead Publishing Ltd., pp. 687–694.

Gilman, J.J. (1959). *Cleavage, Ductility and Tenacity in Crystals.* In Fracture. Wiley, New York.

Golovkin, G.S. (1985). Manufacturing parameters of the formation process for ultimately reinforced organic plastics. *Plastics*, **4**, 31–33 (in Russian).

Goodey, W.J. (1946). Stress diffusion problems. *Aircraft Eng.* June 1946, 195–198; July 1946, 227–234; August 1946, 271–276; September 1946, 313–316; October 1946, 343–346; November 1946, 385–389.

Griffith, A.A. (1920). The phenomenon of rupture and flow in solids. *Philosophical Transactions of the Royal Society*, **A221**, 147–166.

Gunyaev, G.M. (1981). *Structure and Properties of Polymeric Fibrous Composites*. Khimia, Moscow (in Russian).

Hashin, Z. and Rosen, B.W. (1964). The elastic moduli of fiber reinforced materials. *Journal of Applied Mechanics*, **31E**, 223–232.

Jones, R.M. (1999). *Mechanics of Composite Materials*, 2nd edn. Taylor & Francis, Philadelphia, PA.

Kharchenko, E.F. (1999). *High-strength Ultimately Reinforced Organic Plastics*. Moscow (in Russian).

Koltunov, M.A., Pleshkov, L.V., Kanovich, M.Z., Roginskii, S.L. and Natrusov, V.I. (1977). High-strength glass-reinforced plastic shells with radial orientation of the reinforcement. *Polymer Mechanics/Mechanics of Composite Materials*, **13**(6), 928–930.

Kondo, K. and Aoki, T. (1982). Longitudinal shear modulus of unidirectional composites. In *Proc. 4th Int. Conf. on Composite Materials (ICCM-IV), Vol. 1, Progr. in Sci. and Eng. of Composites* (Hayashi, Kawata and Umeka eds.). Tokyo, 1982, pp. 357–364.

Lagace, P.A. (1985). Nonlinear stress–strain behavior of graphite/epoxy laminates. *AIAA Journal*, **223**(10), 1583–1589.

Lee, D.J., Jeong, T.H. and Kim, H.G. (1995). Effective longitudinal shear modulus of unidirectional composites. In *Proc. 10th Int. Conf. on Composite Materials (ICCM-10), Vol. 4, Characterization and Ceramic Matrix Composites*, Canada, 1995, pp. 171–178.

Mikelsons, M.Ya. and Gutans, Yu.A. (1984). Failure of the aluminum–boron plastic in static and cyclic tensile loading. *Mechanics of Composite Materials*, **20**(1), 44–52.

Mileiko, S.T. (1982). Mechanics of metal-matrix fibrous composites. In *Mechanics of Composites* (Obraztsov, I.F. and Vasiliev, V.V. eds.). Mir, Moscow, pp. 129–165.

Peters, S.T. (1998). *Handbook of Composites*. 2nd edn. (S.T. Peters ed.). Chapman & Hall, London.

Roginskii, S.L. and Egorov, N.G. (1966). Effect of prestress on the strength of metal shells reinforced with a glass-reinforced plastic. *Polymer Mechanics/Mechanics of Composite Materials*, **2**(2), 176–178.

Skudra, A.M., Bulavs, F.Ya., Gurvich, M.R. and Kruklinsh, A.A. (1989). *Elements of Structural Mechanics of Composite Truss Systems*. Riga, Zinatne, (in Russian).

Tarnopol'skii, Yu.M. and Roze, A.V. (1969). *Specific Features of Analysis for Structural Elements of Reinforced Plastics*, Riga, Zinatne, (in Russian).

Tarnopol'skii, Yu.M. and Kincis, T.Ya. (1985). *Static Test Methods for Composites*. Van Nostrand Reinhold, New York.

Tikhomirov, P.V. and Yushanov, S.P. (1980). Stress distribution after the fracture of fibers in a unidirectional composite. In *Mechanics of Composite Materials*, Riga, pp. 28–43 (in Russian).

Timoshenko, S.P. and Gere, J.M. (1961). *Theory of Elastic Stability*, 2nd edn. McGraw-Hill, New York.

Van Fo Fy (Vanin), G.A. (1966). Elastic constants and state of stress of glass-reinforced strip. *Journal of Polymer Mechanics*, **2**(4), 368–372.

Vasiliev, V.V. and Tarnopol'skii, Yu.M. (1990). *Composite Materials. Handbook* (V.V. Vasiliev, and Yu.M. Tarnopol'skii eds.). Mashinostroenie, Moscow, (in Russian).

Woolstencroft, D.H., Haresceugh, R.I. and Curtis, A.R. (1982). The compressive behavior of carbon fiber reinforced plastic. In *Proc. 4th Int. Conf. on Composite Materials (ICCM-IV), Vol. 1, Progr. in Sci. and Eng. of Composites* (Hayashi, Kawata and Umeka eds.). Tokyo, 1982, pp. 439–446.

Zabolotskii, A.A. and Varshavskii, V.Ya. (1984). Multireinforced (Hybrid) composite materials. In *Science and Technology Reviews*, Composite Materials, Part 2, Moscow.

Chapter 4

MECHANICS OF A COMPOSITE LAYER

A typical composite laminate consists of individual layers (see Fig. 4.1) which are usually made of unidirectional plies with the same or regularly alternating orientation. A layer can also be made from metal, thermosetting or thermoplastic polymer, or fabric or can have a spatial three-dimensionally reinforced structure. In contrast to a ply as considered in Chapter 3, a layer is generally referred to the global coordinate frame x, y, and z of the structural element rather than to coordinates 1, 2, and 3 associated with the ply orientation. Usually, a layer is much thicker than a ply and has a more complicated structure, but this structure does not change through its thickness, or this change is ignored. Thus, a layer can be defined as a three-dimensional structural element that is uniform in the transverse (normal to the layer plane) direction.

4.1. Isotropic layer

The simplest layer that can be observed in composite laminates is an isotropic layer of metal or thermoplastic polymer that is used to protect the composite material (Fig. 4.2) and to provide tightness. For example, filament-wound composite pressure vessels usually have a sealing metal (Fig. 4.3) or thermoplastic (Fig. 4.4) internal liner, which can also be used as a mandrel for winding. Since the layer is isotropic, we need only one coordinate system and let it be the global coordinate frame as in Fig. 4.5.

4.1.1. Linear elastic model

The explicit form of Hooke's law in Eqs. (2.48) and (2.54) can be written as

$$
\begin{aligned}
\varepsilon_x &= \frac{1}{E}(\sigma_x - \nu\sigma_y - \nu\sigma_z), \quad \gamma_{xy} = \frac{\tau_{xy}}{G} \\
\varepsilon_y &= \frac{1}{E}(\sigma_y - \nu\sigma_x - \nu\sigma_z), \quad \gamma_{xz} = \frac{\tau_{xz}}{G} \\
\varepsilon_z &= \frac{1}{E}(\sigma_z - \nu\sigma_x - \nu\sigma_y), \quad \gamma_{yz} = \frac{\tau_{yz}}{G}
\end{aligned}
\tag{4.1}
$$

Fig. 4.1. Laminated structure of a composite pipe.

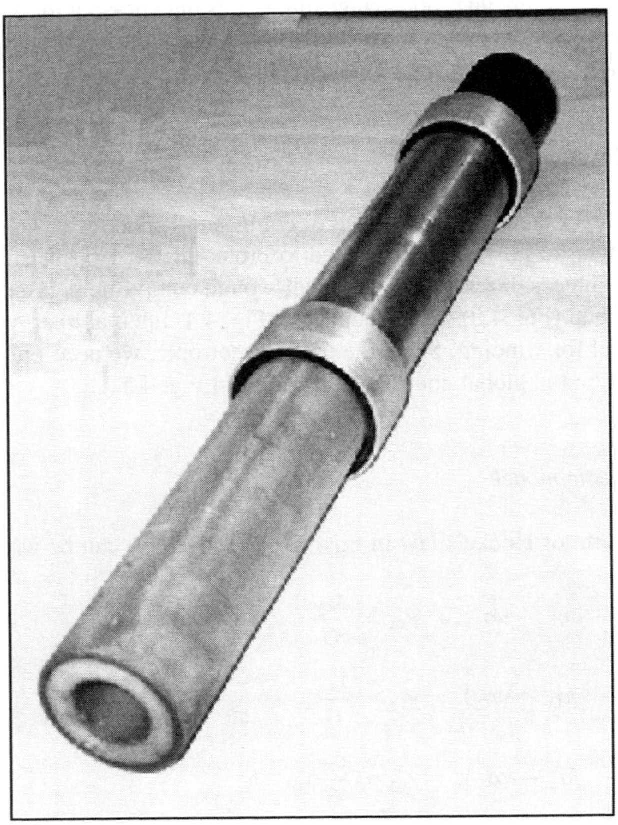

Fig. 4.2. Composite drive shaft with external metal protection layer. Courtesy of CRISM.

Fig. 4.3. Aluminum liner for a composite pressure vessel.

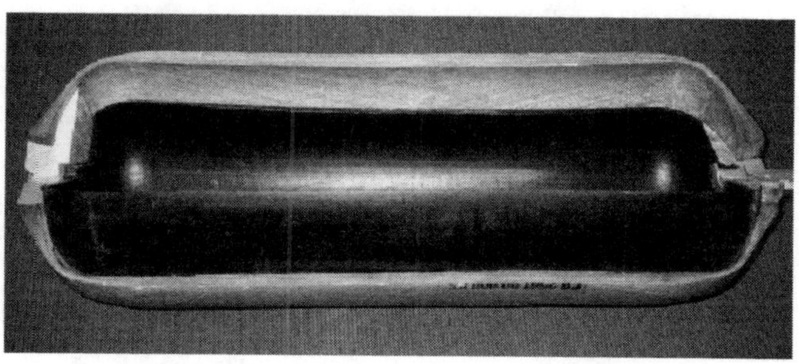

Fig. 4.4. Filament-wound composite pressure vessel with a polyethylene liner. Courtesy of CRISM.

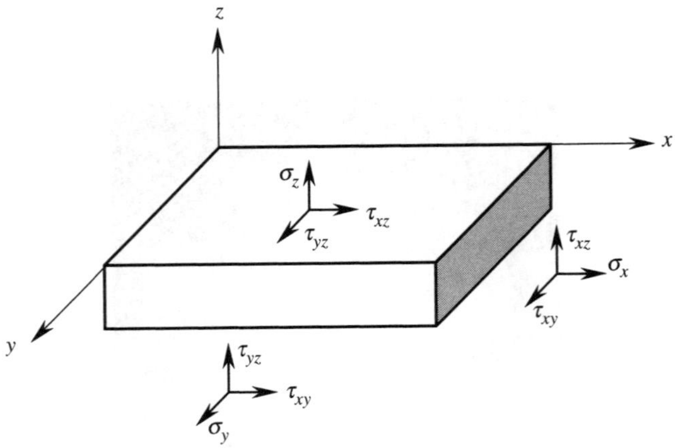

Fig. 4.5. An isotropic layer.

where E is the modulus of elasticity, ν the Poisson's ratio, and G is the shear modulus which can be expressed in terms of E and ν with Eq. (2.57). Adding Eqs. (4.1) for normal strains we get

$$\varepsilon_0 = \frac{1}{K}\sigma_0 \tag{4.2}$$

where

$$\varepsilon_0 = \varepsilon_x + \varepsilon_y + \varepsilon_z \tag{4.3}$$

is the volume deformation. For small strains, the volume $\mathrm{d}V_1$ of an infinitesimal material element after deformation can be found knowing the volume $\mathrm{d}V$ before the deformation and ε_0 as

$$\mathrm{d}V_1 = (1 + \varepsilon_0)\mathrm{d}V$$

Volume deformation is related to the mean stress

$$\sigma_0 = \frac{1}{3}(\sigma_x + \sigma_y + \sigma_z) \tag{4.4}$$

through the volume or bulk modulus

$$K = \frac{E}{3(1 - 2\nu)} \tag{4.5}$$

For $\nu = 1/2$, $K \to \infty$, $\varepsilon_0 = 0$, and $\mathrm{d}V_1 = \mathrm{d}V$ for any stress. Such materials are called incompressible – they do not change their volume under deformation and can change only their shape.

The foregoing equations correspond to the general three-dimensional stress state of a layer. However, working as a structural element of a thin-walled composite laminate, a layer is usually loaded with a system of stresses one of which, namely, transverse normal stress σ_z is much less than the other stresses. Bearing this in mind, we can neglect the terms in Eqs. (4.1) that include σ_z and write these equations in a simplified form

$$\begin{aligned} &\varepsilon_x = \frac{1}{E}(\sigma_x - \nu\sigma_y), \quad \varepsilon_y = \frac{1}{E}(\sigma_y - \nu\sigma_x) \\ &\gamma_{xy} = \frac{\tau_{xy}}{G}, \quad \gamma_{xz} = \frac{\tau_{xz}}{G}, \quad \gamma_{yz} = \frac{\tau_{yz}}{G} \end{aligned} \tag{4.6}$$

or

$$\begin{aligned} &\sigma_x = \overline{E}(\varepsilon_x + \nu\varepsilon_y), \quad \sigma_y = \overline{E}(\varepsilon_y + \nu\varepsilon_x) \\ &\tau_{xy} = G\gamma_{xy}, \quad \tau_{xz} = G\gamma_{xz}, \quad \tau_{yz} = G\gamma_{yz} \end{aligned} \tag{4.7}$$

where $\overline{E} = E/(1 - \nu^2)$.

4.1.2. Nonlinear models

Materials of metal and polymeric layers considered in this section demonstrate linear response only under moderate stresses (see Figs. 1.11 and 1.14). Further loading results in nonlinear behavior, to describe which we need to apply one of the nonlinear material models discussed in Section 1.1.

A relatively simple nonlinear constitutive theory suitable for polymeric layers can be constructed using a nonlinear elastic material model (see Fig. 1.2). In the strict sense, this model can be applied to materials whose stress–strain curves are the same for active loading and unloading. However, normally structural analysis is undertaken only for active loading. If unloading is not considered, an elastic model can be formally used for materials that are not perfectly elastic.

There exist a number of models developed to describe the nonlinear behavior of highly deformable elastomers such as rubber (Green and Adkins, 1960). Polymeric materials used to form isotropic layers of composite laminates admitting, in principal, high strains usually do not demonstrate them in composite structures whose deformation is governed by fibers with relatively low ultimate elongation (1–3%). So, creating the model, we can restrict ourselves to the case of small strains, i.e., to materials whose typical stress–strain diagram is shown in Fig. 4.6.

A natural way is to apply Eqs. (2.41) and (2.42), i.e., (we use tensor notations for stresses and strains introduced in Section 2.9 and the rule of summation over repeated subscripts)

$$dU = \sigma_{ij} d\varepsilon_{ij}, \quad \sigma_{ij} = \frac{\partial U}{\partial \varepsilon_{ij}} \tag{4.8}$$

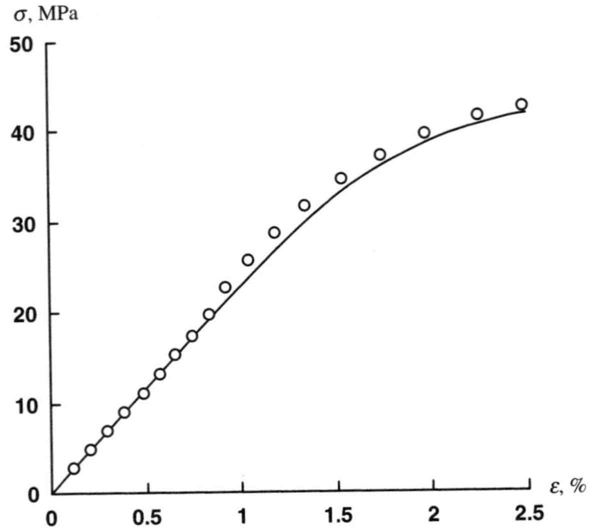

Fig. 4.6. A typical stress–strain diagram (circles) for a polymeric film and its cubic approximation (solid line).

Approximation of elastic potential U as a function of ε_{ij} with some unknown parameters allows us to write constitutive equations directly using the second relation in Eqs. (4.8). However, the polynomial approximation similar to Eq. (2.43), which is the most simple and natural results in a constitutive equation of the type $\sigma = S\varepsilon^n$, in which S is some stiffness coefficient and n is an integer. As can be seen in Fig. 4.7, the resulting stress–strain curve is not typical for the materials under study. Better agreement with nonlinear experimental diagrams presented, e.g., in Fig. 4.6, is demonstrated by the curve specified by the equation $\varepsilon = C\sigma^n$, in which C is some compliance coefficient. To arrive at this form of a constitutive equation, we need to have a relationship similar to the second one in Eqs. (4.8) but allowing us to express strains in terms of stresses. Such relationships exist and are known as Castigliano's formulas. To derive them, introduce the complementary elastic potential U_c in accordance with the following equation

$$dU_c = \varepsilon_{ij} d\sigma_{ij} \tag{4.9}$$

The term 'complementary' becomes clear if we consider a bar in Fig. 1.1 and the corresponding stress–strain curve in Fig. 4.8. The area 0BC below the curve represents U in accordance with the first equation in Eqs. (4.8), whereas the area 0AC above the curve is equal to U_c. As shown in Section 2.9, dU in Eqs. (4.8) is an exact differential. To prove the same for dU_c, consider the following sum

$$dU + dU_c = \sigma_{ij} d\varepsilon_{ij} + \varepsilon_{ij} d\sigma_{ij} = d(\sigma_{ij}\varepsilon_{ij})$$

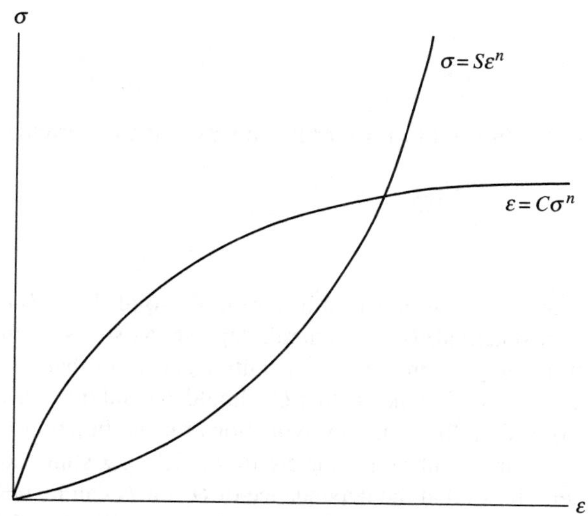

Fig. 4.7. Two forms of approximation of the stress–strain curve.

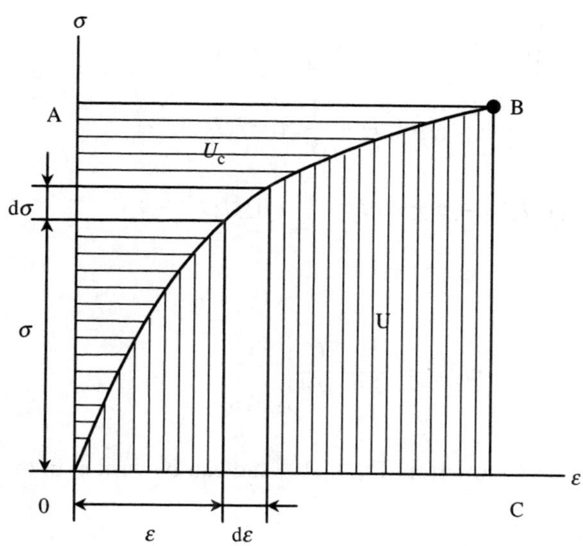

Fig. 4.8. Geometric interpretation of elastic potential, U, and complementary potential, U_c.

which is obviously an exact differential. Since $\mathrm{d}U$ in this sum is also an exact differential, $\mathrm{d}U_\mathrm{c}$ should have the same property and can be expressed as

$$\mathrm{d}U_\mathrm{c} = \frac{\partial U_\mathrm{c}}{\partial \sigma_{ij}} \mathrm{d}\sigma_{ij}$$

Comparing this result with Eq. (4.9), we arrive at Castigliano's formulae

$$\varepsilon_{ij} = \frac{\partial U_\mathrm{c}}{\partial \sigma_{ij}} \tag{4.10}$$

which are valid for any elastic solid (for a linear elastic solid, $U_\mathrm{c} = U$).

The complementary potential, U_c, in general, depends on stresses, but for an isotropic material, Eq. (4.10) should yield invariant constitutive equations that do not depend on the direction of coordinate axes. This means that U_c should depend on stress invariants I_1, I_2, and I_3 in Eqs. (2.13). Using different approximations for the function $\mathrm{U_c}$ (I_1, I_2, I_3), we can construct different classes of nonlinear elastic models. Existing experimental verification of such models shows that the dependence of U_c on I_3 can be neglected. Thus, we can present the complementary potential in a simplified form $\mathrm{U_c}$ (I_1, I_2) and expand this function as a Taylor series as

$$\begin{aligned} U_\mathrm{c} = {} & c_0 + c_{11}I_1 + \frac{1}{2}c_{12}I_1^2 + \frac{1}{3!}c_{13}I_1^3 + \frac{1}{4!}c_{14}I_1^4 + \cdots \\ & + c_{21}I_2 + \frac{1}{2}c_{22}I_2^2 + \frac{1}{3!}c_{23}I_2^3 + \frac{1}{4!}c_{24}I_2^4 + \cdots \\ & + \frac{1}{2}c_{1121}I_1I_2 + \frac{1}{3!}c_{1221}I_1^2I_2 + \frac{1}{3!}c_{1122}I_1I_2^2 + \cdots \\ & + \frac{1}{4!}c_{1321}I_1^3I_2 + \frac{1}{4!}c_{1222}I_1^2I_2^2 + \frac{1}{4!}c_{1123}I_1I_2^3 + \cdots \end{aligned} \tag{4.11}$$

where

$$c_{in} = \left.\frac{\partial^n U_\mathrm{c}}{\partial I_i^n}\right|_{\sigma_{ij}=0}, \quad c_{injm} = \left.\frac{\partial^{n+m} U_\mathrm{c}}{\partial I_i^n \partial I_j^m}\right|_{\sigma_{ij}=0}$$

Constitutive equations follow from Eq. (4.10) and can be written in the form

$$\varepsilon_{ij} = \frac{\partial U_\mathrm{c}}{\partial I_1}\frac{\partial I_1}{\partial \sigma_{ij}} + \frac{\partial U_\mathrm{c}}{\partial I_2}\frac{\partial I_2}{\partial \sigma_{ij}} \tag{4.12}$$

Assuming that for zero stresses $U_\mathrm{c} = 0$ and $\varepsilon_{ij} = 0$ we should take $c_0 = 0$ and $c_{11} = 0$ in Eq. (4.11).

Consider a plane stress state with stresses σ_x, σ_y, τ_{xy} shown in Fig. 4.5. The stress invariants in Eqs. (2.13) to be substituted into Eq. (4.12) are

$$I_1 = \sigma_x + \sigma_y, \quad I_2 = -\sigma_x\sigma_y + \tau_{xy}^2 \tag{4.13}$$

A linear elastic material model is described with Eq. (4.11) if we take

$$U_c = \frac{1}{2}c_{12}I_1^2 + c_{21}I_2 \tag{4.14}$$

Using Eqs. (4.12)–(4.14) and engineering notations for stresses and strains, we arrive at

$$\varepsilon_x = c_{12}(\sigma_x + \sigma_y) - c_{21}\sigma_y, \quad \varepsilon_y = c_{12}(\sigma_x + \sigma_y) - c_{21}\sigma_x, \quad \gamma_{xy} = 2c_{21}\tau_{xy}$$

These equations coincide with the corresponding equations in Eqs. (4.6) if we take

$$c_{12} = \frac{1}{E}, \quad c_{21} = \frac{1+\nu}{E}$$

To describe a nonlinear stress–strain diagram of the type shown in Fig. 4.6, we can generalize Eq. (4.14) as

$$U_c = \frac{1}{2}c_{12}I_1^2 + c_{21}I_2 + \frac{1}{4!}c_{14}I_1^4 + \frac{1}{2}c_{22}I_2^2$$

Then, Eq. (4.12) yields the following cubic constitutive law

$$\varepsilon_x = c_{12}(\sigma_x + \sigma_y) - c_{21}\sigma_y + \frac{1}{6}c_{14}(\sigma_x + \sigma_y)^3 + c_{22}(\sigma_x\sigma_y - \tau_{xy}^2)\sigma_y$$

$$\varepsilon_y = c_{12}(\sigma_x + \sigma_y) - c_{21}\sigma_x + \frac{1}{6}c_{14}(\sigma_x + \sigma_y)^3 + c_{22}(\sigma_x\sigma_y - \tau_{xy}^2)\sigma_x$$

$$\gamma_{xy} = 2\left[c_{21} - c_{22}\left(\sigma_x\sigma_y - \tau_{xy}^2\right)\right]\tau_{xy}$$

The corresponding approximation is shown in Fig. 4.6 with a solid line. Retaining more higher order terms in Eq. (4.11), we can describe the nonlinear behavior of any isotropic polymeric material.

To describe the nonlinear elastic–plastic behavior of metal layers, we should use constitutive equations of the theory of plasticity. There exist two basic versions of this theory – the deformation theory and the flow theory which are briefly described below.

According to the deformation theory of plasticity, the strains are decomposed into two components – elastic strains (with superscript 'e') and plastic strains (superscript 'p'), i.e.,

$$\varepsilon_{ij} = \varepsilon_{ij}^{\mathrm{e}} + \varepsilon_{ij}^{\mathrm{p}} \tag{4.15}$$

We again use the tensor notations of strains and stresses (i.e., ε_{ij} and σ_{ij}) introduced in Section 2.9. Elastic strains are related to stresses by Hooke's law, Eqs. (4.1), which can be written with the aid of Eq. (4.10) in the form

$$\varepsilon_{ij}^{e} = \frac{\partial U_e}{\partial \sigma_{ij}} \tag{4.16}$$

where U_e is the elastic potential that for a linear elastic solid coincides with the complementary potential U_c in Eq. (4.10). An explicit expression for U_e can be obtained from Eq. (2.51) if we change strains for stresses with the aid of Hooke's law, i.e.,

$$U_e = \frac{1}{2E}\left[\sigma_{11}^2 + \sigma_{22}^2 + \sigma_{33}^2 - 2\nu(\sigma_{11}\sigma_{22} + \sigma_{11}\sigma_{33} + \sigma_{22}\sigma_{33})\right] + \frac{1}{2G}\left(\sigma_{12}^2 + \sigma_{13}^2 + \sigma_{23}^2\right) \tag{4.17}$$

Now describing the plastic strains in Eq. (4.15) in a form similar to Eq. (4.16)

$$\varepsilon_{ij}^{p} = \frac{\partial U_p}{\partial \sigma_{ij}} \tag{4.18}$$

where U_p is the plastic potential. To approximate the dependence of U_p on stresses, a special generalized stress characteristic, i.e., the so-called stress intensity σ, is introduced in the classical theory of plasticity as

$$\sigma = \frac{1}{\sqrt{2}}\left[(\sigma_{11} - \sigma_{22})^2 + (\sigma_{22} - \sigma_{33})^2 + (\sigma_{11} - \sigma_{33})^2 + 6\left(\sigma_{12}^2 + \sigma_{13}^2 + \sigma_{23}^2\right)\right]^{\frac{1}{2}} \tag{4.19}$$

Transforming Eq. (4.19) with the aid of Eqs. (2.13), we can reduce it to the following form

$$\sigma = \sqrt{I_1^2 + 3I_2}$$

This means that σ is an invariant characteristic of a stress state, i.e., that it does not depend on the orientation of a coordinate frame. For unidirectional tension as in Fig. 1.1, we have only one nonzero stress, e.g., σ_{11}. Then, Eq. (4.19) yields $\sigma = \sigma_{11}$. In a similar way, the strain intensity ε can be introduced as

$$\varepsilon = \frac{\sqrt{2}}{3}\left[(\varepsilon_{11} - \varepsilon_{22})^2 + (\varepsilon_{22} - \varepsilon_{33})^2 + (\varepsilon_{11} - \varepsilon_{33})^2 + 6\left(\varepsilon_{12}^2 + \varepsilon_{13}^2 + \varepsilon_{23}^2\right)\right]^{\frac{1}{2}} \tag{4.20}$$

The strain intensity is also an invariant characteristic. For uniaxial tension (Fig. 1.1) with stress σ_{11} and strain ε_{11} in the loading direction, we have $\varepsilon_{22} = \varepsilon_{33} = -\nu_p\varepsilon_{11}$, where

ν_p is the elastic–plastic Poisson's ratio which, in general, depends on σ_{11}. For this case, Eq. (4.20) yields

$$\varepsilon = \frac{2}{3}(1+\nu_p)\varepsilon_{11} \tag{4.21}$$

For an incompressible material (see Section 4.1.1), $\nu_p = 1/2$ and $\varepsilon = \varepsilon_{11}$. Thus, the numerical coefficients in Eqs. (4.19) and (4.20) provide $\sigma = \sigma_{11}$ and $\varepsilon = \varepsilon_{11}$ for uniaxial tension of an incompressible material. The stress and strain intensities in Eqs. (4.19) and (4.20) have an important physical meaning. As known from experiments, metals do not demonstrate plastic properties under loading with stresses $\sigma_x = \sigma_y = \sigma_z = \sigma_0$ resulting only in a change of material volume. Under such loading, materials exhibit only elastic volume deformation specified by Eq. (4.2). Plastic strains occur in metals if we change the material shape. For a linear elastic material, the elastic potential U in Eq. (2.51) can be reduced after rather cumbersome transformation with the aid of Eqs. (4.3), (4.4) and (4.19), (4.20) to the following form

$$U = \frac{1}{2}\sigma_0\varepsilon_0 + \frac{1}{2}\sigma\varepsilon \tag{4.22}$$

The first term in the right-hand side part of this equation is the strain energy associated with the volume change, whereas the second term corresponds to the change of material shape. Thus, σ and ε in Eqs. (4.19) and (4.20) are stress and strain characteristics associated with the change of material shape under which it demonstrates the plastic behavior.

In the theory of plasticity, the plastic potential U_p is assumed to be a function of stress intensity σ, and according to Eq. (4.18), the plastic strains are given by

$$\varepsilon_{ij}^{p} = \frac{dU_p}{d\sigma}\frac{\partial\sigma}{\partial\sigma_{ij}} \tag{4.23}$$

Consider further a plane stress state with stresses σ_x, σ_y, and τ_{xy} in Fig. 4.5. For this case, Eq. (4.19) takes the form

$$\sigma = \sqrt{\sigma_x^2 + \sigma_y^2 - \sigma_x\sigma_y + 3\tau_{xy}^2} \tag{4.24}$$

Using Eqs. (4.15)–(4.17), (4.23), and (4.24), we finally arrive at the following constitutive equations

$$\begin{aligned}
\varepsilon_x &= \frac{1}{E}(\sigma_x - \nu\sigma_y) + \omega(\sigma)\left(\sigma_x - \frac{1}{2}\sigma_y\right) \\
\varepsilon_y &= \frac{1}{E}(\sigma_y - \nu\sigma_x) + \omega(\sigma)\left(\sigma_y - \frac{1}{2}\sigma_x\right) \\
\gamma_{xy} &= \frac{1}{G}\tau_{xy} + 3\omega(\sigma)\tau_{xy}
\end{aligned} \tag{4.25}$$

in which

$$\omega(\sigma) = \frac{1}{\sigma}\frac{dU_p}{d\sigma} \tag{4.26}$$

To find $\omega(\sigma)$, we need to specify the dependence of U_c on σ. The most simple and suitable for practical applications is the power approximation

$$U_p = C\sigma^n \tag{4.27}$$

where C and n are some experimental constants. As a result, Eq. (4.26) yields

$$\omega(\sigma) = Cn\sigma^{n-2} \tag{4.28}$$

To determine coefficients C and n, we introduce the basic assumption of the plasticity theory concerning the existence of a universal stress–strain diagram (master curve). According to this assumption, for any particular material, there exists a relationship between stress and strain intensities, i.e., $\sigma = \varphi(\varepsilon)$ (or $\varepsilon = f(\sigma)$), that is one and the same for all loading cases. This fact enables us to find coefficients C and n from a test under uniaxial tension and thus extend the obtained results to an arbitrary state of stress.

Indeed, consider uniaxial tension as in Fig. 1.1 with stress σ_{11}. For this case, $\sigma = \sigma_x$, and Eqs. (4.25) yield

$$\varepsilon_x = \frac{\sigma_x}{E} + \omega(\sigma_x)\sigma_x \tag{4.29}$$

$$\varepsilon_y = -\frac{\nu}{E}\sigma_x - \frac{1}{2}\omega(\sigma_x)\sigma_x \tag{4.30}$$

$$\gamma_{xy} = 0$$

Solving Eq. (4.29) for $\omega(\sigma_x)$, we get

$$\omega(\sigma_x) = \frac{1}{E_s(\sigma_x)} - \frac{1}{E} \tag{4.31}$$

where $E_s = \sigma_x/\varepsilon_x$ is the secant modulus introduced in Section 1.1 (see Fig. 1.4). Using now the existence of the universal diagram for stress intensity σ and taking into account that $\sigma = \sigma_x$ for uniaxial tension we can generalize Eq. (4.31) and write it for an arbitrary state of stress as

$$\omega(\sigma) = \frac{1}{E_s(\sigma)} - \frac{1}{E} \tag{4.32}$$

To determine $E_s(\sigma) = \sigma/\varepsilon$, we need to plot the universal stress–strain curve. For this purpose, we can use an experimental diagram $\sigma_x(\varepsilon_x)$ for the case of uniaxial tension, e.g., the one shown in Fig. 4.9 for an aluminum alloy with a solid line. To plot the universal

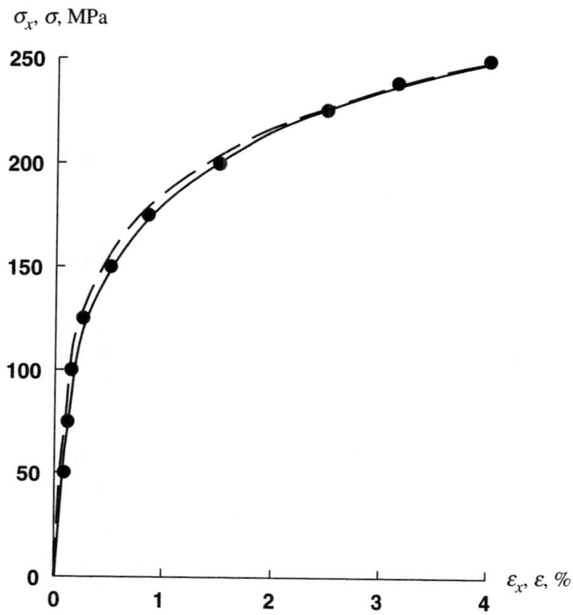

Fig. 4.9. Experimental stress–strain diagram for an aluminum alloy under uniaxial tension (solid line), the universal stress–strain curve (dashed line) and its power approximation (dots).

curve $\sigma(\varepsilon)$, we should put $\sigma = \sigma_x$ and change the scale on the strain axis in accordance with Eq. (4.21). To do this, we need to know the plastic Poisson's ratio ν_p which can be found from $\nu_p = -\varepsilon_y/\varepsilon_x$. Using Eqs. (4.29) and (4.30), we arrive at

$$\nu_p = \frac{1}{2} - \frac{E_s}{E}\left(\frac{1}{2} - \nu\right)$$

It follows from this equation that, $\nu_p = \nu$ if $E_s = E$ and $\nu_p \to 1/2$ for $E_s \to 0$. The dependencies of E_s and ν_p on ε for the aluminum alloy under consideration are presented in Fig. 4.10. With the aid of this figure and Eq. (4.21) in which we should take $\varepsilon_{11} = \varepsilon_x$ we can calculate ε and plot the universal curve shown in Fig. 4.9 with a dashed line. As can be seen, this curve is slightly different from the diagram corresponding to a uniaxial tension. For the power approximation in Eq. (4.27), we get from Eqs. (4.26) and (4.32) the following equations

$$\omega(\sigma) = Cn\sigma^{n-2}, \quad \omega(\sigma) = \frac{\varepsilon}{\sigma} - \frac{1}{E}$$

Matching these results, we find

$$\varepsilon = \frac{\sigma}{E} + Cn\sigma^{n-1} \tag{4.33}$$

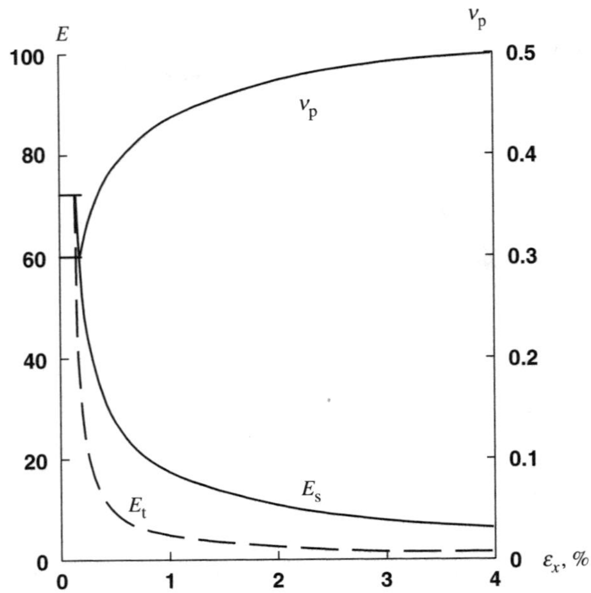

Fig. 4.10. Dependencies of the secant modulus (E_s), tangent modulus (E_t), and the plastic Poisson's ratio (ν_p), on strain for an aluminum alloy.

This is a traditional approximation for a material with a power hardening law. Now, we can find C and n using Eq. (4.33) to approximate the dashed line in Fig. 4.9. The results of this approximation are shown in this figure with dots that correspond to $E = 71.4\,\text{GPa}$, $n = 6$, and $C = 6.23 \times 10^{-15}\,(\text{MPa})^{-5}$.

Thus, constitutive equations of the deformation theory of plasticity are specified by Eqs. (4.25) and (4.32). These equations are valid only for active loading that can be identified by the condition $d\sigma > 0$. Being applied for unloading (i.e., for $d\sigma < 0$), Eqs. (4.25) correspond to nonlinear elastic material with stress–strain diagram shown in Fig. 1.2. For an elastic–plastic material (see Fig. 1.5), the unloading diagram is linear. So, if we reduce the stresses by some decrements $\Delta\sigma_x$, $\Delta\sigma_y$, and $\Delta\tau_{xy}$, the corresponding decrements of strains will be

$$\Delta\varepsilon_x = \frac{1}{E}(\Delta\sigma_x - \nu\Delta\sigma_y), \quad \Delta\varepsilon_y = \frac{1}{E}(\Delta\sigma_y - \nu\Delta\sigma_x), \quad \Delta\gamma_{xy} = \frac{1}{G}\Delta\tau_{xy}$$

Direct application of the nonlinear equations (4.25) substantially hinders the problem of stress–strain analysis because these equations include function $\omega(\sigma)$ in Eq. (4.32) which, in turn, contains the secant modulus $E_s(\sigma)$. For the power approximation corresponding to Eq. (4.33), E_s can be expressed analytically, i.e.,

$$\frac{1}{E_s} = \frac{1}{E} + Cn\sigma^{n-2}$$

However, in many cases E_s is given graphically as in Fig. 4.10 or numerically in the form of a table. Thus, Eqs. (4.25) sometimes cannot be even written in an explicit analytical form. This implies application of numerical methods in conjunction with iterative linearization of Eqs. (4.25).

There exist several methods of such linearization that will be demonstrated using the first equation in Eqs. (4.25), i.e.,

$$\varepsilon_x = \frac{1}{E}(\sigma_x - \nu\sigma_y) + \omega(\sigma)\left(\sigma_x - \frac{1}{2}\sigma_y\right) \tag{4.34}$$

In the method of elastic solutions (Ilyushin, 1948), Eq. (4.34) is used in the following form

$$\varepsilon_x^s = \frac{1}{E}(\sigma_x^s - \nu\sigma_y^s) + \eta_{s-1} \tag{4.35}$$

where s is the number of the iteration step and

$$\eta_{s-1} = \omega(\sigma_{s-1})\left(\sigma_x^{s-1} - \frac{1}{2}\sigma_y^{s-1}\right)$$

For the first step ($s = 1$), we take $\eta_0 = 0$ and solve the problem of linear elasticity with Eq. (4.35) in the form

$$\varepsilon_x^1 = \frac{1}{E}(\sigma_x^1 - \nu\sigma_y^1) \tag{4.36}$$

Finding the stresses, we calculate η_1 and write Eq. (4.35) as

$$\varepsilon_x^2 = \frac{1}{E}(\sigma_x^2 - \nu\sigma_y^2) + \eta_1$$

where the first term is linear, whereas the second term is a known function of coordinates. Thus, we have another linear problem resolving which we find stresses, calculate η_2, and switch to the third step. This process is continued until the strains corresponding to some step become sufficiently close within the stipulated accuracy to the results found at the previous step.

Thus, the method of elastic solutions reduces the initial nonlinear problem to a sequence of linear problems of the theory of elasticity for the same material but with some initial strains that can be transformed into initial stresses or additional loads. This method readily provides a nonlinear solution for any problem that has a linear solution, analytical or numerical. The main shortcoming of the method is its poor convergence. Graphical interpretation of this process for the case of uniaxial tension with stress σ is presented in Fig. 4.11a. This figure shows a simple way to improve the convergence of the process.

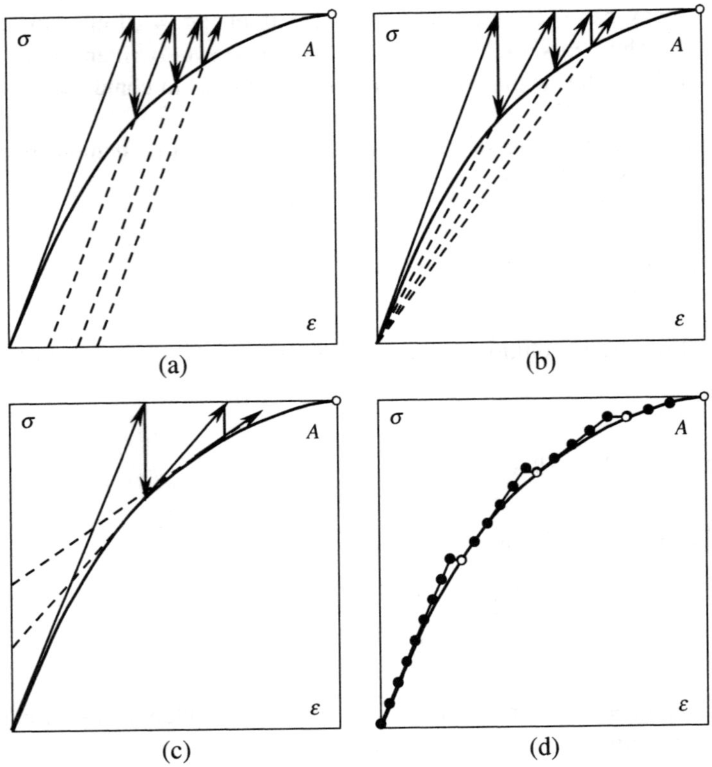

Fig. 4.11. Geometric interpretation of (a) the method of elastic solutions, (b) the method of variable elasticity parameters, (c) Newton's method, and (d) method of successive loading.

If we need to find the strain at the point of the curve that is close to point A, it is not necessary to start the process with initial modulus E. Taking $E' < E$ in Eq. (4.36) we can reach the result with much fewer steps.

According to the method of elastic variables (Birger, 1951), we should present Eq. (4.34) as

$$\varepsilon_x^s = \frac{1}{E}(\sigma_x^s - \nu\sigma_y^s) + \omega(\sigma_{s-1})\left(\sigma_x^s - \frac{1}{2}\sigma_y^s\right) \tag{4.37}$$

In contrast to Eq. (4.35), stresses σ_x^s and σ_y^s in the second term correspond to the current step rather than to the previous one. This enables us to write Eq. (4.37) in a form analogous to Hooke's law, i.e.,

$$\varepsilon_x^s = \frac{1}{E_{s-1}}(\sigma_x^s - \nu_{s-1}\sigma_y^s) \tag{4.38}$$

where

$$E_{s-1}=\left[\frac{1}{E}+\omega(\sigma_{s-1})\right]^{-1}, \quad \nu_{s-1}=E_{s-1}\left[\frac{\nu}{E}+\frac{1}{2}\omega(\sigma_{s-1})\right] \tag{4.39}$$

are elastic variables corresponding to the step with number $s-1$. The iteration procedure is similar to that described above. For the first step we take $E_0 = E$ and $\nu_0 = \nu$ in Eq. (4.38). We then find σ_x^1, σ_y^1, and σ_1, determine E_1, ν_1, switch to the second step and so on. Graphical interpretation of the process is shown in Fig. 4.11b. Convergence of this method is by an order faster than that of the method of elastic solutions. However, elastic variables in the linear constitutive equation of the method, Eq. (4.38), depend on stresses and hence, on coordinates whence the method has obtained its name. This method can be efficiently applied in conjunction with the finite element method according to which the structure is simulated with the system of elements with constant stiffness coefficients. Being calculated for each step with the aid of Eqs. (4.39), these stiffnesses will change only with transition from one element to another, which is as apparent would not practically hinder the calculation procedure for the finite element method.

The iteration process having the best convergence is provided by the classical Newton's method requiring the following form of Eq. (4.34)

$$\varepsilon_x^s=\varepsilon_x^{s-1}+c_{11}^{s-1}\left(\sigma_x^s-\sigma_x^{s-1}\right)+c_{12}^{s-1}\left(\sigma_y^s-\sigma_y^{s-1}\right)+c_{13}^{s-1}\left(\tau_{xy}^s-\tau_{xy}^{s-1}\right) \tag{4.40}$$

where

$$c_{11}^{s-1}=\frac{1}{E}+\omega(\sigma_{s-1})+\left(\sigma_x^{s-1}-\frac{1}{2}\sigma_y^{s-1}\right)\frac{\partial}{\partial\sigma_x^{s-1}}\omega(\sigma_{s-1})$$

$$c_{12}^{s-1}=-\frac{\nu}{E}-\frac{1}{2}\omega(\sigma_{s-1})+\left(\sigma_x^{s-1}-\frac{1}{2}\sigma_y^{s-1}\right)\frac{\partial}{\partial\sigma_y^{s-1}}\omega(\sigma_{s-1})$$

$$c_{13}^{s-1}=\left(\sigma_x^{s-1}-\frac{1}{2}\sigma_y^{s-1}\right)\frac{\partial}{\partial\tau_{xy}^{s-1}}\omega(\sigma_{s-1})$$

Since coefficients c are known from the previous step $(s-1)$, Eq. (4.40) is linear with respect to stresses and strains corresponding to step number s. Graphical interpretation of this method is presented in Fig. 4.11c. In contrast to the methods discussed above, Newton's method has no physical interpretation and being characterized with very high convergence, is rather cumbersome for practical applications.

The iteration methods discussed above are used to solve direct problems of stress analysis, i.e., to find stresses and strains induced by a given load. However, there exists another class of problems requiring us to evaluate the load-carrying capacity of the structure. To solve these problems, we need to trace the evolution of stresses while the load increases from zero to some ultimate value. To do this, we can use the method of successive loading. According to this method, the load is applied with some increments, and for each s-step of loading the strain is determined as

$$\varepsilon_x^s = \varepsilon_x^{s-1} + \frac{1}{E_{s-1}}\left(\Delta\sigma_x^s - \nu_{s-1}\Delta\sigma_y^s\right) \tag{4.41}$$

where E_{s-1} and ν_{s-1} are specified by Eqs. (4.39) and correspond to the previous loading step. Graphical interpretation of this method is shown in Fig. 4.11d. To obtain reliable results, the load increments should be as small as possible, because the error of calculation is cumulative in this method. To avoid this effect, the method of successive loading can be used in conjunction with the method of elastic variables. Being applied after several loading steps (black circles in Fig. 4.11d) the latter method allows us to eliminate the accumulated error and to start again the process of loading from a 'correct' initial state (light circles in Fig. 4.11d).

Returning to the constitutive equations of the deformation theory of plasticity, Eq. (4.25), it is important to note that these equations are algebraic. This means that strains corresponding to some combination of loads are determined by the stresses induced by these loads and do not depend on the history of loading, i.e., on what happened to the material before this combination of loads was reached.

However, existing experimental data show that, in general, strains should depend on the history of loading. This means that constitutive equations should be differential rather than algebraic as they are in the deformation theory. Such equations are provided by the flow theory of plasticity. According to this theory, decomposition in Eq. (4.15) is used for infinitesimal increments of stresses, i.e.,

$$\mathrm{d}\varepsilon_{ij} = \mathrm{d}\varepsilon_{ij}^{\mathrm{e}} + \mathrm{d}\varepsilon_{ij}^{\mathrm{p}} \tag{4.42}$$

Here, increments of elastic strains are related to the increments of stresses by Hooke's law, e.g., for the plane stress state

$$\mathrm{d}\varepsilon_x^{\mathrm{e}} = \frac{1}{E}(\mathrm{d}\sigma_x - \nu\mathrm{d}\sigma_y), \quad \mathrm{d}\varepsilon_y^{\mathrm{e}} = \frac{1}{E}(\mathrm{d}\sigma_y - \nu\mathrm{d}\sigma_x), \quad \mathrm{d}\gamma_{xy} = \frac{1}{G}\mathrm{d}\tau_{xy} \tag{4.43}$$

whereas increments of plastic strains

$$\mathrm{d}\varepsilon_{ij}^{\mathrm{p}} = \frac{\partial U_{\mathrm{p}}}{\partial\sigma_{ij}}\mathrm{d}\lambda$$

are expressed in the form of Eq. (4.18) but include a parameter λ which characterizes the loading process.

Assuming that $U_{\mathrm{p}} = U_{\mathrm{p}}(\sigma)$, where σ is the stress intensity specified by Eq. (4.19) or (4.24), we get

$$\mathrm{d}\varepsilon_{ij}^{\mathrm{p}} = \frac{\mathrm{d}U_{\mathrm{p}}}{\mathrm{d}\sigma}\frac{\partial\sigma}{\partial\sigma_{ij}}\mathrm{d}\lambda$$

The explicit form of these equations for the plane stress state is

$$\begin{aligned} d\varepsilon_x^p &= d\omega(\sigma)\left(\sigma_x - \frac{1}{2}\sigma_y\right) \\ d\varepsilon_y^p &= d\omega(\sigma)\left(\sigma_y - \frac{1}{2}\sigma_x\right) \\ d\gamma_{xy}^p &= 3d\omega(\sigma)\tau_{xy} \end{aligned} \tag{4.44}$$

where

$$d\omega(\sigma) = \frac{1}{\sigma}\frac{dU_p}{d\sigma}d\lambda \tag{4.45}$$

To determine the parameter λ, assume that the plastic potential U_p, being on the one hand a function of σ, can be treated as the work performed by stresses on plastic strains, i.e.,

$$\begin{aligned} dU_p &= \frac{\partial U_p}{\partial \sigma}d\sigma \\ &= \sigma_x d\varepsilon_x^p + \sigma_y d\varepsilon_y^p + \tau_{xy} d\gamma_{xy}^p \end{aligned}$$

Substituting strain increments from Eqs. (4.44) and taking into account Eq. (4.24) for σ, we have

$$\frac{\partial U_p}{\partial \sigma}d\sigma = \sigma^2 d\omega(\sigma)$$

With due regard to Eq. (4.45), we arrive at the following simple and natural relationship $d\lambda = d\sigma/\sigma$. Thus, Eq. (4.45) takes the form

$$d\omega(\sigma) = \frac{d\sigma}{\sigma^2}\frac{dU_p}{d\sigma} \tag{4.46}$$

and Eqs. (4.42)–(4.44) result in the following constitutive equations for the flow theory

$$\begin{aligned} d\varepsilon_x &= \frac{1}{E}(d\sigma_x - \nu d\sigma_y) + d\omega(\sigma)\left(\sigma_x - \frac{1}{2}\sigma_y\right) \\ d\varepsilon_y &= \frac{1}{E}(d\sigma_y - \nu d\sigma_x) + d\omega(\sigma)\left(\sigma_y - \frac{1}{2}\sigma_x\right) \\ d\gamma_{xy} &= \frac{1}{G}d\tau_{xy} + 3d\omega(\sigma)\tau_{xy} \end{aligned} \tag{4.47}$$

As can be seen, in contrast to the deformation theory, stresses govern the increments of plastic strains rather than the strains themselves.

In the general case, irrespective of any particular approximation of plastic potential U_p, we can obtain for function $d\omega(\sigma)$ in Eqs. (4.47) an expression similar to Eq. (4.32). Consider uniaxial tension for which Eqs. (4.47) yield

$$d\varepsilon_x = \frac{d\sigma_x}{E} + d\omega(\sigma_x)\sigma_x$$

Repeating the derivation of Eq. (4.32), we finally have

$$d\omega(\sigma) = \frac{d\sigma}{\sigma}\left(\frac{1}{E_t(\sigma)} - \frac{1}{E}\right) \tag{4.48}$$

where $E_t(\sigma) = d\sigma/d\varepsilon$ is the tangent modulus introduced in Section 1.1 (see Fig. 1.4). The dependence of E_t on strain for an aluminum alloy is shown in Fig. 4.10. For the power approximation for plastic potential

$$U_p = B\sigma^n \tag{4.49}$$

matching Eqs. (4.46) and (4.48), we arrive at the following equation

$$\frac{d\varepsilon}{d\sigma} = \frac{1}{E} + Bn\sigma^{n-2}$$

Upon integration, we get

$$\varepsilon = \frac{\sigma}{E} + \frac{Bn}{n-1}\sigma^{n-1} \tag{4.50}$$

As can be seen, this equation has the same form as Eq. (4.33). The only difference is in the form of coefficients C and B. As in the theory of deformation, Eq. (4.50) can be used to approximate the experimental stress–strain curve and to determine coefficients B and n. Thus, the constitutive equations for the flow theory of plasticity are specified by Eqs. (4.47) and (4.48).

For a plane stress state, introduce the stress space shown in Fig. 4.12 and referred to a Cartesian coordinate frame with stresses as coordinates. In this space, any loading can be presented as a curve specified by the parametric equations $\sigma_x = \sigma_x(p)$, $\sigma_y = \sigma_y(p)$, and $\tau_{xy} = \tau_{xy}(p)$, in which p is the loading parameter. To find strains corresponding to point A on the curve, we should integrate Eqs. (4.47) along this curve, thus taking into account the whole history of loading. In the general case, the obtained result will be different from what follows from Eqs. (4.25) of the deformation theory for point A. However, there exists one loading path (the straight line 0A in Fig. 4.12) that is completely determined by the location of its final point A. This is the so-called proportional loading during which the stresses increase in proportion to parameter p, i.e.,

$$\sigma_x = \sigma_x^0 p, \quad \sigma_y = \sigma_y^0 p, \quad \tau_{xy} = \tau_{xy}^0 p \tag{4.51}$$

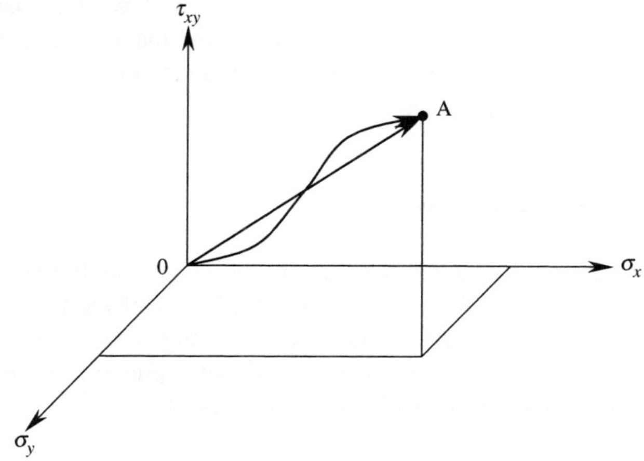

Fig. 4.12. Loading path (0A) in the stress space.

where stresses with superscript '0' can depend on coordinates only. For such loading, $\sigma = \sigma_0 p$, $\mathrm{d}\sigma = \sigma_0 \mathrm{d}p$, and Eqs. (4.46) and (4.49) yield

$$\mathrm{d}\omega(\sigma) = Bn\sigma^{n-3}\mathrm{d}\sigma = Bn\sigma_0^{n-2}p^{n-3}\mathrm{d}p \tag{4.52}$$

Consider, for example, the first equation of Eqs. (4.47). Substituting Eqs. (4.51) and (4.52), we have

$$\mathrm{d}\varepsilon_x = \frac{1}{E}\left(\sigma_x^0 - \nu\sigma_y^0\right)\mathrm{d}p + Bn\sigma_0^{n-2}\left(\sigma_x^0 - \frac{1}{2}\sigma_y^0\right)p^{n-2}\mathrm{d}p$$

This equation can be integrated with respect to p. Using again Eqs. (4.51), we arrive at the constitutive equation of the deformation theory

$$\varepsilon_x = \frac{1}{E}(\sigma_x - \nu\sigma_y) + B\frac{n}{n-1}\sigma^{n-2}\left(\sigma_x - \frac{1}{2}\sigma_y\right)$$

Thus, for a proportional loading, the flow theory reduces to the deformation theory of plasticity. Unfortunately, before the problem is solved and the stresses are found we do not know whether the loading is proportional or not and which particular theory of plasticity should be used. There exists a theorem of proportional loading (Ilyushin, 1948) according to which the stresses increase proportionally and the deformation theory can be used if:

(1) external loads increase in proportion to one loading parameter,
(2) the material is incompressible and its hardening can be described with the power law $\sigma = S\varepsilon^n$.

In practice, both conditions of this theorem are rarely met. However, existing experience shows that the second condition is not very important and that the deformation theory of plasticity can be reliably (but approximately) applied if all the loads acting on the structure increase in proportion to one parameter.

4.2. Unidirectional orthotropic layer

A composite layer with the simplest structure consists of unidirectional plies whose material coordinates, 1, 2, and 3, coincide with coordinates of the layer, x, y, and z, as in Fig. 4.13. An example of such a layer is presented in Fig. 4.14 – the principal material axes of an outer circumferential unidirectional layer of a pressure vessel coincide with global (axial and circumferential) coordinates of the vessel.

4.2.1. Linear elastic model

For the layer under study, the constitutive equations, Eqs. (2.48) and (2.53), yield

$$
\begin{aligned}
\varepsilon_1 &= \frac{\sigma_1}{E_1} - \nu_{12}\frac{\sigma_2}{E_2} - \nu_{13}\frac{\sigma_3}{E_3} \\
\varepsilon_2 &= \frac{\sigma_2}{E_2} - \nu_{21}\frac{\sigma_1}{E_1} - \nu_{23}\frac{\sigma_3}{E_3} \\
\varepsilon_3 &= \frac{\sigma_3}{E_3} - \nu_{31}\frac{\sigma_1}{E_1} - \nu_{32}\frac{\sigma_2}{E_2} \\
\gamma_{12} &= \frac{\tau_{12}}{G_{12}}, \quad \gamma_{13} = \frac{\tau_{13}}{G_{13}}, \quad \gamma_{23} = \frac{\tau_{23}}{G_{23}}
\end{aligned}
\tag{4.53}
$$

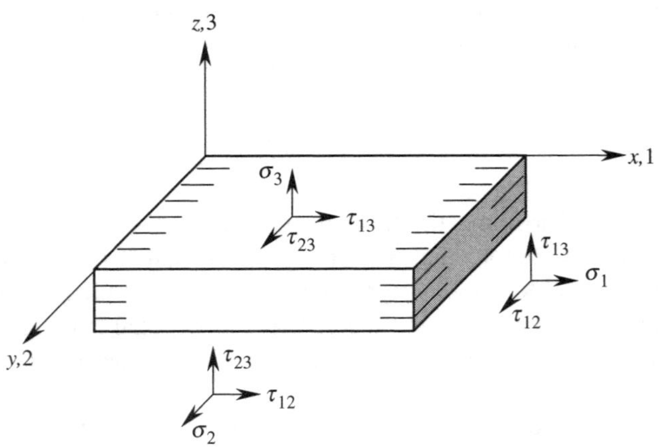

Fig. 4.13. An orthotropic layer.

Fig. 4.14. Filament-wound composite pressure vessel.

where

$$\nu_{12}E_1 = \nu_{21}E_2, \quad \nu_{13}E_1 = \nu_{31}E_3, \quad \nu_{23}E_2 = \nu_{32}E_3$$

The inverse form of Eqs. (4.53) is

$$\begin{aligned}
\sigma_1 &= A_1(\varepsilon_1 + \mu_{12}\varepsilon_2 + \mu_{13}\varepsilon_3) \\
\sigma_2 &= A_2(\varepsilon_2 + \mu_{21}\varepsilon_1 + \mu_{23}\varepsilon_3) \\
\sigma_3 &= A_3(\varepsilon_3 + \mu_{31}\varepsilon_1 + \mu_{32}\varepsilon_2) \\
\tau_{12} &= G_{12}\gamma_{12}, \quad \tau_{13} = G_{13}\gamma_{13}, \quad \tau_{23} = G_{23}\gamma_{23}
\end{aligned} \tag{4.54}$$

where

$$A_1 = \frac{E_1}{D}(1 - \nu_{23}\nu_{32}), \quad A_2 = \frac{E_2}{D}(1 - \nu_{13}\nu_{31}), \quad A_3 = \frac{E_3}{D}(1 - \nu_{12}\nu_{21})$$

$$D = 1 - \nu_{12}\nu_{23}\nu_{31} - \nu_{13}\nu_{21}\nu_{32} - \nu_{13}\nu_{31} - \nu_{12}\nu_{21} - \nu_{23}\nu_{32}$$

$$\mu_{12} = \frac{\nu_{12} + \nu_{13}\nu_{32}}{1 - \nu_{23}\nu_{32}}, \quad \mu_{21} = \frac{\nu_{21} + \nu_{23}\nu_{31}}{1 - \nu_{13}\nu_{31}}$$

$$\mu_{13} = \frac{\nu_{13} + \nu_{12}\nu_{23}}{1 - \nu_{23}\nu_{32}}, \quad \mu_{31} = \frac{\nu_{31} + \nu_{21}\nu_{32}}{1 - \nu_{12}\nu_{21}}$$

$$\mu_{23} = \frac{\nu_{23} + \nu_{13}\nu_{21}}{1 - \nu_{13}\nu_{31}}, \quad \mu_{32} = \frac{\nu_{31} + \nu_{21}\nu_{32}}{1 - \nu_{12}\nu_{21}}$$

As for an isotropic layer considered in Section 4.1, the terms including the transverse normal stress σ_3 can be neglected for a thin layer in Eqs. (4.53) and (4.54), and they can be written in the following simplified forms

$$\varepsilon_1 = \frac{\sigma_1}{E_1} - \nu_{12}\frac{\sigma_2}{E_2}, \quad \varepsilon_2 = \frac{\sigma_2}{E_2} - \nu_{21}\frac{\sigma_1}{E_1}$$
$$\gamma_{12} = \frac{\tau_{12}}{G_{12}}, \quad \gamma_{13} = \frac{\tau_{13}}{G_{13}}, \quad \gamma_{23} = \frac{\tau_{23}}{G_{23}} \tag{4.55}$$

and

$$\sigma_1 = \overline{E}_1(\varepsilon_1 + \nu_{12}\varepsilon_2), \quad \sigma_2 = \overline{E}_2(\varepsilon_2 + \nu_{21}\varepsilon_1)$$
$$\tau_{12} = G_{12}\gamma_{12}, \quad \tau_{13} = G_{13}\gamma_{13}, \quad \tau_{23} = G_{23}\gamma_{23} \tag{4.56}$$

where

$$\overline{E}_{1,2} = \frac{E_{1,2}}{1 - \nu_{12}\nu_{21}}$$

The constitutive equations presented above include elastic constants for a layer that are determined experimentally. For in-plane characteristics E_1, E_2, G_{12}, and ν_{12}, the corresponding test methods are discussed in Chapter 3. The transverse modulus E_3 is usually found by testing the layer under compression in the z-direction. The transverse shear moduli G_{13} and G_{23} can be obtained by various methods, e.g., by inducing pure shear in two symmetric specimens shown in Fig. 4.15 and calculating the shear modulus as $G_{13} = P/(2A\gamma)$, where A is the in-plane area of the specimen.

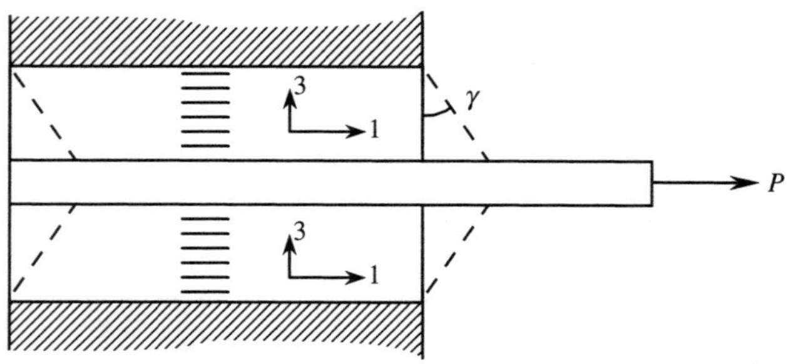

Fig. 4.15. A test to determine transverse shear modulus.

Table 4.1
Transverse shear moduli of unidirectional composites (Herakovich, 1998).

Material	Glass–epoxy	Carbon–epoxy	Aramid–epoxy	Boron–Al
G_{23} (GPa)	4.1	3.2	1.4	49.1

For unidirectional composites, $G_{13} = G_{12}$ (see Table 3.5) whereas typical values of G_{23} are listed in Table 4.1 (Herakovich, 1998).

Poisson's ratios ν_{31} and ν_{32} can be determined by measuring the change in the layer thickness under in-plane tension in directions 1 and 2.

4.2.2. Nonlinear models

Consider Figs. 3.40–3.43 showing typical stress–strain diagrams for unidirectional advanced composites. As can be seen, the materials demonstrate linear behavior only under tension. The curves corresponding to compression are slightly nonlinear, whereas the shear curves are definitely nonlinear. It should be emphasized that this does not mean that the linear constitutive equations presented in Section 4.2.1 are not valid for these materials. First, it should be taken into account that the deformations of properly designed composite materials are controlled by the fibers, and they do not allow the shear strain to reach the values at which the shear stress–strain curve is highly nonlinear. Second, the shear stiffness is usually very small in comparison with the longitudinal one, and so is its contribution to the apparent material stiffness. The material behavior is usually close to linear even if the shear deformation is nonlinear. Thus, a linear elastic model provides, as a rule, a reasonable approximation to the actual material behavior. However, there exist problems to solve in which we need to allow for material nonlinearity and apply one of the nonlinear constitutive theories discussed below.

First, note that material behavior under elementary loading (pure tension, compression, and shear) is specified by experimental stress–strain diagrams of the type shown in Figs. 3.40–3.43, and we do not need any theory. The necessity for a theory occurs if we are to study the interaction of simultaneously acting stresses. Because for the layer under study this interaction usually takes place for in-plane stresses σ_1, σ_2, and τ_{12} (see Fig. 4.13), we consider further the plane state of stress.

In the simplest, but quite useful for practical engineering analysis approach, the stress interaction is ignored completely, and the linear constitutive equations, Eqs. (4.55), are generalized as

$$\varepsilon_1 = \frac{\sigma_1}{E_1^s} - \nu_{12}^s\frac{\sigma_2}{E_2^s}, \quad \varepsilon_2 = \frac{\sigma_2}{E_2^s} - \nu_{21}^s\frac{\sigma_1}{E_1^s}, \quad \gamma_{12} = \frac{\tau_{12}}{G_{12}^s} \tag{4.57}$$

where the superscript 's' indicates the corresponding secant characteristics specified by Eqs. (1.8). These characteristics depend on stresses and are determined using experimental diagrams similar to those presented in Figs. 3.40–3.43. Particularly, diagrams $\sigma_1(\varepsilon_1)$ and $\varepsilon_2(\varepsilon_1)$ plotted under uniaxial longitudinal loading yield $E_1^s(\sigma_1)$ and $\nu_{21}^s(\sigma_1)$, secant moduli $E_2^s(\sigma_2)$ and $G_{12}^s(\tau_{12})$ are determined from experimental curves for $\sigma_2(\varepsilon_2)$ and $\tau_{12}(\gamma_{12})$, respectively, whereas ν_{12}^s is found from the symmetry condition in Eqs. (4.53). In a more rigorous model (Jones, 1977), the secant characteristics of the material in Eqs. (4.57) are also functions but in this case they are functions of strain energy U in Eq. (2.51) rather than of individual stresses. Models of this type provide adequate results for unidirectional composites with moderate nonlinearity.

To describe pronounced nonlinear elastic behavior of a unidirectional layer, we can use Eq. (4.10). Expanding the complementary potential U_c into a Taylor series with respect to stresses, we have

$$U_c = c_0 + c_{ij}\sigma_{ij} + \frac{1}{2}c_{ijkl}\sigma_{ij}\sigma_{kl} + \frac{1}{3!}c_{ijklmn}\sigma_{ij}\sigma_{kl}\sigma_{mn} + \frac{1}{4!}c_{ijklmnpq}\sigma_{ij}\sigma_{kl}\sigma_{mn}\sigma_{pq} + \frac{1}{5!}c_{ijklmnpqrs}\sigma_{ij}\sigma_{kl}\sigma_{mn}\sigma_{pq}\sigma_{rs} + \frac{1}{6!}c_{ijklmnpqrstw}\sigma_{ij}\sigma_{kl}\sigma_{mn}\sigma_{pq}\sigma_{rs}\sigma_{tw} + \cdots \tag{4.58}$$

where

$$c_0 = U_c\ (\sigma_{ij} = 0), \quad c_{ij} = \left.\frac{\partial U_c}{\partial \sigma_{ij}}\right|_{\sigma_{ij}=0}, \quad c_{ijkl} = \left.\frac{\partial^2 U_c}{\partial \sigma_{ij}\partial \sigma_{kl}}\right|_{\sigma_{ij}=0,\ \sigma_{kl}=0}, \text{ etc.}$$

A sixth-order approximation with the terms presented in Eq. (4.58) (where summation over repeated subscripts is implied) allows us to construct constitutive equations including stresses in the fifth power. The coefficients 'c' should be found from experiments with material specimens. Since these coefficients are particular derivatives that do not depend on the sequence of differentiation, the sequence of their subscripts is not important. As a result, the sixth-order polynomial in Eq. (4.58) includes 84 'c'-coefficients. This is clearly far too many for the practical analysis of composite materials. To reduce the number of coefficients, we can first use some general considerations. Namely, assume that $U_c = 0$ and $\varepsilon_{ij} = 0$ if there are no stresses ($\sigma_{ij} = 0$). Then, $c_0 = 0$ and $c_{ij} = 0$. Second, we should take into account that the material under study is orthotropic. This means that normal stresses do not induce shear strain, and shear stresses do not cause normal strains. Third, the direction of shear stresses should influence only shear strains, i.e., shear stresses should have only even powers in constitutive equations for normal strains, whereas the corresponding equation for shear strain should include only odd powers of shear stresses. As a result, the constitutive equations will contain 37 coefficients and take the following form (in new notations for coefficients and stresses)

$$\varepsilon_1 = a_1\sigma_1 + a_2\sigma_1^2 + a_3\sigma_1^3 + a_4\sigma_1^4 + a_5\sigma_1^5 + d_1\sigma_1 + 2d_2\sigma_1\sigma_2 + d_3\sigma_2^2 + 3d_4\sigma_1^2\sigma_2 + d_5\sigma_2^3 + d_6\sigma_1\sigma_2^2 + 4d_7\sigma_1^3\sigma_2 + 3d_8\sigma_1^2\sigma_2^2 + 2d_9\sigma_1\sigma_2^3 + d_{10}\sigma_2^4 + 5d_{11}\sigma_1^4\sigma_2$$

$$+ 4d_{12}\sigma_1^3\sigma_2^2 + 3d_{13}\sigma_1^2\sigma_2^3 + 2d_{14}\sigma_1\sigma_2^4 + d_{15}\sigma_2^5 + k_1\sigma_1\tau_{12}^2 + k_2\sigma_2\tau_{12}^2$$

$$+ 3k_3\sigma_1^2\tau_{12}^2 + 4k_4\sigma_1^3\tau_{12}^2 + 2k_5\sigma_1\tau_{12}^4$$

$$\varepsilon_2 = b_1\sigma_2 + b_2\sigma_2^2 + b_3\sigma_2^3 + b_4\sigma_2^4 + b_5\sigma_2^5 + d_1\sigma_1 + d_2\sigma_1^2 + 2d_3\sigma_1\sigma_2 + d_4\sigma_1^3$$

$$+ 3d_5\sigma_1\sigma_2^2 + d_6\sigma_1^2\sigma_2 + d_7\sigma_1^4 + 2d_8\sigma_1^3\sigma_2 + 3d_9\sigma_1^2\sigma_2^2 + 4d_{10}\sigma_1\sigma_2^3 + d_{11}\sigma_1^5$$

$$+ 2d_{12}\sigma_1^4\sigma_2 + 3d_{13}\sigma_1^3\sigma_2^2 + 2d_{14}\sigma_1^2\sigma_2^3 + 5d_{15}\sigma_1\sigma_2^4 + m_1\sigma_2\tau_{12}^2 + k_2\sigma_1\tau_{12}^2$$

$$+ 3m_2\sigma_2^2\tau_{12}^2 + 4m_3\sigma_2^3\tau_{12}^2 + 2m_4\sigma_2\tau_{12}^4 \tag{4.59}$$

$$\gamma_{12} = c_1\tau_{12} + c_2\tau_{12}^3 + c_3\tau_{12}^5 + k_1\tau_{12}\sigma_1^2 + m_1\tau_{12}\sigma_2^2 + 2k_2\tau_{12}\sigma_1\sigma_2$$

$$+ 2k_3\tau_{12}\sigma_1^3 + 2m_2\tau_{12}\sigma_2^3 + 2k_4\tau_{12}\sigma_1^4 + 4k_5\tau_{12}^3\sigma_1^2 + 2m_3\tau_{12}\sigma_2^4 + 4m_4\tau_{12}^3\sigma_2^2$$

For unidirectional composites, the dependence $\varepsilon_1(\sigma_1)$ is linear which means that we should put $d_2 = \ldots d_{15} = 0$, $k_1 = \ldots k_5 = 0$. Then, the foregoing equations reduce to

$$\varepsilon_1 = a_1\sigma_1 + d_1\sigma_2$$

$$\varepsilon_2 = b_1\sigma_2 + b_2\sigma_2^2 + b_3\sigma_2^3 + b_4\sigma_2^4 + b_5\sigma_2^5 + d_1\sigma_1 + m_1\sigma_2\tau_{12}^2 + 3m_2\sigma_2^2\tau_{12}^2$$

$$+ 4m_3\sigma_2^3\tau_{12}^2 + 2m_4\sigma_2\tau_{12}^4 \tag{4.60}$$

$$\gamma_{12} = c_1\tau_{12} + c_2\tau_{12}^3 + c_3\tau_{12}^5 + m_1\tau_{12}\sigma_2^2 + 2m_2\tau_{12}\sigma_2^3 + 2m_3\tau_{12}\sigma_2^4 + 4m_4\tau_{12}^3\sigma_2^2$$

As an example, consider a specific unidirectional two-matrix fiberglass composite with high in-plane transverse and shear deformation (see Section 4.4.3 for further details). The stress–strain curves corresponding to transverse tension, compression, and in-plane shear are shown in Fig. 4.16. Solid lines correspond to Eqs. (4.60) used to approximate the experimental results (circles in Fig. 4.16). The coefficients a_1 and d_1 in Eqs. (4.60) are found using diagrams $\varepsilon_1(\sigma_1)$ and $\varepsilon_2(\sigma_2)$ which are linear and not shown here. The coefficients $b_1 \ldots b_5$ and c_1, c_2, and c_3 are determined using the least-squares method to approximate curves $\sigma_2^+(\varepsilon_2)$, $\sigma_2^-(\varepsilon_2)$, and $\tau_{12}(\gamma_{12})$. The other coefficients, i.e. $m_1 \ldots m_4$, should be determined with the aid of a more complicated experiment involving loading that induces both stresses σ_2 and τ_{12} acting simultaneously. This experiment is described in Section 4.3.

As follows from Figs. 3.40–3.43, unidirectional composites demonstrate pronounced nonlinearity only under shear. Assuming that the dependence $\varepsilon_2(\sigma_2)$ is also linear, we can reduce Eqs. (4.60) to

$$\varepsilon_1 = a_1\sigma_1 + d_1\sigma_2, \quad \varepsilon_2 = b_1\sigma_2 + d_1\sigma_1, \quad \gamma_{12} = c_1\tau_{12} + c_2\tau_{12}^3 + c_3\tau_{12}^5$$

For practical analysis, an even simpler form of these equations (with $c_3 = 0$) can be used (Hahn and Tsai, 1973).

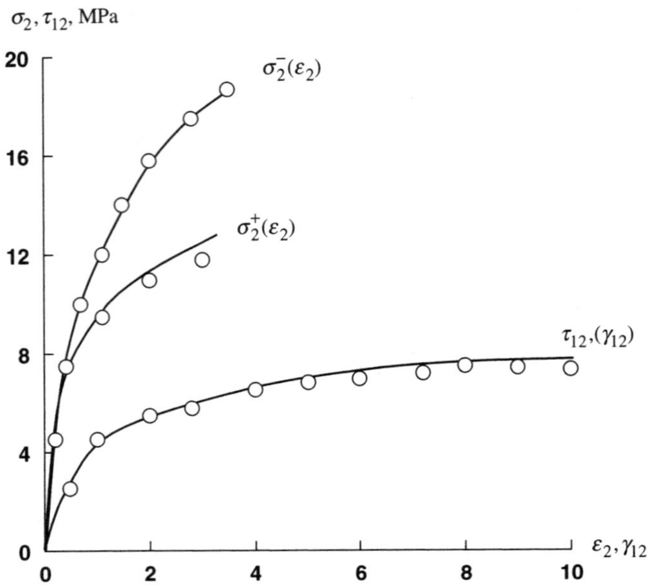

Fig. 4.16. Calculated (solid lines) and experimental (circles) stress–strain diagrams for a two-matrix unidirectional composite under in-plane transverse tension (σ_2^+), compression (σ_2^-) and shear (τ_{12}).

Nonlinear behavior in composite materials can also be described with the aid of the theory of plasticity which can be constructed as a direct generalization of the classical plasticity theory developed for metals and described in Section 4.1.2.

To construct such a theory, we decompose strains in accordance with Eq. (4.15) and use Eqs. (4.16) and (4.18) to determine elastic and plastic strains as

$$\varepsilon_{ij}^{\mathrm{e}} = \frac{\partial U_{\mathrm{e}}}{\partial \sigma_{ij}}, \quad \varepsilon_{ij}^{\mathrm{p}} = \frac{\partial U_{\mathrm{p}}}{\partial \sigma_{ij}} \tag{4.61}$$

where U_{e} and U_{p} are elastic and plastic potentials. For elastic potential, elasticity theory yields

$$U = c_{ijkl}\sigma_{ij}\sigma_{kl} \tag{4.62}$$

where c_{ijkl} are compliance coefficients, and summation over repeated subscripts is implied. The plastic potential is assumed to be a function of stress intensity, σ, which is constructed for a plane stress state as a direct generalization of Eq. (4.24), i.e.,

$$\sigma = a_{ij}\sigma_{ij} + \sqrt{a_{ijkl}\sigma_{ij}\sigma_{kl}} + \sqrt[3]{a_{ijklmn}\sigma_{ij}\sigma_{kl}\sigma_{mn}} + \cdots \tag{4.63}$$

where the coefficients 'a' are material constants characterizing its plastic behavior. Finally, we use the power law in Eq. (4.27) for the plastic potential.

To write constitutive equations for a plane stress state, we return to engineering notations for stresses and strains and use conditions that should be imposed on an orthotropic material and are discussed above in application to Eqs. (4.59). Finally, Eqs. (4.15), (4.27) and (4.61)–(4.63) yield

$$
\begin{aligned}
\varepsilon_1 &= a_1\sigma_1 + d_1\sigma_2 + n\sigma^{n-1}\left[\frac{1}{R_1}(b_{11}\sigma_1 + c_{12}\sigma_2) + \frac{1}{R_2^2}\left(d_{11}\sigma_1^2 + 2e_{12}\sigma_1\sigma_2 + e_{21}\sigma_2^2\right)\right] \\
\varepsilon_2 &= b_1\sigma_2 + d_1\sigma_1 + n\sigma^{n-1}\left[\frac{1}{R_1}(b_{22}\sigma_2 + c_{12}\sigma_1) + \frac{1}{R_2^2}\left(d_{22}\sigma_2^2 + 2e_{21}\sigma_2\sigma_1 + e_{12}\sigma_1^2\right)\right] \\
\gamma_{12} &= c_1\tau_{12} + 2n\sigma^{n-1}\frac{b_{12}}{R_1}\tau_{12}
\end{aligned}
\tag{4.64}
$$

where

$$
\begin{aligned}
\sigma &= R_1 + R_2 \\
R_1 &= \sqrt{b_{11}\sigma_1^2 + b_{22}\sigma_2^2 + b_{12}\tau_{12}^2 + 2c_{12}\sigma_1\sigma_2} \\
R_2 &= \sqrt[3]{d_{11}\sigma_1^3 + d_{22}\sigma_2^3 + 3e_{12}\sigma_1^2\sigma_2 + 3e_{21}\sigma_1\sigma_2^2}
\end{aligned}
$$

Deriving Eqs. (4.64), we use new notations for coefficients and restrict ourselves to the three-term approximation for σ as in Eq. (4.63).

For independent uniaxial loading along the fibers, across the fibers, and in pure shear, Eqs. (4.64) reduce to

$$
\begin{aligned}
\varepsilon_1 &= a_1\sigma_1 + n\left(\sqrt{b_{11}\sigma_1^2} + \sigma_1\sqrt[3]{d_{11}}\right)^{n-1}\left(\sqrt{b_{11}}\frac{\sigma_1}{\sqrt{\sigma_1^2}} + \sqrt[3]{d_{11}}\right) \\
\varepsilon_2 &= b_1\sigma_2 + n\left(\sqrt{b_{22}\sigma_2^2} + \sigma_2\sqrt[3]{d_{22}}\right)^{n-1}\left(\sqrt{b_{22}}\frac{\sigma_2}{\sqrt{\sigma_2^2}} + \sqrt[3]{d_{22}}\right) \\
\gamma_{12} &= \left[c_1 + 2n\sqrt{b_{12}^n}\left(\sqrt{\tau_{12}^2}\right)^{n-1}\right]\tau_{12}
\end{aligned}
\tag{4.65}
$$

If nonlinear material behavior does not depend on the sign of normal stresses, then $d_{11} = d_{22} = 0$ in Eqs. (4.65). In the general case, Eqs. (4.65) allow us to describe materials with high nonlinearity and different behavior under tension and compression.

As an example, consider a boron–aluminum unidirectional composite whose experimental stress–strain diagrams (Herakovich, 1998) are shown in Fig. 4.17 (circles) along with the corresponding approximations (solid lines) plotted with the aid of Eqs. (4.65).

4.3. Unidirectional anisotropic layer

Consider now a unidirectional layer studied in the previous section and assume that its principal material axis 1 makes some angle ϕ with the x-axis of the global coordinate frame (see Fig. 4.18). An example of such a layer is shown in Fig. 4.19.

4.3.1. Linear elastic model

Constitutive equations of the layer under study referred to the principal material coordinates are given by Eqs. (4.55) and (4.56). We need now to derive such equations for the global coordinate frame x, y, and z (see Fig. 4.18). To do this, we should transfer stresses σ_1, σ_2, τ_{12}, τ_{13}, τ_{23} acting in the layer and the corresponding strains ε_1, ε_2, γ_{12}, γ_{13}, γ_{23} into stress and strain components σ_x, σ_y, τ_{xy}, τ_{xz}, τ_{yz} and ε_x, ε_y, γ_{xy}, γ_{xz}, γ_{yz} using Eqs. (2.8), (2.9) and (2.21), (2.27) for coordinate transformation of stresses and strains. According to Fig. 4.18, the directional cosines, Eqs. (2.1), for this transformation are (we take $x' = 1$, $y' = 2$, $z' = 3$)

$$
\begin{aligned}
&l_{x^1x} = c, \quad l_{x^1y} = s, \quad l_{x^1z} = 0 \\
&l_{y^1x} = -s, \quad l_{y^1y} = c, \quad l_{y^1z} = 0 \\
&l_{z^1x} = 0, \quad l_{z^1y} = 0, \quad l_{z^1z} = 1
\end{aligned}
\tag{4.66}
$$

where $c = \cos\phi$ and $s = \sin\phi$. Using Eqs. (2.8) and (2.9), we get

$$
\begin{aligned}
\sigma_1 &= \sigma_x c^2 + \sigma_y s^2 + 2\tau_{xy} cs \\
\sigma_2 &= \sigma_x s^2 + \sigma_y c^2 - 2\tau_{xy} cs \\
\tau_{12} &= (\sigma_y - \sigma_x)cs + \tau_{xy}\left(c^2 - s^2\right) \\
\tau_{13} &= \tau_{xz} c + \tau_{yz} s \\
\tau_{12} &= -\tau_{xz} s + \tau_{yz} c
\end{aligned}
\tag{4.67}
$$

The inverse form of these equations is

$$
\sigma_x = \sigma_1 c^2 + \sigma_2 s^2 - 2\tau_{12} cs
$$

$$
\sigma_y = \sigma_1 s^2 + \sigma_2 c^2 + 2\tau_{12} cs
$$

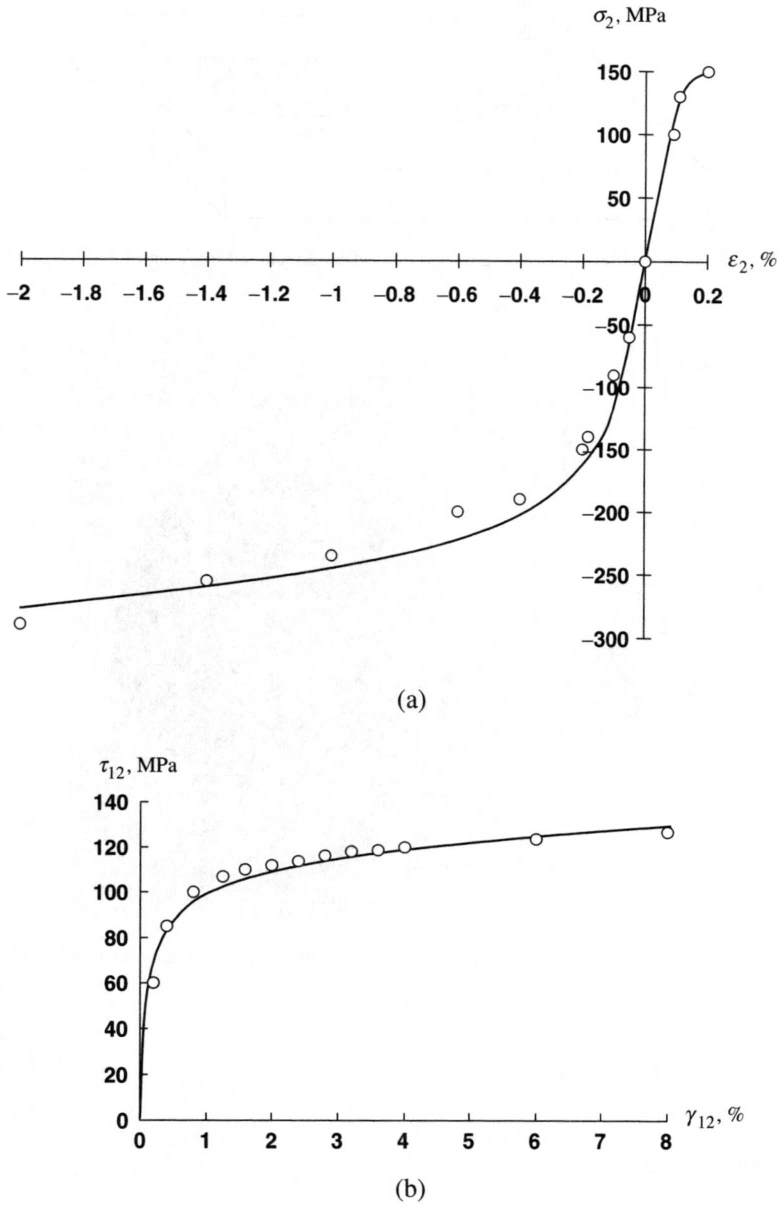

Fig. 4.17. Calculated (solid lines) and experimental (circles) stress–strain diagrams for a boron–aluminum composite under transverse loading (a) and in-plane shear (b).

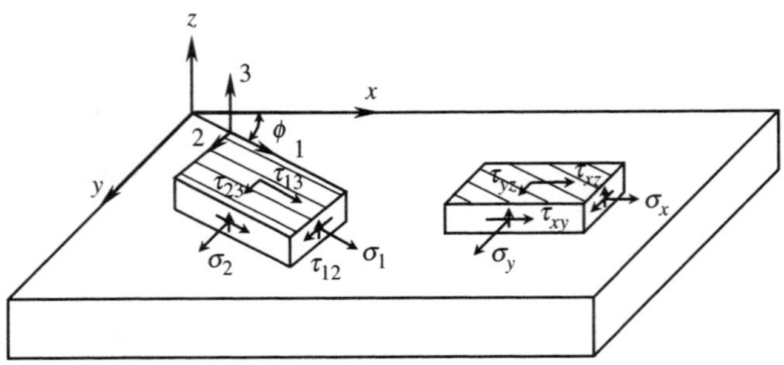

Fig. 4.18. A composite layer consisting of a system of unidirectional plies with the same orientation.

Fig. 4.19. An anisotropic outer layer of a composite pressure vessel. Courtesy of CRISM.

$$\tau_{xy} = (\sigma_1 - \sigma_2)cs + \tau_{12}\left(c^2 - s^2\right) \tag{4.68}$$

$$\tau_{xz} = \tau_{13}c - \tau_{23}s$$

$$\tau_{yz} = \tau_{13}s + \tau_{23}c$$

The corresponding transformation for strains follows from Eqs. (2.21) and (2.27), i.e.,

$$\varepsilon_1 = \varepsilon_x c^2 + \varepsilon_y s^2 + \gamma_{xy} cs$$

$$\varepsilon_2 = \varepsilon_x s^2 + \varepsilon_y c^2 - \gamma_{xy} cs$$

$$\gamma_{12} = 2(\varepsilon_y - \varepsilon_x)cs + \gamma_{xy}\left(c^2 - s^2\right) \tag{4.69}$$
$$\gamma_{13} = \gamma_{xz}c + \gamma_{yz}s$$
$$\gamma_{23} = -\gamma_{xz}s + \gamma_{yz}c$$

or

$$\varepsilon_x = \varepsilon_1 c^2 + \varepsilon_2 s^2 - \gamma_{12}cs$$
$$\varepsilon_y = \varepsilon_1 s^2 + \varepsilon_2 c^2 + \gamma_{12}cs$$
$$\gamma_{xy} = 2(\varepsilon_1 - \varepsilon_2)cs + \gamma_{12}\left(c^2 - s^2\right) \tag{4.70}$$
$$\gamma_{xz} = \gamma_{13}c - \gamma_{23}s$$
$$\gamma_{yz} = \gamma_{13}s + \gamma_{23}c$$

To derive constitutive equations for an anisotropic unidirectional layer, we substitute strains, Eqs. (4.69), into Hooke's law, Eqs. (4.56), and the derived stresses – into Eqs. (4.68). The final result is as follows

$$\sigma_x = A_{11}\varepsilon_x + A_{12}\varepsilon_y + A_{14}\gamma_{xy}$$
$$\sigma_y = A_{21}\varepsilon_x + A_{22}\varepsilon_y + A_{24}\gamma_{xy}$$
$$\tau_{xy} = A_{41}\varepsilon_x + A_{42}\varepsilon_y + A_{44}\gamma_{xy} \tag{4.71}$$
$$\tau_{xz} = A_{55}\gamma_{xz} + A_{56}\gamma_{yz}$$
$$\tau_{yz} = A_{65}\gamma_{xz} + A_{66}\gamma_{yz}$$

The stiffness coefficients in these equations are

$$A_{11} = \overline{E}_1 c^4 + \overline{E}_2 s^4 + 2E_{12}c^2 s^2$$
$$A_{12} = A_{21} = \overline{E}_1 \nu_{12} + (\overline{E}_1 + \overline{E}_2 - 2E_{12})c^2 s^2$$
$$A_{14} = A_{41} = \left[\overline{E}_1 c^2 - \overline{E}_2 s^2 - E_{12}\left(c^2 - s^2\right)\right] cs$$
$$A_{22} = \overline{E}_1 s^4 + \overline{E}_2 c^4 + 2E_{12}c^2 s^2$$
$$A_{24} = A_{42} = \left[\overline{E}_1 s^2 - \overline{E}_2 c^2 + E_{12}\left(c^2 - s^2\right)\right] cs \tag{4.72}$$
$$A_{44} = (\overline{E}_1 + \overline{E}_2 - 2\overline{E}_1 \nu_{12})c^2 s^2 + G_{12}\left(c^2 - s^2\right)^2$$
$$A_{55} = G_{13}c^2 + G_{23}s^2$$
$$A_{56} = A_{65} = (G_{13} - G_{23})cs$$
$$A_{66} = G_{13}s^2 + G_{23}c^2$$

where

$$\overline{E}_{1,2} = \frac{E_{1,2}}{1 - \nu_{12}\nu_{21}}, \quad E_{12} = \overline{E}_1\nu_{12} + 2G_{12}, \quad c = \cos\phi, \quad s = \sin\phi$$

The dependence of stiffness coefficients A_{mn} in Eqs. (4.72) on ϕ has been studied by Tsai and Pagano (see, e.g., Tsai, 1987; Verchery, 1999). Changing the powers of $\sin\phi$ and $\cos\phi$ in Eqs. (4.72) for multiple-angle trigonometric functions, we can reduce these equations to the following form (Verchery, 1999)

$$\begin{aligned}
A_{11} &= S_1 + S_2 + 2S_3\cos 2\phi + S_4\cos 4\phi \\
A_{12} &= -S_1 + S_2 - S_4\cos 4\phi \\
A_{14} &= S_3\sin 2\phi + S_4\sin 4\phi \\
A_{22} &= S_1 + S_2 - 2S_3\cos 2\phi + S_4\cos 4\phi \\
A_{24} &= S_3\sin 2\phi - S_4\sin 4\phi \\
A_{44} &= S_1 - S_4\cos 4\phi \\
A_{55} &= S_5 + S_6\cos 2\phi \\
A_{56} &= 4S_6\sin 2\phi \\
A_{66} &= S_5 - S_6\cos 2\phi
\end{aligned} \tag{4.73}$$

In these equations,

$$\begin{aligned}
S_1 &= \frac{1}{8}\left(A_{11}^0 + A_{22}^0 - 2A_{12}^0 + 4A_{44}^0\right) \\
S_2 &= \frac{1}{4}\left(A_{11}^0 + A_{22}^0 + 2A_{12}^0\right) \\
S_3 &= \frac{1}{4}\left(A_{11}^0 - A_{22}^0\right) \\
S_4 &= \frac{1}{8}\left(A_{11}^0 + A_{22}^0 - 2A_{12}^0 - 4A_{44}^0\right) \\
S_5 &= \frac{1}{2}\left(A_{55}^0 + A_{66}^0\right) \\
S_6 &= \frac{1}{2}\left(A_{55}^0 - A_{66}^0\right)
\end{aligned}$$

where A_n^0 are stiffness coefficients corresponding to $\phi = 0$. It follows from Eqs. (4.72) that,

$$A_{11}^0 = \overline{E}_1, \quad A_{12}^0 = \overline{E}_1 \nu_{12}, \quad A_{14}^0 = A_{24}^0 = A_{56}^0 = 0$$

$$A_{22}^0 = \overline{E}_2, \quad A_{44}^0 = G_{12}, \quad A_{55}^0 = G_{13}, \quad A_{66}^0 = G_{23}$$

As can be seen in Eqs. (4.73), there exist the following differential relationships between tensile and coupling stiffnesses (Verchery and Gong, 1999)

$$\frac{\mathrm{d}A_{11}}{\mathrm{d}\phi} = -4A_{14}, \quad \frac{\mathrm{d}A_{22}}{\mathrm{d}\phi} = 4A_{24}$$

It can be directly checked that Eqs. (4.73) provide three invariant stiffness characteristics whose forms do not depend on ϕ, i.e.,

$$\begin{aligned} &A_{11}(\phi) + A_{22}(\phi) + 2A_{12}(\phi) = A_{11}^0 + A_{22}^0 + 2A_{12}^0 \\ &A_{44}(\phi) - A_{12}(\phi) = A_{44}^0 - A_{12}^0 \\ &A_{55}(\phi) + A_{66}(\phi) = A_{55}^0 + A_{66}^0 \end{aligned} \tag{4.74}$$

Any linear combination of these equations is also an invariant combination of stiffness coefficients.

The inverse form of Eqs. (4.71) can be obtained if we substitute stresses, Eqs. (4.67), into Hooke's law, Eqs. (4.55), and the derived strains in Eqs. (4.70). As a result, we arrive at the following particular form of Eqs. (2.48) and (2.49)

$$\begin{aligned} &\varepsilon_x = \frac{\sigma_x}{E_x} - \nu_{xy}\frac{\sigma_y}{E_y} + \eta_{x,xy}\frac{\tau_{xy}}{G_{xy}}, \quad \varepsilon_y = \frac{\sigma_y}{E_y} - \nu_{yx}\frac{\sigma_x}{E_x} + \eta_{y,xy}\frac{\tau_{xy}}{G_{xy}} \\ &\gamma_{xy} = \frac{\tau_{xy}}{G_{xy}} + \eta_{xy,x}\frac{\sigma_x}{E_x} + \eta_{xy,y}\frac{\sigma_y}{E_y}, \quad \gamma_{xz} = \frac{\tau_{xz}}{G_{xz}} + \lambda_{xz,yz}\frac{\tau_{yz}}{G_{yz}} \\ &\gamma_{yz} = \frac{\tau_{yz}}{G_{yz}} + \lambda_{yz,xz}\frac{\tau_{xz}}{G_{xz}} \end{aligned} \tag{4.75}$$

in which the compliance coefficients are

$$\frac{1}{E_x} = \frac{c^4}{E_1} + \frac{s^4}{E_2} + \left(\frac{1}{G_{12}} - \frac{2\nu_{21}}{E_1}\right)c^2s^2$$

$$\frac{\nu_{xy}}{E_y} = \frac{\nu_{yx}}{E_x} = \frac{\nu_{21}}{E_1} - \left(\frac{1}{E_1} + \frac{1}{E_2} + \frac{\nu_{21}}{E_1} - \frac{1}{G_{12}}\right)c^2s^2$$

$$\frac{\eta_{x,xy}}{G_{xy}} = \frac{\eta_{xy,x}}{E_x} = 2\left[\frac{c^2}{E_1} - \frac{s^2}{E_2} - \left(\frac{1}{2G_{12}} - \frac{\nu_{21}}{E_1}\right)\left(c^2 - s^2\right)\right]cs$$

$$\frac{1}{E_y}=\frac{s^4}{E_1}+\frac{c^4}{E_2}+\left(\frac{1}{G_{12}}-\frac{2\nu_{21}}{E_1}\right)c^2s^2 \tag{4.76}$$

$$\frac{\eta_{y,xy}}{G_{xy}}=\frac{\eta_{xy,y}}{E_y}=2\left[\frac{s^2}{E_1}-\frac{c^2}{E_2}+\left(\frac{1}{2G_{12}}-\frac{\nu_{21}}{E_1}\right)\left(c^2-s^2\right)\right]cs$$

$$\frac{1}{G_{xy}}=4\left(\frac{1}{E_1}+\frac{1}{E_2}+\frac{2\nu_{21}}{E_1}\right)c^2s^2+\frac{1}{G_{12}}\left(c^2-s^2\right)^2$$

$$\frac{1}{G_{xz}}=\frac{c^2}{G_{13}}+\frac{s^2}{G_{23}},\quad \frac{\lambda_{xz,yz}}{G_{yz}}=\frac{\lambda_{yz,xz}}{G_{xz}}=\left(\frac{1}{G_{13}}-\frac{1}{G_{23}}\right)cs,\quad \frac{1}{G_{yz}}=\frac{s^2}{G_{13}}+\frac{c^2}{G_{23}}$$

There exist the following dependencies between the coefficients of Eqs. (4.71) and (4.75)

$$\frac{1}{E_x}=\frac{1}{D_1}\left(A_{22}A_{44}-A_{24}^2\right),\quad \frac{\nu_{xy}}{E_y}=\frac{\nu_{yx}}{E_x}=\frac{1}{D_1}(A_{12}A_{44}-A_{14}A_{24})$$

$$\frac{\eta_{x,xy}}{G_{xy}}=\frac{\eta_{xy,x}}{E_x}=\frac{1}{D_1}(A_{12}A_{24}-A_{22}A_{14}),\quad \frac{1}{E_y}=\frac{1}{D_1}\left(A_{11}A_{44}-A_{14}^2\right)$$

$$\frac{\eta_{y,xy}}{G_{xy}}=\frac{\eta_{xy,y}}{E_y}=\frac{1}{D_1}(A_{12}A_{14}-A_{11}A_{24}),\quad \frac{1}{G_{xy}}=\frac{1}{D_1}\left(A_{11}A_{22}-A_{12}^2\right)$$

$$\frac{1}{G_{xz}}=\frac{A_{66}}{D_2},\quad \frac{1}{G_{yz}}=\frac{A_{55}}{D_2},\quad \frac{\lambda_{xz,yz}}{G_{yz}}=\frac{\lambda_{yz,xz}}{G_{xz}}=-\frac{A_{56}}{D_2}$$

Here,

$$D_1=A_{11}A_{22}A_{44}-A_{11}A_{24}^2-A_{22}A_{14}^2-A_{44}A_{12}^2+2A_{12}A_{14}A_{24}$$

$$D_2=A_{55}A_{66}-A_{56}^2$$

and

$$A_{11}=\frac{1-\eta_{y,xy}\eta_{xy,y}}{D_3E_yG_{xy}},\quad A_{12}=\frac{\nu_{xy}-\eta_{x,xy}\eta_{xy,y}}{D_3E_yG_{xy}}$$

$$A_{14}=-\frac{\eta_{x,xy}+\nu_{xy}\eta_{y,xy}}{D_3E_yG_{xy}},\quad A_{22}=\frac{1-\eta_{x,xy}\eta_{xy,y}}{D_3E_xG_{xy}}$$

$$A_{24}=-\frac{\eta_{y,xy}+\nu_{yx}\eta_{x,xy}}{D_3E_xG_{xy}},\quad A_{44}=\frac{1-\nu_{xy}\nu_{yx}}{D_3E_xE_y}$$

$$A_{55}=\frac{1}{D_4G_{yz}},\quad A_{56}=-\frac{\lambda_{xz,yz}}{D_4G_{yz}},\quad A_{66}=\frac{1}{D_4G_{xz}}$$

where

$$D_3 = \frac{1}{E_x E_y G_{xy}}(1 - \nu_{xy}\nu_{yx} - \eta_{x,\,xy}\eta_{xy,\,x} - \eta_{y,\,xy}\eta_{xy,\,y} - \nu_{xy}\eta_{y,\,xy}\eta_{xy,\,x}$$

$$- \nu_{yx}\eta_{x,\,xy}\eta_{xy,\,y})$$

$$D_4 = \frac{1}{G_{xz}G_{yz}}(1 - \lambda_{xz,\,yz}\lambda_{yz,\,xz})$$

As can be seen in Eqs. (4.71) and (4.75), the layer under study is anisotropic in plane xy because the constitutive equations include shear–extension and shear–shear coupling coefficients η and λ. For $\phi = 0$, the foregoing equations degenerate into Eqs. (4.55) and (4.56) for an orthotropic layer.

The dependencies of stiffness coefficients on the orientation angle for a carbon–epoxy composite with properties listed in Table 3.5 are presented in Figs. 4.20 and 4.21.

Uniaxial tension of the anisotropic layer (the so-called off-axis test of a unidirectional composite) is often used to determine material characteristics that cannot be found in tests with orthotropic specimens or to evaluate constitutive and failure theories. Such a

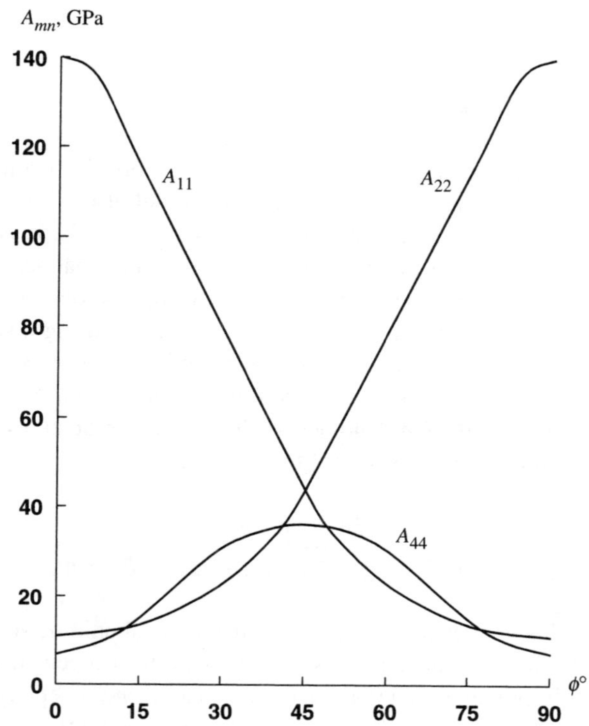

Fig. 4.20. Dependencies of tensile (A_{11}, A_{22}) and shear (A_{44}) stiffnesses of a unidirectional carbon–epoxy layer on the orientation angle.

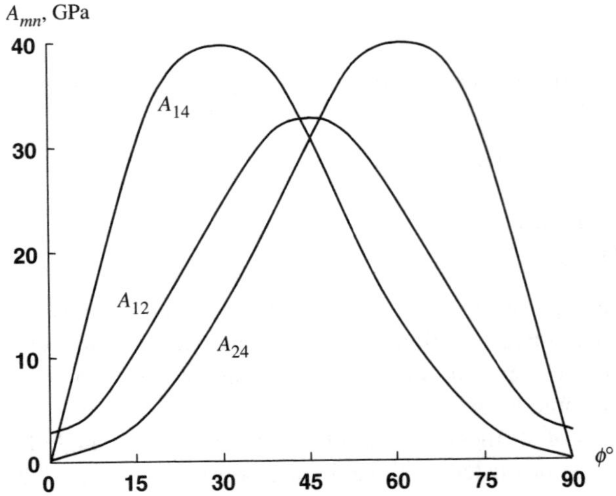

Fig. 4.21. Dependencies of coupling stiffnesses of a unidirectional carbon–epoxy layer on the orientation angle.

test is shown in Fig. 4.22. To study this loading case, we should take $\sigma_y = \tau_{xy} = 0$ in Eqs. (4.75). Then,

$$\varepsilon_x = \frac{\sigma_x}{E_x}, \quad \varepsilon_y = -\nu_{xy}\frac{\sigma_x}{E_x}, \quad \gamma_{xy} = \eta_{xy,\,x}\frac{\sigma_x}{E_x} \tag{4.77}$$

As can be seen in these equations, tension in the x-direction is accompanied not only with transverse contraction, as in orthotropic materials, but also with shear. This results in the deformed shape of the sample shown in Fig. 4.23. This shape is natural because the material stiffness in the fiber direction is much higher than that across the fibers.

Such an experiment, in cases where it can be performed, allows us to determine the in-plane shear modulus, G_{12} in principle material coordinates using a simple tensile test rather than the much more complicated tests described in Section 3.4.3 and shown in Figs. 3.54 and 3.55. Indeed, if we know E_x from the tensile test in Fig. 4.23 and find E_1, E_2, and ν_{21} from tensile tests along and across the fibers (see Sections 3.4.1 and 3.4.2), we can use the first equation of Eqs. (4.76) to determine

$$G_{12} = \frac{\sin^2\phi\cos^2\phi}{(1/E_x) - \left(\cos^4\phi/E_1\right) - \left(\sin^4\phi/E_2\right) + (2\nu_{21}/E_1)\sin^2\phi\cos^2\phi} \tag{4.78}$$

In connection with this, a question arises as to what angle should be substituted into this equation to provide the most accurate result. The answer is given in Fig. 4.24, which displays the strains in principal material coordinates for a carbon–epoxy layer calculated with the aid of Eqs. (4.69) and (4.77). As can be seen in this figure, the most appropriate angle is about 10°. At this angle, the shear strain γ_{12} is much higher than normal strains ε_1 and ε_2, so that material deformation is associated mainly with shear. An off-axis test

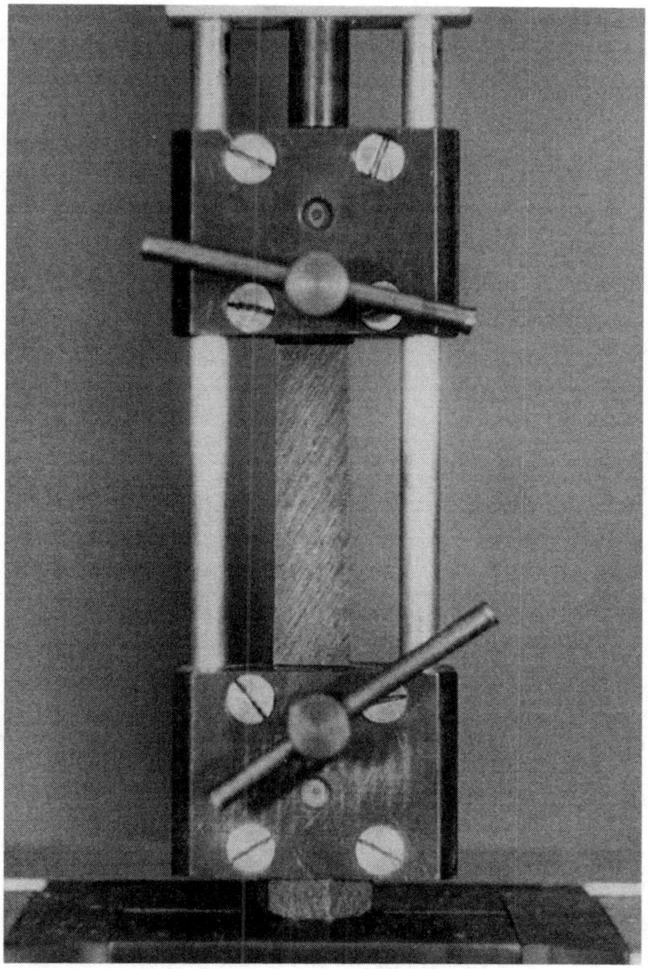

Fig. 4.22. An off-axis test.

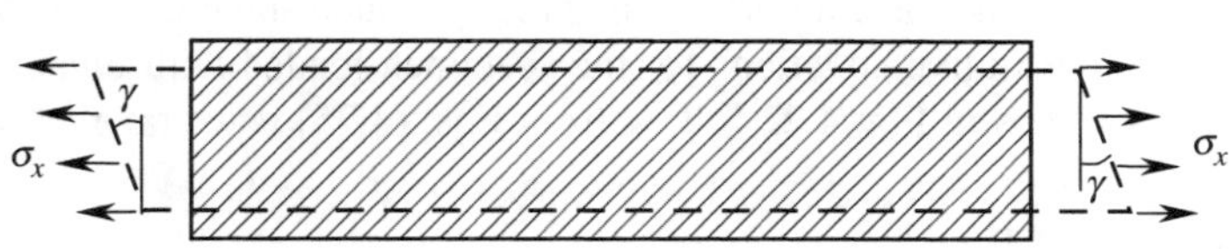

Fig. 4.23. Deformation of a unidirectional layer loaded at an angle to fiber orientation.

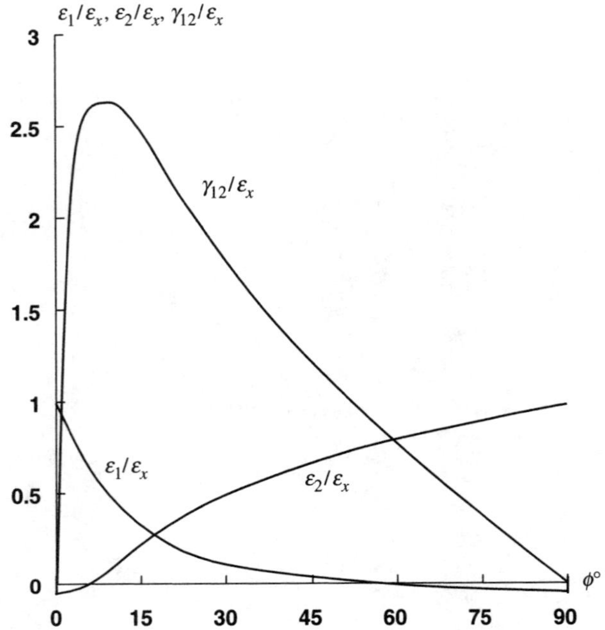

Fig. 4.24. Dependencies of normalized strains in the principle material coordinates on the angle of the off-axis test.

with $\phi = 10°$ can also be used to evaluate material strength in shear $\overline{\tau}_{12}$ (Chamis, 1979). Stresses acting under off-axis tension in the principal material coordinates are statically determinate and can be found directly from Eqs. (4.67) as

$$\sigma_1 = \sigma_x \cos^2 \phi, \quad \sigma_2 = \sigma_x \sin^2 \phi, \quad \tau_{12} = -\sigma_x \sin \phi \cos \phi \tag{4.79}$$

Thus, applying stress σ_x and changing ϕ we can induce proportional loading with different combinations of stresses σ_1, σ_2, and τ_{12} to evaluate putative constitutive or failure theories for a material under study.

However, the test shown in Fig. 4.23 can hardly be performed because the test fixture (see Fig. 4.22) restrains the shear deformation of the specimen and induces a corresponding shear stress. The constitutive equations for the specimen loaded with uniaxial tension as in Fig. 4.23 and fixed as in Fig. 4.22 follow from Eqs. (4.75) if we take $\sigma_y = 0$, i.e.,

$$\varepsilon_x = \frac{\sigma_x}{E_x} + \eta_{x,\,xy} \frac{\tau_{xy}}{G_{xy}} \tag{4.80}$$

$$\gamma_{xy} = \frac{\tau_{xy}}{G_{xy}} + \eta_{xy,\,x} \frac{\sigma_x}{E_x} \tag{4.81}$$

in which elastic constants are specified by Eqs. (4.76). The shear stress, being of a reactive nature, can be found from Eq. (4.81) if we put $\gamma_{xy} = 0$. Then,

$$\tau_{xy} = -\eta_{xy,\,x}\frac{G_{xy}}{E_x}\sigma_x$$

Substituting this result into Eq. (4.80), we arrive at

$$\varepsilon_x = \frac{\sigma_x}{E_x^{\mathrm{a}}} \tag{4.82}$$

Here,

$$E_x^{\mathrm{a}} = \frac{E_x}{1 - \eta_{x,\,xy}\eta_{xy,\,x}} \tag{4.83}$$

is the apparent elastic modulus that can be found from the test shown in Fig. 4.22. As follows from Eq. (4.83), E_x^{a}, in general, does not coincide with E_x as used in Eq. (4.78) for G_{12}.

Thus, measuring σ_x and ε_x we can determine E_x from Eq. (4.82) only under the condition $E_x^{\mathrm{a}} = E_x$, which means that the shear–extension coupling coefficient η must be zero. Applying Eqs. (4.76) and assuming that $\phi \neq 0$ and $\phi \neq 90°$, we arrive at the following condition providing $\eta = 0$

$$\sin^2\phi_0 = \frac{e_1}{e_2} \tag{4.84}$$

in which

$$e_1 = \frac{1+\nu_{21}}{E_1} - \frac{1}{2G_{12}}, \quad e_2 = \frac{1+\nu_{21}}{E_1} + \frac{1+\nu_{12}}{E_2} - \frac{1}{G_{12}}$$

Since $0 \leq \sin^2\phi \leq 1$, there exist two cases in which Eq. (4.84) is valid. The first case corresponds to the following set of inequalities

$$e_1 \geq 0, \quad e_2 > 0, \quad e_2 \geq e_1 \tag{4.85}$$

whereas for the second case,

$$e_1 \leq 0, \quad e_2 < 0, \quad e_2 \leq e_1 \tag{4.86}$$

To be specific, suppose that $E_1 > E_2$. Then, taking into account the symmetry condition $\nu_{12}E_1 = \nu_{21}E_2$ we have

$$\frac{1+\nu_{12}}{E_2} > \frac{1+\nu_{21}}{E_1} \tag{4.87}$$

Consider the first set of inequalities in Eqs. (4.85) and assume that the first of them, which has the following explicit form

$$\frac{1+\nu_{21}}{E_1} \geq \frac{1}{2G_{12}} \tag{4.88}$$

is valid. Then, Eq. (4.87) yields

$$\frac{1+\nu_{12}}{E_2} > \frac{1+\nu_{21}}{E_1} \geq \frac{1}{2G_{12}} \quad \text{or} \quad \frac{1+\nu_{12}}{E_2} > \frac{1}{2G_{12}}$$

Matching this result with the last inequality in Eqs. (4.85) presented in the form

$$\frac{1+\nu_{12}}{E_2} \geq \frac{1}{2G_{12}} \tag{4.89}$$

we can conclude that if the first condition in Eqs. (4.85) is valid, then the last of these conditions is valid too.

Consider the second condition in Eqs. (4.85) and write it in explicit form, i.e.,

$$\frac{1+\nu_{12}}{E_2} + \frac{1+\nu_{21}}{E_1} \geq \frac{1}{G_{12}} \tag{4.90}$$

Transforming Eq. (4.87) and using Eq. (4.89), we have

$$\frac{1+\nu_{12}}{E_2} + \frac{1+\nu_{21}}{E_1} > 2\frac{1+\nu_{21}}{E_1} \geq \frac{1}{G_{12}}$$

which means that the condition in Eq. (4.90) is valid.

So, the set of conditions in Eqs. (4.85) can be reduced to one inequality in Eq. (4.88), which can be written in a final form as

$$G_{12} \geq \frac{E_1}{2(1+\nu_{21})} \tag{4.91}$$

Consider conditions (4.86) and assume that the last of them, which can be presented in the following explicit form

$$\frac{1+\nu_{12}}{E_2} \leq \frac{1}{2G_{12}} \tag{4.92}$$

is valid. Using Eqs. (4.87) and (4.92), we get

$$\frac{1+\nu_{21}}{E_1} < \frac{1+\nu_{12}}{E_2} \leq \frac{1}{2G_{12}} \quad \text{or} \quad \frac{1+\nu_{21}}{E_1} < \frac{1}{2G_{12}}$$

Since the first condition in Eqs. (4.86) can be presented as

$$\frac{1+\nu_{21}}{E_1} \le \frac{1}{2G_{12}}$$

we can conclude that it is satisfied.

Consider the second inequality in Eqs. (4.86) and write it in an explicit form, i.e.,

$$\frac{1+\nu_{21}}{E_1} + \frac{1+\nu_{12}}{E_2} < \frac{1}{G_{12}} \tag{4.93}$$

Using Eqs. (4.87) and (4.92), we get

$$\frac{1+\nu_{21}}{E_1} + \frac{1+\nu_{12}}{E_2} < 2\frac{1+\nu_{12}}{E_2} \le \frac{1}{G_{12}}$$

which means that the condition in Eq. (4.93) is satisfied.

So, the set of conditions in Eqs. (4.86) is reduced to one inequality in Eq. (4.92), which can be written in the following final form

$$G_{12} \le \frac{E_2}{2(1+\nu_{12})} \tag{4.94}$$

Thus, Eq. (4.84) determines the angle ϕ_0 for the orthotropic materials whose mechanical characteristics satisfy the conditions in Eqs. (4.91) or (4.94). Such materials, being loaded at an angle $\phi = \phi_0$, do not experience shear–stretching coupling. The shear modulus can be found from Eq. (4.78) in which $E_x = \sigma_x/\varepsilon_x$, where σ_x and ε_x are the stress and the strain determined in the off-axis tension test shown in Fig. 4.22.

Consider as examples unidirectional composites with typical properties (Table 3.5).

(1) For fiberglass–epoxy composite, we have $E_1 = 60$ GPa, $E_2 = 13$ GPa, $G_{12} = 3.4$ GPa, $\nu_{12} = 0.065$, $\nu_{21} = 0.3$
Calculation in accordance with Eqs. (4.91) and (4.94) yields

$$\frac{E_1}{2(1+\nu_{21})} = 23.08\,\text{GPa}, \qquad \frac{E_2}{2(1+\nu_{12})} = 6.1\,\text{GPa}$$

Thus, the condition in Eq. (4.94) is satisfied, and Eq. (4.84) gives $\phi_0 = 54.31°$.

(2) For aramid–epoxy composite, $E_1 = 95$ GPa, $E_2 = 5.1$ GPa, $G_{12} = 1.8$ GPa, $\nu_{12} = 0.018$, $\nu_{21} = 0.34$

$$\frac{E_1}{2(1+\nu_{21})} = 36.45\,\text{GPa}, \qquad \frac{E_2}{2(1+\nu_{12})} = 2.5\,\text{GPa, and } \phi_0 = 61.45°$$

(3) For carbon–epoxy composite with $E_1 = 140$ GPa, $E_2 = 11$ GPa, $G_{12} = 5.5$ GPa, $\nu_{12} = 0.021$, $\nu_{21} = 0.27$, we have

$$\frac{E_1}{2(1+\nu_{21})} = 55.12\,\text{GPa}, \qquad \frac{E_2}{2(1+\nu_{12})} = 5.39\,\text{GPa}$$

As can be seen, the conditions in Eqs. (4.91) and (4.94) are not satisfied, and angle ϕ_0 does not exist for this material.

As can be directly checked with the aid of Eqs. (4.76), there exists the following relationship between the elastic constants of anisotropic materials (Verchery and Gong, 1999)

$$\frac{d}{d\phi}\left(\frac{1}{E_x}\right) = -2\frac{\eta_{x,\,xy}}{G_{xy}}$$

This equation means that $\eta_{x,\,xy} = 0$ for materials whose modulus E_x reaches the extremum value in the interval $0 < \phi < 90°$. The dependencies of E_x/E_1 on ϕ for the materials considered above as examples, are shown in Fig. 4.25.

As can be seen, curves 1 and 2 corresponding to glass and aramid composites reach the minimum value at $\phi_0 = 54.31°$ and $\phi_0 = 61.45°$, respectively, whereas curve 3 for carbon composite does not have a minimum at $0 < \phi < 90°$.

The dependence $E_x(\phi)$ with the minimum value of E_x reached at $\phi = \phi_0$, where $0 < \phi < 90°$, is typical for composites reinforced in two orthogonal directions. For example, for a fabric composite having $E_1 = E_2$ and $\nu_{12} = \nu_{21}$, Eq. (4.84) yields the well-known result $\phi_0 = 45°$. For a typical fiberglass fabric composite with $E_1 = 26$ GPa, $E_2 = 22$ GPa, $G_{12} = 7.2$ GPa, $\nu_{12} = 0.11$, $\nu_{21} = 0.13$, we have

$$\frac{E_1}{2(1+\nu_{21})} = 11.5 \text{ GPa}, \quad \frac{E_2}{2(1+\nu_{12})} = 9.9 \text{ GPa, and } \phi_0 = 49.13°$$

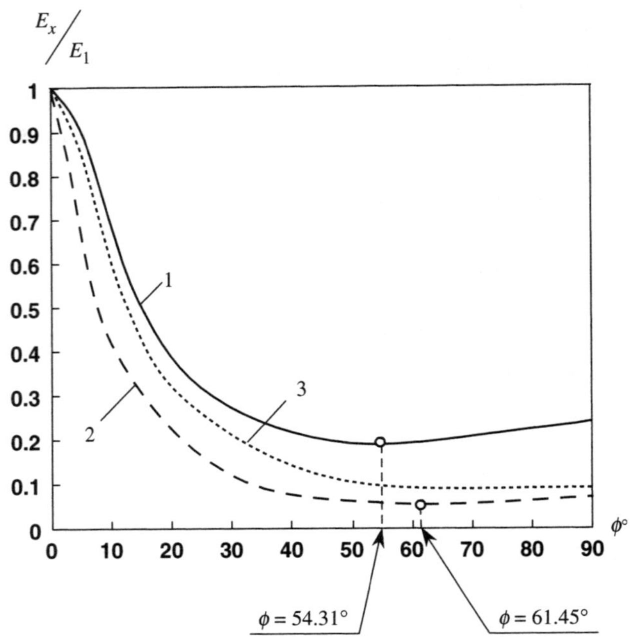

Fig. 4.25. Dependencies of E_x/E_1 on ϕ for fiberglass (1), aramid (2) and carbon (3) epoxy composites.

In conclusion, it should be noted that the actual application of Eq. (4.78) is hindered by the fact that the angle ϕ_0 specified by Eq. (4.84) depends on G_{12}, which is not known and needs to be determined from Eq. (4.78). To find G_{12}, we actually need to perform several tests for several values of G_{12} in the vicinity of the expected value and the corresponding values of ϕ_0 following from Eq. (4.84) and to select the correct value of G_{12}, which satisfies in conjunction with the corresponding value of ϕ_0, both equations – Eqs. (4.78) and (4.84) (Morozov and Vasiliev, 2003).

Consider the general case of an off-axis test (see Fig. 4.22) for a composite specimen with an arbitrary fiber orientation angle ϕ (see Fig. 4.26). To describe this test, we need to study the coupled problem for an anisotropic strip in which shear is induced by tension but is restricted at the strip ends by the jaws of a test frame as in Figs. 4.22 and 4.26. As follows from Fig. 4.26, the action of the grip can be simulated if we apply a bending moment M and a transverse force V such that the rotation of the strip ends (γ in Fig. 4.23) will become zero. As a result, bending normal and shear stresses appear in the strip that can be analyzed with the aid of composite beam theory (Vasiliev, 1993).

To derive the corresponding equations, introduce the conventional assumptions of beam theory according to which axial, u_x, and transverse, u_y, displacements can be presented as

$$u_x = u(x) + y\theta, \quad u_y = v(x)$$

where u and θ are the axial displacement and the angle of rotation of the strip cross section x = constant and v is the strip deflection in the xy-plane (see Fig. 4.26). The strains corresponding to these displacements follow from Eqs. (2.22), i.e.,

$$\begin{aligned} \varepsilon_x &= \frac{\partial u_x}{\partial x} = u' + y\theta' = \varepsilon + y\theta' \\ \gamma_{xy} &= \frac{\partial u_x}{\partial y} + \frac{\partial u_y}{\partial x} = \theta + v' \end{aligned} \tag{4.95}$$

where $(\ )' = d(\ )/dx$ and ε is the elongation of the strip axis. These strains are related to stresses by Eqs. (4.75) which reduce to

$$\begin{aligned} \varepsilon_x &= \frac{\sigma_x}{E_x} + \eta_{x,\,xy}\frac{\tau_{xy}}{G_{xy}} \\ \gamma_{xy} &= \frac{\tau_{xy}}{G_{xy}} + \eta_{xy,\,x}\frac{\sigma_x}{E_x} \end{aligned} \tag{4.96}$$

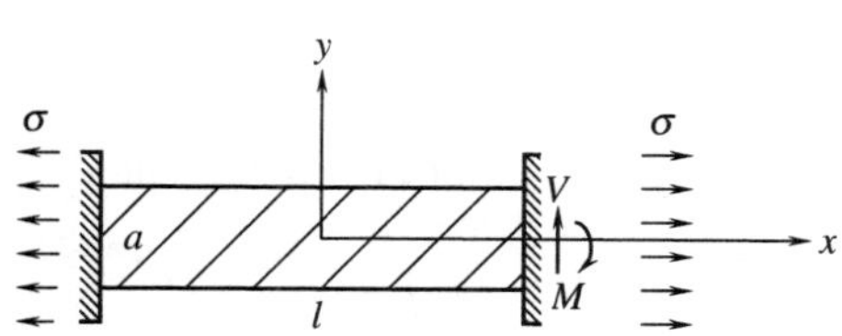

Fig. 4.26. Off-axis tension of a strip fixed at the ends.

The inverse form of these equations is

$$\sigma_x = B_{11}\varepsilon_x + B_{14}\gamma_{xy}, \quad \tau_{xy} = B_{41}\varepsilon_x + B_{44}\gamma_{xy} \tag{4.97}$$

where

$$\begin{aligned} B_{11} &= \frac{E_x}{1-\eta_{x,\,xy}\eta_{xy,x}}, \quad B_{44} = \frac{G_{xy}}{1-\eta_{x,\,xy}\eta_{xy,\,x}} \\ B_{14} &= B_{41} = -\frac{E_x\eta_{x,\,xy}}{1-\eta_{x,\,xy}\eta_{xy,\,x}} = -\frac{G_{xy}\eta_{xy,\,x}}{1-\eta_{x,\,xy}\eta_{xy,\,x}} \end{aligned} \tag{4.98}$$

Now, decompose the strip displacements, strains, and stresses into two components corresponding to

(1) free tension (see Fig. 4.23), and

(2) bending.

For free tension, we have $\tau_{xy} = 0$ and $v = 0$. So, Eqs. (4.95) and (4.96) yield

$$\begin{aligned} \varepsilon_x^{(1)} &= \varepsilon_1 + y\theta_1', \quad \gamma_{xy}^{(1)} = \theta_1 \\ \varepsilon_x^{(1)} &= \frac{\sigma_x^{(1)}}{E_x}, \quad \gamma_{xy}^{(1)} = \eta_{xy,\,x}\frac{\sigma_x^{(1)}}{E_x} \end{aligned} \tag{4.99}$$

Here, $\varepsilon_1 = u_1'$ and $\sigma_x^{(1)} = \sigma = F/ah$, where F is the axial force applied to the strip, a the strip width, and h is its thickness. Since $\sigma_x^{(1)} =$ constant, Eqs. (4.99) give

$$\theta_1 = \eta_{xy,\,x}\frac{\sigma}{E_x} = \text{constant}, \quad \varepsilon_x^1 = \varepsilon_1 = \frac{\sigma}{E_x} = \frac{F}{ah} \tag{4.100}$$

Adding components corresponding to bending (with index 2), we can write the total displacements and strains as

$$\begin{aligned} u_x &= u_1 + u_2 + y(\theta_1 + \theta_2), \quad u_y = v_2 \\ \varepsilon_x &= \varepsilon_1 + \varepsilon_2 + y\theta_2', \quad \gamma_{xy} = \theta_1 + \theta_2 + v_2' \end{aligned}$$

The total stresses can be expressed with the aid of Eqs. (4.97), i.e.,

$$\begin{aligned} \sigma_x &= B_{11}\left(\varepsilon_1 + \varepsilon_2 + y\theta_2'\right) + B_{14}\left(\theta_1 + \theta_2 + v_2'\right) \\ \tau_{xy} &= B_{41}\left(\varepsilon_1 + \varepsilon_2 + y\theta_2'\right) + B_{44}\left(\theta_1 + \theta_2 + v_2'\right) \end{aligned}$$

Transforming these equations with the aid of Eqs. (4.98) and (4.100), we arrive at

$$\begin{aligned} \sigma_x &= \sigma + B_{11}\left(\varepsilon_2 + y\theta_2'\right) + B_{14}\left(\theta_2 + v_2'\right) \\ \tau_{xy} &= B_{41}\left(\varepsilon_2 + y\theta_2'\right) + B_{44}(\theta_2 + v_2') \end{aligned} \tag{4.101}$$

These stresses are statically equivalent to the axial force P, the bending moment M, and the transverse force V, which can be introduced as

$$P = h\int_{-a/2}^{a/2} \sigma_x \mathrm{d}y, \quad M = h\int_{-a/2}^{a/2} \sigma_x y \mathrm{d}y, \quad V = h\int_{-a/2}^{a/2} \tau_{xy} \mathrm{d}y$$

Substitution of Eqs. (4.101) and integration yields

$$P = ah\left[\sigma + B_{11}\varepsilon_2 + B_{14}\left(\theta_2 + v_2'\right)\right], \tag{4.102}$$

$$M = B_{11}h\frac{a^3}{12}\theta_2' \tag{4.103}$$

$$V = ah\left[B_{41}\varepsilon_2 + B_{44}\left(\theta_2 + v_2'\right)\right] \tag{4.104}$$

These forces and moments should satisfy the equilibrium equation that follows from Fig. 4.27, i.e.,

$$P' = 0, \quad V' = 0, \quad M' = V \tag{4.105}$$

Solution of the first equation is $P = F = \sigma ah$. Then, Eq. (4.102) gives

$$\varepsilon_2 = -\frac{B_{14}}{B_{11}}\left(\theta_2 + v_2'\right) \tag{4.106}$$

The second equation of Eqs. (4.105) shows that $V = C_1$, where C_1 is a constant of integration. Then, substituting this result into Eq. (4.104) and eliminating ε_2 with the aid of Eq. (4.106), we have

$$\theta_2 + v_2' = \frac{C_1}{ah\overline{B}_{44}} \tag{4.107}$$

where $\overline{B}_{44} = B_{44} - B_{14}B_{41}$.

Taking in the third equation of Eqs. (4.105) $V = C_1$ and substituting M from Eq. (4.103), we arrive at the following equation for θ_2

$$\theta_2'' = \frac{12C_1}{a^3hB_{11}}$$

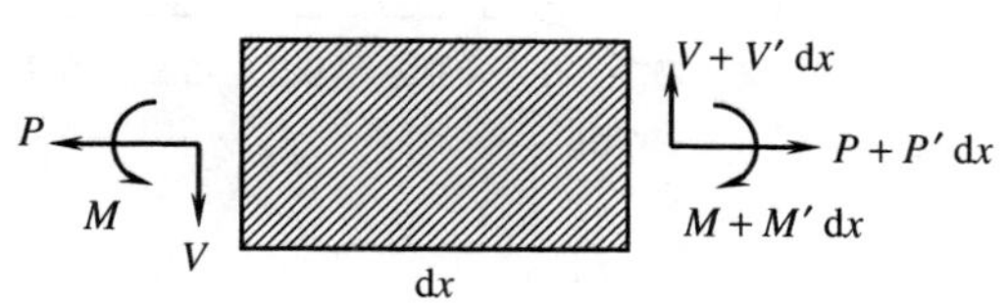

Fig. 4.27. Forces and moments acting on the strip element.

Integration yields

$$\theta_2 = \frac{6C_1}{a^3 h B_{11}} + C_2 x + C_3$$

The total angle of rotation $\theta = \theta_1 + \theta_2$, where θ_1 is specified by Eqs. (4.100), should be zero at the ends of the strip, i.e., $\theta(x = \pm l/2) = 0$. Satisfying these conditions, we have

$$\theta_2 = \frac{3C_1}{a^3 h B_{11}}\left(2x^2 - \frac{l^2}{2}\right) - \eta_{xy,\,x}\frac{\sigma}{E_x} \tag{4.108}$$

Substitution into Eq. (4.107) and integration allows us to find the deflection

$$v_2 = \frac{C_1 x}{ah\overline{B}_{44}} - \frac{3C_1 x}{a^3 h B_{11}}\left(\frac{2x^2}{3} - \frac{l^2}{2}\right) + \eta_{xy,\,x}\frac{\sigma_0 x}{E_x} + C_4 \tag{4.109}$$

This expression includes two constants, C_1 and C_4, which can be determined from the boundary conditions $v_2(x = \pm l/2) = 0$. The final result, following from Eqs. (4.100), (4.108), and (4.109), is

$$v = l\eta_{xy,\,x}\frac{\sigma\overline{x}}{E_x}\left[1 - \frac{B_{11} + \overline{l}^2\overline{B}_{44}\left(3/2 - 2\overline{x}^2\right)}{B_{11} + \overline{l}^2\overline{B}_{44}}\right]$$

$$\theta = \eta_{xy,\,x}\frac{3\sigma\overline{l}^2}{E_x}\frac{\overline{B}_{44}\left(2\overline{x}^2 - 1/2\right)}{B_{11} + \overline{l}^2\overline{B}_{44}} \tag{4.110}$$

where $\overline{l} = l/a$ and $\overline{x} = x/l$. The deflection of a carbon–epoxy strip having $\phi = 45°$ and $\overline{l} = 10$ is shown in Fig. 4.28.

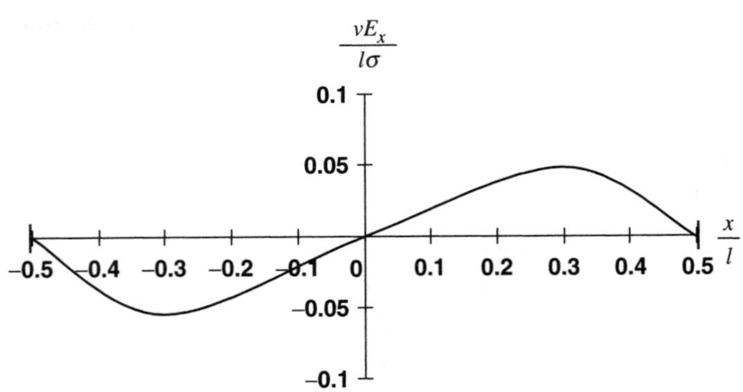

Fig. 4.28. Normalized deflection of a carbon–epoxy strip ($\phi = 45°$, $\overline{l} = 10$).

Now, we can write the relationship between modulus E_x corresponding to the ideal test shown in Fig. 4.23 and apparent modulus E_x^{a} that can be found from the real test shown in Figs. 4.22 and 4.26. Using Eqs. (4.98), (4.100), (4.106), and (4.110), we finally get

$$\sigma = E_x^{\mathrm{a}}\varepsilon$$

where

$$E_x^{\mathrm{a}} = \frac{E_x}{1 - \left\{E_x\eta_{x,\,xy}\eta_{xy,\,x} \big/ \left[E_x + \bar{l}^2 G_{xy}(1 - \eta_{x,\,xy}\eta_{xy,\,x})\right]\right\}}$$

Consider two limiting cases. For an infinitely long strip ($\bar{l} \to \infty$), we have $E_x^{\mathrm{a}} = E_x$. This result corresponds to the case of free shear deformation specified by Eqs. (4.77). For an infinitely short strip ($\bar{l} \to 0$), taking into account Eqs. (4.98), we get

$$E_x^{\mathrm{a}} = \frac{E_x}{1 - \eta_{x,\,xy}\eta_{xy,\,x}} = B_{11}$$

In accordance with Eq. (4.97), this result corresponds to a restricted shear deformation ($\gamma_{xy} = 0$). For a strip with finite length, $E_x < E_x^{\mathrm{a}} < B_{11}$. The dependence of the normalized apparent modulus on the length-to-width ratio for a 45° carbon–epoxy layer is shown in Fig. 4.29. As can be seen, the difference between E_x^{a} and E_x becomes less than 5% for $l > 3a$.

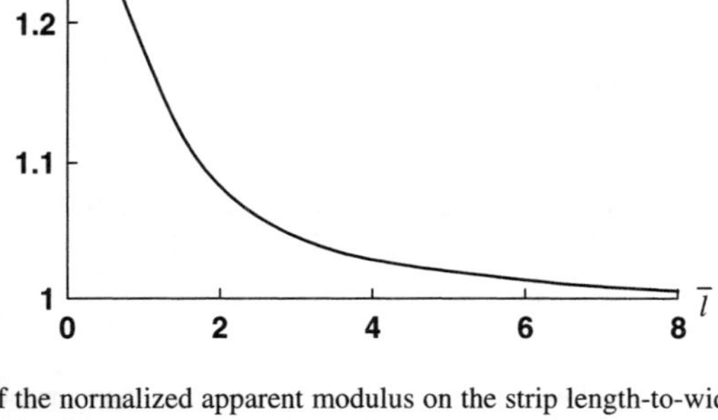

Fig. 4.29. Dependence of the normalized apparent modulus on the strip length-to-width ratio for a 45° carbon–epoxy layer.

4.3.2. Nonlinear models

Nonlinear deformation of an anisotropic unidirectional layer can be studied rather straightforwardly because stresses σ_1, σ_2, and τ_{12} in the principal material coordinates (see Fig. 4.18) are statically determinate and can be found using Eqs. (4.67). Substituting these stresses into the nonlinear constitutive equations, Eqs. (4.60) or Eqs. (4.64), we can express strains ε_1, ε_2, and γ_{12} in terms of stresses σ_x, σ_y, and τ_{xy}. Further substitution of the strains ε_1, ε_2, and γ_{12} into Eqs. (4.70) yields constitutive equations that link strains ε_x, ε_y, and γ_{xy} with stresses σ_x, σ_y, and τ_{xy} thus allowing us to find strains in the global coordinates x, y, and z if we know the corresponding stresses.

As an example of the application of a nonlinear elastic material model described by Eqs. (4.60), consider a two-matrix fiberglass composite (see Section 4.4.3) whose stress–strain curves in the principal material coordinates are presented in Fig. 4.16. These curves allowed us to determine coefficients 'b' and 'c' in Eqs. (4.60). To find the coupling coefficients 'm,' we use a 45° off-axis test. Experimental results (circles) and the corresponding approximation (solid line) are shown in Fig. 4.30. Thus, the constructed model can be used now to predict material behavior under tension at any other (different from 0, 45, and 90°) angle (the corresponding results are given in Fig. 4.31 for 60°) or to study more complicated material structures and loading cases (see Section 4.5).

As an example of the application of the elastic–plastic material model specified by Eq. (4.64), consider a boron–aluminum composite whose stress–strain diagrams in principal material coordinates are shown in Fig. 4.17. Theoretical and experimental curves (Herakovich, 1998) for 30 and 45° off-axis tension of this material are presented in Fig. 4.32.

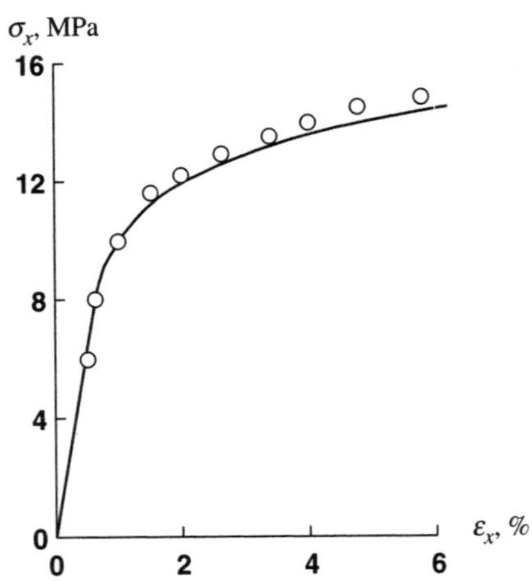

Fig. 4.30. Calculated (solid line) and experimental (circles) stress–strain diagram for 45° off-axis tension of a two-matrix unidirectional composite.

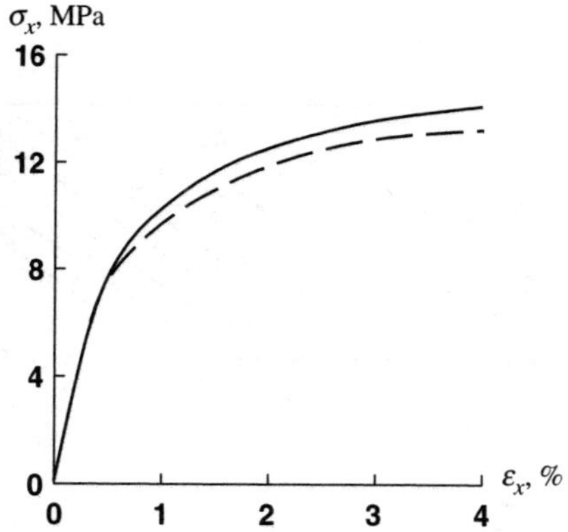

Fig. 4.31. Theoretical (solid line) and experimental (dashed line) stress–strain diagrams for 60° off-axis tension of a two-matrix unidirectional composite.

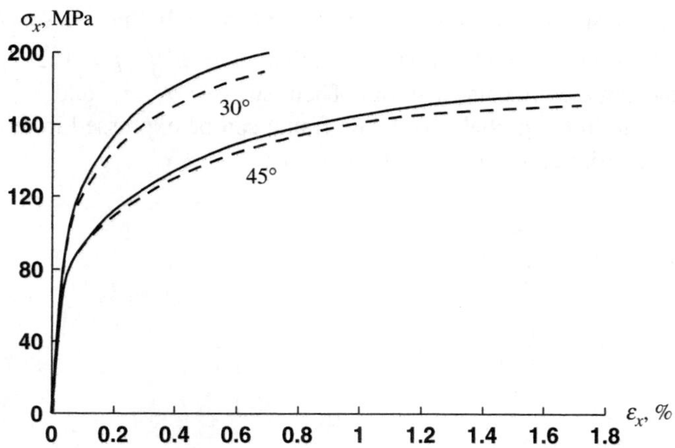

Fig. 4.32. Theoretical (solid lines) and experimental (dashed lines) stress–strain diagrams for 30° and 45° off-axis tension of a boron–aluminum composite.

4.4. Orthogonally reinforced orthotropic layer

The simplest layer reinforced in two directions is the so-called cross-ply layer that consists of alternating plies with 0 and 90° orientations with respect to the global coordinate frame x, y, and z as in Fig. 4.33. Actually, this is a laminated structure, but being formed with a number of plies, it can be treated as a homogeneous orthotropic layer (see Section 5.4.2).

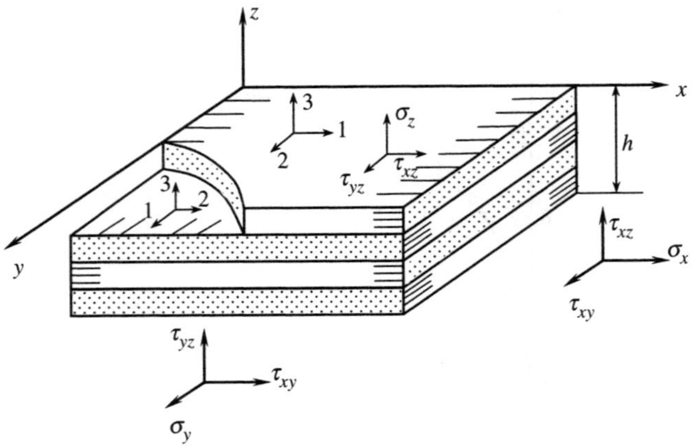

Fig. 4.33. A cross-ply layer.

4.4.1. Linear elastic model

Let the layer consist of m longitudinal (0°) plies with thicknesses $h_0^{(i)}$ ($i = 1, 2, 3, \ldots, m$) and n transverse (90°) plies with thicknesses $h_{90}^{(j)}$ ($j = 1, 2, 3, \ldots, n$) made from one and the same composite material. Then, stresses σ_x, σ_y, and τ_{xy} that comprise the plane stress state in the global coordinate frame can be expressed in terms of stresses in the principal material coordinates of the plies as

$$\sigma_x h = \sum_{i=1}^{m} \sigma_1^{(i)} h_0^{(i)} + \sum_{j=1}^{n} \sigma_2^{(j)} h_{90}^{(j)}$$

$$\sigma_y h = \sum_{i=1}^{m} \sigma_2^{(i)} h_0^{(i)} + \sum_{j=1}^{n} \sigma_1^{(j)} h_{90}^{(j)} \qquad (4.111)$$

$$\tau_{xy} h = \sum_{i=1}^{m} \tau_{12}^{(i)} h_0^{(i)} + \sum_{j=1}^{n} \tau_{12}^{(j)} h_{90}^{(j)}$$

Here, h is the total thickness of the layer (see Fig. 4.33), i.e.,

$$h = h_0 + h_{90}$$

where

$$h_0 = \sum_{i=1}^{m} h_0^{(i)}, \quad h_{90} = \sum_{j=1}^{n} h_{90}^{(j)}$$

are the total thicknesses of the longitudinal and transverse plies.

The stresses in the principal material coordinates of the plies are related to the corresponding strains by Eqs. (3.59) or Eqs. (4.56)

$$
\begin{aligned}
\sigma_1^{(i,j)} &= \overline{E}_1\left(\varepsilon_1^{(i,j)} + \nu_{12}\varepsilon_2^{(i,j)}\right) \\
\sigma_2^{(i,j)} &= \overline{E}_2\left(\varepsilon_2^{(i,j)} + \nu_{21}\varepsilon_1^{(i,j)}\right) \\
\tau_{12}^{(i,j)} &= G_{12}\gamma_{12}^{(i,j)}
\end{aligned}
\tag{4.112}
$$

in which, as earlier $\overline{E}_{1,2} = E_{1,2}/(1-\nu_{12}\nu_{21})$ and $E_1\nu_{12} = E_2\nu_{21}$. Now assume that the deformation of all the plies is the same as that of the deformation of the whole layer, i.e., that

$$
\varepsilon_1^{(i)} = \varepsilon_2^{(j)} = \varepsilon_x, \quad \varepsilon_2^{(i)} = \varepsilon_1^{(j)} = \varepsilon_y, \quad \gamma_{12}^{(i)} = \gamma_{12}^{(j)} = \gamma_{xy}
$$

Then, substituting the stresses, Eqs. (4.112), into Eqs. (4.111), we arrive at the following constitutive equations

$$
\begin{aligned}
\sigma_x &= A_{11}\varepsilon_x + A_{12}\varepsilon_y \\
\sigma_y &= A_{21}\varepsilon_x + A_{22}\varepsilon_y \\
\tau_{xy} &= A_{44}\gamma_{xy}
\end{aligned}
\tag{4.113}
$$

in which the stiffness coefficients are

$$
\begin{aligned}
&A_{11} = \overline{E}_1\overline{h}_0 + \overline{E}_2\overline{h}_{90}, \quad A_{22} = \overline{E}_1\overline{h}_{90} + \overline{E}_2\overline{h}_0 \\
&A_{12} = A_{21} = \overline{E}_1\nu_{12} = \overline{E}_2\nu_{21}, \quad A_{44} = G_{12}
\end{aligned}
\tag{4.114}
$$

and

$$
\overline{h}_0 = \frac{h_0}{h}, \quad \overline{h}_{90} = \frac{h_{90}}{h}
$$

The inverse form of Eqs. (4.113) is

$$
\varepsilon_x = \frac{\sigma_x}{E_x} - \nu_{xy}\frac{\sigma_y}{E_y}, \quad \varepsilon_y = \frac{\sigma_y}{E_y} - \nu_{yx}\frac{\sigma_x}{E_x}, \quad \gamma_{xy} = \frac{\tau_{xy}}{G_{xy}}
\tag{4.115}
$$

where

$$
\begin{aligned}
&E_x = A_{11} - \frac{A_{12}^2}{A_{22}}, \quad E_y = A_{22} - \frac{A_{12}^2}{A_{11}}, \quad G_{xy} = A_{44} \\
&\nu_{xy} = \frac{A_{12}}{A_{11}}, \quad \nu_{yx} = \frac{A_{12}}{A_{22}}
\end{aligned}
\tag{4.116}
$$

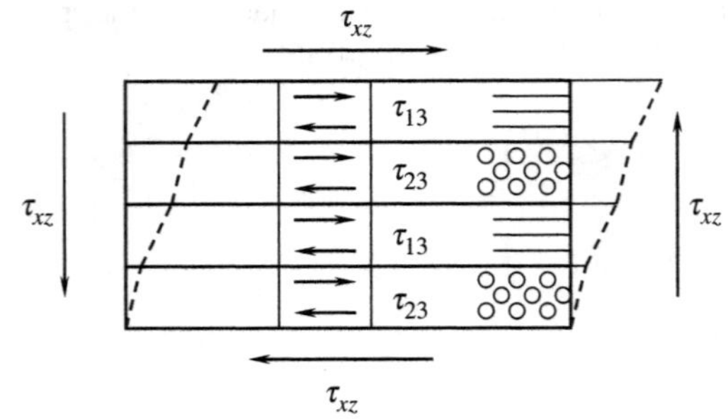

Fig. 4.34. Pure transverse shear of a cross-ply layer.

To determine the transverse shear moduli G_{xz} and G_{yz}, consider, e.g., pure shear in the xz-plane (see Fig. 4.34). It follows from the equilibrium conditions for the plies that

$$\tau_{13}^{(i)} = \tau_{23}^{(j)} = \tau_{xz}, \quad \tau_{23}^{(i)} = \tau_{13}^{(j)} = \tau_{yz} \tag{4.117}$$

The total shear strains can be found as

$$\begin{aligned}\gamma_{xz} &= \frac{1}{h}\left(\sum_{i=1}^{m}\gamma_{13}^{(i)}h_0 + \sum_{j=1}^{n}\gamma_{23}^{(j)}h_{90}\right)\\ \gamma_{yz} &= \frac{1}{h}\left(\sum_{i=1}^{m}\gamma_{23}^{(i)}h_0 + \sum_{j=1}^{n}\gamma_{13}^{(j)}h_{90}\right)\end{aligned} \tag{4.118}$$

where in accordance with Eqs. (4.56)

$$\gamma_{13}^{(i,\,j)} = \frac{\tau_{13}^{(i,\,j)}}{G_{13}}, \quad \gamma_{23}^{(i,\,j)} = \frac{\tau_{23}^{(i,\,j)}}{G_{23}} \tag{4.119}$$

Substituting Eqs. (4.119) into Eqs. (4.118) and using Eqs. (4.117), we arrive at

$$\gamma_{xz} = \frac{\tau_{xz}}{G_{xz}}, \quad \gamma_{yz} = \frac{\tau_{yz}}{G_{yz}}$$

where

$$\frac{1}{G_{xz}} = \frac{\bar{h}_0}{G_{13}} + \frac{\bar{h}_{90}}{G_{23}}, \quad \frac{1}{G_{yz}} = \frac{\bar{h}_0}{G_{23}} + \frac{\bar{h}_{90}}{G_{13}}$$

4.4.2. Nonlinear models

The nonlinear behavior of a cross-ply layer associated with nonlinear material response under loading in the principal material coordinates (see, e.g., Figs. 4.16 and 4.17) can be described using nonlinear constitutive equations, Eqs. (4.60) or Eqs. (4.64) instead of linear equations (4.113).

However, this layer can demonstrate nonlinearity which is entirely different from that studied in the previous sections. This nonlinearity is observed in the cross-ply layer composed of linear elastic plies and is caused by microcracking of the matrix.

To study this phenomenon, consider a cross-ply laminate consisting of three plies as in Fig. 4.35. Equilibrium conditions yield the following equations

$$
\begin{aligned}
&2\left(\sigma_{x1}\overline{h}_1 + \sigma_{x2}\overline{h}_2\right) = \sigma \\
&2\left(\sigma_{y1}\overline{h}_1 + \sigma_{y2}\overline{h}_2\right) = 0
\end{aligned}
\tag{4.120}
$$

in which

$$\overline{h}_1 = h_1/h, \quad \overline{h}_2 = h_2/h, \quad h = 2(h_1 + h_2)$$

The constitutive equations are

$$
\begin{aligned}
&\sigma_{x1,2} = \overline{E}_{1,2}(\varepsilon_x + \nu_{12,21}\varepsilon_y) \\
&\sigma_{y1,2} = \overline{E}_{2,1}(\varepsilon_y + \nu_{21,12}\varepsilon_x)
\end{aligned}
\tag{4.121}
$$

in which $\overline{E}_{1,2} = E_{1,2}/(1 - \nu_{12}\nu_{21})$. We assume that strains ε_x and ε_y do not change through the laminate thickness. Substituting Eqs. (4.121) in Eqs. (4.120), we can find strains and then stresses using again Eqs. (4.121). The final result is

$$\sigma_{x1,2} = \frac{\sigma E_{1,2}\left[E_2\overline{h}_1 + E_1\overline{h}_2 - E_{1,2}\nu_{12,21}^2(\overline{h}_1 + \overline{h}_2)\right]}{2\left[(E_1\overline{h}_1 + E_2\overline{h}_2)(E_2\overline{h}_1 + E_1\overline{h}_2) - E_1^2\nu_{12}^2(h_1 + h_2)^2\right]}$$

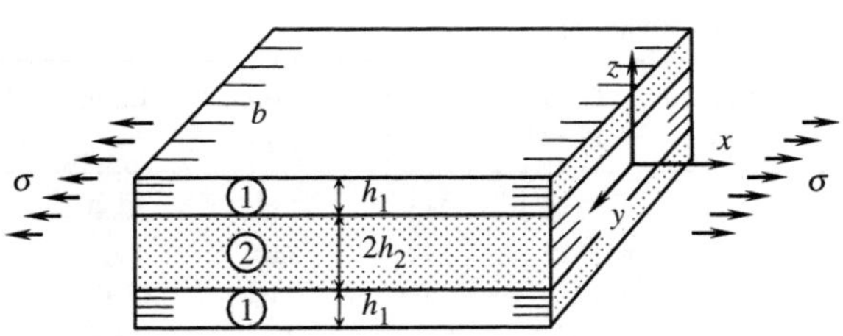

Fig. 4.35. Tension of a cross-ply laminate.

To simplify the analysis, neglect Poisson's effect, i.e., taking $\nu_{12} = \nu_{21} = 0$. Then

$$\sigma_{x1} = \sigma_1^0 = \frac{\sigma E_1}{2(E_1\bar{h}_1 + E_2\bar{h}_2)}, \quad \sigma_{x2} = \sigma_2^0 = \frac{\sigma E_2}{2(E_1\bar{h}_1 + E_2\bar{h}_2)} \tag{4.122}$$

Consider, for example, the case $\bar{h}_1 = \bar{h}_2 = 0.5$ and find the ultimate stresses corresponding to the failure of longitudinal plies or to the failure of the transverse ply. Putting $\sigma_1^0 = \bar{\sigma}_1^+$ and $\sigma_2^0 = \bar{\sigma}_2^+$, we get

$$\bar{\sigma}_x^{(1)} = \bar{\sigma}_1^+ \left(1 + \frac{E_2}{E_1}\right), \quad \bar{\sigma}_x^{(2)} = \bar{\sigma}_2^+ \left(1 + \frac{E_1}{E_2}\right)$$

The results of calculation for the composites listed in Table 3.5 are presented in Table 4.2.

As can be seen, $\bar{\sigma}_x^{(1)} \gg \bar{\sigma}_x^{(2)}$. This means that the first failure occurs in the transverse ply under stress

$$\sigma = \bar{\sigma} = 2\bar{\sigma}_2^+ \left(\bar{h}_2 + \frac{E_1}{E_2}\bar{h}_1\right) \tag{4.123}$$

This stress does not cause failure of the whole laminate because the longitudinal plies can carry the load, but the material behavior becomes nonlinear. Actually, the effect under consideration is the result of the difference between the ultimate elongations of the unidirectional plies along and across the fibers. From the data presented in Table 4.2 we can see that for all the materials listed in this table $\bar{\varepsilon}_1 \gg \bar{\varepsilon}_2$. As a result, transverse plies drawn under tension by longitudinal plies that have much higher stiffness and elongation fail because their ultimate elongation is smaller. This failure is accompanied with a system of cracks parallel to the fibers which can be observed not only in cross-ply layers but also in many other laminates that include unidirectional plies experiencing transverse tension caused by interaction with the adjacent plies (see Fig. 4.36).

Now assume that the acting stress $\sigma \geq \bar{\sigma}$, where $\bar{\sigma}$ is specified by Eq. (4.123) and corresponds to the load causing the first crack in the transverse ply as in Fig. 4.37. To study

Table 4.2
Ultimate stresses causing the failure of longitudinal $\left(\bar{\sigma}_x^{(1)}\right)$ or transverse $\left(\bar{\sigma}_x^{(2)}\right)$ plies and deformation characteristics of typical advanced composites.

σ(MPa); ε (%)	Glass–epoxy	Carbon–epoxy	Carbon–PEEC	Aramid–epoxy	Boron–epoxy	Boron–Al	Carbon–Carbon	Al_2O_3–Al
$\bar{\sigma}_x^{(1)}$	2190	2160	2250	2630	1420	2000	890	1100
$\bar{\sigma}_x^{(2)}$	225	690	1125	590	840	400	100	520
$\bar{\varepsilon}_1$	3	1.43	1.5	2.63	0.62	0.50	0.47	0.27
$\bar{\varepsilon}_2$	0.31	0.45	0.75	0.2	0.37	0.1	0.05	0.13
$\bar{\varepsilon}_1/\bar{\varepsilon}_2$	9.7	3.2	2	13.1	1.68	5	9.4	2.1

Fig. 4.36. Cracks in the circumferential layer of the failed pressure vessel induced by transverse (for the vessel, axial) tension of the layer.

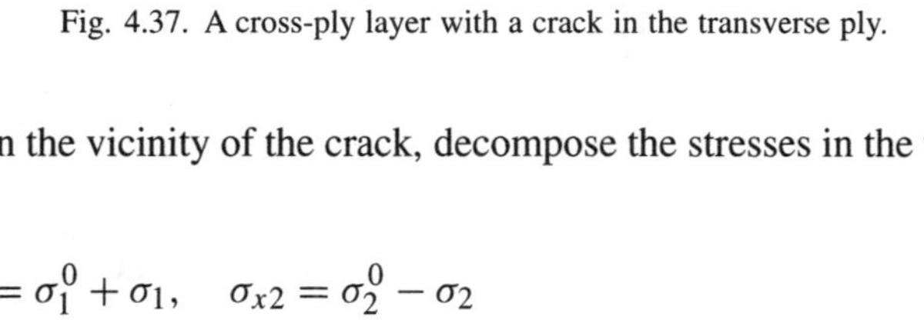

Fig. 4.37. A cross-ply layer with a crack in the transverse ply.

the stress state in the vicinity of the crack, decompose the stresses in the three plies shown in Fig. 4.37 as

$$\sigma_{x1} = \sigma_{x3} = \sigma_1^0 + \sigma_1, \quad \sigma_{x2} = \sigma_2^0 - \sigma_2 \tag{4.124}$$

and assume that the crack induces also transverse through-the-thickness shear and normal stresses

$$\tau_{xzi} = \tau_i, \quad \sigma_{zi} = s_i, \quad i = 1, 2, 3 \tag{4.125}$$

The stresses σ_1^0 and σ_2^0 in Eqs. (4.124) are specified by Eqs. (4.122) with $\sigma = \overline{\sigma}$, corresponding to the acting stress under which the first crack appears in the transverse ply. Stresses σ_1 and σ_2 should be self-balanced, i.e.,

$$\sigma_1 \overline{h}_1 = \sigma_2 \overline{h}_2 \tag{4.126}$$

The total stresses in Eqs. (4.124) and (4.125) should satisfy equilibrium equations, Eq. (2.5), which yield for the problem under study

$$\frac{\partial \sigma_{xi}}{\partial x} + \frac{\partial \tau_{xzi}}{\partial z} = 0, \quad \frac{\partial \sigma_{zi}}{\partial z} + \frac{\partial \tau_{xzi}}{\partial x} = 0 \tag{4.127}$$

where $i = 1, 2, 3$.

To simplify the problem, suppose that the additional stresses σ_1 and σ_2 do not depend on z, i.e., that they are uniformly distributed through the thickness of longitudinal plies. Then, Eqs. (4.127), upon substitution of Eqs. (4.124) and (4.125), can be integrated with respect to z. The resulting stresses should satisfy the following boundary and interface conditions (see Fig. 4.37)

$$\tau_1(z = h_1 + h_2) = 0 \qquad s_1(z = h_1 + h_2) = 0$$

$$\tau_1(z = h_2) = \tau_2(z = h_2) \qquad s_1(z = h_2) = s_2(z = h_2)$$

$$\tau_2(z = -h_2) = \tau_3(z = -h_2) \qquad s_2(z = -h_2) = s_3(z = -h_2)$$

$$\tau_3(z = -h_1 - h_2) = 0 \qquad s_3(z = -h_1 - h_2) = 0$$

Finally, using Eq. (4.126) to express σ_1 in terms of σ_2, we arrive at the following stress distribution (Vasiliev et al., 1970)

$$\begin{gathered}
\sigma_{x1} = \sigma_{x3} = \sigma_1^0 + \sigma_2(x)\frac{h_2}{h_1}, \quad \sigma_{x2} = \sigma_2^0 - \sigma_2(x) \\
\tau_1 = -\frac{h_2}{h_1}\sigma_2'(x)z_1, \quad \tau_2 = \sigma_2'(x)z, \quad \tau_3 = -\frac{h_2}{h_1}\sigma_2'(x)z_2 \\
s_1 = \frac{h_2}{2h_1}\sigma_2''(x)z_1^2, \quad s_2 = -\frac{1}{2}\sigma_2''(x)\left(z^2 - h_1h_2 - h_2^2\right), \quad s_3 = \frac{h_2}{2h_1}\sigma_2''(x)z_2^2
\end{gathered} \tag{4.128}$$

where

$$z_1 = z - h_1 - h_2, \quad z_2 = z + h_1 + h_2, \text{ and } (\ )' = \mathrm{d}(\ )/\mathrm{d}x$$

Thus, we need to find only one unknown function: $\sigma_2(x)$. To do this, we can use the principle of minimum strain energy (see Section 2.11.2) according to which the function $\sigma_2(x)$ should deliver the minimum value of

$$W_\sigma = \frac{1}{2}\int_0^l \sum_{i=1}^{3} \int_{h_i} \left(\frac{\sigma_{xi}^2}{E_{xi}} + \frac{\sigma_{zi}^2}{E_{zi}} - 2\frac{\nu_{xzi}}{E_{zi}}\sigma_{xi}\sigma_{zi} + \frac{\tau_{xzi}^2}{G_{xzi}}\right)\mathrm{d}x \tag{4.129}$$

where $E_{x1} = E_{x3} = E_1$, $E_{x2} = E_2$, $E_{zi} = E_2$, $G_{xz1} = G_{xz3} = G_{13}$, $G_{xz2} = G_{23}$, $\nu_{xz1} = \nu_{xz3} = \nu_{13}$, $\nu_{xz2} = \nu_{23}$ and $E_1, E_2, G_{13}, G_{23}, \nu_{13}, \nu_{23}$ are elastic constants of a unidirectional ply. Substituting stresses, Eqs. (4.128), into the functional in Eq. (4.129), integrating with respect to z, and using the traditional procedure of variational calculus providing $\delta W_\sigma = 0$, we arrive at the following equation for $\sigma_2(x)$

$$\frac{\mathrm{d}^4\sigma_2}{\mathrm{d}x^4} - 2a^2\frac{\mathrm{d}^2\sigma_2}{\mathrm{d}x^2} + b^4\sigma_2 = 0$$

in which

$$a^2 = \frac{1}{d}\left[\frac{h_2}{3G_{23}} + \frac{h_1}{3G_{13}} - \frac{\nu_{23}}{E_2}\left(h_1 + \frac{2}{3}h_2\right) + \frac{h_1\nu_{23}}{3E_2}\right], \quad b^4 = \frac{1}{d}\left(\frac{1}{h_1E_1} + \frac{1}{h_2E_2}\right)$$

$$d = \frac{1}{2E_2}\left[\frac{1}{5}\left(h_1^3 + h_2^3\right) - \frac{2}{3}h_2^2(h_1+h_2) + h_2(h_1+h_2)^2\right]$$

The general solution for this equation is

$$\sigma_2 = \mathrm{e}^{-k_1x}(c_1 \sin k_2x + c_2 \cos k_2x) + \mathrm{e}^{k_1x}(c_3 \sin k_2x + c_4 \cos k_2x) \tag{4.130}$$

where

$$k_1 = \sqrt{\frac{1}{2}(a^2 + b^2)}, \quad k_2 = \sqrt{\frac{1}{2}(b^2 - a^2)}$$

Suppose that the strip shown in Fig. 4.37 is infinitely long in the x-direction. Then, we should have $\sigma_1 \to 0$ and $\sigma_2 \to 0$ for $x \to \infty$ in Eqs. (4.124). This means that we should put $c_3 = c_4 = 0$ in Eq. (4.130). The other two constants, c_1 and c_2, should be determined from the conditions on the crack surface (see Fig. 4.37), i.e.,

$$\sigma_{x2}(x = 0) = 0, \quad \tau_{xz2}(x = 0) = 0$$

Satisfying these conditions, we obtain the following expressions for stresses

$$\begin{aligned}
\sigma_{x1} &= \sigma_{x3} = \sigma_1^0 + \sigma_2^0\frac{h_2}{h_1}\mathrm{e}^{-k_1x}\left(\frac{k_1}{k_2}\sin k_2x + \cos k_2x\right)\\
\sigma_{x2} &= \sigma_2^0\left[1 - \mathrm{e}^{-k_1x}\left(\frac{k_1}{k_2}\sin k_2x + \cos k_2x\right)\right]\\
\tau_{xz2} &= -\frac{\sigma_2^0}{k_2}\left(k_1^2 + k_2^2\right)z\mathrm{e}^{-k_1x}\sin k_2x\\
\sigma_{z2} &= -\frac{\sigma_2^0}{2k_2}\left(k_1^2 + k_2^2\right)[z^2 - h_2(h_1 + h_2)]\mathrm{e}^{-k_1x}(k_1 \sin k_2x - k_2 \cos k_2x)
\end{aligned} \tag{4.131}$$

As an example, consider a glass–epoxy sandwich layer with the following parameters: $h_1 = 0.365\,\mathrm{mm}$, $h_2 = 0.735\,\mathrm{mm}$, $E_1 = 56\,\mathrm{GPa}$, $E_2 = 17\,\mathrm{GPa}$,

$G_{13} = 5.6\,\text{GPa}$, $G_{23} = 6.4\,\text{GPa}$, $\nu_{13} = 0.095$, $\nu_{23} = 0.35$, $\overline{\sigma}_2^+ = 25.5\,\text{MPa}$. The distributions of stresses normalized to the acting stress σ are presented in Fig. 4.38. As can be seen, there is a stress concentration in the longitudinal plies in the vicinity of the crack, whereas the stress in the transverse ply, being zero on the crack surface, practically reaches σ_2^0 at a distance of about 4 mm (or about twice the thickness of the laminate) from the crack. The curves have the expected forms for this problem of stress diffusion. However, analysis of the second equation of Eqs. (4.131) allows us to reveal an interesting phenomenon which can be demonstrated if we increase the vertical scale of the graph in the vicinity of points A and B (see Fig. 4.38). It follows from this analysis that stress σ_{x2} becomes equal to σ_2^0 at point A with coordinate

$$x_A = \frac{1}{k_2}\left[\pi - \tan^{-1}\left(\frac{k_2}{k_1}\right)\right]$$

and reaches a maximum value at point B with coordinate $x_B = \pi/k_2$. This maximum value

$$\sigma_{x2}^{\max} = \sigma_2^0\left(1 + e^{-\pi\frac{k_1}{k_2}}\right)$$

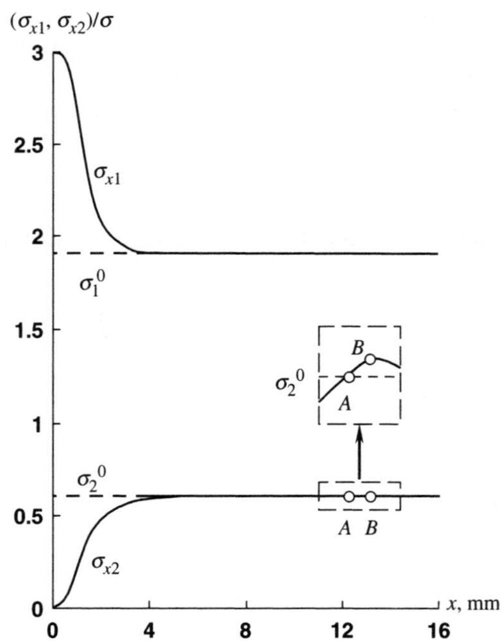

Fig. 4.38. Variation of normalized normal stresses in longitudinal (σ_{x1}) and transverse (σ_{x2}) plies with a distance from the crack.

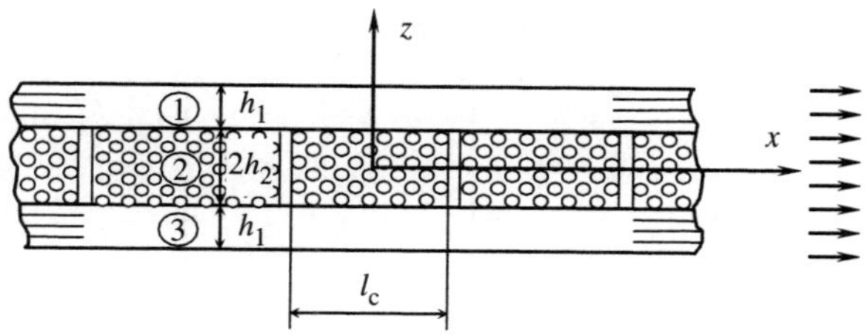

Fig. 4.39. A system of cracks in the transverse ply.

is higher than stress σ_2^0 which causes failure of the transverse ply. This means that a single crack cannot exist. When stress σ_2^0 reaches its ultimate value $\overline{\sigma}_2^+$, a regular system of cracks located at a distance of $l_c = \pi/k_2$ from one another appears in the transverse ply (see Fig. 4.39). For the example considered above, $l_c = 12.6\,\mathrm{mm}$.

To study the stress state of a layer with cracks shown in Fig. 4.39, we can use solution (4.130) but should write it in a different form, i.e.,

$$\begin{aligned}\sigma_2 &= C_1 \sinh k_1x \sin k_2x + C_2 \sinh k_1x \cos k_2x \\ &\quad + C_3 \cosh k_1x \sin k_2x + C_4 \cosh k_1x \cos k_2x \end{aligned} \tag{4.132}$$

Since the stress state of an element $-l_c/2 \leq x \leq l_c/2$ is symmetric with respect to coordinate x, we should put $C_2 = C_3 = 0$ and find constants C_1 and C_4 from the following boundary conditions

$$\sigma_{x2}(x = l_c/2) = 0, \quad \tau_{xz2}(x = l_c/2) = 0 \tag{4.133}$$

where $l_c = \pi/k_2$.

The final expressions for stresses are

$$\begin{aligned}
\sigma_{x1} &= \sigma_{x3} = \sigma_1^0 + \sigma_2^0 \frac{h_2}{h_1 c}\left(\frac{k_1}{k_2}\cosh k_1x \cos k_2x + \sinh k_1x \sin k_2x\right) \\
\sigma_{x2} &= \sigma_2^0\left[1 - \frac{1}{c}\left(\frac{k_1}{k_2}\cosh k_1x \cos k_2x + \sinh k_1x \sin k_2x\right)\right] \\
\tau_{xz2} &= \frac{\sigma_2^0}{k_2 c}\left(k_1^2 + k_2^2\right) z \sinh k_1x \cos k_2x \\
\sigma_{z2} &= -\frac{\sigma_2^0}{2k_2 c}\left(k_1^2 + k_2^2\right)\left[z^2 - h_2(h_1 + h_2)\right](k_1 \cosh k_1x \cos k_2x - k_2 \sinh k_1x \sin k_2x)
\end{aligned} \tag{4.134}$$

in which $c = \sinh(\pi k_1/2k_2)$.

For the layer considered above as an example, stress distributions corresponding to $\sigma = \overline{\sigma} = 44.7\,\mathrm{MPa}$ are shown in Figs. 4.40 and 4.41. Under further loading ($\sigma > \overline{\sigma}$), two

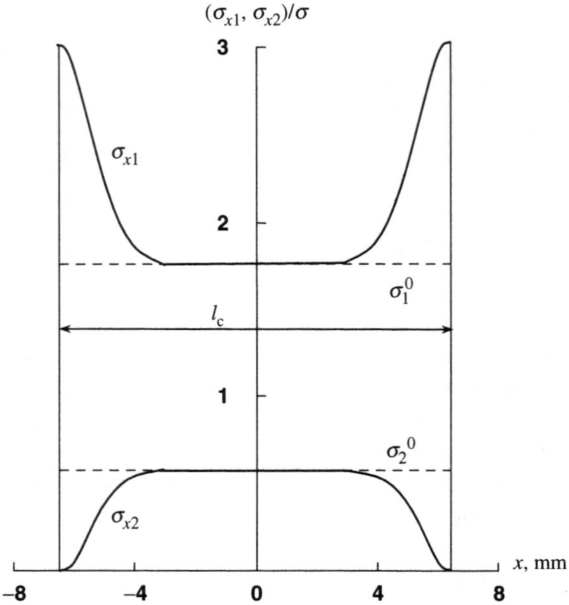

Fig. 4.40. Distribution of normalized stresses in longitudinal (σ_{x1}) and transverse (σ_{x2}) plies between the cracks.

modes of the layer failure are possible. The first one is the formation of another transverse crack separating the block with length l_c in Fig. 4.39 into two pieces. The second one is a delamination in the vicinity of the crack caused by stresses τ_{xz} and σ_z (see Fig. 4.41). Usually, the first situation takes place because stresses τ_{xz} and σ_z are considerably lower than the corresponding ultimate stresses, whereas the maximum value of σ_{x2} is close to the ultimate stress $\sigma_2^0 = \overline{\sigma}_2^+$. Indeed, the second equation of Eqs. (4.134) yields

$$\sigma_{x2}^{\max} = \sigma_{x2}(x = 0) = \sigma_2^0(1 - k)$$

where $k = k_1/(k_2 c)$. For the foregoing example, $k = 3.85 \times 10^{-4}$. So, $\sigma_{x2}^{\max}$ is so close to σ_2^0 that we can presume that under practically the same load, another crack occurs in the central cross section $x = 0$ of the central block in Fig. 4.39 (as well as in all the other blocks). Thus, the distance between the cracks becomes $l_c = \pi/2k_2$ (6.4 mm for the example under study). The corresponding stress distribution can be determined with the aid of Eqs. (4.128) and (4.132), and boundary conditions (4.133) in which we should take $l_c = \pi/2k_2$. The next crack will again appear at the block center and this process will be continued until failure of the longitudinal plies.

To plot the stress–strain diagram of the cross-ply layer with allowance for the cracks in the transverse ply, we introduce the mean longitudinal strain

$$\varepsilon_x = \frac{2}{h_2 l_c} \int_0^{l_c/2} \mathrm{d}x \int_0^{h_2} \varepsilon_{x2} \mathrm{d}z$$

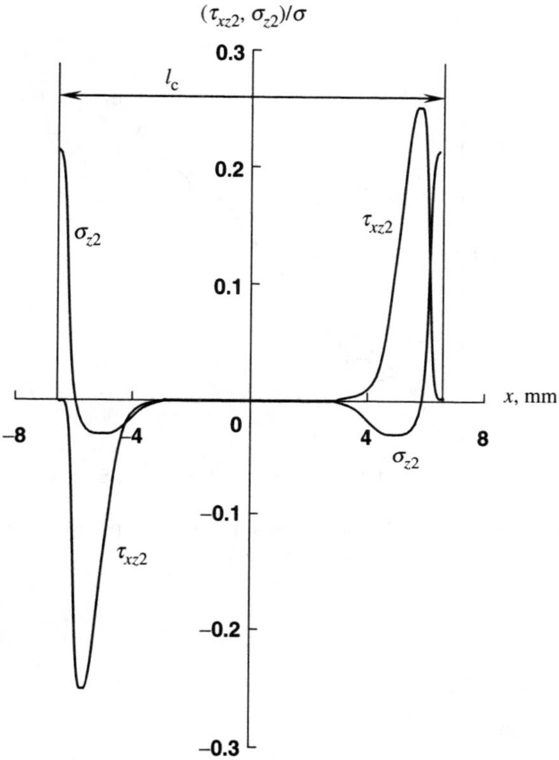

Fig. 4.41. Distribution of normalized shear (τ_{xz2}) and transverse normal stresses (σ_{z2}) at the ply interface ($z = h_2$) between the cracks.

where

$$\varepsilon_{x2} = \frac{1}{E_2}(\sigma_{x2} - \nu_{23}\sigma_{z2})$$

For a layer with the properties given above, such a diagram is shown in Fig. 4.42 with a solid line and is in good agreement with experimental results (circles). The formation of cracks is accompanied with horizontal jumps and reduction in material stiffness. The stress–strain diagram for the transverse layer that is formally singled out of the diagram in Fig. 4.42 is presented in Fig. 4.43.

To develop a nonlinear phenomenological model of the cross-ply layer, we need to approximate the diagram in Fig. 4.43. As follows from this figure and numerous experiments, the most suitable and simple approximation is that shown by the dashed line. It implies that the ply is linear elastic until its transverse stress σ_2 reaches its ultimate value $\overline{\sigma}_2^+$, and after that $\sigma_2 = \overline{\sigma}_2^+$, i.e., σ_2 remains constant up to failure of the longitudinal plies. This means that under transverse tension, a unidirectional ply is in a state of

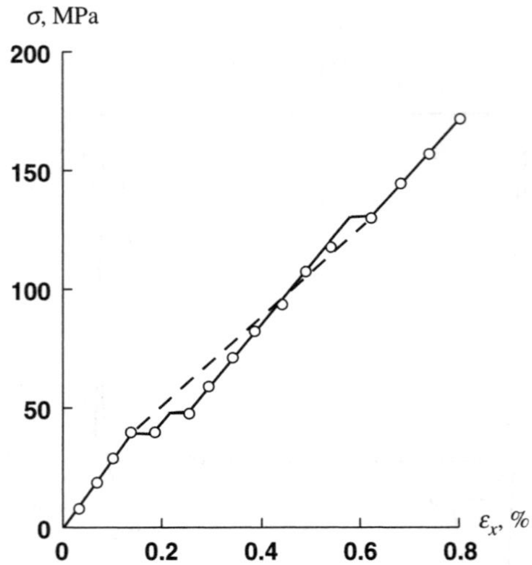

Fig. 4.42. Stress–strain diagram for a glass–epoxy cross-ply layer: ○ experiment; —— theoretical prediction; - - - model.

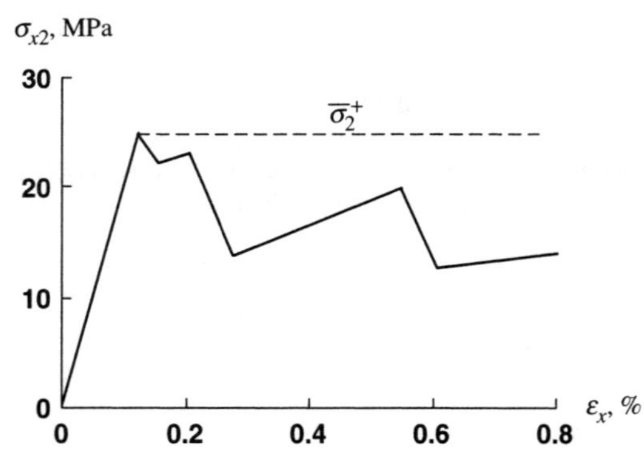

Fig. 4.43. Stress–strain diagram for a transverse ply.

permanent failure and takes from the longitudinal plies the necessary load to support this state (Vasiliev and Elpatievskii, 1967). The stress–strain diagram of the cross-ply layer corresponding to this model is shown in Fig. 4.42 with a dashed line.

Now consider a general plane stress state with stresses σ_x, σ_y, and τ_{xy} as in Fig. 4.44. As can be seen, stress σ_x induces cracks in the inner ply, stress σ_y causes cracks in the outer orthogonal plies, whereas shear stress τ_{xy} can give rise to cracks in all the plies. The ply model that generalizes the model introduced above for a uniaxial tension is demonstrated in Fig. 4.45. To determine strains corresponding to a given combination of stresses σ_x, σ_y, and τ_{xy}, we can use the following procedure.

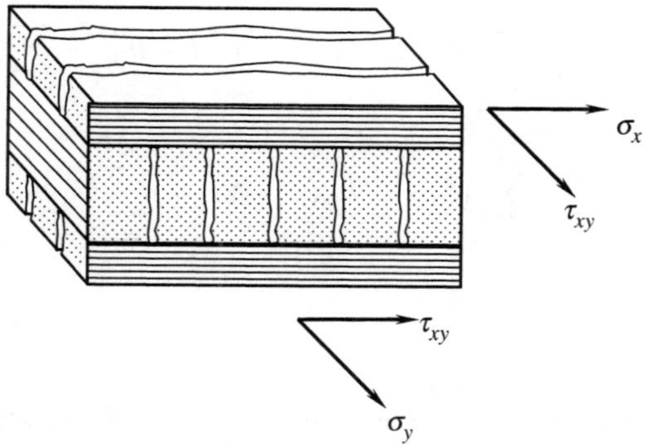

Fig. 4.44. A cross-ply layer in a plane stress state.

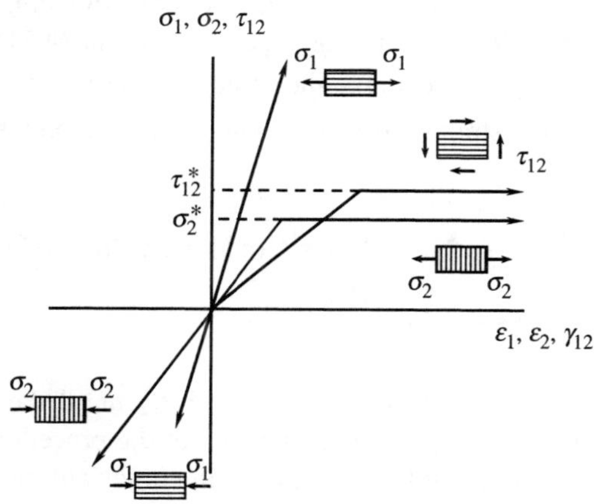

Fig. 4.45. Stress–strain diagrams of a unidirectional ply simulating its behavior in the laminate and allowing for cracks in the matrix.

(1) For the first stage of loading (before the cracks appear), the strains are calculated with the aid of Eqs. (4.114) and (4.115) providing $\varepsilon_x^{(1)}(\sigma)$, $\varepsilon_y^{(1)}(\sigma)$, and $\gamma_{xy}^{(1)}(\sigma)$, where $\sigma = (\sigma_x, \sigma_y, \tau_{xy})$ is the given combination of stresses. Using Eqs. (4.112), we find stresses σ_1, σ_2 and τ_{12} in principal material coordinates for all the plies.
(2) We determine the combination of stresses σ_{1k}^*, σ_{2k}^*, and τ_{12k}^* which induce the first failure of the matrix in some ply and indicate the number of this ply, say k, applying the appropriate strength criterion (see Section 6.2). Then, the corresponding stresses $\sigma^* = (\sigma_x^*, \sigma_y^*, \tau_{xy}^*)$ and strains $\varepsilon_x^{(1)}(\sigma^*)$, $\varepsilon_y^{(1)}(\sigma^*)$, and $\gamma_{xy}^{(1)}(\sigma^*)$ are calculated.
(3) To proceed, i.e., to study the material behavior for $\sigma > \sigma^*$, we need to consider two possible cases for the layer stiffnesses. For this purpose, we should write Eqs. (4.114) for stiffness coefficients in a more general form, i.e.,

$$A_{11} = \sum_{i=1}^{m} \overline{E}_1^{(i)} \overline{h}_0^{(i)} + \sum_{j=1}^{n} \overline{E}_2^{(j)} \overline{h}_{90}^{(j)}, \quad A_{22} = \sum_{i=1}^{m} \overline{E}_2^{(i)} \overline{h}_0^{(i)} + \sum_{j=1}^{n} \overline{E}_1^{(j)} \overline{h}_{90}^{(j)}$$

$$A_{12} = \sum_{i=1}^{m} \nu_{12}^{(i)} \overline{E}_1^{(i)} \overline{h}_0^{(i)} + \sum_{j=1}^{n} \nu_{12}^{(j)} \overline{E}_1^{(j)} \overline{h}_{90}^{(j)}, \quad A_{44} = \sum_{i=1}^{m} G_{12}^{(i)} \overline{h}_0^{(i)} + \sum_{j=1}^{n} G_{12}^{(j)} \overline{h}_{90}^{(j)} \tag{4.135}$$

where $\overline{h}_0^{(i)} = h_0^{(i)}/h$ and $\overline{h}_{90}^{(j)} = h_{90}^{(j)}/h$.

(a) If $\sigma_{2k} > 0$ in the kth ply, it can work only along the fibers, and we should calculate the stiffnesses of the degraded layer taking $E_2^k = 0$, $G_{12}^k = 0$, and $\nu_{12}^k = 0$ in Eqs. (4.135).

(b) If $\sigma_{2k} < 0$ in the kth ply, it cannot work only in shear, so we should take $G_{12}^k = 0$ in Eqs. (4.135).

Thus, we find coefficients $A_{st}^{(2)}$ $(st = 11, 12, 22, 44)$ corresponding to the second stage of loading (with one degraded ply). Using Eqs. (4.116) and (4.115) we can determine $E_x^{(2)}$, $E_y^{(2)}$, $G_{xy}^{(2)}$, $\nu_{xy}^{(2)}$, $\nu_{yx}^{(2)}$ and express the strains in terms of stresses, i.e., $\varepsilon_x^{(2)}(\sigma)$, $\varepsilon_y^{(2)}(\sigma)$, $\gamma_{xy}^{(2)}(\sigma)$. The final strains corresponding to the second stage of loading are calculated as

$$\varepsilon_x^{\mathrm{f}} = \varepsilon_x^{(1)}(\sigma^*) + \varepsilon_x^{(2)}(\sigma - \sigma^*), \quad \varepsilon_y^{\mathrm{f}} = \varepsilon_y^{(1)}(\sigma^*) + \varepsilon_y^{(2)}(\sigma - \sigma^*)$$

$$\gamma_{xy}^{\mathrm{f}} = \gamma_{xy}^{(1)}(\sigma^*) + \gamma_{xy}^{(2)}(\sigma - \sigma^*)$$

To study the third stage, we should find σ_1, σ_2, and τ_{12} in all the plies, except the kth one, identify the next degraded ply and repeat step 3 of the procedure which is continued up to failure of the fibers. The resulting stress–strain curves are multi-segmented broken lines with straight segments and kinks corresponding to degradation of particular plies.

The foregoing procedure was described for a cross-ply layer consisting of plies with different properties. For the layer made of one and the same material, there are only three stages of loading – first, before the plies degradation, second, after the degradation of the longitudinal or the transverse ply only, and third, after the degradation of all the plies.

As a numerical example, consider a carbon–epoxy cylindrical pressure vessel consisting of axial plies with total thickness h_0 and circumferential plies with total thickness h_{90}. The vessel has the following parameters: radius $R = 500$ mm, total thickness of the wall $h = 7.5$ mm, $h_0 = 2.5$ mm, $h_{90} = 5$ mm. The mechanical characteristics of a carbon–epoxy unidirectional ply are $E_1 = 140$ GPa, $E_2 = 11$ GPa, $\nu_{12} = 0.0212$, $\nu_{21} = 0.27$, $\overline{\sigma}_1^+ = 2000$ MPa, $\overline{\sigma}_2^+ = 50$ MPa. Axial, σ_x, and circumferential, σ_y, stresses are expressed as (see Fig. 4.46)

$$\sigma_x = \frac{pR}{2h}, \quad \sigma_y = \frac{pR}{h} \tag{4.136}$$

where p is the internal pressure.

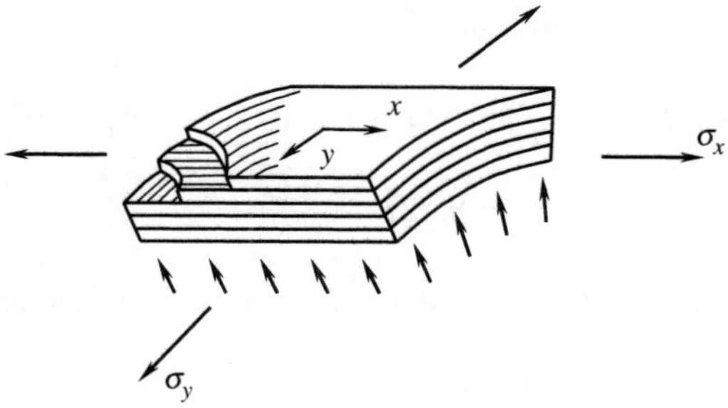

Fig. 4.46. Element of a composite pressure vessel.

Using Eqs. (4.114) and (4.116), we calculate first the stiffness coefficients. The result is as follows

$$A_{11} = 54.1\,\text{GPa}, \quad A_{12} = 3\,\text{GPa}, \quad A_{22} = 97.1\,\text{GPa}$$
$$E_x = 54\,\text{GPa}, \quad E_y = 97\,\text{GPa}, \quad \nu_{xy} = 0.055, \quad \nu_{yx} = 0.031 \tag{4.137}$$

Substituting stresses, Eqs. (4.136) into the constitutive equations, Eqs. (4.115), we obtain

$$\varepsilon_x^{(1)}(p) = \frac{pR}{h}\left(\frac{1}{2E_x} - \frac{\nu_{xy}}{E_y}\right) = 0.58 \times 10^{-3} p$$

$$\varepsilon_y^{(1)}(p) = \frac{pR}{h}\left(\frac{1}{E_y} - \frac{\nu_{yx}}{2E_x}\right) = 0.66 \times 10^{-3} p$$

where p is measured in mega pascals. For axial plies, $\varepsilon_x = \varepsilon_{1,0}$ and $\varepsilon_y = \varepsilon_{2,0}$. The corresponding stresses are

$$\sigma_{1,0}^{(1)}(p) = \overline{E}_1(\varepsilon_{1,0} + \nu_{12}\varepsilon_{2,0}) = 83.2p, \quad \sigma_{2,0}^{(1)}(p) = \overline{E}_2(\varepsilon_{2,0} + \nu_{21}\varepsilon_{1,0}) = 9.04p$$

For circumferential plies, $\varepsilon_x = \varepsilon_{2,90}, \quad \varepsilon_y = \varepsilon_{1,90}$ and

$$\sigma_{1,90}^{(1)}(p) = \overline{E}_1(\varepsilon_{1,90} + \nu_{12}\varepsilon_{2,90}) = 94.15p, \quad \sigma_{2,90}^{(1)}(p) = \overline{E}_2(\varepsilon_{2,90} + \nu_{21}\varepsilon_{1,90}) = 8.4p$$

As can be seen, $\sigma_{2,0}^{(1)} > \sigma_{2,90}^{(1)}$. This means that the cracks appear first in the axial plies under the pressure p^* that can be found from the equation $\sigma_{2,0}^{(1)}(p^*) = \overline{\sigma}_2^+$. The result is $p^* = 5.53\,\text{MPa}$.

To study the second stage of loading for $p > p^*$, we should put $E_2 = 0$, and $\nu_{12} = 0$ in Eqs. (4.135) for the axial plies. Then, the stiffness coefficients and elastic constants become

$$A_{11} = 54.06\,\text{GPa}, \quad A_{12} = 2\,\text{GPa}, \quad A_{22} = 93.4\,\text{GPa}$$

$$E_x = 54\,\text{GPa}, \quad E_y = 93.3\,\text{GPa}, \quad \nu_{xy} = 0.037, \quad \nu_{yx} = 0.021$$

The strains and stresses in the plies are

$$\varepsilon_x^{(2)}(p) = 0.59 \times 10^{-3}p, \quad \varepsilon_y^{(2)} p) = 0.7 \times 10^{-3}p, \quad \sigma_{1,0}^{(2)}(p) = 82.6p$$

$$\sigma_{1,90}^{(2)} p) = 99.8p, \quad \sigma_{2,90}^{(2)}(p) = 8.62p$$

The total transverse stress in the circumferential plies can be calculated as

$$\sigma_{2,90} = \sigma_{2,90}^{(1)}(p^*) + 8.62(p - p^*)$$

Using the condition $\sigma_{2,90}(p^{**}) = \overline{\sigma}_2^+$, we find the pressure $p^{**} = 5.95\,\text{MPa}$ at which cracks appear in the matrix of the circumferential plies.

For $p \geq p^{**}$, we should take $E_2 = 0$ and $\nu_{12} = 0$ for all the plies. Then

$$\begin{aligned}
&A_{11} = 46.2\,\text{GPa}, \quad A_{12} = 0, \quad A_{22} = 93.4\,\text{GPa} \\
&E_x = 46.2\,\text{GPa}, \quad E_y = 93.4\,\text{GPa}, \quad \nu_{xy} = \nu_{yx} = 0 \\
&\varepsilon_x^{(3)}(p) = 0.72 \times 10^{-3}p, \quad \varepsilon_y^{(3)}(p) = 0.71 \times 10^{-3}p \\
&\sigma_{1,0}^{(3)}(p) = 100.8p, \quad \sigma_{1,90}^{(3)}(p) = 99.4p
\end{aligned} \tag{4.138}$$

The total stresses acting along the fibers are

$$\sigma_{1,0}(p) = \sigma_{1,0}^{(1)}(p^*) + \sigma_{1,0}^{(2)}(p^{**} - p^*) + \sigma_{1,0}^{(3)}(p - p^{**}) = 100.8p - 105$$

$$\sigma_{1,90}(p) = \sigma_{1,90}^{(1)}(p^*) + \sigma_{1,90}^{(2)}(p^{**} - p^*) + \sigma_{1,90}^{(3)}(p - p^{**}) = 99.4p - 28.9$$

To determine the ultimate pressure, we can use two possible strength conditions – for axial fibers and for circumferential fibers. The criterion $\sigma_{1,0}(p) = \overline{\sigma}_1^+$ yields $p = 20.9\,\text{MPa}$, whereas the criterion $\sigma_{1,90}(p) = \overline{\sigma}_1^+$ gives $p = 20.4\,\text{MPa}$. Thus, the burst pressure governed by failure of the fibers in the circumferential plies, is $\overline{p} = 20.4\,\text{MPa}$.

The strains can be calculated for all three stages of loading using the following equations

- for $p \leq p^*$

$$\varepsilon_{x,y}(p) = \varepsilon_{x,y}^{(1)}(p)$$

- for $p^* < p \leq p^{**}$

$$\varepsilon_{x,y}(p) = \varepsilon_{x,y}^{(1)}(p^*) + \varepsilon_{x,y}^{(2)}(p - p^*)$$

- for $p^{**} < p \leq \overline{p}$

$$\varepsilon_{x,y}(p) = \varepsilon_{x,y}^{(1)}(p^*) + \varepsilon_{x,y}^{(2)}(p^{**} - p^*) + \varepsilon_{x,y}^{(3)}(p - p^{**})$$

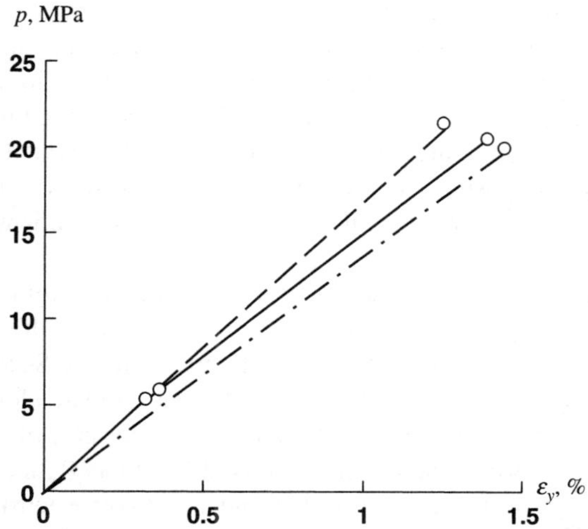

Fig. 4.47. Dependence of the axial and the circumferential strains of the carbon–epoxy pressure vessel on pressure: ——— model allowing for cracks in the matrix; - - - - model ignoring cracks in the matrix; – · – · – model ignoring the matrix.

For the pressure vessel under study, the dependency of the circumferential strain on pressure is shown in Fig. 4.47 (solid line). The circles correspond to failure of the matrix and fibers.

For comparison, consider two limiting cases. First, assume that no cracks occur in the matrix, and the material stiffness is specified by Eqs. (4.137). The corresponding diagram is shown in Fig. 4.47 with a dashed line. Second, suppose that the load is taken by the fibers only, i.e., use the monotropic model of a ply introduced in Section 3.3. Then, the material stiffnesses are given by Eqs. (4.138). The corresponding result is also presented in Fig. 4.47. It follows from this figure that all three models give close results for the burst pressure (which is expected since $\overline{\sigma}_2^+ \ll \overline{\sigma}_1^+$), but different strains.

4.4.3. Two-matrix composites

The problem of the analysis of a cracked cross-ply composite laminate has been studied by Tsai and Azzi (1966), Vasiliev and Elpatievckii (1967), Vasiliev et al. (1970), Hahn and Tsai (1974), Reifsnaider (1977), Hashin (1987), and many other authors. In spite of this, the topic is still receiving repeated attention in the literature (Lungren and Gudmundson, 1999). Taking into account that matrix degradation leads to reduction of material stiffness and fatigue strength, absorption of moisture and many other consequences that are difficult to predict but are definitely undesirable, it is surprising how many efforts have been undertaken to study this phenomenon rather than try to avoid it. At first glance, the problem looks simple – all we need is to synthesize unidirectional composite whose ultimate elongations along and across the fibers, i.e., $\overline{\varepsilon}_1$ and $\overline{\varepsilon}_2$ are the same. Actually, the problem is even simpler, because $\bar{\varepsilon}_2$ can be less than $\bar{\varepsilon}_1$ by a factor that is equal to the safety factor of the structure. This means that matrix degradation can occur but at the load that exceeds the operational level (the safety factor is the ratio of the failure load to the operational load and can vary from 1.25 up to 3 or more depending on the application of a particular composite structure). Returning to Table 4.2, in which $\bar{\varepsilon}_1$ and $\bar{\varepsilon}_2$ are given for typical advanced composites, we can see that $\bar{\varepsilon}_1 > \bar{\varepsilon}_2$ for all the materials and that for polymeric matrices the problem could be, in principle, solved if we could increase $\bar{\varepsilon}_2$ up to about 1%.

Two main circumstances hinder the direct solution of this problem. The first is that being locked between the fibers, the matrix does not show the high elongation that it has under uniaxial tension and behaves as a brittle material (see Section 3.4.2). To study this effect, epoxy resins were modified to have different ultimate elongations. The corresponding curves are presented in Fig. 4.48 (only the initial part of curve 4 is shown in this figure, the ultimate elongation of this resin is 60%). Fiberglass composites that have been fabricated with these resins were tested under transverse tension. As can be seen in Fig. 4.49, the desired value of $\bar{\varepsilon}_2$ (that is about 1%) is reached if the matrix elongation is about 60%. However, the stiffness of this matrix is relatively low, and the second circumstance arises – matrix material with low stiffness cannot provide sufficient stress diffusion in the vicinity of damaged or broken fibers (see Section 3.2.3). As a result, the main material characteristic – its longitudinal tensile strength – decreases. Experimental results corresponding to composites with resins 1, 2, 3, and 4 are presented in Fig. 4.50. Thus, a significant increase in transverse elongation is accompanied with an unacceptable drop in longitudinal strength (see also Chiao, 1979).

One of the possible ways for synthesizing composite materials with high transverse elongation and high longitudinal strength is to combine two matrix materials – one with

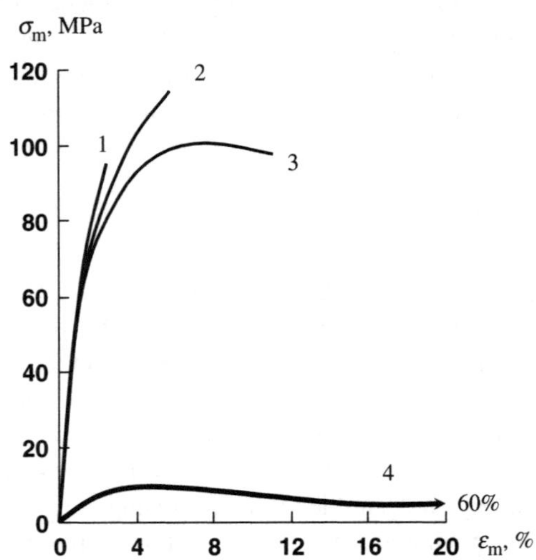

Fig. 4.48. Stress–strain curves for epoxy matrices modified for various ultimate elongations.

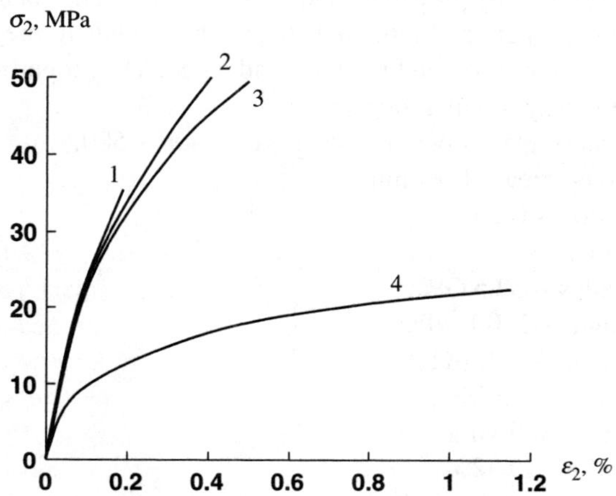

Fig. 4.49. Stress–strain curves for transverse tension of unidirectional fiberglass composites with various epoxy matrices (numbers on the curves correspond to Fig. 4.48).

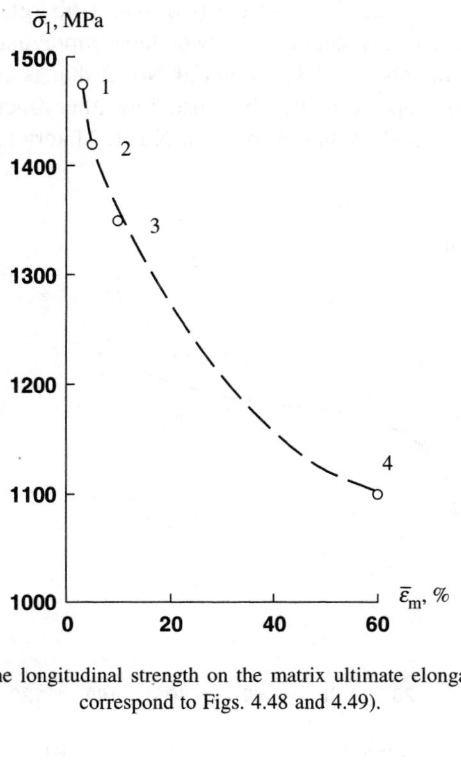

Fig. 4.50. Dependence of the longitudinal strength on the matrix ultimate elongation (numbers on the curve correspond to Figs. 4.48 and 4.49).

high stiffness to bind the fibers and the other with high elongation to provide the appropriate transverse deformability (Vasiliev and Salov, 1984). The manufacturing process involves two-stage impregnation. At the first stage, a fine tow is impregnated with a high-stiffness epoxy resin (of the type 2 in Fig. 4.48) and cured. The properties of the composite fiber fabricated in this way are as follows

- number of elementary glass fibers in the cross section – 500;
- mean cross-sectional area – 0.15 mm^2;
- fiber volume fraction – 0.75;
- density – 2.2 g/cm^3;
- longitudinal modulus – 53.5 GPa;
- longitudinal strength – 2100 MPa;
- longitudinal elongation – 4.5%;
- transverse modulus – 13.5 GPa;
- transverse strength – 400 MPa;
- transverse elongation – 0.32%.

At the second stage, a tape formed of composite fibers is impregnated with a highly deformable epoxy matrix whose stress–strain diagram is presented in Fig. 4.51. The microstructure of the resulting two-matrix unidirectional composite is shown in Fig. 4.52 (the dark areas are cross sections of composite fibers, the magnification is not sufficient to see the elementary glass fibers). Stress–strain diagrams corresponding to transverse tension, compression, and in-plane shear of this material are presented in Fig. 4.16.

The main mechanical characteristics of the two-matrix fiberglass composite are listed in Table 4.3 (material No. 1). As can be seen, two-stage impregnation results in relatively low fiber volume content (about 50%). Material No. 2 that is composed of composite fibers and a conventional epoxy matrix has also low fiber fraction, but its transverse elongation is 10 times lower than that of material No. 1. Material No. 3 is a conventional

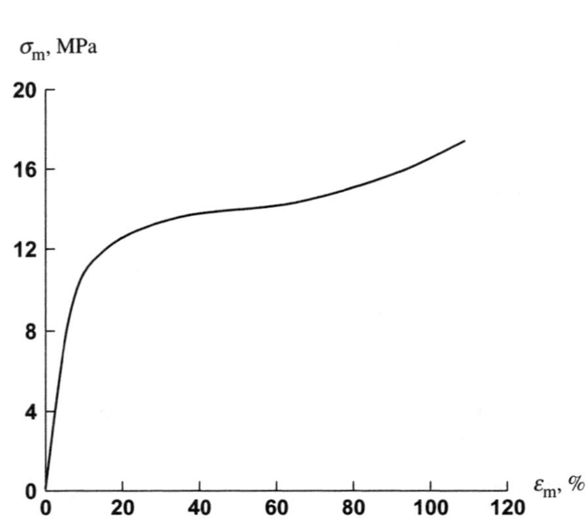

Fig. 4.51. Stress–strain diagram of a deformable epoxy matrix.

Fig. 4.52. Microstructure of a unidirectional two-matrix composite.

Table 4.3
Properties of glass–epoxy unidirectional composites.

No.	Material components	Fiber volume fraction	Longitudinal strength $\overline{\sigma}_1^+$ (MPa)	Ultimate transverse strain $\overline{\varepsilon}_2^+$ (%)	Density ρ (g/cm^3)	Specific strength $\overline{\sigma}_1^+/\rho \times 10^3$ (m)
1	Composite fibers and deformable matrix	0.51	1420	3.0	1.83	77.6
2	Composite fibers and high-stiffness matrix	0.52	1430	0.3	1.88	76.1
3	Glass fibers and high-stiffness matrix	0.67	1470	0.2	2.07	71.0
4	Glass fibers and deformable matrix	0.65	1100	1.2	2.02	54.4

glass–epoxy composite that has the highest longitudinal strength and the lowest transverse strain. Comparing materials No. 1 and No. 3, we can see that although the fiber volume fraction of the two-matrix composite is lower by 24%, its longitudinal strength is less than that of a traditional composite by 3.4% only (because the composite fibers are not damaged in the processing of composite materials), whereas its specific strength is a bit higher (due to its lower density). Material No. 4 demonstrates that direct application of a highly deformable matrix allows us to increase transverse strains but results in a 23% reduction in longitudinal specific strength.

Thus, two-matrix glass–epoxy composites have practically the same longitudinal strength as conventional materials but their transverse elongation is greater by an order of magnitude.

Comparison of a conventional cross-ply glass–epoxy layer and a two-matrix one is presented in Fig. 4.53. Line 1 corresponds to a traditional material and has, typical for this material, a kink corresponding to matrix failure in the transverse plies (see also Fig. 4.37). A theoretical diagram was plotted using the procedure described above. Line 2 corresponds to a two-matrix composite and was plotted using Eqs. (4.60). As can be seen, there is no kink on the stress–strain diagram. To prove that no cracks appear in the matrix

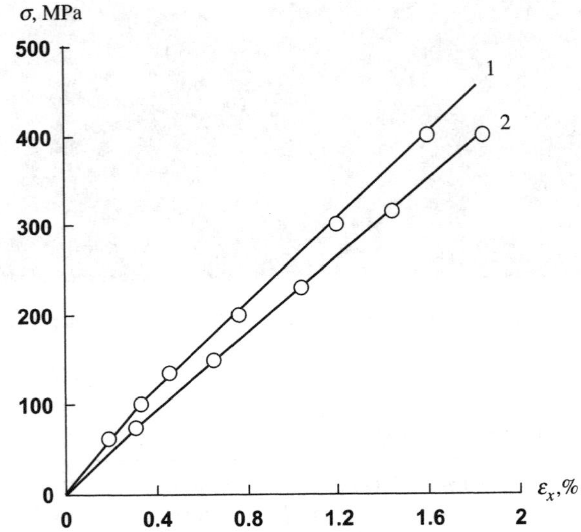

Fig. 4.53. Stress–strain diagrams of a conventional (1) and two-matrix (2) cross-ply glass–epoxy layers under tension: ——— theoretical prediction; ○ experiment.

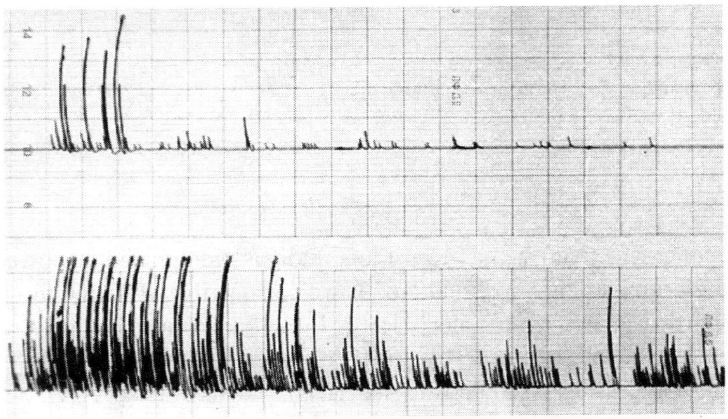

Fig. 4.54. Intensity of acoustic emission for a cross-ply two-matrix composite (above) and a conventional fiberglass composite (below).

of this material under loading, the intensity of acoustic emission was recorded during loading. The results are shown in Fig. 4.54.

Composite fibers of two-matrix materials can also be made from fine carbon or aramid tows, and the deformable thermosetting resin can be replaced with a thermoplastic matrix (Vasiliev et al., 1997). The resulting hybrid thermoset–thermoplastic unidirectional composite is characterized by high longitudinal strength and transverse strain exceeding 1%. Having high strength, composite fibers are not damaged in the process of laying-up or winding, and the tapes formed from these fibers are readily impregnated even with high-viscosity thermoplastic polymers.

4.4.4. *Composites with controlled cracks*

Now we return to the conventional composites discussed in Section 4.4.2. Since the transverse ultimate elongation of a ply, $\overline{\varepsilon}_2$, is less than the corresponding longitudinal elongation, $\overline{\varepsilon}_1$, (see Table 4.2), the stress σ in Eq. (4.123) induces a system of cracks in the matrix of the transverse ply as in Fig. 4.39. As has been already noted, these cracks do not cause laminate failure because its strength is controlled by the longitudinal plies. What is actually not desirable is matrix failure in the process of laminate loading. So, since the cracks shown in Fig. 4.39 will occur anyway at some stress σ, suppose that the material has these cracks before loading, i.e., that the transverse ply consists of individual strips with width l_c as in Fig. 4.39. The problem is to find the width l_c for which no other cracks will appear in the transverse ply up to failure of the fibers in the longitudinal plies.

Consider the solution in Eq. (4.132), take $C_2 = C_3 = 0$ and find the constants C_1 and C_4 from the boundary conditions in Eqs. (4.133) in which l_c is some unknown width. The resulting expression for the stress in the transverse ply is

$$\begin{aligned}\sigma_{x2} = \sigma_2^0 \Bigg\{ 1 - \frac{1}{k_1 \sin\lambda_2 \cos\lambda_2 + k_2 \sinh\lambda_1 \cosh\lambda_1} & [(k_2 \cosh\lambda_1 \sin\lambda_2 \\ & - k_1 \sinh\lambda_1 \cos\lambda_2) \sinh k_1 x \sin k_2 x \\ & + (k_1 \cosh\lambda_1 \sin\lambda_2 + k_2 \sinh\lambda_1 \cos\lambda_2) \cosh k_1 x \cos k_2 x] \Bigg\}\end{aligned}$$

in which $\lambda_1 = k_1 l_c/2$ and $\lambda_2 = k_2 l_c/2$. The maximum stress acts at $x = 0$ (see Fig. 4.40) and can be presented as

$$\sigma_2^{\mathrm{m}} = \sigma_2^0 [1 - F(l_c)] \tag{4.139}$$

where

$$F(l_c) = \frac{2(k_1 \cosh\lambda_1 \sin\lambda_2 + k_2 \sinh\lambda_1 \cos\lambda_2)}{k_1 \sin 2\lambda_2 + k_2 \sinh 2\lambda_1} \tag{4.140}$$

The stress σ_2^0 in Eq. (4.139) is specified by the second equation of Eqs. (4.122). Taking into account the first equation, we have

$$\sigma_2^0 = \frac{E_2}{E_1}\sigma_1^0$$

where σ_1^0 is the stress in the longitudinal plies. So, Eq. (4.139) can be written as

$$\sigma_2^{\rm m} = \frac{E_2}{E_1}\sigma_1^0\left[1 - F(l_{\rm c})\right] \tag{4.141}$$

Now suppose that $\sigma_1 = \overline{\sigma}_1$, i.e., that the longitudinal stress reaches the corresponding ultimate value. The cracks in the matrix of the transverse ply do not appear if $\sigma_2^{\rm m} \leq \overline{\sigma}_2$, where $\overline{\sigma}_2$ is the transverse tensile strength of the ply. Then, Eq. (4.141) yields

$$F(l_{\rm c}) \geq t \tag{4.142}$$

where $t = 1 - \dfrac{E_1\overline{\sigma}_2}{E_2\overline{\sigma}_1}$.

As an example, consider a cross-ply (see Fig. 4.35) carbon–epoxy composite with the following parameters

$$E_1 = 140\,\text{GPa}, \quad E_2 = 11\,\text{GPa}, \quad G_{13} = 5.5\,\text{GPa}, \quad G_{23} = 4.1\,\text{GPa},$$

$$\nu_{23} = 0.3, \quad \overline{\sigma}_1 = 2000\,\text{MPa}, \quad \overline{\sigma}_2 = 50\,\text{MPa}$$

for which $t = 0.68$. Introduce normalized thicknesses of the plies as

$$\overline{h}_1 = \frac{2h_1}{h}, \quad \overline{h}_2 = \frac{2h_2}{h}$$

where $h = 2(h_1 + h_2)$ (see Fig. 4.37). Let $\overline{h}_1 = 1 - \alpha$, $\overline{h}_2 = \alpha$, where the parameter α specifies the relative thickness of the transverse ply. The dependencies of the coefficients $\overline{k}_1 = k_1/h$ and $\overline{k}_2 = k_2/h$ (in which k_1 and k_2 are given in the notations to Eq. (4.130)) on the parameter α are shown in Fig. 4.55. The dependence of function F in Eq. (4.140) on the normalized distance between the cracks $\overline{l}_{\rm c} = l_{\rm c}/h$ is presented in Fig. 4.56 for $\alpha = 0.1, 0.5$, and 0.9. The intersections of the horizontal line $F = t = 0.68$ give the values of $\overline{l}_{\rm c}$ for which no new cracks appear in the transverse ply up to the fibers' failure. The final dependence of $\overline{l}_{\rm c}$ on α is shown in Fig. 4.57. As can be seen, $l_{\rm c}$ varies from about 2 up to 4 thicknesses of the laminate. For $h_1 = h_2 = \delta$, where $\delta = 0.15\,$mm is the thickness of the unidirectional ply, we get $h = 4\delta$, $\alpha = 0.5$, and $l_{\rm c} = 1.9\,$mm. A yarn of such width is typical for carbon fabrics made of $3K$ carbon tows. Experiments with such fabric composites show that the tensile stress–strain diagram of the material is linear up to failure, and no cracks are observed in the matrix.

4.5. Angle-ply orthotropic layer

The angle-ply layer is a combination of an even number of alternating plies with angles $+\phi$ and $-\phi$ as in Fig. 4.58. The structure of this layer is typical for the process of filament winding (see Fig. 4.59). As for the cross-ply layer considered in the previous

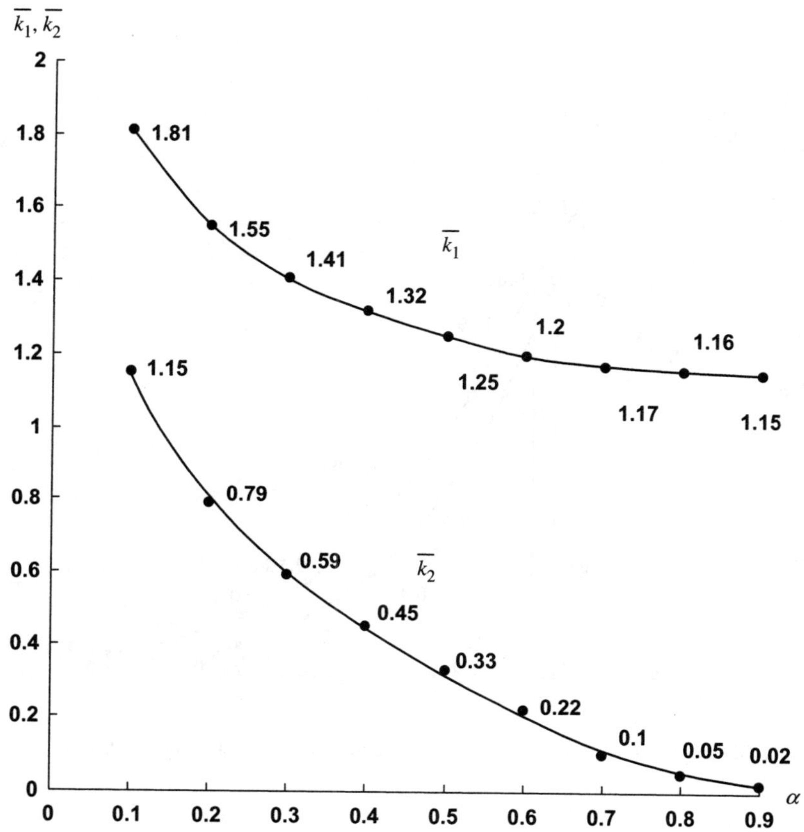

Fig. 4.55. Dependencies of the coefficients $\overline{k}_1$ and $\overline{k}_2$ on the relative thickness of the transverse ply α.

section, an angle-ply layer is actually a laminate, but for a large number of plies it can be approximately treated as a homogeneous orthotropic layer (see Section 5.4.3).

4.5.1. Linear elastic model

Consider two symmetric systems of unidirectional anisotropic plies (see Section 4.3) consisting of the same number of plies, made of one and the same material and having alternating angles $+\phi$ and $-\phi$. Then, the total stresses σ_x, σ_y, and τ_{xy} acting on the layer can be expressed in terms of the corresponding stresses acting in the $+\phi$ and $-\phi$ plies as

$$\sigma_x h = \sigma_x^+ \frac{h}{2} + \sigma_x^- \frac{h}{2}, \quad \sigma_y h = \sigma_y^+ \frac{h}{2} + \sigma_y^- \frac{h}{2}, \quad \tau_{xy} h = \tau_{xy}^+ \frac{h}{2} + \tau_{xy}^- \frac{h}{2} \tag{4.143}$$

where h is the total thickness of the layer. Stresses with superscripts '+' and '−' are related to strains ε_x, ε_y, and γ_{xy} (which are presumed to be the same for all the plies)

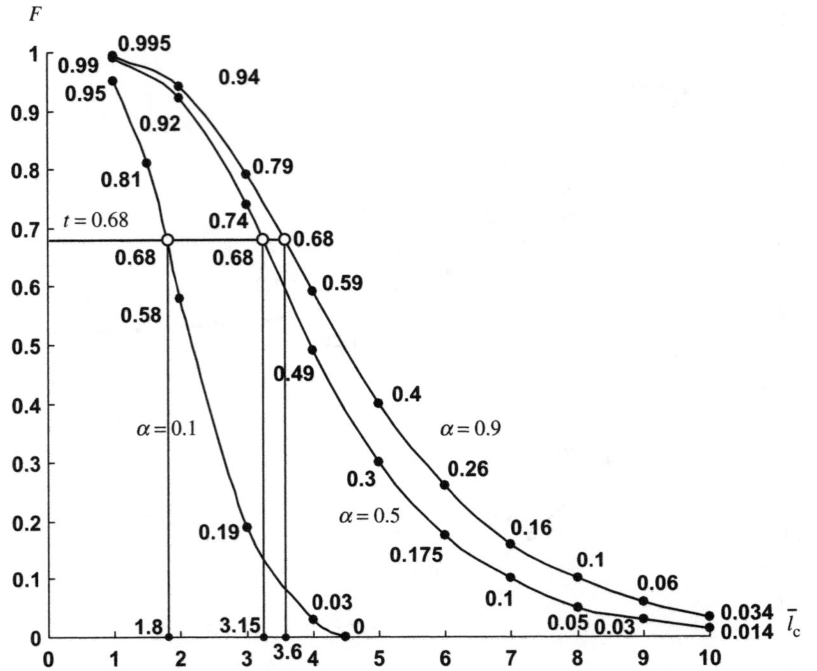

Fig. 4.56. Dependencies of function F on the normalized width of the strip for $\alpha = 0.1$, 0.5, and 0.9.

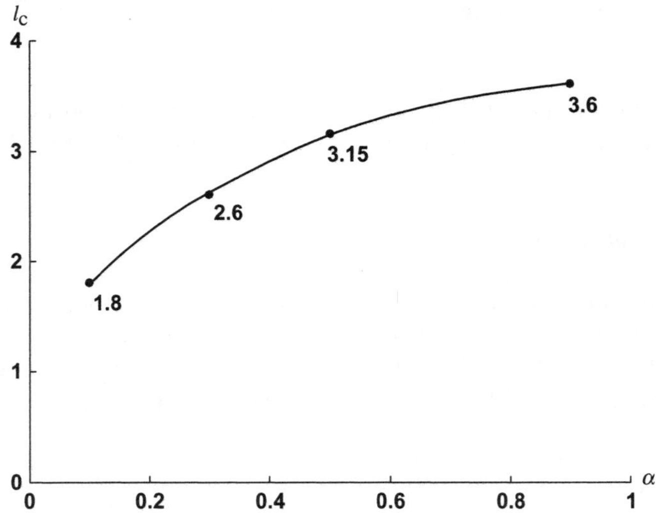

Fig. 4.57. Normalized width of the strip $\bar{l}_c$ as a function of the relative thickness of transverse ply α.

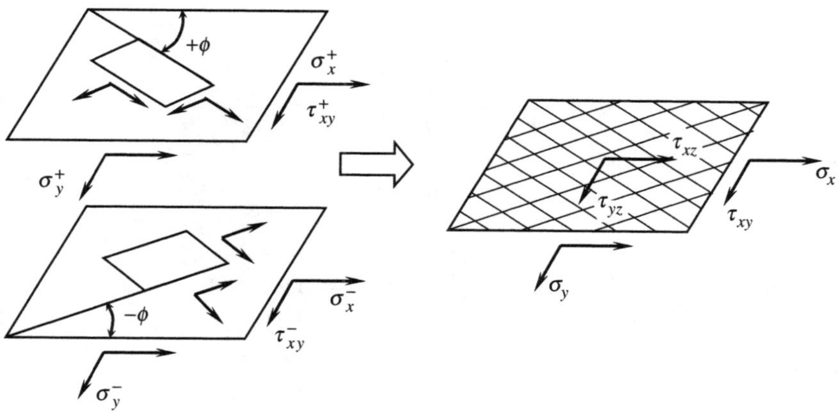

Fig. 4.58. Two symmetric plies forming an angle-ply layer.

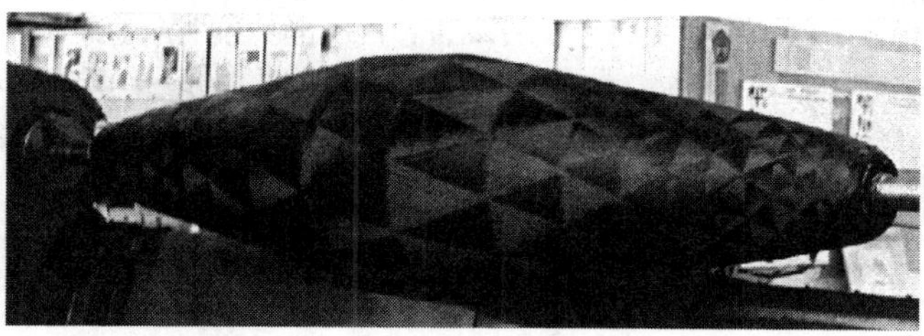

Fig. 4.59. Angle-ply layer of a filament-wound shell. Courtesy of CRISM.

by Eqs. (4.71), i.e.,

$$\sigma_x^{\pm} = A_{11}^{\pm}\varepsilon_x + A_{12}^{\pm}\varepsilon_y + A_{14}^{\pm}\gamma_{xy}, \quad \sigma_y^{\pm} = A_{21}^{\pm}\varepsilon_x + A_{22}^{\pm}\varepsilon_y + A_{24}^{\pm}\gamma_{xy},$$
$$\tau_{xy}^{\pm} = A_{41}^{\pm}\varepsilon_x + A_{42}^{\pm}\varepsilon_y + A_{44}^{\pm}\gamma_{xy} \tag{4.144}$$

in which $A_{11}^+ = A_{11}^- = A_{11}$, $A_{12}^+ = A_{12}^- = A_{12}$, $A_{22}^+ = A_{22}^- = A_{22}$, $A_{14}^+ = -A_{14}^- = A_{14}$, $A_{24}^+ = -A_{24}^- = A_{24}$, $A_{44}^+ = A_{44}^- = A_{44}$, where A_{mn} (mn = 11, 12, 22, 14, 24, 44) are specified by Eqs. (4.72). Substituting Eqs. (4.144) into Eqs. (4.143), we arrive at the following constitutive equations for an angle-ply layer

$$\sigma_x = A_{11}\varepsilon_x + A_{12}\varepsilon_y$$
$$\sigma_y = A_{21}\varepsilon_x + A_{22}\varepsilon_y \tag{4.145}$$
$$\tau_{xy} = A_{44}\gamma_{xy}$$

The inverse form of these equations is

$$\varepsilon_x = \frac{\sigma_x}{E_x} - \nu_{xy}\frac{\sigma_y}{E_y}, \quad \varepsilon_y = \frac{\sigma_y}{E_y} - \nu_{yx}\frac{\sigma_x}{E_x}, \quad \gamma_{xy} = \frac{\tau_{xy}}{G_{xy}} \tag{4.146}$$

where

$$E_x = A_{11} - \frac{A_{12}^2}{A_{22}}, \quad E_y = A_{22} - \frac{A_{12}^2}{A_{11}}, \quad G_{xy} = A_{44},$$
$$\nu_{xy} = \frac{A_{12}}{A_{11}}, \quad \nu_{yx} = \frac{A_{12}}{A_{22}} \tag{4.147}$$

It follows from Eqs. (4.145) and (4.146) that the layer under study is orthotropic.

Now derive constitutive equations relating transverse shear stresses τ_{xz} and τ_{yz} and the corresponding shear strains γ_{xz} and γ_{yz}. Let the angle-ply layer be loaded by stress τ_{xz}. Then for all the plies, $\tau_{xz}^+ = \tau_{xz}^- = \tau_{xz}$ and because the layer is orthotropic, $\gamma_{xz}^+ = \gamma_{xz}^- = \gamma_{xz}$, $\gamma_{yz}^+ = \gamma_{yz}^- = \gamma_{yz} = 0$. In a similar way, applying stress τ_{yz} we have $\tau_{yz}^+ = \tau_{yz}^- = \tau_{yz}$, $\gamma_{yz}^+ = \gamma_{yz}^- = \gamma_{yz}$, $\gamma_{xz}^+ = \gamma_{xz}^- = \gamma_{xz} = 0$. Writing the last two constitutive equations of Eqs. (4.71) for these two cases, we arrive at

$$\tau_{xz} = A_{55}\gamma_{xz}, \quad \tau_{yz} = A_{66}\gamma_{yz} \tag{4.148}$$

where the stiffness coefficients A_{55} and A_{66} are specified by Eqs. (4.72).

The dependencies of E_x and G_{xy} on ϕ, plotted using Eqs. (4.147), are shown in Fig. 4.60 with solid lines. The theoretical curve for E_x is in very good agreement with experimental data shown with circles (Lagace, 1985). For comparison, the same moduli are presented for the $+\phi$ anisotropic layer considered in Section 4.3.1. As can be seen, $E_x(\pm\phi) \geq E_x^+$. To explain this effect, consider uniaxial tension of both layers in the x-direction. Whereas tension of the $+\phi$ and $-\phi$ individual plies shown in Fig. 4.61 is accompanied with shear strain, the system of these plies does not demonstrate shear under tension and, as a result, has higher stiffness. Working as plies of a symmetric angle-ply layer, individual anisotropic $+\phi$ and $-\phi$ plies are loaded not only with a normal stress σ_x that is applied to the layer, but also with shear stress τ_{xy} that restricts the shear of individual plies (see Fig. 4.61). In order to find the reactive shear stress, which is balanced between the plies, we can use Eqs. (4.75). Taking $\sigma_y = 0$, we can simulate the stress–strain state of the ply in the angle-ply layer putting $\gamma_{xy} = 0$. Then, the third equation yields

$$\tau_{xy} = -\eta_{xy,x}^+ \frac{G_{xy}^+}{E_x^+}\sigma_x$$

Superscript '+' indicates that elastic constants correspond to an individual $+\phi$ ply. Substituting this shear stress into the first equation of Eqs. (4.75), we arrive at $\sigma_x = E_x\varepsilon_x$,

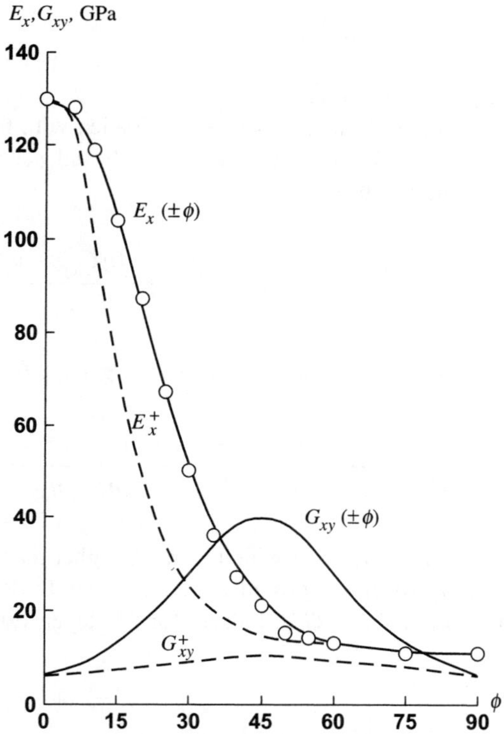

Fig. 4.60. Dependencies of the moduli of a carbon–epoxy layer on the orientation angle: ——— orthotropic angle-ply $\pm\phi$ layer; - - - - anisotropic $+\phi$ layer; ○ experiment for an angle-ply layer.

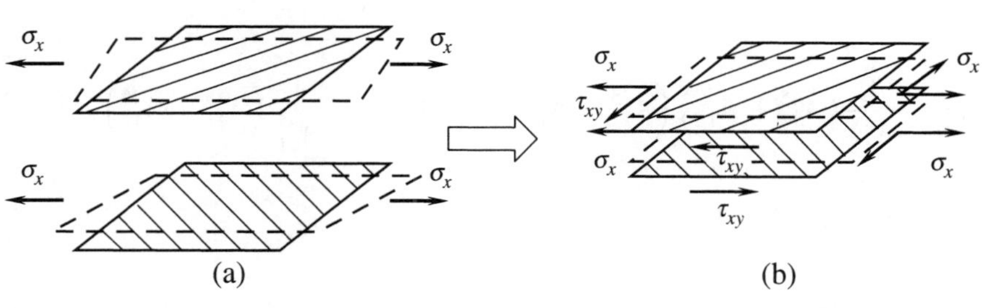

Fig. 4.61. Deformation and stresses induced in individual plies (a) and bonded symmetric plies (b) by uniaxial tension.

where

$$E_x = \frac{E_x^+}{1 - \eta_{x,xy}^+ \eta_{xy,x}^+} = \frac{E_x^+}{1 - \frac{G_{xy}^+}{E_x^+}(\eta_{xy,x}^+)^2} \tag{4.149}$$

is the modulus of the $\pm\phi$ angle-ply layer.

Under pure shear of an angle-ply layer, its plies are loaded with the additional normal stresses. These stresses can be found if we take $\varepsilon_x = 0$ and $\varepsilon_y = 0$ in the first two equations of Eqs. (4.75). The result is

$$\sigma_x = -\tau_{xy} \frac{E_x^+(\eta_{x,xy}^+ - \nu_{xy}^+ \eta_{y,xy}^+)}{G_{xy}^+(1 - \nu_{xy}^+ \nu_{yx}^+)}, \quad \sigma_y = -\tau_{xy} \frac{E_y^+(\eta_{y,xy}^+ - \nu_{yx}^+ \eta_{x,xy}^+)}{G_{xy}^+(1 - \nu_{xy}^+ \nu_{yx}^+)}$$

Substituting these expressions into the third equation, we get $\tau_{xy} = G_{xy}\gamma_{xy}$, where

$$G_{xy} = \frac{G_{xy}^+(1 - \nu_{xy}^+ \nu_{yx}^+)}{1 - \nu_{xy}^+ \nu_{yx}^+ - \eta_{x,xy}^+ \eta_{xy,x}^+ - \eta_{y,xy}^+ \eta_{xy,y}^+ - \nu_{xy}^+ \eta_{y,xy}^+ \eta_{xy,x}^+ - \nu_{yx}^+ \eta_{x,xy}^+ \eta_{xy,y}^+}$$

is the shear modulus of an angle-ply layer which is much higher than G_{xy}^+ (see Fig. 4.60).

Tension of $\pm 45°$angle-ply specimen provides a simple way to determine the in-plane shear modulus of a unidirectional ply, G_{12}. Indeed, for this layer, Eqs. (4.72) and (4.147) yield

$$A_{11}^{45} = A_{22}^{45} = \frac{1}{4}(\overline{E}_1 + \overline{E}_2 + 2\overline{E}_1\nu_{12} + 4G_{12})$$

$$A_{12}^{45} = \overline{E}_1\nu_{12} + \frac{1}{4}(\overline{E}_1 + \overline{E}_2 - 2\overline{E}_1\nu_{12} - 4G_{12})$$

and

$$E_{45} = \frac{1}{A_{11}^{45}}\left(A_{11}^{45} + A_{12}^{45}\right)\left(A_{11}^{45} - A_{12}^{45}\right), \quad 1 + \nu_{45} = \frac{1}{A_{11}^{45}}\left(A_{11}^{45} + A_{12}^{45}\right)$$

Taking into account that $A_{11}^{45} - A_{12}^{45} = 2G_{12}$, we have

$$G_{12} = \frac{E_{45}}{2(1 + \nu_{45})} \tag{4.150}$$

Thus, to find G_{12}, we can test a $\pm 45°$ specimen under tension, measure ε_x and ε_y, determine $E_{45} = \sigma_x/\varepsilon_x$, $\nu_{45} = -\varepsilon_y/\varepsilon_x$, and use Eq. (4.150) rather than perform the cumbersome tests described in Section 3.

4.5.2. Nonlinear models

To describe nonlinear behavior of an angle-ply layer associated with material nonlinearity in its plies, we can use nonlinear constitutive equations, Eqs. (4.60) or (4.64) instead of Hooke's law. Indeed, assuming that the ply behavior is linear under tension or compression along the fibers, we can write these equations in the following general form

$$\varepsilon_1 = c_{11}\sigma_1 + c_{12}\sigma_2, \quad \varepsilon_2 = c_{12}\sigma_1 + c_{22}\sigma_2 + \omega_2(\sigma_2, \tau_{12}), \quad \gamma_{12} = c_{44}\tau_{12} + \omega_{12}(\sigma_2, \tau_{12})$$

Functions ω_2 and ω_{12} include all the nonlinear terms. The inverse form of these equations is

$$\begin{aligned} &\sigma_1 = C_{11}\varepsilon_1 + C_{12}\varepsilon_2 - C_{12}\omega_2, \quad \sigma_2 = C_{12}\varepsilon_1 + C_{22}\varepsilon_2 - C_{22}\omega_2, \\ &\tau_{12} = C_{44}\gamma_{12} - C_{44}\omega_{12} \end{aligned} \tag{4.151}$$

in which

$$C_{11} = \frac{c_{22}}{c}, \quad C_{22} = \frac{c_{11}}{c}, \quad C_{44} = \frac{1}{c_{44}}, \quad C_{12} = -\frac{c_{12}}{c}, \quad c = c_{11}c_{22} - c_{12}^2$$

Repeating the derivation of Eqs. (4.145) but using this time Eqs. (4.151) as the constitutive equations for the ply, we arrive at

$$\sigma_x = A_{11}\varepsilon_x + A_{12}\varepsilon_y - A_{11}^{\omega}, \quad \sigma_y = A_{21}\varepsilon_x + A_{22}\varepsilon_y - A_{22}^{\omega}, \quad \tau_{xy} = A_{44}\gamma_{xy} - A_{44}^{\omega}$$

where $s = \sin\phi$ and $c = \cos\phi$.

$$\begin{aligned} &A_{11}^{\omega} = \left(C_{22}s^2 + C_{12}c^2\right)\omega_2 - 2C_{44}cs\omega_{12}, \quad A_{22}^{\omega} = \left(C_{22}c^2 + C_{12}s^2\right)\omega_2 + 2C_{44}cs\omega_{12}, \\ &A_{44}^{\omega} = (C_{12} - C_{22})cs\omega_2 + C_{44}\left(c^2 - s^2\right)\omega_{12} \end{aligned}$$

These equations can be used in conjunction with the method of elastic solutions described in Section 4.1.2.

As an example, consider the two-matrix glass–epoxy composite described in Section 4.4.3 (see also Figs. 4.16, 4.30, and 4.31). Theoretical (solid lines) and experimental (dashed lines) stress–strain diagrams for ±30, ±45, and ±75° angle-ply layers under tension along the x-axis are shown in Fig. 4.62.

Angle-ply layers demonstrate a specific type of material nonlinearity – structural nonlinearity that can occur in the layers composed of linear elastic plies due to the change of the plies' orientations caused by loading. Since this effect manifests itself at high strains, consider a geometrically nonlinear problem of the ply deformation. This deformation can be described with the longitudinal, ε_1, transverse, ε_2, and shear, γ_{12}, strains that follow from Fig. 4.63 and can be expressed as

$$\varepsilon_1 = \frac{1}{ds_1}(ds_1' - ds_1), \quad \varepsilon_2 = \frac{1}{ds_2}(ds_2' - ds_2), \quad \gamma_{12} = \frac{\pi}{2} - \psi \tag{4.152}$$

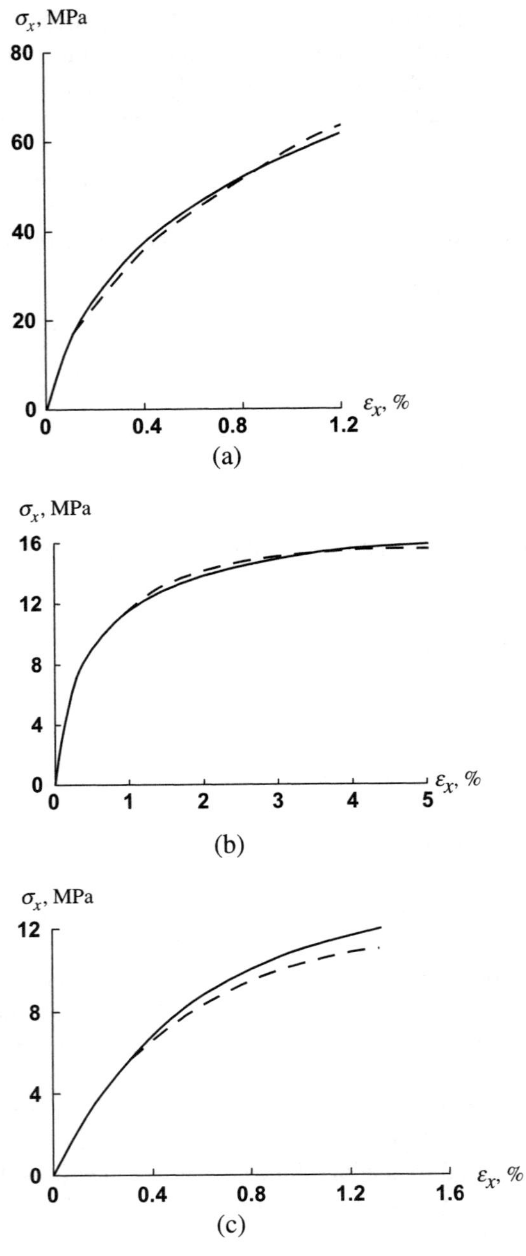

Fig. 4.62. Theoretical (solid lines) and experimental (dashed lines) stress–strain diagrams for ±30° (a), ±45° (b), and ±75° (c) angle-ply two-matrix composites under uniaxial tension.

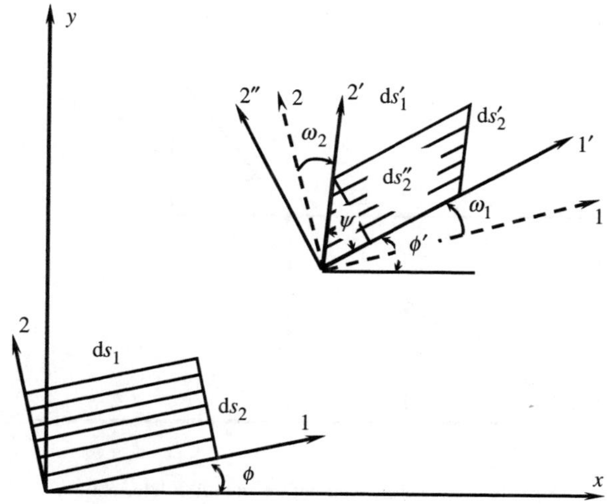

Fig. 4.63. Ply element before and after deformation.

In addition to this, we introduce strain ε_2'' in the direction normal to the fibers

$$\varepsilon_2'' = \frac{1}{ds_2}(ds_2'' - ds_2) \tag{4.153}$$

and the angle of rotation of the element as a solid in the 12-plane

$$\omega_{12} = \frac{1}{2}(\omega_1 - \omega_2)$$

where $\omega_1 = \phi' - \phi,\ \omega_2 = \frac{\pi}{2} + \phi - (\phi' + \psi)$ are the angles of rotation of axes $1'$ and $2'$ (see Fig. 4.63). Thus,

$$\omega_{12} = \phi' - \phi + \frac{\psi}{2} - \frac{\pi}{4} \tag{4.154}$$

Consider some arbitrary element ds_α, shown in Fig. 4.64, and introduce its strain

$$ds_\alpha = \frac{1}{ds_\alpha}(ds_\alpha' - ds_\alpha) \tag{4.155}$$

Repeating the derivation described in Section 2.5, we have

$$ds_\alpha^2 = dx^2 + dy^2$$

$$(ds_\alpha')^2 = (dx')^2 + (dy')^2 = (dx + du_x)^2 + (dy + du_y)^2$$

$$= (1 + \varepsilon_x)^2 dx^2 + (1 + \varepsilon_y)^2 dy^2 + 2\varepsilon_{xy} dx dy$$

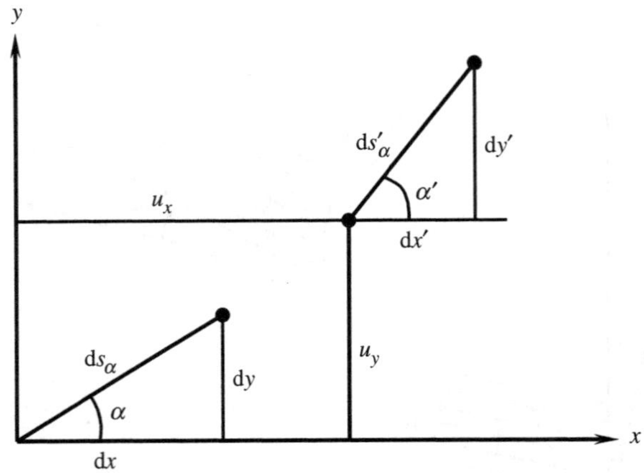

Fig. 4.64. Linear element before and after deformation.

where

$$(1+\varepsilon_x)^2 = 1 + 2\left[\frac{\partial u_x}{\partial x} + \frac{1}{2}\left(\frac{\partial u_x}{\partial x}\right)^2 + \frac{1}{2}\left(\frac{\partial u_y}{\partial x}\right)^2\right]$$

$$(1+\varepsilon_y)^2 = 1 + 2\left[\frac{\partial u_y}{\partial y} + \frac{1}{2}\left(\frac{\partial u_x}{\partial y}\right)^2 + \frac{1}{2}\left(\frac{\partial u_y}{\partial y}\right)^2\right] \tag{4.156}$$

$$\varepsilon_{xy} = \frac{\partial u_x}{\partial y} + \frac{\partial u_y}{\partial x} + \frac{\partial u_x}{\partial x}\frac{\partial u_x}{\partial y} + \frac{\partial u_y}{\partial x}\frac{\partial u_y}{\partial y}$$

Using Eq. (4.155), we arrive at

$$(1+\varepsilon_\alpha)^2 = (1+\varepsilon_x)^2\cos^2\alpha + (1+\varepsilon_y)^2\sin^2\alpha + \varepsilon_{xy}\sin 2\alpha \tag{4.157}$$

where $\cos\alpha = dx/ds_\alpha$ and $\sin\alpha = dy/ds_\alpha$.
In a similar way, we can find the angle α' after the deformation, i.e.,

$$\sin\alpha' = \frac{dy'}{ds_\alpha} = \frac{1}{1+\varepsilon_\alpha}\left[\left(1+\frac{\partial u_y}{\partial y}\right)\sin\alpha + \frac{\partial u_y}{\partial x}\cos\alpha\right]$$

$$\cos\alpha' = \frac{dx'}{ds_\alpha} = \frac{1}{1+\varepsilon_\alpha}\left[\left(1+\frac{\partial u_x}{\partial x}\right)\cos\alpha + \frac{\partial u_x}{\partial y}\sin\alpha\right] \tag{4.158}$$

Now return to the ply element in Fig. 4.63. Taking $\alpha = \phi$ in Eqs. (4.157) and (4.158), we obtain

$$
\begin{aligned}
(1+\varepsilon_1)^2 &= (1+\varepsilon_x)^2\cos^2\phi + (1+\varepsilon_y)^2\sin^2\phi + \varepsilon_{xy}\sin 2\phi \\
\sin\phi' &= \frac{1}{1+\varepsilon_1}\left[\left(1+\frac{\partial u_y}{\partial y}\right)\sin\phi + \frac{\partial u_y}{\partial x}\cos\phi\right] \\
\cos\phi' &= \frac{1}{1+\varepsilon_1}\left[\left(1+\frac{\partial u_x}{\partial x}\right)\cos\phi + \frac{\partial u_x}{\partial y}\sin\phi\right]
\end{aligned}
\tag{4.159}
$$

Putting $\alpha = \frac{\pi}{2} + \phi$, we have

$$
\begin{aligned}
(1+\varepsilon_2)^2 &= (1+\varepsilon_x)^2\sin^2\phi + (1+\varepsilon_y)^2\cos^2\phi - 2\varepsilon_{xy}\sin 2\phi \\
\sin(\phi'+\psi) &= \frac{1}{1+\varepsilon_2}\left[\left(1+\frac{\partial u_y}{\partial y}\right)\cos\phi - \frac{\partial u_y}{\partial x}\sin\phi\right] \\
\cos(\phi'+\psi) &= \frac{1}{1+\varepsilon_2}\left[-\left(1+\frac{\partial u_x}{\partial x}\right)\cos\phi + \frac{\partial u_x}{\partial y}\sin\phi\right]
\end{aligned}
\tag{4.160}
$$

Using the last equation of Eqs. (4.152), we can find the shear strain as $\sin\gamma_{12} = \cos\psi$. After some rearrangement, with the aid of Eqs. (4.159) and (4.160), we arrive at

$$
\sin\gamma_{12} = \frac{1}{(1+\varepsilon_1)(1+\varepsilon_2)}\left\{\left[(1+\varepsilon_y)^2 - (1+\varepsilon_x)^2\right]\sin\phi\cos\phi + \varepsilon_{xy}\cos 2\phi\right\}
\tag{4.161}
$$

For $\phi = 0$, axes 1 and 2 coincide, respectively, with axes x and y (see Fig. 4.63), and Eq. (4.161) yields

$$
\sin\gamma_{xy} = \frac{\varepsilon_{xy}}{(1+\varepsilon_x)(1+\varepsilon_y)}
\tag{4.162}
$$

Using this result to express ε_{xy}, we can write Eqs. (4.159)–(4.161) in the following final form

$$
\begin{aligned}
(1+\varepsilon_1)^2 &= (1+\varepsilon_x)^2\cos^2\phi + (1+\varepsilon_y)^2\sin^2\phi + (1+\varepsilon_x)(1+\varepsilon_y)\sin\gamma_{xy}\sin 2\phi \\
(1+\varepsilon_2)^2 &= (1+\varepsilon_x)^2\sin^2\phi + (1+\varepsilon_y)^2\cos^2\phi - (1+\varepsilon_x)(1+\varepsilon_y)\sin\gamma_{xy}\sin 2\phi \\
\sin\gamma_{12} &= \frac{1}{(1+\varepsilon_1)(1+\varepsilon_2)}\left\{\left[(1+\varepsilon_y)^2 - (1+\varepsilon_x)^2\right]\sin\phi\cos\phi\right. \\
&\qquad \left. +(1+\varepsilon_x)(1+\varepsilon_y)\sin\gamma_{xy}\cos 2\phi\right\}
\end{aligned}
\tag{4.163}
$$

It follows from Fig. 4.63 and the last equation of Eqs. (4.152), that $ds_2'' = ds_2' \sin\psi = ds_2' \cos\gamma_{12}$. So, in accordance with Eqs. (4.152) and (4.153),

$$1 + \varepsilon_2'' = (1 + \varepsilon_2)\cos\gamma_{12}$$

Using Eqs. (4.163) to transform this equation we get

$$1 + \varepsilon_2'' = \frac{(1 + \varepsilon_x)(1 + \varepsilon_y)}{1 + \varepsilon_1}\cos\gamma_{xy} \tag{4.164}$$

To express ϕ' in terms of ϕ and strains referred to the global coordinate frame x, y, consider Eq. (4.154). After rather cumbersome transformation with the aid of Eqs. (4.159) and (4.160), we obtain

$$\sin 2\omega_{12} = \frac{1}{(1+\varepsilon_1)(1+\varepsilon_2)}\left\{\left(\frac{\partial u_y}{\partial x} - \frac{\partial u_x}{\partial y} + \frac{\partial u_y}{\partial x}\frac{\partial u_y}{\partial y} - \frac{\partial u_x}{\partial x}\frac{\partial u_x}{\partial y}\right)\cos^2 2\phi\right.$$
$$+\left(\frac{\partial u_y}{\partial x} - \frac{\partial u_x}{\partial y} + \frac{\partial u_x}{\partial x}\frac{\partial u_y}{\partial x} - \frac{\partial u_x}{\partial y}\frac{\partial u_y}{\partial y}\right)\sin^2 2\phi$$
$$\left.+\frac{1}{4}\left[\left(\frac{\partial u_x}{\partial x} - \frac{\partial u_y}{\partial y}\right)^2 - \left(\frac{\partial u_y}{\partial x} + \frac{\partial u_x}{\partial y}\right)^2\right]\sin 4\phi\right\}$$

Taking $\phi = 0$, we can write rotation angle ω_z around the z-axis of the global coordinate frame, i.e.,

$$\sin 2\omega_z = \frac{1}{(1+\varepsilon_x)(1+\varepsilon_y)}\left(\frac{\partial u_y}{\partial x} - \frac{\partial u_x}{\partial y} + \frac{\partial u_y}{\partial x}\frac{\partial u_y}{\partial y} - \frac{\partial u_x}{\partial x}\frac{\partial u_x}{\partial y}\right) \tag{4.165}$$

Consider now Eqs. (4.156), (4.162), and (4.165) which form a set of four algebraic equations with respect to the derivatives of the displacements. Omitting the solution procedure, we can write the final outcome as

$$\frac{\partial u_x}{\partial x} = (1+\varepsilon_x)\cos\left(\frac{\gamma_{xy}}{2} + \omega_z\right) - 1, \quad \frac{\partial u_x}{\partial y} = (1+\varepsilon_y)\sin\left(\frac{\gamma_{xy}}{2} - \omega_z\right),$$
$$\frac{\partial u_y}{\partial x} = (1+\varepsilon_x)\sin\left(\frac{\gamma_{xy}}{2} + \omega_z\right), \quad \frac{\partial u_y}{\partial y} = (1+\varepsilon_y)\cos\left(\frac{\gamma_{xy}}{2} - \omega_z\right) - 1$$

Substituting these expressions into Eqs. (4.159), we have

$$\sin\phi' = \frac{1}{1+\varepsilon_1}\left[(1+\varepsilon_x)\sin\left(\frac{\gamma_{xy}}{2} + \omega_z\right)\cos\phi + (1+\varepsilon_y)\cos\left(\frac{\gamma_{xy}}{2} - \omega_z\right)\sin\phi\right.$$
$$\cos\phi' = \frac{1}{1+\varepsilon_1}\left[(1+\varepsilon_x)\cos\left(\frac{\gamma_{xy}}{2} + \omega_z\right)\cos\phi + (1+\varepsilon_y)\sin\left(\frac{\gamma_{xy}}{2} - \omega_z\right)\sin\phi\right. \tag{4.166}$$

The derived nonlinear equations, Eqs. (4.163), generalize Eqs. (4.69) for the case of large strains, whereas Eqs. (4.166) allow us to find the fiber orientation angle after deformation.

The equilibrium equations, Eqs. (4.68), retain their form but should be written for the deformed state, i.e.,

$$
\begin{aligned}
\sigma_x &= \sigma_1' \cos^2\phi' + \sigma_2'' \sin^2\phi' - \tau_{12}' \sin 2\phi' \\
\sigma_y &= \sigma_1' \sin^2\phi' + \sigma_2'' \cos^2\phi' + \tau_{12}' \sin 2\phi' \\
\tau_{xy} &= (\sigma_1' - \sigma_2'') \sin\phi' \cos\phi' + \tau_{12}' \cos 2\phi'
\end{aligned}
\tag{4.167}
$$

where σ_1', σ_2'', and τ_{12}' are stresses referred to coordinate frame $1'2''$ (see Fig. 4.63) and to the current thickness of the ply.

Consider a problem of uniaxial tension of a $\pm\phi$ angle-ply layer with stress σ_x. For this case, $\gamma_{xy} = 0$, $\omega_z = 0$, and Eqs. (4.163), (4.164), (4.166) take the form

$$
\begin{aligned}
(1+\varepsilon_1)^2 &= (1+\varepsilon_x)^2 \cos^2\phi + (1+\varepsilon_y)^2 \sin^2\phi \\
(1+\varepsilon_2)^2 &= (1+\varepsilon_x)^2 \sin^2\phi + (1+\varepsilon_y)^2 \cos^2\phi \\
\sin\gamma_{12} &= \frac{\sin\phi\cos\phi}{(1+\varepsilon_1)(1+\varepsilon_2)}[(1+\varepsilon_y)^2 - (1+\varepsilon_x)^2] \\
1+\varepsilon_2'' &= \frac{(1+\varepsilon_x)(1+\varepsilon_y)}{1+\varepsilon_1} \\
\sin\phi' &= \frac{1+\varepsilon_y}{1+\varepsilon_1}\sin\phi, \quad \cos\phi' = \frac{1+\varepsilon_x}{1+\varepsilon_1}\cos\phi
\end{aligned}
$$

For composite materials, the longitudinal strain ε_1 is usually small, and these equations can be further simplified as follows

$$
\begin{aligned}
\varepsilon_1 &= \varepsilon_x \cos^2\phi + \varepsilon_y \sin^2\phi + \frac{1}{2}\left(\varepsilon_x^2 \cos^2\phi + \varepsilon_y^2 \sin^2\phi\right) \\
(1+\varepsilon_2)^2 &= (1+\varepsilon_x)^2 \sin^2\phi + (1+\varepsilon_y)^2 \cos^2\phi \\
\sin\gamma_{12} &= \frac{1}{1+\varepsilon_2}\left[(1+\varepsilon_y)^2 - (1+\varepsilon_x)^2\right]\sin\phi\cos\phi \\
1+\varepsilon_2'' &= (1+\varepsilon_x)(1+\varepsilon_y) \\
\tan\phi' &= \frac{1+\varepsilon_y}{1+\varepsilon_x}\tan\phi
\end{aligned}
\tag{4.168}
$$

As an example, consider a specially synthesized highly deformable composite material made from glass composite fibers and thermoplastic matrix as discussed in

Section 4.4.3. Neglecting interaction of strains, we take constitutive equations for the unidirectional ply as

$$\sigma_1' = \frac{E_1\varepsilon_1}{1+\varepsilon_2''}, \quad \sigma_2' = \omega_2(\varepsilon_2''), \quad \tau_{12}' = \omega_{12}(\gamma_{12}) \tag{4.169}$$

where E_1 in the first equation is the longitudinal elasticity modulus, whereas ε_2'' in the denominator takes account of the decrease of the ply stiffness due to the increase in the fiber spacing. The constant E_1 and functions ω_2 and ω_{12} are determined from the experimental stress–strain diagrams for 0, 90, and ±45° specimens that are shown in Fig. 4.65. The results of calculations with the aid of Eqs. (4.167)–(4.169) are presented together with the corresponding experimental data in Fig. 4.66.

The foregoing equations comprise the analytical background for a promising manufacturing process allowing us to fabricate composite parts with complicated shapes by deforming partially cured preforms of simple shapes made by winding or laying-up (see, e.g., Cherevatsky, 1999). An example of such a part is presented in Fig. 4.67. The curved composite pipe shown in this figure was fabricated from a straight cylinder that was partially cured, loaded with pre-assigned internal pressure and end forces and moments, and cured completely in this state. The desired deformation of the part under loading is provided by the appropriate change of the fibers' orientation angles governed by Eqs. (4.163), (4.166), and (4.167).

Angle-ply layers can also demonstrate nonlinear behavior caused by the matrix cracking described in Section 4.4.2. To illustrate this type of nonlinearity, consider carbon–epoxy ±15, ±30, ±45, ±60, and ±75° angle-ply specimens studied experimentally by Lagace (1985). The unidirectional ply has the following mechanical properties: $E_1 = 131$ GPa, $E_2 = 11$ GPa, $G_{12} = 6$ GPa, $\nu_{21} = 0.28$, $\overline{\sigma}_1^+ = 1770$ MPa, $\overline{\sigma}_2^+ = 54$ MPa,

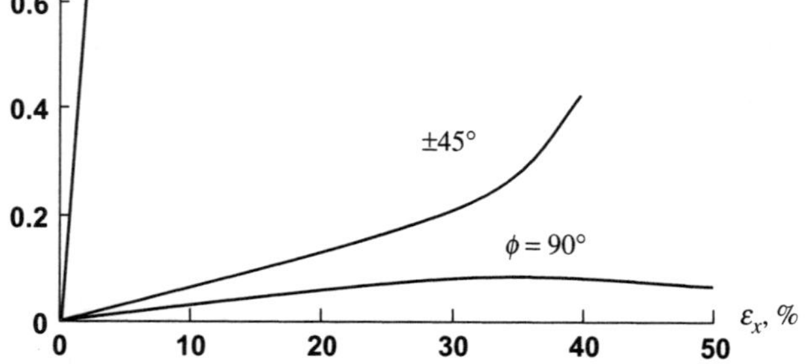

Fig. 4.65. Experimental stress–strain diagrams for 0, ±45, and 90° angle-ply layers.

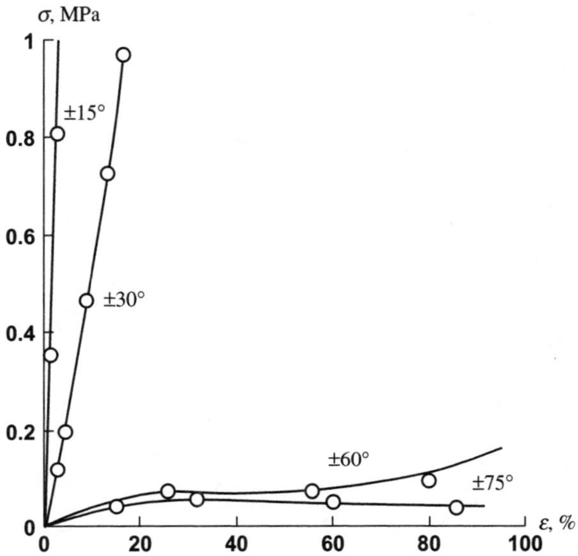

Fig. 4.66. Calculated (circles) and experimental (solid lines) stress–strain diagrams for ±15, ±30, ±60, and ±75° angle-ply layers.

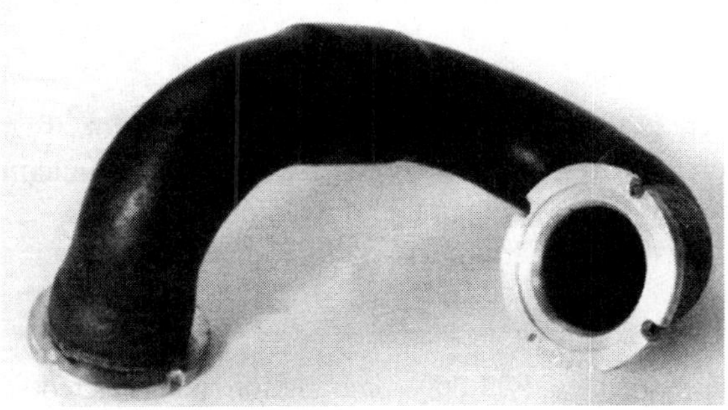

Fig. 4.67. A curved angle-ply pipe made by deformation of a filament-wound cylinder.

$\overline{\sigma}_2^- = 230\,\text{MPa}$, and $\overline{\tau}_{12} = 70\,\text{MPa}$. The dependencies $\sigma_1(\varepsilon_1)$ and $\sigma_2(\varepsilon_2)$ are linear, whereas for in-plane shear, the stress–strain diagram is not linear and is shown in Fig. 4.68. To take into account the material nonlinearity associated with shear, we use the constitutive equation derived in Section 4.2.2, i.e.,

$$\gamma_{12} = c_1\tau_{12} + c_2\tau_{12}^3$$

in which $c_1 = 1/G_{12}$ and $c_2 = 5.2 \times 10^{-8}(\text{MPa})^{-3}$.

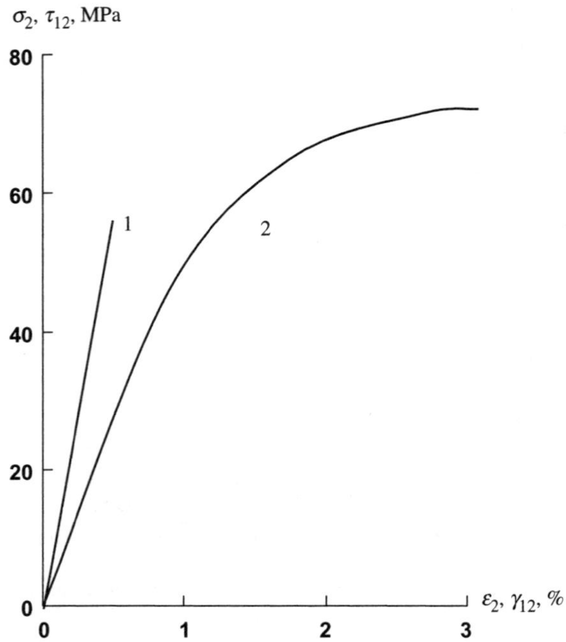

Fig. 4.68. Experimental stress–strain diagrams for transverse tension (1) and in-plane shear (2) of a carbon–epoxy unidirectional ply.

The specimens were tested in uniaxial tension in the x-direction. To calculate the applied stress σ_x that causes failure of the matrix, we use the simplest maximum stress strength criterion (see Chapter 6) which ignores the interaction of stresses, i.e.,

$$-\overline{\sigma}_2^- \leq \sigma_2 \leq \overline{\sigma}_2^+, \quad |\tau_{12}| \leq \overline{\tau}_{12}$$

Nonlinear behavior associated with ply degradation is predicted applying the procedure described in Section 4.4.2. Stress–strain diagrams are plotted using the method of successive loading (see Section 4.1.2).

Consider a $\pm 15°$ angle-ply layer. Point 1 on the theoretical diagram, shown in Fig. 4.69, corresponds to cracks in the matrix caused by shear. These cracks do not result in complete failure of the matrix because the transverse normal stress σ_2 is compressive (see Fig. 4.70) and does not reach $\overline{\sigma}_2^-$ before the failure of fibers under tension (point 2 on the diagram). As can be seen, the theoretical prediction of the material stiffness is quite good, whereas the predicted material strength (point 2) is much higher than the experimental (dot on the solid line). The reasons for this are discussed in the next section.

The theoretical diagram corresponding to the $\pm 30°$ layer (see Fig. 4.69) also has two specific points. Point 1 again corresponds to cracks in the matrix induced by the shear stress τ_{12}, whereas point 2 indicates complete failure of the matrix caused by the compressive stress σ_2 which reaches $\overline{\sigma}_2^-$ at this point. After the matrix fails, the fibers of an angle-ply

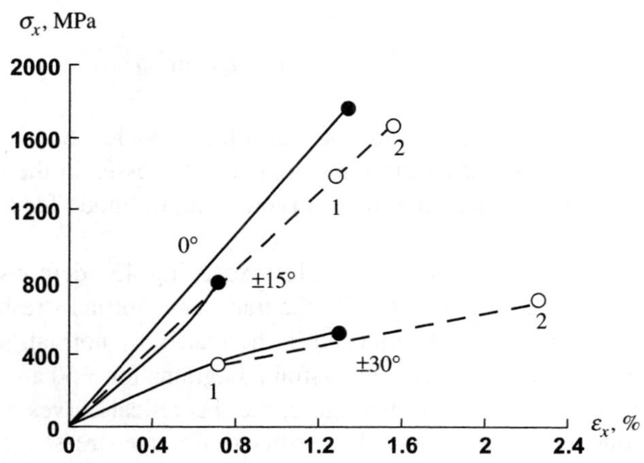

Fig. 4.69. Experimental (solid lines) and calculated (dashed lines) stress–strain diagrams for 0, ±15, and ±30° angle-ply carbon–epoxy layers.

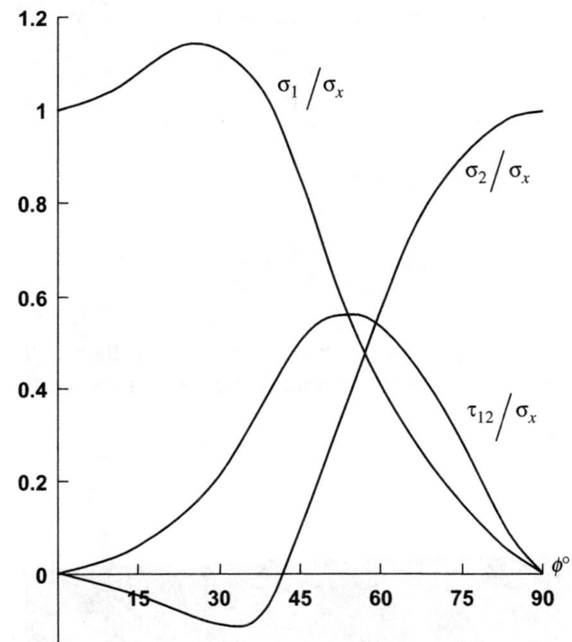

Fig. 4.70. Dependencies of the normalized stresses in the plies on the ply orientation angle.

layer cannot take the load. Indeed, putting $E_2 = G_{12} = \nu_{12} = 0$ in Eqs. (4.72), we obtain the following stiffness coefficients

$$A_{11} = E_1 \cos^4 \phi, \quad A_{22} = E_1 \sin^4 \phi, \quad A_{12} = E_1 \sin^2 \phi \cos^2 \phi$$

With these coefficients, the first equation of Eqs. (4.147) yields $E_x = 0$, which means that the system of fibers becomes a mechanism, and the stresses in the fibers, no matter how high they are, cannot balance the load. A typical failure mode for a $\pm 30°$ angle-ply specimen is shown in Fig. 4.71.

Angle-ply layers with fiber orientation angles exceeding 45° demonstrate a different type of behavior. As can be seen in Fig. 4.70, the transverse normal stress σ_2 is tensile for $\phi \geq 45°$. This means that the cracks induced in the matrix by normal, σ_2, or shear, τ_{12}, stresses cause failure of the layer. The stress–strain diagrams for ± 60 and $\pm 75°$ layers are shown in Fig. 4.72. As follows from this figure, the theoretical curves are linear and are close to the experimental ones, whereas the predicted ultimate stresses (circles) are again higher than the experimental values (dots).

Now consider the $\pm 45°$ angle-ply layer which demonstrates a very specific behavior. For this layer, the transverse normal stress, σ_2, is tensile but not high (see Fig. 4.70), and the cracks in the matrix are caused by the shear stress, τ_{12}. According to the ply model we use, to predict material response after the cracks appeared, we should take $G_{12} = 0$ in the stiffness coefficients. Then, Eqs. (4.72) yield

$$A_{11} = A_{12} = A_{22} = \frac{1}{4}(\overline{E}_1 + \overline{E}_2) + \frac{1}{2}\overline{E}_1 \nu_{12}$$

whereas Eqs. (4.146) and (4.147) give

$$\varepsilon_x = \frac{A_{22}\sigma_x}{A_{11}A_{22} - A_{12}^2}, \quad \varepsilon_y = -\frac{A_{12}\sigma_x}{A_{11}A_{22} - A_{12}^2}$$

The denominator of both expressions is zero, so it looks as though the material becomes a mechanism and should fail under the load that causes cracks in the matrix. However,

Fig. 4.71. A failure mode of $\pm 30°$ angle-ply specimen.

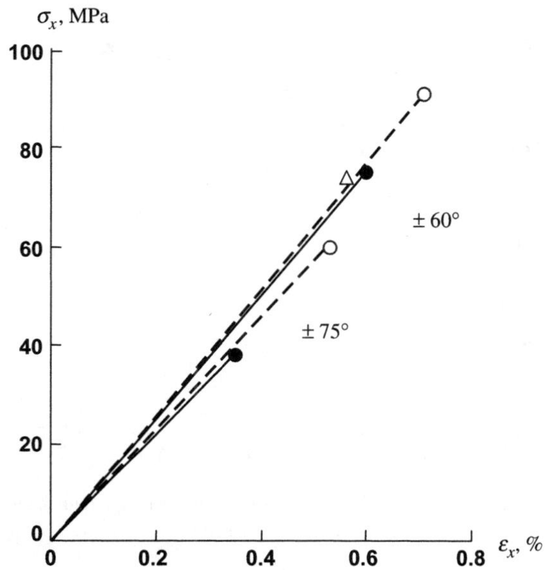

Fig. 4.72. Experimental (solid lines) and calculated (dashed lines) stress–strain diagrams for ±60 and ±75° angle-ply carbon–epoxy layers.

this is not the case. To explain why, consider the last equation of Eqs. (4.168), i.e.,

$$\tan\phi' = \frac{1+\varepsilon_y}{1+\varepsilon_x}\tan\phi$$

For the layer under study, $\tan\phi = 1$, $\varepsilon_y < 0$, $\varepsilon_x > 0$, so $\tan\phi' < 1$ and $\phi' < 45°$. However, in the plies with $\phi < 45°$ the transverse normal stresses, σ_2, become compressive (see Fig. 4.70) and close the cracks. Thus, the load exceeding the level at which the cracks appear due to shear locks the cracks and induces compression across the fibers thus preventing material failure. Since ϕ' is only slightly less than 45°, the material stiffness, E_x, is very low and slightly increases with the increase in strains and decrease of ϕ'. For the material under study, the calculated and experimental diagrams are shown in Fig. 4.73. The circle on the theoretical curve indicates the stress σ_x that causes cracks in the matrix. More pronounced behavior of this type is demonstrated by the glass–epoxy composites whose stress–strain diagram is presented with curve 1 in Fig. 4.74 (Alfutov and Zinoviev, 1982). A specific plateau on the curve and material hardening at high strain are the result of the angle variation as is also shown in Fig. 4.74 (line 2).

4.5.3. *Free-edge effects*

As shown in the previous section, there is a significant difference between the predicted and measured strength of an angle-ply specimen loaded in tension. This difference

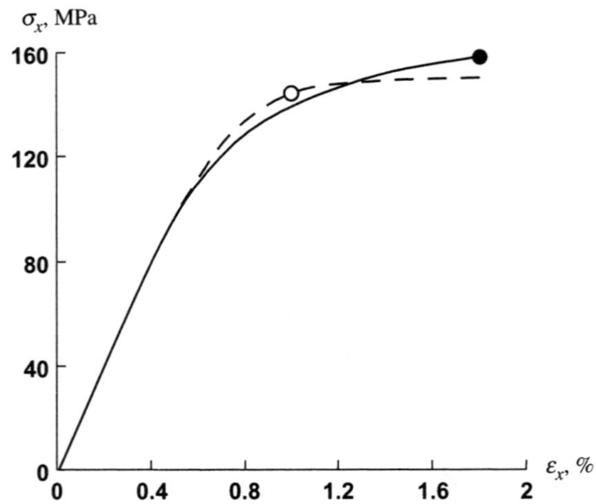

Fig. 4.73. Experimental (solid line) and calculated (dashed line) stress–strain diagrams for ±45° angle-ply carbon–epoxy layer.

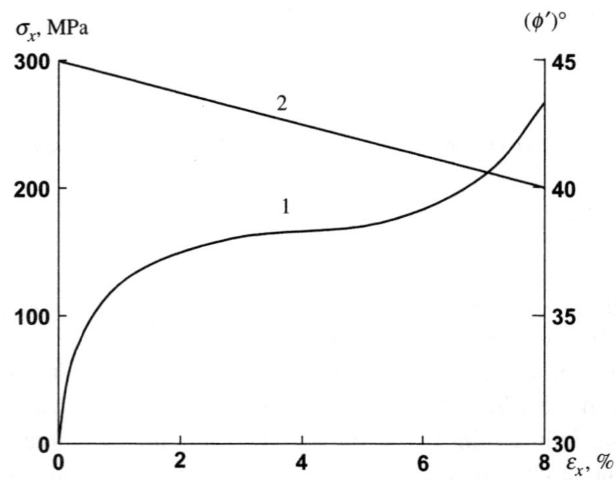

Fig. 4.74. Experimental dependencies of stress (1) and ply orientation angle (2) on strain for ±45° angle-ply glass–epoxy composite.

is associated with the stress concentration that occurs in the vicinity of the specimen longitudinal edges and was not taken into account in the analysis.

To study a free-edge effect in an angle-ply specimen, consider a strip whose initial width a is much smaller than the length l. Under tension with longitudinal stress σ, symmetric plies with orientation angles $+\phi$ and $-\phi$ tend to deform as shown in Fig. 4.75. As can be seen, deformation of the plies in the y-direction is the same, whereas the deformation in the x-direction tends to be different. This means that symmetric plies forming the

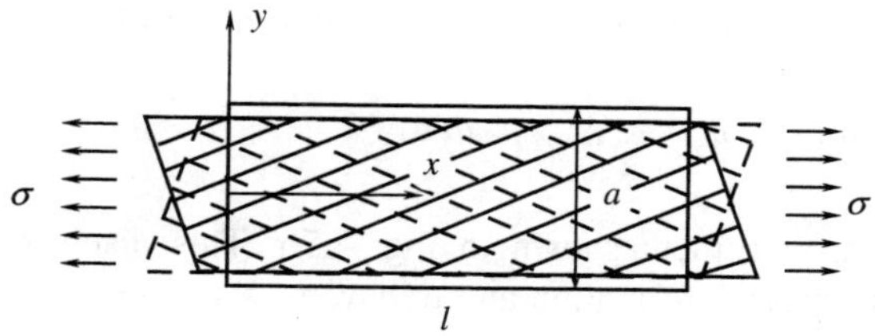

Fig. 4.75. Deformation of symmetric plies under tension.

angle-ply layer interact through the interlaminar shear stress τ_{xz} acting between the plies in the longitudinal direction. To describe the ply interaction, introduce the model shown in Fig. 4.76 according to which the in-plane stresses in the plies are applied to their middle surfaces, whereas transverse shear stresses act in some hypothetical layers introduced between these surfaces.

To simplify the problem, we further assume that the transverse stress can be neglected, i.e., $\sigma_y = 0$, and that the axial strain in the middle part of the long strip is constant, i.e., $\varepsilon_x = \varepsilon =$ constant. Thus, the constitutive equations, Eqs. (4.75), for a $+\phi$ ply have the form

$$\varepsilon_x = \frac{\sigma_x}{E_x^+} + \eta_{x,xy}^+ \frac{\tau_{xy}}{G_{xy}^+} \tag{4.170}$$

$$\varepsilon_y = -\nu_{yx}^+ \frac{\sigma_x}{E_x^+} + \eta_{y,xy}^+ \frac{\tau_{xy}}{G_{xy}^+} \tag{4.171}$$

$$\gamma_{xy} = \eta_{xy,x}^+ \frac{\sigma_x}{E_x^+} + \frac{\tau_{xy}}{G_{xy}^+} \tag{4.172}$$

where the elastic constants for an individual ply are specified by Eqs. (4.76). The strain-displacement equations, Eqs. (2.22), for the problem under study are

$$\varepsilon_x = \varepsilon = \frac{\partial u_x}{\partial x}, \quad \varepsilon_y = \frac{\partial u_y}{\partial y}, \quad \gamma_{xy} = \frac{\partial u_x}{\partial y} + \frac{\partial u_y}{\partial x} \tag{4.173}$$

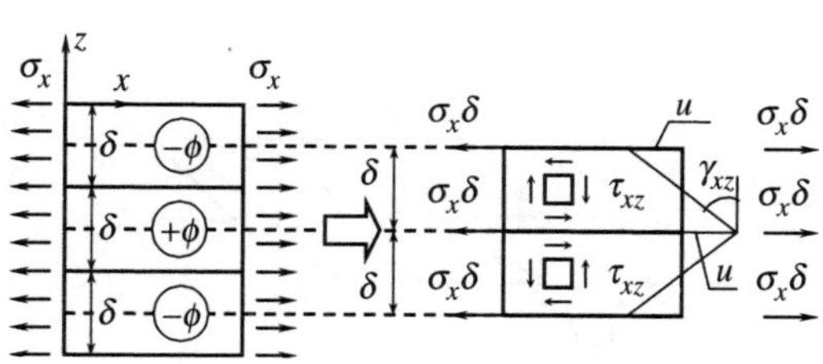

Fig. 4.76. A model simulating the plies interaction.

Integration of the first equation yields for the $+\phi$ and $-\phi$ plies

$$u_x^{+\phi} = \varepsilon \cdot x + u(y), \quad u_x^{-\phi} = \varepsilon \cdot x - u(y) \tag{4.174}$$

where $u(y)$ is the displacement shown in Fig. 4.76. This displacement results in the following transverse shear deformation and transverse shear stress

$$\gamma_{xz} = \frac{2}{\delta} u(y), \quad \tau_{xz} = G_{xz}\gamma_{xz} \tag{4.175}$$

where G_{xz} is the transverse shear modulus of the ply specified by Eqs. (4.76). Consider the equilibrium state of $+\phi$ ply element shown in Fig. 4.77. Equilibrium equations can be written as

$$\delta \frac{\partial \tau_{xy}}{\partial x} = 0, \quad \delta \frac{\partial \tau_{xy}}{\partial y} - 2\tau_{xz} = 0 \tag{4.176}$$

The first of these equations shows that τ_{xy} does not depend on x. Since the axial stress, σ_x, in the middle part of a long specimen also does not depend on x, Eqs. (4.171) and (4.173) allow us to conclude that ε_y, and hence u_y, do not depend on x. As a result, the last equation of Eqs. (4.173) yields in conjunction with the first equation of Eqs. (4.174)

$$\gamma_{xy} = \frac{\partial u_x}{\partial y} = \frac{du}{dy}$$

Using this expression and substituting ε from Eq. (4.170) in Eq. (4.172), we arrive at

$$\tau_{xy} = \frac{G_{xy}^{+}}{1-\eta}\left(\frac{du}{dy} - \eta_{xy,x}\varepsilon\right) \tag{4.177}$$

where $\eta = \eta_{x,xy}^{+}\eta_{xy,x}^{+}$.

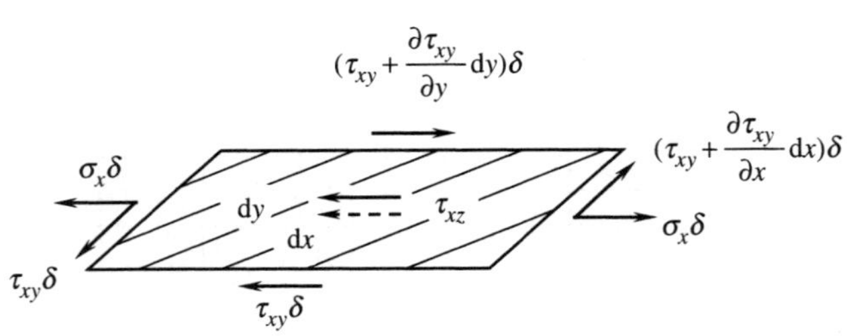

Fig. 4.77. Forces acting on the infinitesimal element of a ply.

Substitution of Eqs. (4.175) and (4.177) into the second equation of Eqs. (4.176) provides the following governing equation for the problem under study

$$\frac{d^2u}{dy^2} - k^2u = 0 \tag{4.178}$$

in which

$$k^2 = \frac{4G_{xz}(1-\eta)}{G_{xy}^+\delta^2}$$

Using the symmetry conditions, we can present the solution of Eq. (4.178) as

$$u = C\sinh ky$$

The constant C can be determined from the boundary conditions for the free longitudinal edges of the specimen (see Fig. 4.75) according to which $\tau_{xy}(y = \pm a/2) = 0$. Satisfying these conditions and using Eqs. (4.170), (4.171), (4.175), and (4.177), we finally obtain

$$\begin{aligned}
\varepsilon_x &= \varepsilon \\
\varepsilon_y &= \frac{\varepsilon}{1-\eta}\left[\eta_{y,xy}^+\eta_{xy,x}^+\left(\frac{\cosh\lambda\overline{y}}{\cosh\lambda} - 1\right) + \nu_{yx}^+\left(\eta\frac{\cosh\lambda\overline{y}}{\cosh\lambda} - 1\right)\right] \\
\gamma_{xy} &= \varepsilon\eta_{xy,x}^+\frac{\cosh\lambda\overline{y}}{\cosh\lambda} \\
\sigma_x &= \varepsilon E_x^+\left[1 - \frac{\eta}{1-\eta}\left(\frac{\cosh\lambda\overline{y}}{\cosh\lambda} - 1\right)\right] \\
\tau_{xy} &= \frac{G_{xy}^+\eta_{xy,x}^+}{1-\eta}\left(\frac{\cosh\lambda\overline{y}}{\cosh\lambda} - 1\right) \\
\tau_{xz} &= \frac{2\varepsilon}{k\delta}G_{xz}\eta_{xy,x}^+\frac{\sinh\lambda\overline{y}}{\cosh\lambda}
\end{aligned} \tag{4.179}$$

where

$$\lambda = \frac{ka}{2} = \frac{a}{\delta}\sqrt{(1-\eta)\frac{G_{xz}}{G_{xy}^+}}, \quad \overline{y} = \frac{2y}{a} \tag{4.180}$$

The axial stress, σ_x, should provide the stress resultant equal to σa (see Fig. 4.75), i.e.,

$$\int_{-a/2}^{a/2}\sigma_x dy = \sigma a$$

This condition allows us to determine the axial strain as

$$\varepsilon = \frac{\sigma}{E_x}$$

where

$$E_x = E_x^+\left[1 + \frac{\eta}{1-\eta}\left(1 - \frac{1}{\lambda}\tanh\lambda\right)\right] \tag{4.181}$$

is the apparent modulus of an angle-ply specimen.

Consider two limiting cases. First, suppose that $G_{xz} = 0$, i.e., that the plies are not bonded. Then, $\lambda = 0$ and because

$$\lim_{\lambda\to 0} \frac{1}{\lambda}\tanh\lambda = 1$$

$E_x = E_x^+$. Second, assume that $G_{xz} \to \infty$, i.e., that the interlaminar shear stiffness is infinitely high. Then $\lambda \to \infty$ and Eq. (4.181) yields

$$E_x = \frac{E_x^+}{1-\eta} \tag{4.182}$$

This result coincides with Eq. (4.149), which specifies the modulus of an angle-ply layer.

For finite values of G_{xz}, the parameter λ in Eqs. (4.180) is rather large because it includes the ratio of the specimen width, a, to the ply thickness, δ, which is, usually, a large number. Taking into account that $\tanh\lambda \le 1$, we can neglect the last term in Eq. (4.181) in comparison with unity. Thus, this equation reduces to Eq. (4.182). This means that tension of angle-ply specimens allows us to measure material stiffness with good accuracy despite the fact that the fibers are cut on the longitudinal edges of the specimens.

However, this is not true for the strength. The distribution of stresses over the half-width of the carbon–epoxy specimen with the properties given above and $a/\delta = 20$, $\phi = 45°$ is shown in Fig. 4.78. The stresses σ_x, τ_{xy}, and τ_{xz} were calculated with the aid of Eqs. (4.179), whereas stresses σ_1, σ_2, and τ_{12} in the principal material directions of the plies were found using Eqs. (4.69) for the corresponding strains and Hooke's law for the plies. As can be seen in Fig. 4.78, there exists a significant concentration of stress σ_2 that causes cracks in the matrix. Moreover, the interlaminar shear stress τ_{xz} that appears in the vicinity of the specimen edge can induce delamination of the specimen. The maximum value of stress σ_2 is

$$\sigma_2^{\max} = \sigma_2(\overline{y} = 1) = \overline{E}_2\varepsilon\,[(1 - \nu_{21}\nu_{yx}^+)\sin^2\phi$$
$$+ (\nu_{21} - \nu_{yx})\cos^2\phi - (1 - \nu_{21})\eta_{xy,x}^+ \sin\phi\cos\phi]$$

Using the modified strength condition, i.e., $\sigma_2^{\max} = \overline{\sigma}_2^+$ to evaluate the strength of $\pm 60°$ specimen, we arrive at the result shown with a triangular symbol in Fig. 4.72. As can

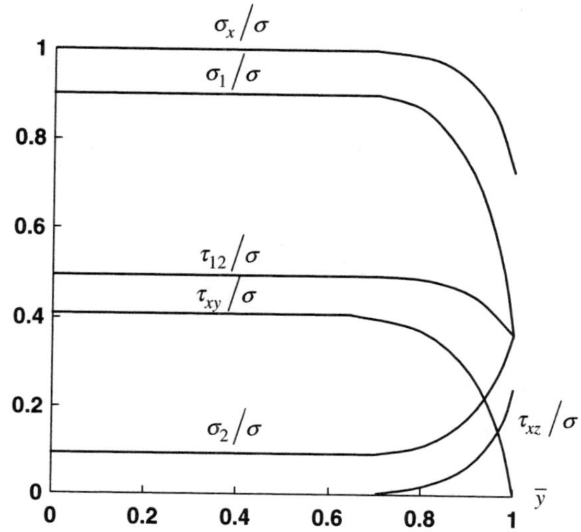

Fig. 4.78. Distribution of normalized stresses over the width of a ±45° angle-ply carbon–epoxy specimen.

be seen, the allowance for the stress concentration results is in fair agreement with the experimental strength (dot).

Thus, the strength of angle-ply specimens is reduced by the free-edge effects, which causes a dependence of the observed material strength on the width of the specimen. Such dependence is shown in Fig. 4.79 for 105-mm diameter and 2.5-mm-thick fiberglass rings made by winding at ±35°angles with respect to the axis and loaded with internal pressure by two half-disks as in Fig. 3.46 (Fukui et al., 1966).

It should be emphasized that the free-edge effect occurs in specimens only and does not show itself in composite structures which, being properly designed, must not have free edges of such a type.

4.6. Fabric layers

Textile preforming plays an important role in composite technology providing glass, aramid, carbon (see Fig. 4.80), and hybrid fabrics that are widely used as reinforcing materials. The main advantages of woven composites are their cost efficiency and high processability, particularly, in lay-up manufacturing of large-scale structures (see Figs. 4.81 and 4.82). However, on the other hand, processing of fibers and their bending in the process of weaving results in substantial reduction of material strength and stiffness. As can be seen in Fig. 4.83, in which a typical woven structure is shown the warp (lengthwise) and fill (crosswise) yarns forming the fabric make angle $\alpha \geq 0$ with the plane of the fabric layer.

To demonstrate how this angle influences material stiffness, consider tension of the structure shown in Fig. 4.83 in the warp direction. The apparent modulus of elasticity can

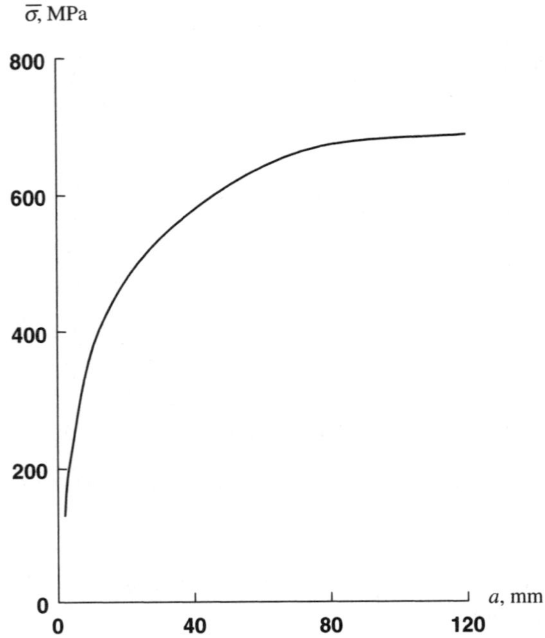

Fig. 4.79. Experimental dependence of strength of a ±35° angle-ply layer on the width of the specimen.

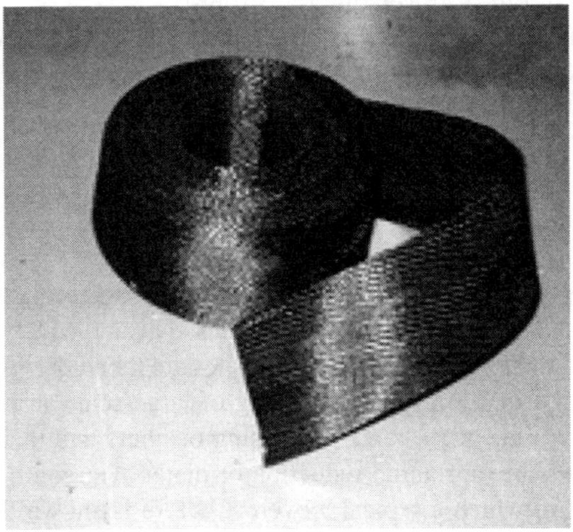

Fig. 4.80. A carbon fabric tape.

Fig. 4.81. A composite body of a boat made of fiberglass fabric by lay-up method. Courtesy of CRISM.

Fig. 4.82. A composite leading edge of an aeroplane wing made of carbon fabric by lay-up method. Courtesy of CRISM.

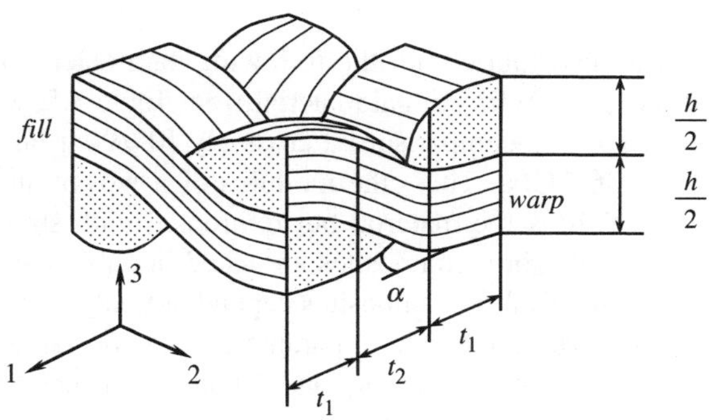

Fig. 4.83. Unit cell of a fabric structure.

be expressed as

$$E_a A_a = E_f A_f + E_w A_w \tag{4.183}$$

where $A_a = h(2t_1 + t_2)$ is the apparent cross-sectional area and

$$A_f = \frac{h}{2}(2t_1 + t_2), \quad A_w = \frac{h}{4}(4t_1 + t_2)$$

are the areas of the fill and warp yarns in the cross section. Substitution into Eq. (4.183) yields

$$E_a = \frac{1}{2}\left[E_f + \frac{E_w(4t_1 + t_2)}{2(2t_1 + t_2)}\right]$$

Since the fibers of the fill yarns are orthogonal to the loading direction, we can take $E_f = E_2$, where E_2 is the transverse modulus of a unidirectional composite. The compliance of the warp yarn can be decomposed into two parts corresponding to t_1 and t_2 in Fig. 4.83, i.e.,

$$\frac{2t_1 + t_2}{E_w} = \frac{2t_1}{E_1} + \frac{t_2}{E_\alpha}$$

where E_1 is the longitudinal modulus of a unidirectional composite, whereas E_α can be determined with the aid of the first equation of Eqs. (4.76) if we change ϕ for α, i.e.,

$$\frac{1}{E_\alpha} = \frac{\cos^4\alpha}{E_1} + \frac{\sin^4\alpha}{E_2} + \left(\frac{1}{G_{12}} - \frac{2\nu_{21}}{E_1}\right)\sin^2\alpha\cos^2\alpha \tag{4.184}$$

The final result is as follows

$$E_a = \frac{E_2}{2} + \frac{E_1(4t_1 + t_2)}{4\left\{2t_1 + t_2\left[\cos^4\alpha + \frac{E_1}{E_2}\sin^4\alpha + \left(\frac{E_1}{G_{12}} - 2\nu_{21}\right)\sin^2\alpha\cos^2\alpha\right]\right\}} \tag{4.185}$$

For example, consider a glass fabric with the following parameters: $\alpha = 12°$, $t_2 = 2t_1$. Taking elastic constants for a unidirectional material from Table 3.5, we get for the fabric composite $E_a = 23.5\,\text{GPa}$. For comparison, a cross-ply [0°/90°] laminate made of the same material has $E = 36.5\,\text{GPa}$. Thus, the modulus of a woven structure is lower by 37% than the modulus of the same material but reinforced with straight fibers. Typical mechanical characteristics of fabric composites are listed in Table 4.4.

The stiffness and strength of fabric composites depend not only on the yarns and matrix properties, but also on the material structural parameters, i.e., on fabric count and weave. The fabric count specifies the number of warp and fill yarns per inch (25.4 mm), whereas the weave determines how the warp and the fill yarns are interlaced. Typical weave patterns are shown in Fig. 4.84 and include plain, twill, and triaxial woven fabrics. In the

Table 4.4
Typical properties of fabric composites.

Property	Glass fabric–epoxy	Aramid fabric–epoxy	Carbon fabric–epoxy
Fiber volume fraction	0.43	0.46	0.45
Density (g/cm^3)	1.85	1.25	1.40
Longitudinal modulus (GPa)	26	34	70
Transverse modulus (GPa)	22	34	70
Shear modulus (GPa)	7.2	5.6	5.8
Poisson's ratio	0.13	0.15	0.09
Longitudinal tensile strength (MPa)	400	600	860
Longitudinal compressive strength (MPa)	350	150	560
Transverse tensile strength (MPa)	380	500	850
Transverse compressive strength (MPa)	280	150	560
In-plane shear strength (MPa)	45	44	150

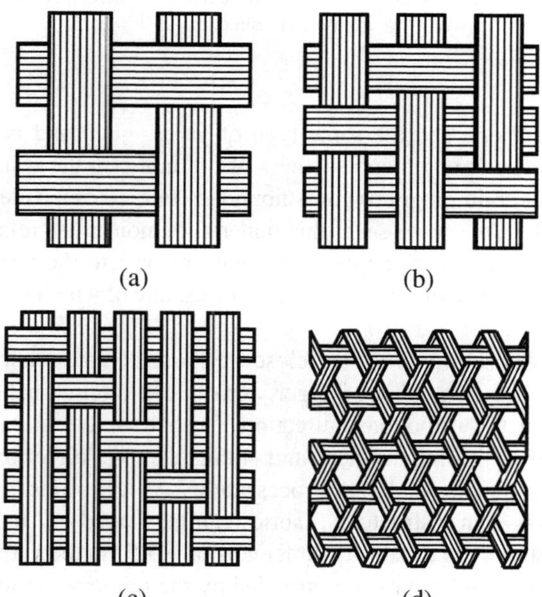
(a) (b)
(c) (d)

Fig. 4.84. Plain (a), twill (b) and (c), and triaxial (d) woven fabrics.

plain weave (see Fig. 4.84a) which is the most common and the oldest, the warp yarn is repeatedly woven over the fill yarn and under the next fill yarn. In the twill weave, the warp yarn passes over and under two or more fill yarns (as in Fig. 4.84b and c) in a regular way.

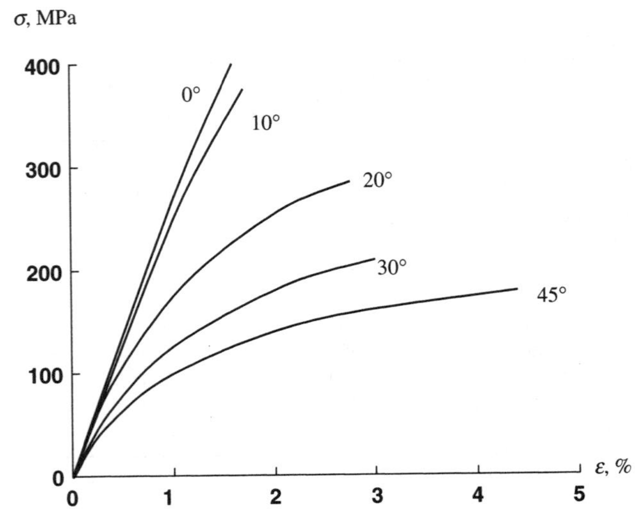

Fig. 4.85. Stress–strain curves for fiberglass fabric composite loaded in tension at different angles with respect to the warp direction.

Being formed from one and the same type of yarns, plain and twill weaves provide approximately the same strength and stiffness of the fabric in the warp and the fill directions. Typical stress–strain diagrams for a fiberglass fabric composite of such a type are presented in Fig. 4.85. As can be seen, this material demonstrates relatively low stiffness and strength under tension at an angle of 45° with respect to the warp or fill directions. To improve these properties, multiaxial woven fabrics, one of which is shown in Fig. 4.84d, can be used.

Fabric materials whose properties are closer to those of unidirectional composites are made by weaving a greater number of larger yarns in the longitudinal direction and fewer and smaller yarns in the orthogonal direction. Such a weave is called unidirectional. It provides materials with high stiffness and strength in one direction, which is specific for unidirectional composites and high processability typical of fabric composites.

Being fabricated as planar structures, fabrics can be shaped on shallow surfaces using the material's high stretching capability under tension at 45° to the yarns' directions. Many more possibilities for such shaping are provided by the implementation of knitted fabrics whose strain to failure exceeds 100%. Moreover, knitting allows us to shape the fibrous preform in accordance with the shape of the future composite part. There exist different knitting patterns, some of which are shown in Fig. 4.86. Relatively high curvature of the yarns in knitted fabrics, and possible fiber breakage in the process of knitting, result in materials whose strength and stiffness are less than those of woven fabric composites, but whose processability is better, and the cost is lower. Typical stress–strain diagrams for composites reinforced by knitted fabrics are presented in Fig. 4.87.

Material properties close to those of woven composites are provided by braided structures which, being usually tubular in form, are fabricated by mutual intertwining,

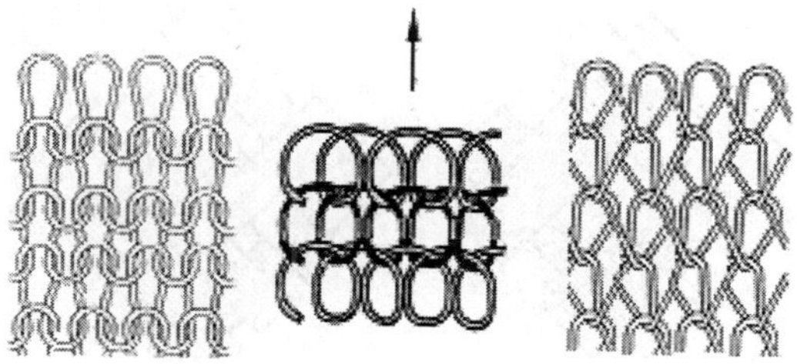

Fig. 4.86. Typical knitted structures.

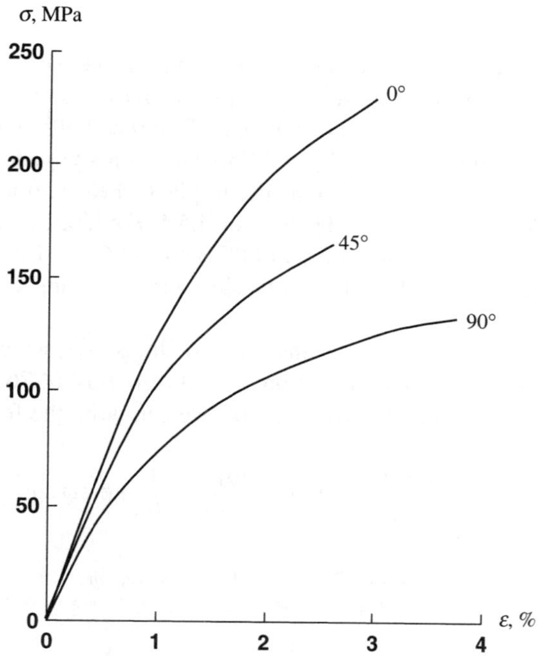

Fig. 4.87. Typical stress–strain curves for fiberglass-knitted composites loaded in tension at different angles with respect to direction indicated by the arrow Fig. 4.86.

or twisting of yarns around each other. Typical braided structures are shown in Fig. 4.88. The biaxial braided fabrics in Fig. 4.88 can incorporate longitudinal yarns forming a triaxial braid whose structure is similar to that shown in Fig. 4.84d. Braided preforms are characterized with very high processability providing near net-shape manufacturing of tubes and profiles with various cross-sectional shapes.

Although microstructural models of the type shown in Fig. 4.83 which lead to equations similar to Eq. (4.185) have been developed to predict the stiffness and even strength characteristics of fabric composites (e.g., Skudra et al., 1989), for practical design and analysis, these characteristics are usually determined by experimental methods. The elastic

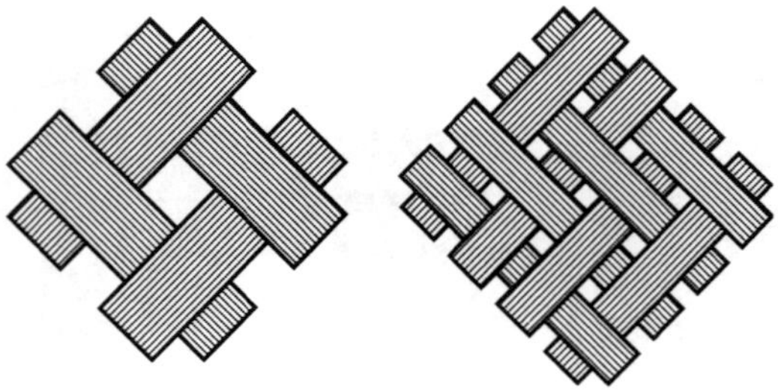

(a) (b)

Fig. 4.88. Diamond (a) and regular (b) braided fabric structures.

constants entering the constitutive equations written in principal material coordinates, e.g., Eqs. (4.55), are determined by testing strips cut out of fabric composite plates at different angles with respect to the orthotropy axes. The 0 and 90° specimens are used to determine moduli of elasticity E_1 and E_2 and Poisson's ratios ν_{12} and ν_{21} (or parameters for nonlinear stress–strain curves), whereas the in-plane shear stiffness can be obtained with the aid of off-axis tension described in Section 4.3.1. For fabric composites, the elastic constants usually satisfy conditions in Eqs. (4.85) and (4.86), and there exists the angle ϕ specified by Eq. (4.84) such that off-axis tension under this angle is not accompanied with shear–extension coupling.

Since Eq. (4.84) specifying ϕ includes the shear modulus G_{12}, which is not known, we can transform the results presented in Section 4.3.1. Using Eqs. (4.76) and assuming that there is no shear–extension coupling ($\eta_{x,xy} = 0$), we can write the following equations

$$\frac{1}{E_x} = \frac{1+\nu_{21}}{E_1}\cos^4\phi + \frac{1+\nu_{12}}{E_2}\sin^4\phi - \frac{\nu_{21}}{E_1} + \frac{1}{G_{12}}\sin^2\phi\cos^2\phi$$

$$\frac{\nu_{yx}}{E_x} = \frac{\nu_{21}}{E_1} - \left(\frac{1+\nu_{21}}{E_1} + \frac{1+\nu_{12}}{E_2} - \frac{1}{G_{12}}\right)\sin^2\phi\cos^2\phi \tag{4.186}$$

$$\frac{1+\nu_{21}}{E_1}\cos^2\phi - \frac{1+\nu_{12}}{E_2}\sin^2\phi - \frac{1}{2G_{12}}\cos 2\phi = 0$$

Summing up the first two of these equations, we get

$$\frac{1+\nu_{yx}}{E_x} = \left(\frac{1+\nu_{21}}{E_1}\cos^2\phi - \frac{1+\nu_{12}}{E_2}\sin^2\phi\right)\cos 2\phi + \frac{2}{G_{12}}\sin^2\phi\cos^2\phi$$

Using the third equation, we arrive at the following remarkable result

$$G_{12} = \frac{E_x}{2(1+\nu_{yx})} \tag{4.187}$$

similar to the corresponding formula for isotropic materials, Eq. (2.57). It should be emphasized that Eq. (4.187) is valid for off-axis tension in the x-direction making some special angle ϕ with the principal material axis 1. This angle is given by Eq. (4.84). Another form of this expression follows from the last equation of Eqs. (4.186) and (4.187), i.e.,

$$\sin^2\phi = \frac{\left[(1+\nu_{yx})/E_x\right] - \left[(1+\nu_{21})/E_1\right]}{2\left[(1+\nu_{yx})/E_x\right] - (1+\nu_{21})/E_1 - (1+\nu_{12})/E_2} \tag{4.188}$$

For fabric composites whose stiffness in the warp and the fill directions is the same ($E_1 = E_2$), Eq. (4.188) yields $\phi = 45°$.

4.7. Lattice layer

A layer with a relatively low density and high stiffness can be obtained with a lattice structure which can be made by a winding modified in such a way that the tapes are laid onto preceding tapes and not beside them, as in conventional filament winding (see Fig. 4.89). The lattice layer can be the single layer of the structure as in Fig. 4.90, or can be combined with a skin as in Fig. 4.91. As a rule, lattice structures have the form of cylindrical or conical shells in which the lattice layer is formed with two systems of ribs – a symmetric system of helical ribs and a system of circumferential ribs (see Figs. 4.90 and 4.91). However, there exist lattice structures with three systems of ribs as in Fig. 4.92.

In general, a lattice layer can consist of k symmetric systems of ribs making angles $\pm\phi_j (j = 1, 2, 3 \ldots k)$ with the x-axis and having geometric parameters shown in Fig. 4.93. Particularly, the lattice layer presented in this figure has $k = 2, \phi_1 = \phi$, and $\phi_2 = 90°$.

Fig. 4.89. Winding of a lattice layer. Courtesy of CRISM.

Fig. 4.90. Carbon–epoxy lattice spacecraft fitting in the assemble fixture. Courtesy of CRISM.

Fig. 4.91. Interstage composite lattice structure. Courtesy of CRISM.

Fig. 4.92. A composite lattice shear web structure.

Since the lattice structure is formed with dense and regular systems of ribs, the ribs can be smeared over the layer surface when modeled, which is thus simulated with a continuous layer having some effective (apparent) stiffnesses. Taking into account that the ribs work in their axial directions only, neglecting the ribs' torsion and bending in the plane of the lattice layer, and using Eqs. (4.72), we get

$$A_{11} = \sum_{j=1}^{k} B_j \cos^4 \phi_j, \quad A_{22} = \sum_{j=1}^{k} B_j \sin^4 \phi_j,$$

$$A_{12} = A_{21} = A_{44} = \sum_{j=1}^{k} B_j \sin^2 \phi_j \cos^2 \phi_j, \tag{4.189}$$

$$A_{44} = \sum_{j=1}^{k} C_j \cos^2 \phi_j, \quad A_{55} = \sum_{j=1}^{k} C_j \sin^2 \phi_j$$

Here, $B_j = E_j \delta_j / a_j$ and $C_j = G_j \delta_j / a_j$, where E_j and G_j are the modulus of elasticity and the shear modulus of the ribs' materials, δ_j are the ribs' widths, and a_j are the ribs' spacings (see Fig. 4.93).

4.8. Spatially reinforced layers and bulk materials

The layers considered in the previous sections and formed of unidirectionally reinforced plies and tapes (Sections 4.2–4.5 and 4.7) or fabrics reinforced in the layer plane

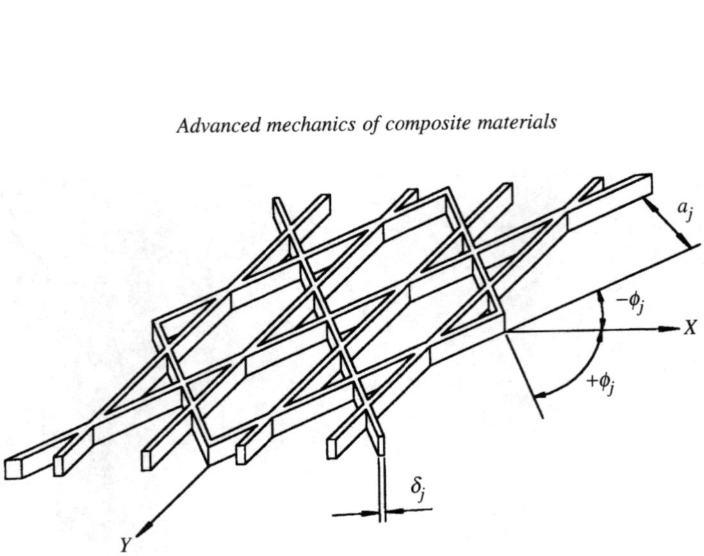

Fig. 4.93. Geometric parameters of a lattice structure.

(Section 4.6) suffer from a serious shortcoming – their transverse (normal to the layer plane) stiffness and strength are substantially lower than the corresponding in-plane characteristics. To improve the material properties under tension or compression in the z-direction and in shear in the xz- and the yz-planes (see, e.g., Fig. 4.18), the material should be additionally reinforced with fibers or yarns directed along the z-axis or making some angles (less than a right angle) with this axis.

A simple and natural way of such triaxial reinforcement is provided by the implementation of three-dimensionally woven or braided fabrics. Three-dimensional weaving or braiding is a variant of the corresponding planar process wherein some yarns are going in the thickness direction. An alternative method involves assembling elementary fabric layers or unidirectional plies into a three-dimensionally reinforced structure by sewing or stitching. Depending on the size of the additional yarn and frequency of sewing or stitching, the transverse mechanical properties of the two-dimensionally reinforced composite can be improved to a greater or lesser extent. A third way is associated with the introduction of composite or metal pins parallel to the z-axis that can be inserted in the material before or after it is cured. A similar effect can be achieved by the so-called needle punching. The needles puncture the fabric, break the fibers that compose the yarns, and direct the broken fibers through the layer thickness. Short fibers (or whiskers) may also be introduced into the matrix with which the fabrics or the systems of fibers are impregnated.

Another class of spatially reinforced composites, used mainly in carbon–carbon technology, is formed by bulk materials multi-dimensionally reinforced with fine rectilinear yarns composed of carbon fibers bound with a polymeric or carbon matrix. The basic structural element of these materials is a parallelepiped shown in Fig. 4.94. The simplest spatial structure is the so-called 3D (three-dimensionally reinforced) in which reinforcing elements are directed along the ribs AA_1, AB, and AD of the basic parallelepiped in Fig. 4.94. This structure is shown in Fig. 4.95 (Vasiliev and Tarnopol'skii, 1990). A more complicated 4D structure with reinforcing elements directed along the diagonals AC_1, A_1C, BD_1, and B_1D (see Fig. 4.94) is shown in Fig. 4.96 (Tarnopol'skii et al., 1987). An example of this structure is presented in Fig. 1.22. A cross section of a 5D structure reinforced along

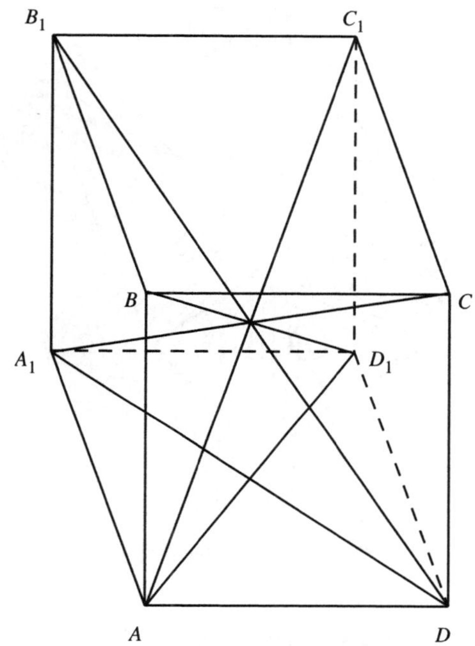

Fig. 4.94. The basic structural element of multi-dimensionally reinforced materials.

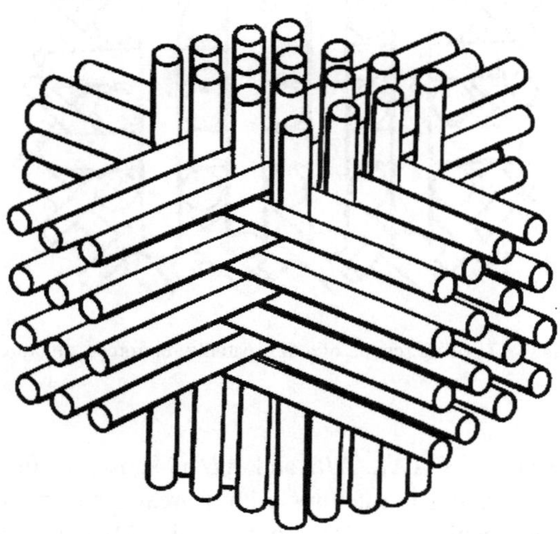

Fig. 4.95. 3D spatially reinforced structure.

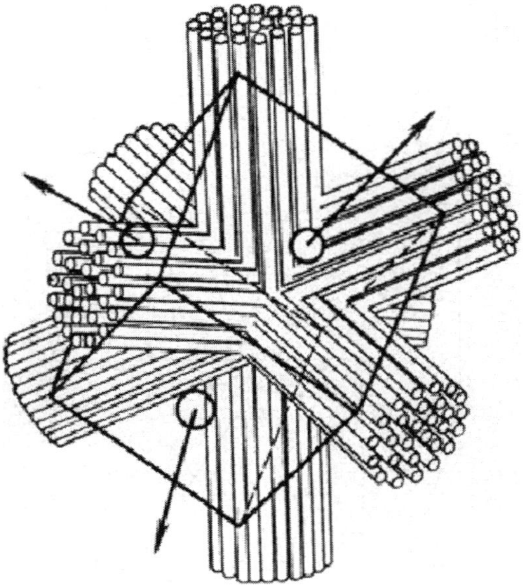

Fig. 4.96. 4D spatially reinforced structure.

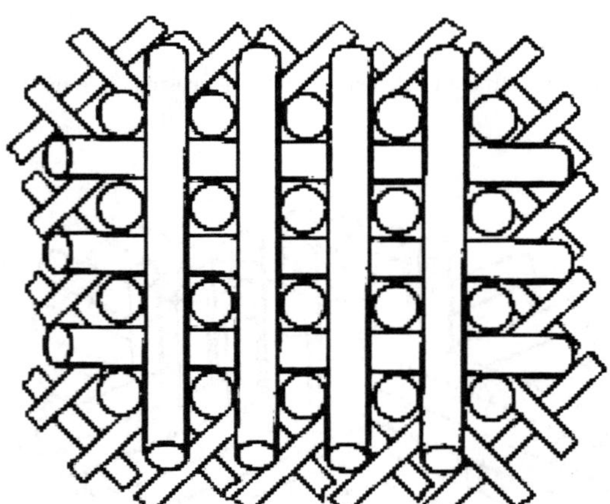

Fig. 4.97. Cross section of a 5D spatially reinforced structure.

diagonals AD_1, A_1D and ribs AA_1, AB, and AD is shown in Fig. 4.97 (Vasiliev and Tarnopol'skii, 1990). There exist structures with a greater number of reinforcing directions. For example, combination of a 4D structure (Fig. 4.96) with reinforcements along the ribs *AB* and *AD* (see Fig. 4.94) results in a 6D structure; addition of reinforcements in the direction of the rib AA_1 gives a 7D structure, and so on up to 13D which is the most complicated of the spatial structures under discussion.

The mechanical properties of multi-dimensional composite structures can be qualitatively predicted with the microstructural models discussed, e.g., by Tarnopol'skii et al. (1992). However, for practical applications these characteristics are usually obtained by experimental methods. Being orthotropic in the global coordinates of the structure x, y, and z, spatially reinforced composites are described within the framework of a phenomenological model ignoring their microctructure by three-dimensional constitutive equations analogous to Eqs. (4.53) or Eqs. (4.54) in which 1 should be changed for x, 2 for y, and 3 for z. These equations include nine independent elastic constants. Stiffness coefficients in the basic plane, i.e., E_x, E_y, G_{xy}, and ν_{xy}, are determined using traditional tests developed for unidirectional and fabric composites as discussed in Sections 3.4, 4.2, and 4.6. The transverse modulus E_z and the corresponding Poisson's ratios ν_{xz} and ν_{yz} can be determined using material compression in the z-direction. Transverse shear moduli G_{xz} and G_{yz} can be calculated using the results of a three-point beam bending test shown in Fig. 4.98. A specimen cut out of the material is loaded with force P, and the deflection at the central point, w, is measured. According to the theory of composite beams (Vasiliev, 1993)

$$w = \frac{Pl^3}{4bh^3E_x}\left(1 + \frac{h^2E_x}{l^2G_{xz}}\right)$$

Knowing P, the corresponding w and modulus E_x (or E_y), we can calculate G_{xz} (or G_{yz}). It should be noted that for reliable calculation the beam should be rather short, because for high ratios of l/h the second term in parenthesis is small in comparison with unity.

The last spatially reinforced structure that is considered here is formed by a unidirectional composite material whose principal material axes 1, 2, and 3 make some angles with the global structural axes x, y, and z (see Fig. 4.99). In the principal material coordinates, the constitutive equations have the form of Eqs. (4.53) or Eqs. (4.54). Introducing directional cosines l_{xi}, l_{yi}, and l_{zi} which are cosines of the angles that the i-axis ($i = 1, 2, 3$) makes with axes x, y, and z, respectively, applying Eqs. (2.8), (2.9), and (2.31) to transform stresses and strains in coordinates 1, 2, and 3 to stresses and strains referred to coordinates x, y, and z, and using the procedure described in Section 4.3.1, we finally

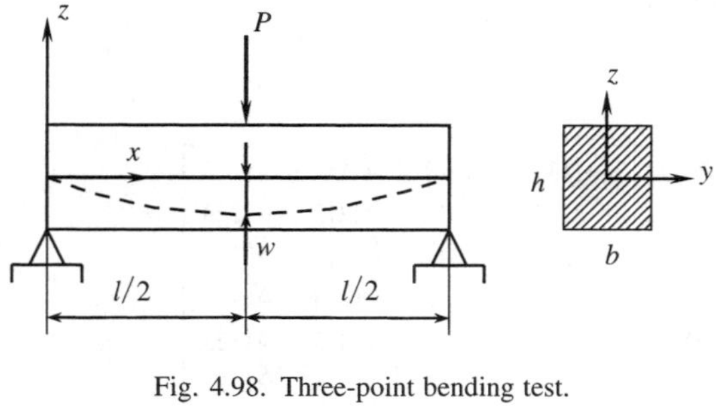

Fig. 4.98. Three-point bending test.

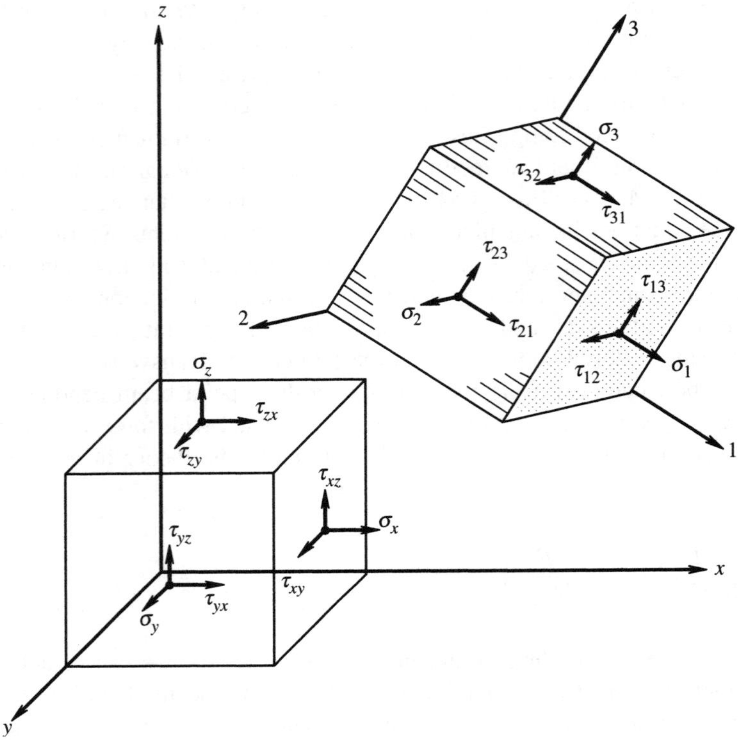

Fig. 4.99. Material elements referred to the global structural coordinate frame x, y, and z and to the principal material axes 1, 2, and 3.

arrive at the following constitutive equations in the global structural coordinate frame

$$\begin{Bmatrix} \sigma_x \\ \sigma_y \\ \sigma_z \\ \tau_{xy} \\ \tau_{xz} \\ \tau_{yz} \end{Bmatrix} = [S] \begin{Bmatrix} \varepsilon_x \\ \varepsilon_y \\ \varepsilon_z \\ \gamma_{xy} \\ \gamma_{xz} \\ \gamma_{yz} \end{Bmatrix} \tag{4.190}$$

in which

$$[S] = \begin{bmatrix} S_{1111} & S_{1122} & S_{1133} & S_{1112} & S_{1113} & S_{1123} \\ & S_{2222} & S_{2233} & S_{2212} & S_{2213} & S_{2223} \\ & & S_{3333} & S_{3312} & S_{3313} & S_{3323} \\ & & & S_{1212} & S_{1213} & S_{1223} \\ & \text{sym} & & & S_{1313} & S_{1323} \\ & & & & & S_{2323} \end{bmatrix}$$

is the stiffness matrix where

$$S_{1111} = A_1 l_{x1}^4 + A_2 l_{x2}^4 + A_3 l_{x3}^4 + 2A_{112} l_{x1}^2 l_{x2}^2 + 2A_{113} l_{x1}^2 l_{x3}^2 + 2A_{223} l_{x2}^2 l_{x3}^2 \quad (1,2,3)$$

$$\begin{aligned} S_{1122} = {} & A_1 l_{x1}^2 l_{y1}^2 + A_2 l_{x2}^2 l_{y2}^2 + A_3 l_{x3}^2 l_{y3}^2 + A_1 \mu_{12} \left(l_{x1}^2 l_{y2}^2 + l_{x2}^2 l_{y1}^2 \right) \\ & + A_1 \mu_{13} \left(l_{x1}^2 l_{y3}^2 + l_{x3}^2 l_{y1}^2 \right) + A_2 \mu_{23} \left(l_{x2}^2 l_{y3}^2 + l_{x3}^2 l_{y3}^2 \right) \\ & + 4(G_{12} l_{x1} l_{x2} l_{y1} l_{y2} + G_{13} l_{x1} l_{x3} l_{y1} l_{y3} + G_{23} l_{x2} l_{x3} l_{y2} l_{y3}) \quad (1,2,3) \end{aligned}$$

$$\begin{aligned} S_{1112} = {} & A_1 l_{x1}^3 l_{y1} + A_2 l_{x2}^3 l_{y2} + A_3 l_{x3}^3 l_{y3} + A_{112}(l_{x1} l_{y2} + l_{x2} l_{y1}) l_{x1} l_x^2 \\ & + A_{113}(l_{x1} l_{y3} + l_{x3} l_{y1}) l_{x1} l_{x3} + A_{223}(l_{x2} l_{y3} + l_{x3} l_{y2}) l_{x2} l_{x3} \quad (1,2,3) \end{aligned}$$

$$\begin{aligned} S_{1113} = {} & A_1 l_{x3}^3 l_{z1} + A_2 l_{x2}^3 l_{z2} + A_3 l_{x3}^3 l_{z3} + A_{112}(l_{x1} l_{z2} + l_{x2} l_{z1}) l_{x1} l_{x2} \\ & + A_{113}(l_{x1} l_{z2} + l_{x3} l_{z1}) l_{x1} l_{x3} + A_{223}(l_{x2} l_{z3} + l_{x3} l_{z2}) l_{x2} l_{x3} \quad (1,2,3) \end{aligned}$$

$$\begin{aligned} S_{1123} = {} & A_1 l_{x1}^2 l_{y1} l_{z1} + A_2 l_{x2}^2 l_{y2} l_{z2} + A_3 l_{x3}^2 l_{y3} l_{z3} \\ & + A_1 \mu_{12} \left(l_{x1}^2 l_{y2} l_{z2} + l_{x2}^2 l_{y1} l_{z1} \right) + A_1 \mu_{13} \left(l_{x1}^2 l_{y3} l_{z3} + l_{x3}^2 l_{y1} l_{z1} \right) \\ & + A_2 \mu_{23} (l_{x2}^2 l_{y3} l_{z3} + l_{x3}^2 l_{y2} l_{z2}) + 2\,[G_{12}(l_{y1} l_{z2} + l_{z1} l_{y2}) l_{x1} l_{x2} \\ & + G_{13}(l_{y1} l_{z2} + l_{z1} l_{y2}) l_{x1} l_{x3} + G_{23}(l_{y3} l_{z2} + l_{y2} l_{z3}) l_{x2} l_{x3}] \quad (1,2,3) \end{aligned}$$

$$\begin{aligned} S_{1212} = {} & A_1 l_{x1}^2 l_{y1}^2 + A_2 l_{x2}^2 l_{y2}^2 + A_3 l_{x3}^2 l_{y3}^2 \\ & + 2(A_1 \mu_{12} l_{x1} l_{x2} l_{y1} l_{y2} + A_1 \mu_{13} l_{x1} l_{x3} l_{y1} l_{y3} + A_2 \mu_{23} l_{x2} l_{x3} l_{y2} l_{y3}) \\ & + G_{12}(l_{x1} l_{y2} + l_{x2} l_{y1})^2 + G_{13}(l_{x1} l_{y3} + l_{x3} l_{y1})^2 + G_{23}(l_{x2} l_{y3} + l_{x3} l_{y2})^2 \\ & \quad (1,2,3) \end{aligned}$$

$$\begin{aligned} S_{1213} = {} & A_1 l_{x1}^2 l_{y1} l_{z1} + A_2 l_{x2}^2 l_{y2} l_{z2} + A_3 l_{x3}^2 l_{y3} l_{z3} \\ & + A_1 \mu_{12}(l_{y1} l_{z2} + l_{y2} l_{z1}) l_{x1} l_{x2} + A_1 \mu_{13}(l_{y1} l_{z3} + l_{y2} l_{z1}) l_{x1} l_{x3} \\ & + A_2 \mu_{23}(l_{y2} l_{z3} + l_{y3} l_{z2}) l_{x1} l_{x3} + G_{12}(l_{x1} l_{y2} + l_{x2} l_{y1})(l_{x1} l_{y2} + l_{x2} l_{z1}) \\ & + G_{13}(l_{x1} l_{y3} + l_{x3} l_{y1})(l_{x1} l_{z3} + l_{x3} l_{z2}) \\ & + G_{23}(l_{x2} l_{y3} + l_{x3} l_{y2})(l_{x2} l_{z3} + l_{x3} l_{z2}) \quad (1,\ 2,\ 3) \end{aligned} \qquad (4.191)$$

It should be noted that stiffness coefficients are symmetric with respect to the couples of subscripts ($S_{ijkl} = S_{klij}$) and that notation (1, 2, 3) means that performing permutation, i.e., changing 1 for 2, 2 for 3, and 3 for 1, we can use Eqs. (4.191) to write the expressions for all the stiffness coefficients entering Eq. (4.190). The coefficients A_i and μ_{ij} in Eqs. (4.191) are given in the notations to Eqs. (4.54) and

$$A_{112} = A_1\mu_{12} + 2G_{12}, \quad A_{113} = A_1\mu_{13} + 2G_{13}, \quad A_{223} = A_2\mu_{23} + 2G_{23}$$

Resolving Eqs. (4.190) for strains, we arrive at Eq. (2.48) with the following coefficients for the compliance matrix in Eq. (2.49)

$$\begin{aligned}
\frac{1}{E_x} &= \frac{l_{x1}^4}{E_1} + \frac{l_{y1}^4}{E_2} + \frac{l_{z1}^4}{E_3} + C_{122}l_{x1}^2 l_{y1}^2 + C_{133}l_{x1}^2 l_{z1}^2 + C_{233}l_{y1}^2 l_{z1}^2 \quad (1, 2, 3), \quad (x, y, z) \\
\frac{\nu_{xy}}{E_y} &= \frac{\nu_{yx}}{E_x} = \frac{\nu_{12}}{E_2}\left(l_{x1}^2 l_{y2}^2 + l_{x2}^2 l_{y1}^2\right) + \frac{\nu_{13}}{E_3}\left(l_{x1}^2 l_{z2}^2 + l_{x2}^2 l_{z1}^2\right) \\
&\quad + \frac{\nu_{23}}{E_3}\left(l_{y1}^2 l_{z2}^2 + l_{y2}^2 l_{z1}^2\right) - \frac{l_{x1}^2 l_{x2}^2}{E_1} - \frac{l_{y1}^2 l_{y2}^2}{E_2} - \frac{l_{z1}^2 l_{z2}^2}{E_3} \\
&\quad - \frac{1}{G_{12}} l_{x1}l_{x2}l_{y1}l_{y2} - \frac{1}{G_{13}} l_{x1}l_{x2}l_{z1}l_{z2} - \frac{1}{G_{23}} l_{y1}l_{y2}l_{z1}l_{z2} \quad (1, 2, 3), \quad (x, y, z)
\end{aligned} \tag{4.192}$$

$$\begin{aligned}
\frac{\eta_{x,xy}}{G_{xy}} &= \frac{\eta_{xy,x}}{E_x} = 2\left(\frac{l_{x1}^3 l_{x2}}{E_1} + \frac{l_{y1}^3 l_{y2}}{E_2} + \frac{l_{z1}^3 l_{z2}}{E_3}\right) \\
&\quad + C_{122}(l_{x1}l_{y2} + l_{x2}l_{y1})l_{x1}l_{y1} + C_{133}(l_{x1}l_{z2} + l_{x2}l_{z1})l_{x1}l_{z1} \\
&\quad + C_{233}(l_{y1}l_{z2} + l_{y2}l_{z1})l_{y1}l_{z1} \quad (1, 2, 3), \quad (x, y, z),
\end{aligned}$$

$$\begin{aligned}
\frac{\eta_{x,xz}}{G_{xz}} &= \frac{\eta_{xz,x}}{E_x} = 2\left(\frac{l_{x1}^3 l_{x3}}{E_1} + \frac{l_{y1}^3 l_{y3}}{E_2} + \frac{l_{z1}^3 l_{z3}}{E_3}\right) \\
&\quad + C_{122}(l_{x1}l_{y3} + l_{y1}l_{x3})l_{x1}l_{y1} + C_{133}(l_{x1}l_{z3} + l_{x3}l_{z1})l_{x1}l_{z1} \\
&\quad + C_{233}(l_{y1}l_{z3} + l_{y3}l_{z1})l_{y1}l_{z1} \quad (1, 2, 3), \quad (x, y, z),
\end{aligned}$$

$$\begin{aligned}
\frac{\eta_{x,yz}}{G_{yz}} &= \frac{\eta_{yz,x}}{E_x} = 2\left[\frac{l_{x1}^2 l_{x2}l_{x3}}{E_1} + \frac{l_{y1}^2 l_{y2}l_{y3}}{E_2} + \frac{l_{z1}^2 l_{z2}l_{z3}}{E_3}\right. \\
&\quad - \frac{\nu_{12}}{E_2}\left(l_{x1}^2 l_{y2}l_{y3} + l_{y1}^2 l_{x2}l_{x3}\right) - \frac{\nu_{13}}{E_3}\left(l_{x1}^2 l_{z2}l_{z3} + l_{z1}^2 l_{x2}l_{x3}\right)
\end{aligned}$$

$$-\frac{\nu_{23}}{E_3}\left(l_{y1}^2 l_{z2} l_{z3} + l_{z1}^2 l_{y2} l_{y3}\right)\Bigg] + \frac{l_{x1} l_{y1}}{G_{12}}(l_{x2} l_{y3} + l_{x3} l_{y2})$$

$$+\frac{l_{x1} l_{z1}}{G_{13}}(l_{x2} l_{z3} + l_{x3} l_{z2}) + \frac{l_{y1} l_{z1}}{G_{23}}(l_{y2} l_{z3} + l_{y3} l_{z2}) \quad (1,\ 2,\ 3), \quad (x,\ y,\ z)$$

$$\frac{1}{G_{xy}} = 4\Bigg[\frac{l_{x1}^2 l_{x2}^2}{E_1} + \frac{l_{y1}^2 l_{y2}^2}{E_2} + \frac{l_{z1}^2 l_{z2}^2}{E_3} - 2\left(\frac{\nu_{12}}{E_2} l_{x1} l_{x2} l_{y1} l_{y2} + \frac{\nu_{13}}{E_3} l_{x1} l_{x2} l_{z1} l_{z2}\right.$$

$$\left.+\frac{\nu_{23}}{E_3} l_{y1} l_{y2} l_{z1} l_{z2}\right)\Bigg] + \frac{1}{G_{12}}(l_{x1} l_{y2} + l_{x2} l_{y1})^2 + \frac{1}{G_{13}}(l_{x1} l_{z2} + l_{x2} l_{z1})^2$$

$$+\frac{1}{G_{23}}(l_{y1} l_{z2} + l_{y2} l_{z1})^2 \quad (1,\ 2,\ 3),\ (x,\ y,\ z)$$

$$\frac{\lambda_{xy,\,xz}}{G_{xz}} = \frac{\lambda_{xz,\,xy}}{G_{xy}} = 4\Bigg[\frac{l_{x1}^2 l_{x2} l_{x3}}{E_1} + \frac{l_{y1}^2 l_{y2} l_{y3}}{E_2} + \frac{l_{z1}^2 l_{z2} l_{z3}}{E_3}$$

$$-\frac{\nu_{12}}{E_2}(l_{x3} l_{y2} + l_{x2} l_{y3}) l_{x1} l_{y1} - \frac{\nu_{13}}{E_3}(l_{x2} l_{z3} + l_{z2} l_{x3}) l_{x1} l_{z1}$$

$$-\frac{\nu_{23}}{E_3}(l_{y2} l_{z3} + l_{y3} l_{z2}) l_{y1} l_{z1}\Bigg] + \frac{1}{G_{12}}(l_{x1} l_{y3} + l_{x3} l_{y1})(l_{x1} l_{y2} + l_{x2} l_{y1})$$

$$+\frac{1}{G_{13}}(l_{x1} l_{z3} + l_{x3} l_{z1})(l_{x1} l_{z2} + l_{x2} l_{z1})$$

$$+\frac{1}{G_{23}}(l_{y1} l_{z3} + l_{y3} l_{z1})(l_{y1} l_{z2} + l_{y2} l_{z1}) \quad (1,\ 2,\ 3),\ (x,\ y,\ z)$$

in which

$$C_{122} = \frac{1}{G_{12}} - \frac{2\nu_{12}}{E_2}, \quad C_{133} = \frac{1}{G_{13}} - \frac{2\nu_{13}}{E_3}, \quad C_{233} = \frac{1}{G_{23}} - \frac{2\nu_{23}}{E_3}$$

Consider a special spatial structure (Pagano and Whitford, 1985) formed by a fabric composite in which the plies reinforced at angle ϕ (warp direction) with respect to the x-axis make angles α and β with the x-axis and the y-axis, respectively, as in Fig. 4.100. The directional cosines for this structure are

$$l_{x1} = \cos\lambda\cos\psi, \quad l_{x2} = -\sin\lambda\cos\psi$$

$$l_{x3} = -\sin\psi, \quad l_{y1} = \sin\lambda\cos\beta - \cos\lambda\sin\beta\sin\psi$$

$$l_{y2} = \cos\lambda\cos\beta + \sin\lambda\sin\beta\sin\psi, \quad l_{y3} = -\sin\beta\cos\psi$$

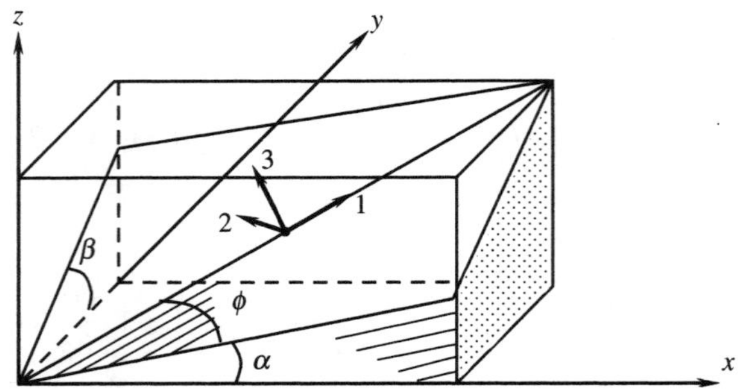

Fig. 4.100. Orientation angles in a spatial composite structure.

$$l_{z1} = \sin\lambda \sin\beta + \cos\lambda \cos\beta \sin\psi$$

$$l_{z2} = \cos\lambda \sin\beta - \sin\lambda \cos\beta \sin\psi, \quad l_{z3} = \cos\beta \cos\psi$$

where

$$\lambda = \phi + \tan^{-1}(\tan\beta \sin\psi), \quad \psi = \tan^{-1}(\tan\alpha \cos\beta)$$

The dependencies of elastic constants E_x, E_y, G_{xz}, and G_{yz} calculated with the aid of Eqs. (4.192) for the material with $E_1 = 12.9\,\text{GPa}$, $E_2 = 5.2\,\text{GPa}$, $E_3 = 3\,\text{GPa}$, $G_{12} = G_{13} = 1.5\,\text{GPa}$, $G_{23} = 1$ GPa, $\nu_{21} = 0.15$, $\nu_{31} = 0.2$, and $\nu_{32} = 0.2$ are presented in Fig. 4.101 (Vasiliev and Morozov, 1988).

For planar structures ($\alpha = \beta = 0$), Eqs. (4.191) and (4.192) generalize Eqs. (4.72) and (4.76) for a three-dimensional stress state of a layer.

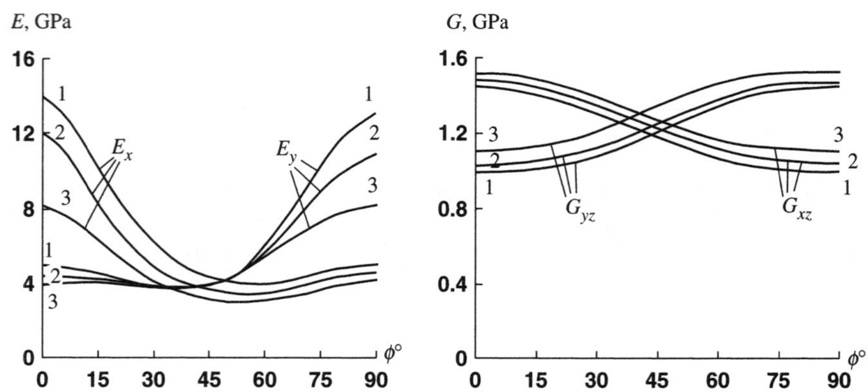

Fig. 4.101. Dependencies of the elastic constants of a spatially reinforced composite on the orientation angles: $1 - \alpha = \beta = 0°$, $2 - \alpha = \beta = 8°$, $3 - \alpha = \beta = 16°$.

4.9. References

Alfutov, N.A. and Zinoviev, P.A. (1982). Deformation and failure of fibrous composites with brittle polymeric matrix under plane stress. In *Mechanics of Composites* (I.F. Obraztsov and V.V. Vasiliev eds.). Mir, Moscow, pp. 166–185.

Birger, I.A. (1951). General solutions of some problems of the plasticity theory. *Prikladnaya Matematika i Mekhanika*, **15**(6), 765–770 (in Russian).

Chamis, C.C. (1979). Impetus of composite mechanics on test methods for fiber composites. In *Proc. 1st USA–USSR Symp. Fracture of Composite Materials*, Riga, USSR, 4–7 Sept. 1978 (G.C. Sih and V.P. Tamuzh eds.). Sijthoff & Noordhoff, The Netherlands, pp. 329–348.

Cherevatsky, A.S. (1999). Manufacturing technology of wound structures by transformation of wound preforms. In *Proc. 12th Int. Conf. on Composite Materials (ICCM-12)*, Paris, France, 5–9 July 1999 (CD-ROM), 10 p.

Chiao, T.T. (1979). Some interesting mechanical behaviors of fiber composite materials. In *Proc. 1st USA–USSR Symp. Fracture of Composite Materials*, Riga, USSR, 4–7 Sept. 1978 (G.C. Sih and V.P. Tamuzh eds.). Sijthoff & Noordhoff, Alphen a/d Rijn., pp. 385–392.

Fukui, S., Kawata, K., Kobayashi, R., Takana, N., Hashimoto, S., Otani, N., and Hondo, A. (1966). Some theoretical and experimental studies on the width variation effects for the filament- wound cylinders. In *Proc. 6th Int. Symp. Space Technol. and Sci.*, Tokyo, pp. 467–470.

Green, A.E. and Adkins, J.E. (1960). *Large Elastic Deformations and Nonlinear Continuum Mechanics*. Oxford University Press, London.

Hahn, H.T. and Tsai, S.W. (1973). Nonlinear elastic behavior of unidirectional composite laminae. *Journal of Composite Materials*, **7**, 102–118.

Hahn, H.T. and Tsai, S.W. (1974). On the behavior of composite laminates after initial failures. *Journal of Composite Materials*, **8**, 288–305.

Hashin, Z. (1987). Analysis of orthogonally cracked laminates under tension. *Journal of Applied Mechanics*, **54**, 872–879.

Herakovich, C.T. (1998). *Mechanics of Fibrous Composites*. Wiley, New York.

Ilyushin, A.A. (1948). *Plasticity*. Gostekhizdat, Moscow (in Russian).

Jones, R.M. (1977). A nonsymmetric compliance matrix approach to nonlinear multimodulus orthotropic materials. *AIAA Journal*, **15**(10), 1436–1443.

Lagace, P.A. (1985). Nonlinear stress-strain behavior of graphite/epoxy laminates. *AIAA Journal*, **23**(10), 1583–1589.

Lungren, J.-E. and Gudmundson, P. (1999). Influence of matrix cracks and swelling on moisture absorption in cross-ply GFRP laminates. In *Proc. 12th Int. Conf. on Composite Materials* (*ICCM-12*), Paris, France, 5–9 July, (CD-ROM), 9 p.

Morozov, E.V. and Vasiliev, V.V. (2003). Determination of the shear modulus of orthotropic materials from off-axis tension tests. *Composite Structures*, **62**, 379–382.

Pagano, N.J. and Whitford, L.E. (1985). On the solution for the elastic response of involute bodies. *Composite Science and Technology*, **22**(4), 295–317.

Reifsnaider, K.L. (1977). Some fundamental aspects of the fatigue and fracture responses of composite materials. In *Proc. 14th Annual Meeting of Society of Engineering Science*, Nov. 14–16, Bethlehem, PA, pp. 373–384.

Skudra, A.M., Bulavs, F.Ya., Gurvich, M.R., and Kruklinsh, A.A. (1989). *Elements of Structural Mechanics of Composite Truss Systems*. Zinatne, Riga (in Russian).

Tarnopol'skii, Yu.M., Zhigun, I.G. and Polyakov, V.A. (1987). *Spatially Reinforced Composite Materials – Handbook*. Mashinostroenie, Moscow (in Russian).

Tarnopol'skii, Yu.M., Zhigun, I.G. and Polyakov, V.A. (1992). *Spatially Reinforced Composites*. Technomic, PA.

Tsai, S.W. (1987). *Composite Design*, 3rd edn. Think Composites, Dayton.

Tsai, S.W. and Azzi, V.D. (1966). Strength of laminated composite materials. *AIAA Journal*, **4**(2), 296–301.

Vasiliev, V.V. (1993). *Mechanics of Composite Structures*, Taylor & Francis, Washington.

Vasiliev, V.V. and Elpatievskii, A.N. (1967). Deformation of tape-wound cylindrical shells under internal pressure. *Polymer Mechanics/Mechanics of Composite Materials*, **3**(5), 604–607.

Vasiliev, V.V. and Morozov, E.V. (1988). Applied theory of spatially reinforced composite shells. *Mechanics of Composite Materials*, **24**(3), 393–400.

Vasiliev, V.V. and Salov, V.A. (1984). Development and examination of a two-matrix glass-fiber composite with high transverse strain. *Mechanics of Composite Materials*, **20**(4), 1984, 463–467.

Vasiliev, V.V., Dudchenko, A.A., and Elpatievskii, A.N. (1970). Analysis of the tensile deformation of glass-reinforced plastics. *Polymer Mechanics/Mechanics of Composite Materials*, **6**(1), 127–130.

Vasiliev, V.V. and Tarnopol'skii, Yu.M. (eds.). (1990). *Composite Materials. Handbook*. Mashinostroenie, Moscow (in Russian).

Vasiliev, V.V., Salov, V.A., and Salov, O.V. (1997). Load-Carrying Shell of Revolution Mode of Composite Materials, *Patent of Russian Federation*, No. 209197.

Verchery, G. (1999). Designing with anisotropy. Part 1: Methods and general results for laminates. In *Proc. 12th Int. Conf. on Composite Materials* (*ICCM-12*), Paris, France, 5–9 July 1999, (CD-ROM), 11 p.

Verchery, G. and Gong, X.J. (1999). Pure tension with off-axis tests for orthotropic laminates. In *Proc. 12th Int. Conf. on Composite Materials* (*ICCM-12*), Paris, France, 5–9 July, (CD-ROM), 7 p.

Chapter 5

MECHANICS OF LAMINATES

A typical composite structure consists of a system of layers bonded together. The layers can be made of different isotropic or anisotropic materials, and have different structures (see Chapter 4), thicknesses, and mechanical properties. In contrast to typical layers which are described in Chapter 4 and whose basic properties are determined experimentally, the laminate characteristics are usually calculated using the information concerning the number of layers, their stacking sequence, geometric and mechanical properties, which must be known. A finite number of layers can be combined to form so many different laminates that the concept of studying them using experimental methods does not seem realistic. Whereas the most complicated typical layer is described with nine stiffness coefficients A_{mn} (mn = 11, 22, 12, 14, 24, 44, 55, 56, 66), some of which can be calculated, the laminate is characterized by 21 coefficients and demonstrates coupling effects that are difficult to simulate in experiments.

Thus, the topic of this chapter is to provide equations allowing us to predict the behavior of a laminate as a system of layers with given properties. The only restriction that is imposed on the laminate as an element of a composite structure concerns its total thickness, which is assumed to be much smaller than the other dimensions of the structure.

5.1. Stiffness coefficients of a generalized anisotropic layer

For the sake of brevity, consider first a thin homogeneous layer, which is anisotropic in the xy-plane and whose mechanical properties are some functions of the normal coordinate z (see Fig. 5.1). Coordinate axes x and y belong to some plane which is referred to as a reference plane such that $z = 0$ on this plane and $-e \leq z \leq s$ for the layer under study. There exist some special locations of the reference plane discussed below, but in this section its coordinates e and s are not specified. We introduce two assumptions both based on the fact that thickness $h = e + s$ is small.

First, it is assumed that the layer thickness, h, does not change under the action of stresses shown in Fig. 5.1. Actually, the thickness does change, but because it is small, this change is negligible. This means that there is no strain in the z-direction, and in

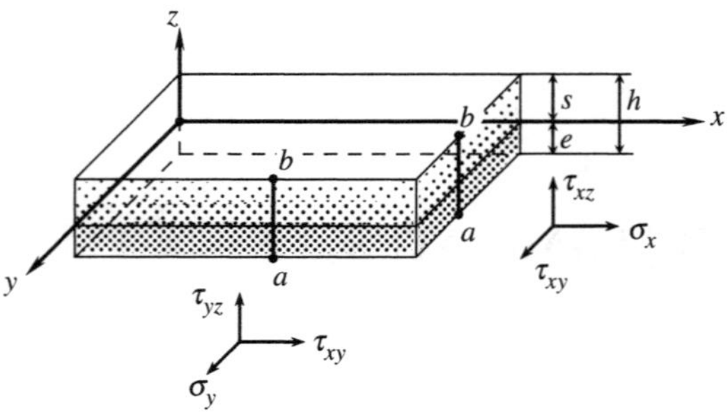

Fig. 5.1. An element of a generalized layer.

accordance with Eqs. (2.22),

$$\varepsilon_z = \frac{\partial u_z}{\partial z} = 0, \quad u_z = w(x, y) \tag{5.1}$$

Here, $w(x, y)$ is the so-called normal deflection which is a translational displacement of a normal element a–b (see Fig. 5.1) as a solid in the z-direction.

Second, we suppose that in-plane displacements u_x and u_y are linear functions of the thickness coordinate z, i.e.,

$$\begin{aligned} u_x(x, y, z) &= u(x, y) + z\theta_x(x, y) \\ u_y(x, y, z) &= v(x, y) + z\theta_y(x, y) \end{aligned} \tag{5.2}$$

where u and v are the displacements of the points of the reference plane $z = 0$ or, which is the same, the translational displacements of the normal element a–b (see Fig. 5.1) as a solid in the x- and y-direction, whereas θ_x and θ_y are the angles of rotations (usually referred to as 'rotations') of the normal element a–b in the xz- and yz-planes. Geometric interpretation of the first expression in Eqs. (5.2) is presented in Fig. 5.2.

In-plane strains of the layer, ε_x, ε_y, and γ_{xy}, can be found using Eqs. (2.22), (5.1), and (5.2) as

$$\begin{aligned} \varepsilon_x &= \frac{\partial u_x}{\partial x} = \varepsilon_x^0 + z\kappa_x \\ \varepsilon_y &= \frac{\partial u_y}{\partial y} = \varepsilon_y^0 + z\kappa_y \\ \gamma_{xy} &= \frac{\partial u_x}{\partial y} + \frac{\partial u_y}{\partial x} = \gamma_{xy}^0 + z\kappa_{xy} \end{aligned} \tag{5.3}$$

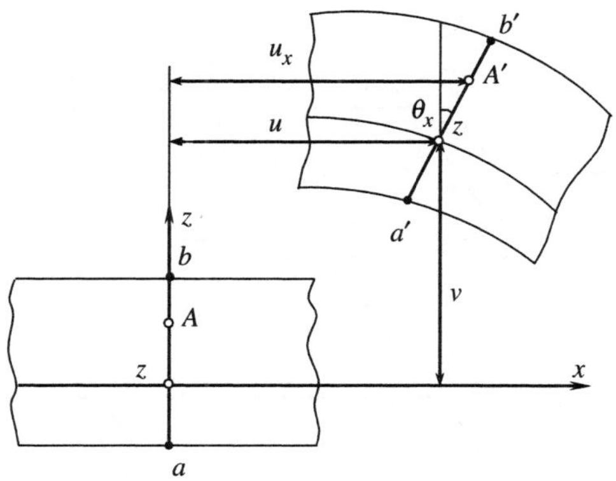

Fig. 5.2. Decomposition of displacement u_x of point A into translational (u) and rotation ($z\theta_x$) components.

where

$$\varepsilon_x^0 = \frac{\partial u}{\partial x}, \quad \varepsilon_y^0 = \frac{\partial v}{\partial y}, \quad \gamma_{xy}^0 = \frac{\partial u}{\partial y} + \frac{\partial v}{\partial x},$$

$$\kappa_x = \frac{\partial \theta_x}{\partial x}, \quad \kappa_y = \frac{\partial \theta_y}{\partial y}, \quad \kappa_{xy} = \frac{\partial \theta_x}{\partial y} + \frac{\partial \theta_y}{\partial x}$$

These generalized strains correspond to the following four basic deformations of the layer shown in Fig. 5.3:

- in-plane tension or compression $\left(\varepsilon_x^0, \varepsilon_y^0\right)$,
- in-plane shear $\left(\gamma_{xy}^0\right)$,
- bending in the xz- and yz-planes (κ_x, κ_y), and
- twisting (κ_{xy}).

The constitutive equations for an anisotropic layer, Eqs. (4.71), upon substitution of Eqs. (5.3), yield

$$\begin{aligned}
\sigma_x &= A_{11}\varepsilon_x^0 + A_{12}\varepsilon_y^0 + A_{14}\gamma_{xy}^0 + z(A_{11}\kappa_x + A_{12}\kappa_y + A_{14}\kappa_{xy}) \\
\sigma_y &= A_{21}\varepsilon_x^0 + A_{22}\varepsilon_y^0 + A_{24}\gamma_{xy}^0 + z(A_{21}\kappa_x + A_{22}\kappa_y + A_{24}\kappa_{xy}) \\
\tau_{xy} &= A_{41}\varepsilon_x^0 + A_{42}\varepsilon_y^0 + A_{44}\gamma_{xy}^0 + z(A_{41}\kappa_x + A_{42}\kappa_y + A_{44}\kappa_{xy})
\end{aligned} \tag{5.4}$$

where $A_{mn} = A_{nm}$ are the stiffness coefficients of the material that will depend, in general, on the coordinate z.

It follows from Eqs. (5.4), that the stresses depend on six generalized strains ε, γ, and κ which are functions of coordinates x and y only. To derive the constitutive equations for the layer under study, we introduce the corresponding force functions as stress resultants

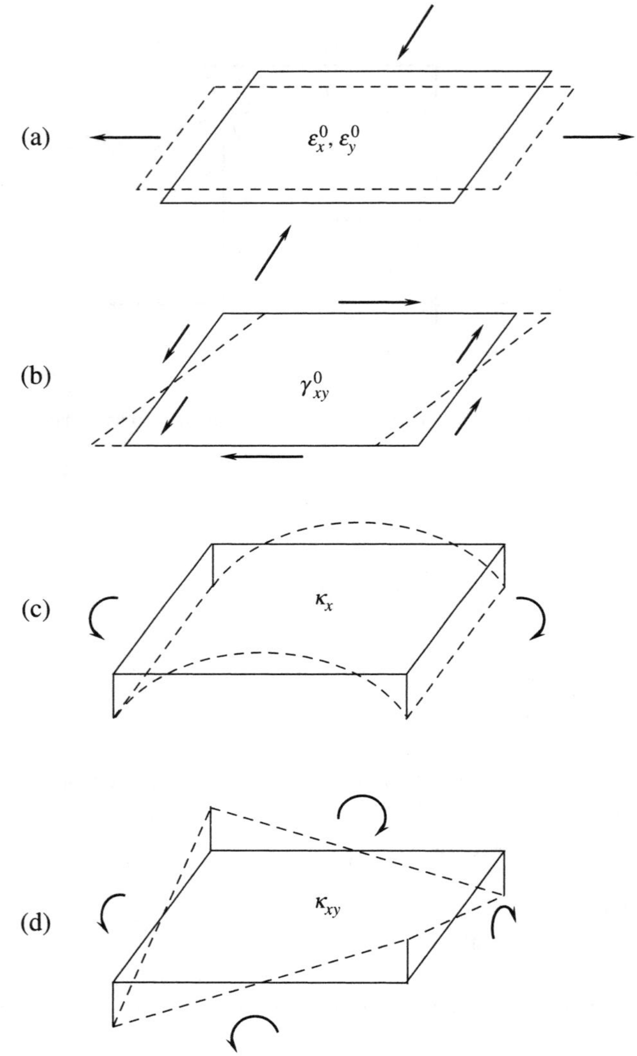

Fig. 5.3. Basic deformations of the layer: (a) in-plane tension and compression $\left(\varepsilon_x^0, \varepsilon_y^0\right)$; (b) in-plane shear $\left(\gamma_{xy}^0\right)$; (c) bending ($\kappa_x$); (d) twisting ($\kappa_{xy}$).

and couples shown in Fig. 5.4 and specified as (see also Fig. 5.1)

$$N_x = \int_{-e}^{s} \sigma_x \mathrm{d}z, \quad N_y = \int_{-e}^{s} \sigma_y \mathrm{d}z, \quad N_{xy} = \int_{-e}^{s} \tau_{xy} \mathrm{d}z,$$

$$M_x = \int_{-e}^{s} \sigma_x z \mathrm{d}z, \quad M_y = \int_{-e}^{s} \sigma_y z \mathrm{d}z, \quad M_{xy} = \int_{-e}^{s} \tau_{xy} z \mathrm{d}z$$

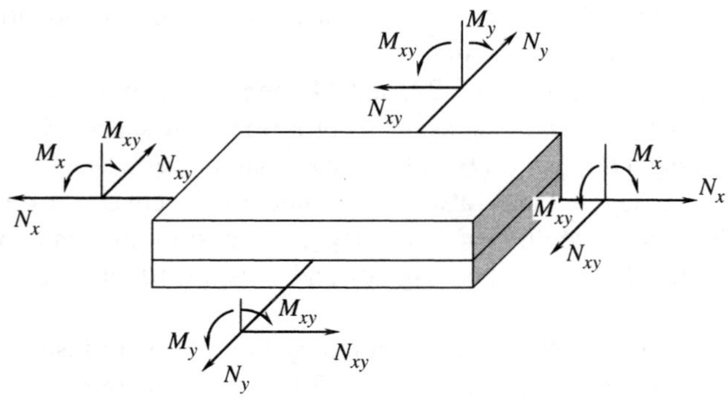

Fig. 5.4. Stress resultants and couples applied to the reference plane of the layer.

Substituting the stresses, Eqs. (5.4), into these equations, we arrive at constitutive equations that relate stress resultants and couples to the corresponding generalized strains, i.e.,

$$
\begin{aligned}
N_x &= B_{11}\varepsilon_x^0 + B_{12}\varepsilon_y^0 + B_{14}\gamma_{xy}^0 + C_{11}\kappa_x + C_{12}\kappa_y + C_{14}\kappa_{xy} \\
N_y &= B_{21}\varepsilon_x^0 + B_{22}\varepsilon_y^0 + B_{24}\gamma_{xy}^0 + C_{21}\kappa_x + C_{22}\kappa_y + C_{24}\kappa_{xy} \\
N_{xy} &= B_{41}\varepsilon_x^0 + B_{42}\varepsilon_y^0 + B_{44}\gamma_{xy}^0 + C_{41}\kappa_x + C_{42}\kappa_y + C_{44}\kappa_{xy} \\
M_x &= C_{11}\varepsilon_x^0 + C_{12}\varepsilon_y^0 + C_{14}\gamma_{xy}^0 + D_{11}\kappa_x + D_{12}\kappa_y + D_{14}\kappa_{xy} \\
M_y &= C_{21}\varepsilon_x^0 + C_{22}\varepsilon_y^0 + C_{24}\gamma_{xy}^0 + D_{21}\kappa_x + D_{22}\kappa_y + D_{24}\kappa_{xy} \\
M_{xy} &= C_{41}\varepsilon_x^0 + C_{42}\varepsilon_y^0 + C_{44}\gamma_{xy}^0 + D_{41}\kappa_x + D_{42}\kappa_y + D_{44}\kappa_{xy}
\end{aligned}
\tag{5.5}
$$

These equations include membrane stiffness coefficients

$$B_{mn} = B_{nm} = \int_{-e}^{s} A_{mn}\mathrm{d}z \tag{5.6}$$

which specify the layer stiffness under in-plane deformation (Fig. 5.3a and b), bending stiffness coefficients

$$D_{mn} = D_{nm} = \int_{-e}^{s} A_{mn}z^2\mathrm{d}z \tag{5.7}$$

which are associated with the layer bending and twisting (Fig. 5.3c and d), and membrane–bending coupling coefficients

$$C_{mn} = C_{nm} = \int_{-e}^{s} A_{mn}z\mathrm{d}z \tag{5.8}$$

through which in-plane stress resultants are related to bending deformations, and stress couples are linked with in-plane strains.

Coefficients with subscripts 11, 12, 22, and 44 compose the basic set of the layer stiffnesses associated with in-plane extension, contraction, and shear $(B_{11}, B_{12}, B_{22}, B_{44})$, bending and twisting $(D_{11}, D_{12}, D_{22}, D_{44})$, and coupling effects $(C_{11}, C_{12}, C_{22}, C_{44})$. For an anisotropic layer there also exists coupling between extension (a) and shear (b) in Fig. 5.3 (coefficients B_{14}, B_{24}), extension (a) and twisting (d) in Fig. 5.3 (coefficients C_{14}, C_{24}), bending (c) and twisting (d) in Fig. 5.3 (coefficients D_{14}, D_{24}).

The forces and moments N and M specified by Eqs. (5.5) are resultants and couples of in-plane stresses σ_x, σ_y, and τ_{xy} (see Fig. 5.1). However, there are also transverse shear stresses τ_{xz} and τ_{yz} which should be expressed in terms of the corresponding shear strains. Unfortunately, we cannot apply for this purpose the direct approach that was used above to derive Eqs. (5.5). This different approach involves strain–displacement equations, Eqs. (2.22),

$$\gamma_{xz} = \frac{\partial u_x}{\partial z} + \frac{\partial u_z}{\partial x}, \quad \gamma_{yz} = \frac{\partial u_y}{\partial z} + \frac{\partial u_z}{\partial y} \tag{5.9}$$

in conjunction with Hooke's law

$$\tau_{xz} = A_{55}\gamma_{xz} + A_{56}\gamma_{yz}, \quad \tau_{yz} = A_{65}\gamma_{xz} + A_{66}\gamma_{yz} \tag{5.10}$$

or

$$\gamma_{xz} = a_{55}\tau_{xz} + a_{56}\tau_{yz}, \quad \gamma_{yz} = a_{65}\tau_{xz} + a_{66}\tau_{yz} \tag{5.11}$$

where A_{mn} and a_{mn} are stiffness and compliance coefficients, respectively. The problem is associated with Eqs. (5.2) which specify only approximate dependence of displacements u_x and u_y on coordinate z (the actual distribution of u_x and u_y through the layer thickness is not known) and must not be differentiated with respect to z. So we cannot substitute Eqs. (5.2) into Eqs. (5.9) which include derivatives of u_x and u_y with respect to z. To see what can happen if we violate this well-known mathematical restriction, consider a sandwich laminate shown in Fig. 5.5. It can be seen that while linear approximation of $u(z)$ (dashed line) looks reasonable, the derivatives of the actual displacements and the approximate ones have little in common.

To derive constitutive equations for transverse shear, consider Fig. 5.6. The actual distribution of shear stresses τ_{xz} and τ_{yz} across the layer thickness is not known, but we can assume that it is not important. Indeed, as follows from Eqs. (5.1), elements *a–b* (see Fig. 5.6) along which the shear stresses act are absolutely rigid. This means (in accordance with the corresponding theorem of Statics of Solids) that the displacements of these elements in the z-direction depend only on the resultants of the shear

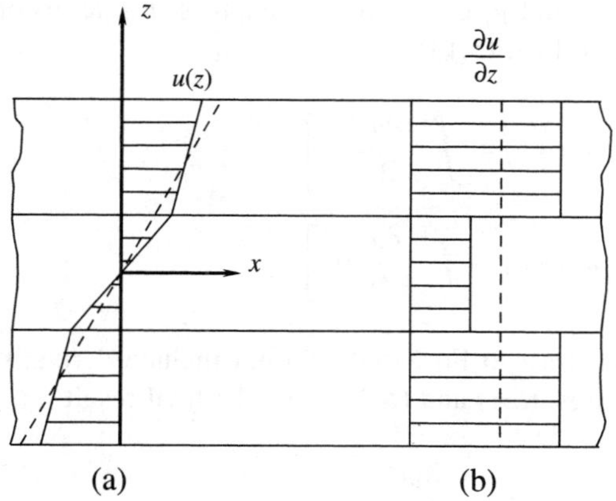

Fig. 5.5. Actual (solid lines) and approximate (dashed lines) distributions of a displacement (a) and its derivative (b) through the thickness of a sandwich laminate.

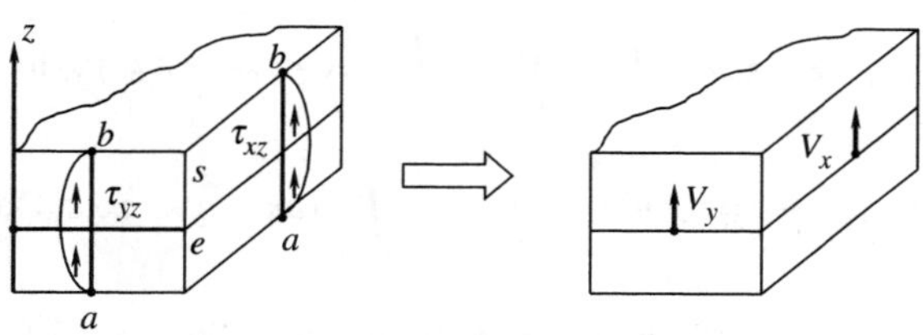

Fig. 5.6. Reduction of transverse shear stresses to stress resultants (transverse shear forces).

stresses, i.e., on transverse shear forces

$$V_x = \int_{-e}^{s} \tau_{xz} \mathrm{d}z, \quad V_y = \int_{-e}^{s} \tau_{yz} \mathrm{d}z \tag{5.12}$$

Since the particular distributions of τ_{xz} and τ_{yz} do not influence the displacements, we can introduce some average stresses having the same resultants as the actual ones, i.e.,

$$\tau_x = \frac{V_x}{h} = \frac{1}{h}\int_{-e}^{s} \tau_{xz} \mathrm{d}z, \quad \tau_y = \frac{V_y}{h} = \frac{1}{h}\int_{-e}^{s} \tau_{yz} \mathrm{d}z$$

However, according to Eqs. (5.11), shear strains are linear combinations of shear stresses. So, we can use the same law to introduce average shear strains as

$$\gamma_x = \frac{1}{h}\int_{-e}^{s} \gamma_{xz} \mathrm{d}z, \quad \gamma_y = \frac{1}{h}\int_{-e}^{s} \gamma_{yz} \mathrm{d}z \tag{5.13}$$

Average shear strains γ_x and γ_y can be readily expressed in terms of displacements if we substitute Eqs. (5.9) into Eqs. (5.13), i.e.,

$$\gamma_x = \frac{1}{h}\left[u_x(s) - u_x(-e) + \int_{-e}^{s} \frac{\partial u_z}{\partial x} \mathrm{d}z\right]$$

$$\gamma_y = \frac{1}{h}\left[u_y(s) - u_x(-e) + \int_{-e}^{s} \frac{\partial u_z}{\partial y} \mathrm{d}z\right]$$

These equations, in contrast to Eqs. (5.9), do not include derivatives with respect to z. So, we can substitute Eqs. (5.1) and (5.2) to get the final result

$$\gamma_x = \theta_x + \frac{\partial w}{\partial x}, \quad \gamma_y = \theta_y + \frac{\partial w}{\partial y} \tag{5.14}$$

Consider Eqs. (5.10) and (5.11). Integrating them over the layer thickness and using Eqs. (5.12) and (5.13), we get

$$V_x = \int_{-e}^{s} (A_{55}\gamma_{xz} + A_{56}\gamma_{yz})\mathrm{d}z, \quad V_y = \int_{-e}^{s} (A_{65}\gamma_{xz} + A_{66}\gamma_{yz})\mathrm{d}z$$

$$\gamma_x = \frac{1}{h}\int_{-e}^{s} (a_{55}\tau_{xz} + a_{56}\tau_{yz})\mathrm{d}z, \quad \gamma_y = \frac{1}{h}\int_{-e}^{s} (a_{65}\tau_{xz} + a_{66}\tau_{yz})\mathrm{d}z$$

Since the actual distribution of stresses and strains according to the foregoing reasoning is not significant, we can change them for the corresponding average stresses and strains:

$$V_x = S_{55}\gamma_x + S_{56}\gamma_y, \quad V_y = S_{65}\gamma_x + S_{66}\gamma_y \tag{5.15}$$

$$\gamma_x = s_{55}V_x + s_{56}V_y, \quad \gamma_y = s_{65}V_x + s_{66}V_y \tag{5.16}$$

where

$$S_{mn} = S_{nm} = \int_{-e}^{s} A_{mn}\mathrm{d}z \tag{5.17}$$

$$s_{mn} = s_{nm} = \frac{1}{h^2}\int_{-e}^{s} a_{mn}\mathrm{d}z \tag{5.18}$$

It should be emphasized that Eqs. (5.16) are not the inverse form of Eqs. (5.15). Indeed, solving Eqs. (5.16), using Eqs. (5.18), and taking into account that

$$a_{55} = \overline{A}_{66}, \quad a_{56} = -\overline{A}_{56}, \quad a_{66} = \overline{A}_{55},$$

$$\overline{A}_{mn} = \frac{A_{mn}}{A_{55}A_{66} - A_{56}^2}$$

we arrive at Eqs. (5.15) in which S_{mn} should be changed to

$$\overline{S}_{mn} = \frac{h^2 \int_{-e}^{s} \overline{A}_{mn} \mathrm{d}z}{\left(\int_{-e}^{s} \overline{A}_{55} \mathrm{d}z\right)\left(\int_{-e}^{s} \overline{A}_{66} \mathrm{d}z\right) - \left(\int_{-e}^{s} \overline{A}_{56} \mathrm{d}z\right)^2} \tag{5.19}$$

These expressions, in general, do not coincide with Eqs. (5.17).

Thus, the constitutive equations for transverse shear are specified by Eqs. (5.15), and there exist two, in general different, approximate forms of stiffness coefficients – Eqs. (5.17) and (5.19). The fact that equations obtained in this way are approximate is quite natural because the assumed displacement field, Eqs. (5.1) and (5.2), is also approximate.

To compare two possible forms of constitutive equations for transverse shear, consider for the sake of brevity an orthotropic layer for which

$$A_{56} = 0, \quad a_{56} = 0, \quad A_{55} = G_{xz}, \quad A_{66} = G_{yz},$$

$$a_{55} = \overline{A}_{66} = \frac{1}{G_{xz}}, \quad a_{66} = \overline{A}_{55} = \frac{1}{G_{yz}}$$

For transverse shear in the xz-plane, Eqs. (5.15) yield

$$V_x = S_{55}\gamma_x \quad \text{or} \quad V_x = \overline{S}_{55}\gamma_x \tag{5.20}$$

in which, in accordance with Eq. (5.17)

$$S_{55} = \int_{-e}^{s} G_{xz} \mathrm{d}z \tag{5.21}$$

whereas Eq. (5.19) yields

$$\overline{S}_{55} = \frac{h^2}{\int_{-e}^{s} \frac{\mathrm{d}z}{G_{xz}}} \tag{5.22}$$

If the shear modulus does not depend on z, both equations, Eq. (5.21) and (5.22), give the same result $S_{55} = \overline{S}_{55} = G_{xz}h$. The same, of course, holds true for the transverse shear in the yz-plane.

Using the energy method applied in Section 3.3, we can show that the Eqs. (5.21) and (5.22) provide the upper and the lower bounds for the exact transverse shear stiffness. Indeed, consider a strip with unit width experiencing transverse shear induced by force V_x as in Fig. 5.7. Assume that Eq. (5.20) links the actual force V_x with the exact angle $\gamma_x = \Delta/l$ through the exact shear stiffness S_{55}^e which we do not know, and which we

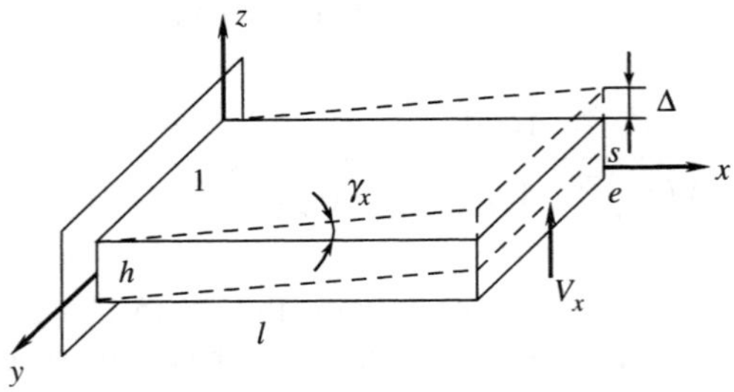

Fig. 5.7. Transverse shear of a strip with unit width.

would like to evaluate. To do this, we can use the two variational principles described in Section 2.11. According to the principle of minimum total potential energy

$$T_{\text{ext}} \leq T_{\text{adm}} \tag{5.23}$$

where

$$T_{\text{ext}} = U_{\text{ext}}^{\varepsilon} - A_{\text{ext}}, \quad T_{\text{adm}} = U_{\text{adm}}^{\varepsilon} - A_{\text{adm}}$$

are the total energies of the exact state and some admissible kinematic state expressed in terms of the strain energy, U, and work A performed by force V_x on displacement Δ (see Fig. 5.7). For both states

$$A_{\text{ext}} = A_{\text{adm}} = V_x \Delta$$

and condition (5.23) reduces to

$$U_{\text{ext}}^{\varepsilon} \leq U_{\text{adm}}^{\varepsilon} \tag{5.24}$$

For the exact state, with due regard to Eq. (5.20), we get

$$U_{\text{ext}} = \frac{l}{2} V_x \gamma_x = \frac{l}{2} S_{55}^e \gamma_x^2 \tag{5.25}$$

For the admissible state, we should use the following general equation

$$U = \frac{1}{2} \int_0^l \mathrm{d}x \int_{-e}^{s} \tau_{xz} \gamma_{xz} \mathrm{d}z = \frac{1}{2} \int_0^l \mathrm{d}x \int_{-e}^{s} G_{xz} \gamma_{xz}^2 \mathrm{d}z = U^{\varepsilon}$$

and admit some approximation for γ_{xz}. The simplest one is $\gamma_{xz} = \gamma_x$, so that

$$U_{\text{adm}}^{\varepsilon} = \frac{l}{2} \gamma_x^2 \int_{-e}^{s} G_{xz} \mathrm{d}z \tag{5.26}$$

Then, Eqs. (5.24)–(5.26) yield

$$S_{55}^e \le \int_{-e}^{s} G_{xz} \mathrm{d}z$$

Comparing this inequality with Eq. (5.21), we can conclude that this equation specifies the upper bound for S_{55}^e.

To determine the lower bound, we should apply the principle of minimum strain energy, according to which

$$U_{\text{ext}} \le U_{\text{adm}}^{\sigma} \tag{5.27}$$

where

$$U_{\text{ext}} = \frac{l}{2} V_x \gamma_x = \frac{l}{2} \cdot \frac{V_x^2}{S_{55}^e}$$

For the admissible state we should apply

$$U = \frac{1}{2} \int_0^l \mathrm{d}x \int_{-e}^{s} \tau_{xz} \gamma_{xz} \mathrm{d}z = \frac{1}{2} \int_0^l \mathrm{d}x \int_{-e}^{s} \frac{\tau_{xz}^2}{G_{xz}} \mathrm{d}z = U^{\sigma}$$

and use some admissible distribution for τ_{xz}. The simplest approximation is $\tau_{xz} = V_x/h$ so that

$$U_{\text{adm}}^{\sigma} = \frac{l}{2h^2} V_x^2 \int_{-e}^{s} \frac{\mathrm{d}z}{G_{xz}}$$

Substitution in the condition (5.27) yields

$$S_{55}^e \ge \frac{h^2}{\int_{-e}^{s} \frac{\mathrm{d}z}{G_{xz}}}$$

Thus, Eq. (5.22) provides the lower bound for S_{55}^e, and the exact stiffness satisfies the following inequality

$$\frac{h^2}{\int_{-e}^{s} \frac{\mathrm{d}z}{G_{xz}}} \le S_{55}^e \le \int_{-e}^{s} G_{xz} \mathrm{d}z$$

It should be emphasized that S_{55}^e in this analysis is not the actual shear stiffness coefficient of the laminate. It is the exact value of the stiffness coefficient which can be found using the exact stress and strain fields following from three-dimensional elasticity equations.

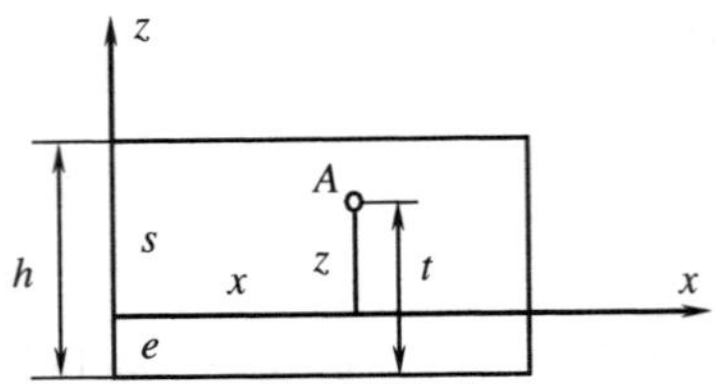

Fig. 5.8. Coordinates of an arbitrary point A.

So, constitutive equations for the generalized layer under study are specified by Eqs. (5.5) and (5.15). Stiffness coefficients, which are given by Eqs. (5.6)–(5.8), and (5.17) or (5.19), can be written in a form more suitable for calculations. To do this, introduce the new coordinate $t = z + e$ such that $0 \leq t \leq h$ (see Fig. 5.8). Transforming the integrals to this new variable, we have

$$B_{mn} = I_{mn}^{(0)}, \quad C_{mn} = I_{mn}^{(1)} - eI_{mn}^{(0)}, \quad D_{mn} = I_{mn}^{(2)} - 2eI_{mn}^{(1)} + e^2 I_{mn}^{(0)} \tag{5.28}$$

where $mn = 11, 12, 22, 14, 24, 44$ and

$$I_{mn}^{(r)} = \int_0^h A_{mn} t^r \mathrm{d}t, \quad r = 0, 1, 2 \tag{5.29}$$

The transverse shear stiffnesses, Eqs. (5.17) and (5.19), take the form

$$S_{mn} = I_{mn}^{(0)} \tag{5.30}$$

and

$$\overline{S}_{mn} = \frac{h^2 \overline{I}_{mn}^{(0)}}{\overline{I}_{55}^{(0)} \overline{I}_{66}^{(0)} - \left(\overline{I}_{56}^{(0)}\right)^2} \tag{5.31}$$

where $mn = 55, 56, 66$ and

$$\overline{I}_{mn}^{(0)} = \int_0^h \overline{A}_{mn} \mathrm{d}t \tag{5.32}$$

The coefficients $\overline{A}_{mn}$ are specified by the expression given in notations to Eq. (5.19).

5.2. Stiffness coefficients of a homogeneous layer

Consider a layer whose material stiffness coefficients A_{mn} do not depend on coordinate z. Then

$$I_{mn}^{(r)} = \frac{A_{mn}}{r+1} h^{r+1}, \quad \overline{I}_{mn}^{(0)} = \overline{A}_{mn} h \tag{5.33}$$

and Eqs. (5.28), (5.30), and (5.31) yield the following stiffness coefficients for the layer

$$\begin{aligned} &B_{mn} = A_{mn} h, \quad C_{mn} = A_{mn}\left(\frac{h}{2} - e\right), \\ &D_{mn} = A_{mn}\left(\frac{h^3}{3} - eh + e^2\right), \quad S_{mn} = A_{mn} h \end{aligned} \tag{5.34}$$

Both Eqs. (5.30) and (5.31) give the same result for S_{mn}. It follows from the second of Eqs. (5.34), that the membrane–bending coupling coefficients C_{mn} become equal to zero if we take $e = h/2$, i.e., if the reference plane coincides with the middle-plane of the layer shown in Fig. 5.9. In this case, Eqs. (5.5) and (5.15) take the following de-coupled form

$$\begin{aligned} &N_x = B_{11}\varepsilon_x^0 + B_{12}\varepsilon_y^0 + B_{14}\gamma_{xy}^0, \quad N_y = B_{21}\varepsilon_x^0 + B_{22}\varepsilon_y^0 + B_{24}\gamma_{xy}^0, \\ &N_{xy} = B_{41}\varepsilon_x^0 + B_{42}\varepsilon_y^0 + B_{44}\gamma_{xy}^0, \\ &M_x = D_{11}\kappa_x + D_{12}\kappa_y + D_{14}\kappa_{xy}, \quad M_y = D_{21}\kappa_x + D_{22}\kappa_y + D_{24}\kappa_{xy}, \\ &M_{xy} = D_{41}\kappa_x + D_{42}\kappa_y + D_{44}\kappa_{xy}, \\ &V_x = S_{55}\gamma_x + S_{56}\gamma_y, \quad V_y = S_{65}\gamma_x + S_{66}\gamma_y \end{aligned} \tag{5.35}$$

As can be seen, we have arrived at three independent groups of constitutive equations, i.e., for in-plane stressed state of the layer, bending and twisting, and transverse shear.

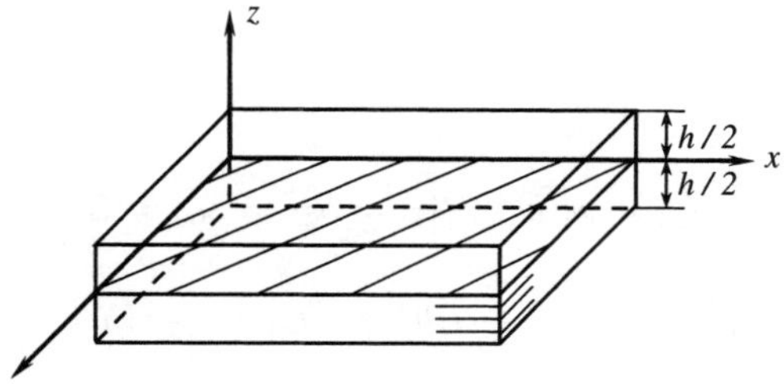

Fig. 5.9. Middle-plane of a laminate.

The stiffness coefficients, Eqs. (5.34), become

$$B_{mn} = A_{mn}h, \quad D_{mn} = \frac{A_{mn}}{12}h^3, \quad S_{mn} = A_{mn}h \tag{5.36}$$

For an orthotropic layer, there are no in-plane stretching–shear coupling ($B_{14} = B_{24} = 0$) and transverse shear coupling ($S_{56} = 0$). Then, Eqs. (5.35) reduce to

$$\begin{aligned}
&N_x = B_{11}\varepsilon_x^0 + B_{12}\varepsilon_y^0, \quad N_y = B_{21}\varepsilon_x^0 + B_{22}\varepsilon_y^0, \quad N_{xy} = B_{44}\gamma_{xy}^0 \\
&M_x = D_{11}\kappa_x + D_{12}\kappa_y, \quad M_y = D_{21}\kappa_x + D_{22}\kappa_y, \quad M_{xy} = D_{44}\kappa_{xy} \\
&V_x = S_{55}\gamma_x, \quad V_y = S_{66}\gamma_y
\end{aligned} \tag{5.37}$$

In terms of engineering elastic constants, the material stiffness coefficients of an orthotropic layer can be expressed as

$$A_{11} = \overline{E}_x, \quad A_{12} = \nu_{xy}\overline{E}_x, \quad A_{22} = \overline{E}_y, \quad A_{44} = G_{xy}, \quad A_{55} = G_{xz}, \quad A_{66} = G_{yz} \tag{5.38}$$

where $\overline{E}_{x,y} = E_{x,y}/(1 - \nu_{xy}\nu_{yx})$. Then, Eqs. (5.36) yield

$$\begin{aligned}
&B_{11} = \overline{E}_x h, \quad B_{12} = \nu_{xy}\overline{E}_x h, \quad B_{22} = \overline{E}_y h, \quad B_{44} = G_{xy}h \\
&D_{11} = \frac{1}{12}\overline{E}_x h^3, \quad D_{12} = \frac{\nu_{xy}}{12}\overline{E}_x h^3, \quad D_{22} = \frac{1}{12}\overline{E}_y h^3, \quad D_{44} = \frac{1}{12}G_{xy}h^3 \\
&S_{55} = G_{xz}h, \quad S_{66} = G_{yz}h
\end{aligned} \tag{5.39}$$

Finally, for an isotropic layer, we have

$$E_x = E_y = E, \quad \nu_{xy} = \nu_{yx} = \nu, \quad G_{xy} = G_{xz} = G_{yz} = G = \frac{E}{2(1+\nu)}$$

and

$$\begin{aligned}
&B_{11} = B_{22} = \overline{E}h, \quad B_{12} = \nu\overline{E}h, \quad B_{44} = S_{55} = S_{66} = Gh \\
&D_{11} = D_{22} = \frac{1}{12}\overline{E}h^3, \quad D_{12} = \frac{\nu}{12}\overline{E}h^3, \quad D_{44} = \frac{1}{12}Gh^3
\end{aligned} \tag{5.40}$$

where $\overline{E} = E/(1 - \nu^2)$.

5.3. Stiffness coefficients of a laminate

Consider the general case, i.e., a laminate consisting of an arbitrary number of layers with different thicknesses h_i and stiffnesses $A_{mn}^{(i)}$ $(i = 1, 2, 3, \ldots, k)$. The location of an arbitrary ith layer of the laminate is specified by the coordinate t_i, which is the distance from the bottom plane of the laminate to the top plane of the ith layer (see Fig. 5.10). Assuming that the material stiffness coefficients do not change within thickness of the layer, and using piece-wise integration, we can write parameter I_{mn} in Eqs. (5.29) and (5.32) as

$$I_{mn}^{(r)} = \frac{1}{r+1}\sum_{i=1}^{k} A_{mn}^{(i)}\left(t_i^{r+1} - t_{i-1}^{r+1}\right), \quad \overline{I}_{mn}^{(0)} = \sum_{i=1}^{k} \overline{A}_{mn}^{(i)}(t_i - t_{i-1}) \tag{5.41}$$

where $r = 0, 1, 2$ and $t_0 = 0$, $t_k = h$ (see Fig. 5.10). For thin layers, Eqs. (5.41) can be reduced to the following form, which is more suitable for calculations

$$I_{mn}^{(0)} = \sum_{i=1}^{k} A_{mn}^{(i)} h_i, \quad \overline{I}_{mn}^{(0)} = \sum_{i=1}^{k} \overline{A}_{mn}^{(i)} h_i,$$

$$I_{mn}^{(1)} = \frac{1}{2}\sum_{i=1}^{k} A_{mn}^{(i)} h_i (t_i + t_{i-1}), \tag{5.42}$$

$$I_{mn}^{(2)} = \frac{1}{3}\sum_{i=1}^{k} A_{mn}^{(i)} h_i \left(t_i^2 + t_i t_{i-1} + t_{i-1}^2\right)$$

in which $h_i = t_i - t_{i-1}$ is the thickness of the ith layer.

The membrane, coupling, and bending stiffness coefficients of the laminate are specified by Eqs. (5.28) and (5.42).

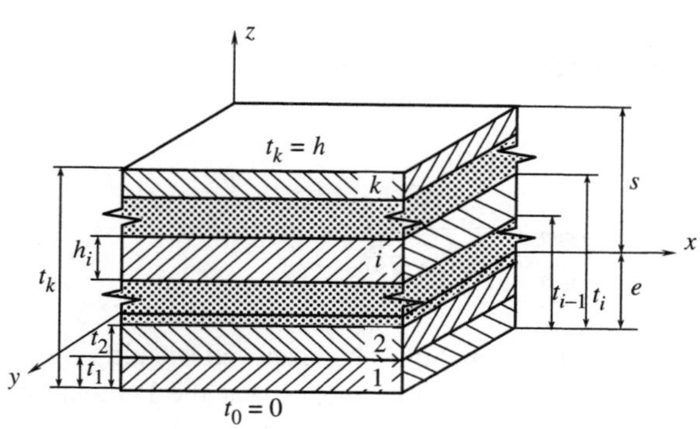

Fig. 5.10. Structure of the laminate.

Consider transverse shear stiffnesses that have two different forms determined by Eqs. (5.30) and (5.31) in which

$$I_{mn}^{(0)} = \sum_{i=1}^{k} A_{mn}^{(i)} h_i, \quad \overline{I}_{mn}^{(0)} = \sum_{i=1}^{k} \overline{A}_{mn}^{(i)} h_i \tag{5.43}$$

A particular case, important for practical applications, is an orthotropic laminate for which Eqs. (5.5) take the form

$$\begin{aligned}
N_x &= B_{11}\varepsilon_x^0 + B_{12}\varepsilon_y^0 + C_{11}\kappa_x + C_{12}\kappa_y \\
N_y &= B_{21}\varepsilon_x^0 + B_{22}\varepsilon_y^0 + C_{12}\kappa_x + C_{22}\kappa_y \\
N_{xy} &= B_{44}\gamma_{xy}^0 + C_{44}\kappa_{xy} \\
M_x &= C_{11}\varepsilon_x^0 + C_{12}\varepsilon_y^0 + D_{11}\kappa_x + D_{12}\kappa_y \\
M_y &= C_{21}\varepsilon_x^0 + C_{22}\varepsilon_y^0 + D_{21}\kappa_x + D_{22}\kappa_y \\
M_{xy} &= C_{44}\gamma_{xy}^0 + D_{44}\kappa_{xy}
\end{aligned} \tag{5.44}$$

Here, membrane, coupling, and bending stiffnesses, B_{mn}, C_{mn}, and D_{mn}, are specified by Eqs. (5.28), i.e.,

$$B_{mn} = I_{mn}^{(0)}, \quad C_{mn} = I_{mn}^{(1)} - eI_{mn}^{(0)}, \quad D_{mn} = I_{mn}^{(2)} - 2eI_{mn}^{(1)} + e^2 I_{mn}^{(0)} \tag{5.45}$$

where $mn = 11, 12, 22, 44$.

Transverse shear forces V_x and V_y are specified by equations similar to Eqs. (5.20)

$$V_x = S_{55}\gamma_x, \quad V_y = S_{66}\gamma_y$$

in which the corresponding stiffness coefficients, Eqs. (5.30) and (5.31) reduce to $(mn = 55, 66)$

$$S_{mn} = \sum_{i=1}^{k} A_{mn}^{(i)} h_i, \quad \overline{S}_{mm} = \frac{h^2}{\sum_{i=1}^{k} \frac{h_i}{A_{mm}^{(i)}}} \tag{5.46}$$

Laminates composed of unidirectional plies have special stacking-sequence notations. For example, notation $[0_2^\circ/+45^\circ/-45^\circ/90_2^\circ]$ means that the laminate consists of 0° layer having two plies, ±45° angle-ply layer, and 90° layer also having two plies. Notation $[0^\circ/90^\circ]_5$ means that the laminate has five cross-ply layers.

5.4. Symmetric laminates

Symmetric laminates are composed of layers that are symmetrically arranged with respect to the laminate's middle plane as shown in Fig. 5.11. Introduce the layer coordinate z_i, (see Fig. 5.11). Since for any layer which is above the middle surface $z = 0$ and has the coordinate z_i there is a similar layer which is located under the middle surface and has the coordinate $(-z_i)$, the integration over the laminate thickness can be performed from $z = 0$ to $z = h/2$ (see Fig. 5.11). Then, the integrals for B_{mn} and D_{mn} similar to Eqs. (5.6) and (5.7) must be doubled, whereas the integral for C_{mn} similar to Eqs. (5.8) is equal to zero. Thus, the stiffness coefficients entering Eqs. (5.5) become

$$B_{mn} = 2\int_0^{h/2} A_{mn}\mathrm{d}z, \quad D_{mn} = 2\int_0^{h/2} A_{mn}z^2\mathrm{d}z, \quad C_{mn} = 0 \tag{5.47}$$

For a symmetric laminate shown in Fig. 5.11, we get

$$\begin{aligned} B_{mn} &= 2\sum_{i=1}^{k/2} A_{mn}^{(i)}(z_i - z_{i-1}) = 2\sum_{i=1}^{k/2} A_{mn}^{(i)}h_i \\ C_{mn} &= 0 \\ D_{mn} &= \frac{2}{3}\sum_{i=1}^{k/2} A_{mn}^{(i)}\left(z_i^3 - z_{i-1}^3\right) = \frac{2}{3}\sum_{i=1}^{k/2} A_{mn}^{(i)}h_i\left(z_i^2 + z_i z_{i-1} + z_{i-1}^2\right) \end{aligned} \tag{5.48}$$

where $h_i = z_i - z_{i-1}$. The transverse shear stiffness coefficients are given by Eqs. (5.30) and (5.31) in which

$$I_{mn}^{(0)} = 2\sum_{i=1}^{k} A_{mn}^{(i)}h_i, \quad \overline{I}_{mn}^{(0)} = 2\sum_{i=1}^{k/2} \overline{A}_{mn}^{(i)}h_i, \quad \overline{A}_{mn}^{(i)} = \frac{A_{mn}^{(i)}}{A_{55}^{(i)}A_{66}^{(i)} - \left(A_{56}^{(i)}\right)^2} \tag{5.49}$$

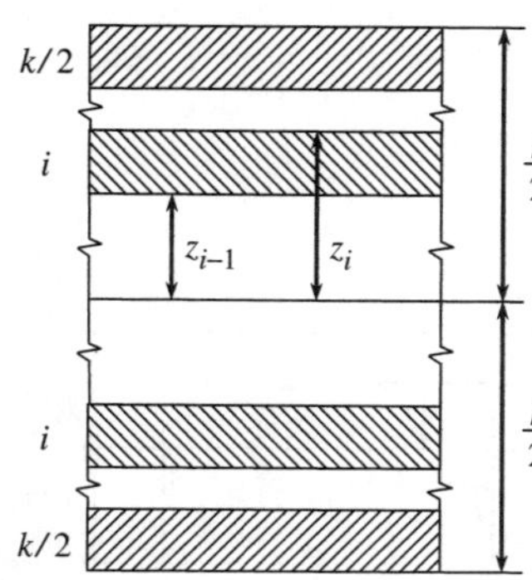

Fig. 5.11. Layer coordinates of a symmetric laminate.

To indicate symmetric laminates, a contracted stacking-sequence notation is used, e.g., $[0°/90°/45°]_s$ instead of $[0°/90°/45°/45°/90°/0°]$. Symmetric laminates are characterized by a specific feature – their bending stiffness is higher than the bending stiffness of any asymmetric laminate composed of the same layers. To show this property of symmetric laminates, consider Eqs. (5.28) and (5.29) and apply them to calculate stiffness coefficients with some combination of subscripts, e.g., $m = 1$ and $n = 1$. Since the coordinate of the reference plane, e, is an arbitrary parameter, we can find it from the condition $C_{11} = 0$. Then,

$$e = \frac{I_{11}^{(1)}}{I_{11}^{(0)}} \tag{5.50}$$

and

$$D_{11} = I_{11}^{(2)} - \left[\frac{\left(I_{11}^{(1)}\right)^2}{I_{11}^{(0)}} \right] \tag{5.51}$$

Introduce a new coordinate for an arbitrary point A in Fig. 5.12 as $z = t-(h/2)$. Changing t to z, we can present Eq. (5.29) in the form

$$I_{11}^{(r)} = \int_{-h/2}^{h/2} A_{11} \left(\frac{h}{2} + z \right)^r \mathrm{d}z$$

Substituting these integrals into Eqs. (5.50) and (5.51), we have

$$e = \frac{h}{2} + \frac{J_{11}^{(1)}}{J_{11}^{(0)}} \tag{5.52}$$

and

$$D_{11} = J_{11}^{(2)} - \left[\frac{\left(J_{11}^{(1)}\right)^2}{J_{11}^{(0)}} \right] \tag{5.53}$$

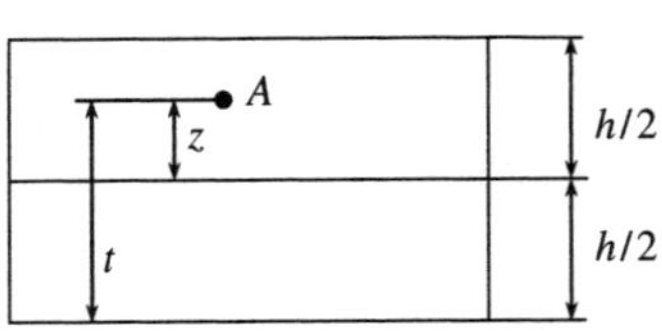

Fig. 5.12. Coordinate of point A referred to the middle plane.

where

$$J_{11}^{(r)} = \int_{-h/2}^{h/2} A_{11} z^r \mathrm{d}z \tag{5.54}$$

and $r = 0, 1, 2$.

Now decompose A_{11} as a function of z into symmetric and antisymmetric components, i.e.,

$$A_{11}(z) = A_{11}^{\mathrm{s}}(z) + A_{11}^{\mathrm{a}}(z)$$

Then, Eq. (5.54) yields

$$J_{11}^{(0)} = \int_{-h/2}^{h/2} A_{11}^{\mathrm{s}} \mathrm{d}z, \quad J_{11}^{(1)} = \int_{-h/2}^{h/2} A_{11}^{\mathrm{a}} z \mathrm{d}z, \quad J_{11}^{(2)} = \int_{-h/2}^{h/2} A_{11}^{\mathrm{s}} z^2 \mathrm{d}z$$

As can be seen from Eq. (5.53), D_{11} reaches its maximum value if $J_{11}^{(1)} = 0$ or $A_{11}^{\mathrm{a}} = 0$ and $A_{11} = A_{11}^{\mathrm{s}}$. In this case, Eq. (5.52) gives $e = h/2$.

Thus, symmetric laminates provide the maximum bending stiffness for a given number and mechanical properties of layers and, being referred to the middle-plane, do not have membrane–bending coupling effects. This essentially simplifies the behavior of the laminate under loading and constitutive equations which have the form specified by Eqs. (5.35).

5.5. Engineering stiffness coefficients of orthotropic laminates

It follows from Eqs. (5.28) that the laminate stiffness coefficients depend, in the general case, on the coordinate of the reference surface e. By changing e, we can change the bending stiffness coefficient D_{mn}. Naturally, the result of the laminate analysis undertaken with the aid of the constitutive equations, Eqs. (5.5) does not depend on the particular pre-assigned value of the coordinate e because of the coupling coefficients C_{mn} which also depend on e. To demonstrate this, consider an orthotropic laminated element loaded with axial forces N and bending moments M uniformly distributed over the element width as in Fig. 5.13. Suppose that the element displacement does not depend on coordinate y. Then, taking $N_x = N$, $M_x = M$, $\varepsilon_y^0 = 0$ and $\kappa_y = 0$ in Eqs. (5.44), we get

$$N = B_{11}\varepsilon_x^0 + C_{11}\kappa_x, \quad M = C_{11}\varepsilon_x^0 + D_{11}\kappa_x \tag{5.55}$$

where, in accordance with Eqs. (5.28),

$$B_{11} = I_{11}^{(0)}, \quad C_{11} = I_{11}^{(1)} - eI_{11}^{(0)}, \quad D_{11} = I_{11}^{(2)} - 2eI_{11}^{(1)} + e^2 I_{11}^{(0)} \tag{5.56}$$

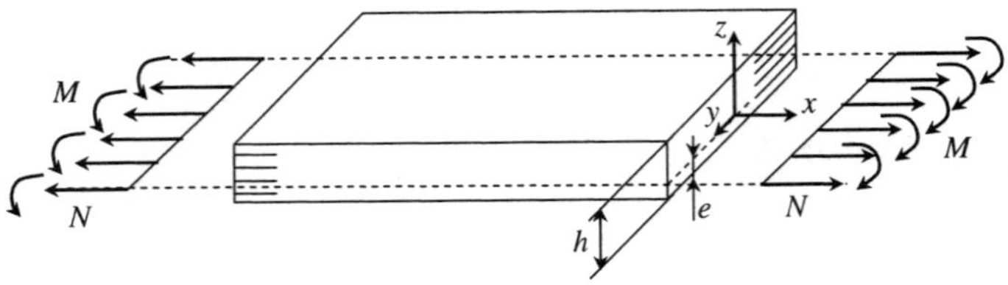

Fig. 5.13. Laminated element under tension and bending.

Here, as follows from Eqs. (5.41)

$$I_{11}^{(r)} = \frac{1}{r+1}\sum_{i=1}^{k} A_{11}^{(i)}\left(t_i^{r+1} - t_{i-1}^{r+1}\right) \tag{5.57}$$

($r = 0,\ 1,\ 2$) are coefficients which do not depend on the coordinate of the reference plane e. It is important to emphasize that forces N in Fig. 5.13 act in the reference plane $z = 0$, and the strain ε_x^0 in Eqs. (5.55) is the strain of the reference plane.

Solving Eqs. (5.55) for ε_x^0 and κ_x, we have

$$\varepsilon_x^0 = \frac{1}{D_1}(D_{11}N - C_{11}M), \quad \kappa_x = \frac{1}{D_1}(B_{11}M - C_{11}N) \tag{5.58}$$

where

$$D_1 = B_{11}D_{11} - C_{11}^2 \tag{5.59}$$

Substituting B, D, and C from Eqs. (5.56), we find

$$D_1 = I_{11}^{(0)} I_{11}^{(2)} - \left(I_{11}^{(1)}\right)^2$$

As can be seen, the parameter D_1 does not depend on e.

Consider now the same element but loaded with forces P applied to the middle plane of the element as in Fig. 5.14. As follows from Fig. 5.15 showing the element cross section, the forces and the moments in Fig. 5.13 induced by the forces in Fig. 5.14 are

$$N = P, \quad M = P\left(\frac{h}{2} - e\right) \tag{5.60}$$

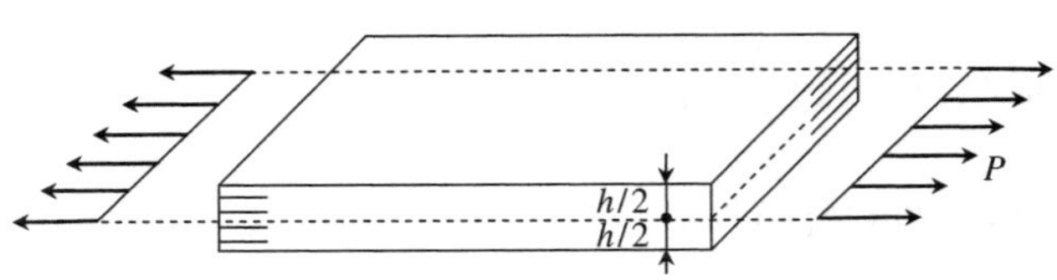

Fig. 5.14. Laminated element under tension.

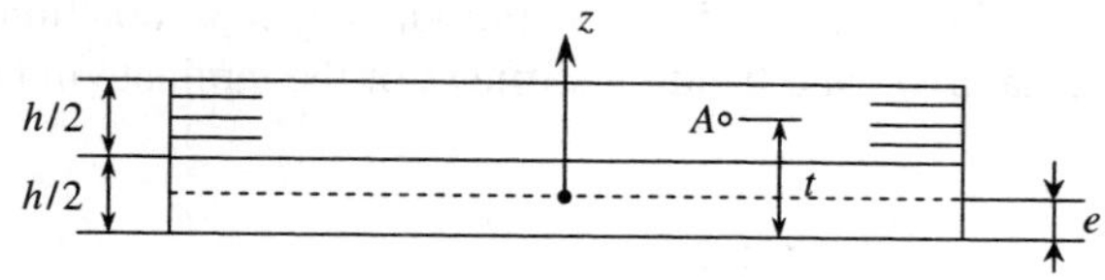

Fig. 5.15. Cross section of the element.

Substitution of Eqs. (5.60) into Eqs. (5.58) yields

$$\varepsilon_x^0 = \frac{P}{D_1}\left[I_{11}^{(2)} - eI_{11}^{(1)} - \frac{h}{2}\left(I_{11}^{(1)} - eI_{11}^{(0)}\right)\right] \tag{5.61}$$

$$\kappa_x = \frac{P}{D_1}\left(\frac{h}{2}I_{11}^{(0)} - I_{11}^{(1)}\right) \tag{5.62}$$

It follows from Eq. (5.62), that κ_x does not depend on e, which is expected because the curvature induced by forces P in Fig. 5.14 is the same for all the planes z = constant of the element. However, Eq. (5.61) includes e which is also expected because ε_x^0 is the strain in the plane $z = 0$ located at the distance e from the lower plane of the element (see Fig. 5.15). Let us find the strain ε_x^t at some arbitrary point A of the cross section for which $z = t - e$ (see Fig. 5.15). Using the first equation of Eqs. (5.3), we have

$$\varepsilon_x^t = \varepsilon_x^0 + (t - e)\,\kappa_x = \frac{P}{D_1}\left[I_{11}^{(2)} - \frac{h}{2}\left(tI_{11}^{(0)} - I_{11}^{(1)}\right) - tI_{11}^{(1)}\right]$$

This equation includes the coordinate of point A and does not depend on e. Thus, taking an arbitrary coordinate of the reference plane, and applying Eqs. (5.56) for the stiffness coefficients, we arrive at values of C_{11} and D_{11}, the combination of which provides the final result that does not depend on e. However, the derived stiffness coefficient D_{11} is not the actual bending stiffness of the laminate which cannot depend on e.

To determine the actual stiffness of the laminate, return to Eqs. (5.58) for ε_x^0 and κ_x. Suppose that $C_{11} = 0$, which means that the laminate has no bending–stretching coupling effects. Then, Eq. (5.59) yields $D = B_{11}D_{11}$ and Eqs. (5.58) become

$$\varepsilon_x^0 = \frac{N}{B_{11}}, \quad \kappa_x = \frac{M}{D_{11}} \tag{5.63}$$

It is obvious that now B_{11} is the actual axial stiffness and D_{11} is the actual bending stiffness of the laminate. However, Eqs. (5.63) are valid only if $C_{11} = 0$. Using the second equation of Eqs. (5.56), we get

$$e = \frac{I_{11}^{(1)}}{I_{11}^{(0)}} \tag{5.64}$$

Substituting this result into Eqs. (5.56) and introducing new notations $B_x = B_{11}$ and $D_x = D_{11}$ for the actual axial and bending stiffness of the laminate in the x-direction, we arrive at

$$B_x = I_{11}^{(0)}, \quad D_x = I_{11}^{(2)} - \frac{\left(I_{11}^{(1)}\right)^2}{I_{11}^{(0)}} \tag{5.65}$$

Here, coefficients $I_{11}^{(r)}$ $(r = 0, 1, 2)$ are specified by Eqs. (5.57). The corresponding stiffnesses in the y-direction (see Fig. 5.13) are determined from similar equations, i.e.,

$$B_y = I_{22}^{(0)}, \quad D_y = I_{22}^{(2)} - \frac{\left(I_{22}^{(1)}\right)^2}{I_{22}^{(0)}} \tag{5.66}$$

in which

$$I_{22}^{(r)} = \frac{1}{r+1}\sum_{i=1}^{k} A_{22}^{(i)}\left(t_i^{r+1} - t_{i-1}^{r+1}\right)$$

For symmetric laminates, as discussed in Section 5.4, $C_{mn} = 0$ and coefficients D_{mn} in Eqs. (5.48) specify the actual bending stiffnesses of the laminate, i.e.,

$$\begin{aligned} D_x &= \frac{2}{3}\sum_{i=1}^{k/2} A_{11}^{(i)} h_i \left(z_i^2 + z_i z_{i-1} + z_{i-1}^2\right) \\ D_y &= \frac{2}{3}\sum_{i=1}^{k/2} A_{22}^{(i)} h_i \left(z_i^2 + z_i z_{i-1} + z_{i-1}^2\right) \end{aligned} \tag{5.67}$$

where coordinates z_i and z_{i-1} are shown in Fig. 5.11. Note, that if the number of layers k is not even, the central layer is divided by the plane $z = 0$ into two identical layers, so k becomes even.

To find the shear stiffness, consider the element in Fig. 5.13 but loaded with shear forces, S, and twisting moments H, uniformly distributed along the element edges as shown in Fig. 5.16. It should be recalled that forces and moments are applied to the element reference plane $z = 0$ (see Fig. 5.13). Taking $N_{xy} = S$ and $M_{xy} = H$ in the corresponding Eqs. (5.44), we get

$$S = B_{44}\gamma_{xy}^0 + C_{44}\kappa_{xy}, \quad H = C_{44}\gamma_{xy}^0 + D_{44}\kappa_{xy} \tag{5.68}$$

in which, in accordance with Eqs. (5.28) and (5.41),

$$B_{44} = I_{44}^{(0)}, \quad C_{44} = I_{44}^{(1)} - eI_{44}^{(0)}, \quad D_{44} = I_{44}^{(2)} - 2eI_{44}^{(1)} + e^2 I_{44}^{(0)} \tag{5.69}$$

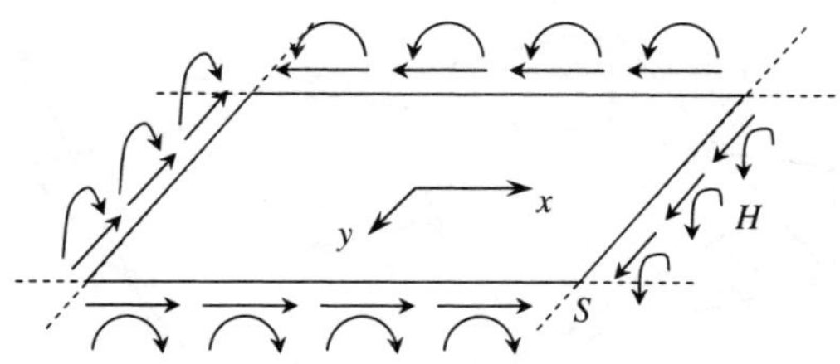

Fig. 5.16. Shear and torsion of the element.

where

$$I_{44}^{(r)} = \frac{1}{r+1}\sum_{i=1}^{k} A_{44}^{(i)}\left(t_i^{r+1} - t_{i-1}^{r+1}\right)$$

The solution of Eqs. (5.68) is

$$\gamma_{xy}^0 = \frac{1}{D_4}(D_{44}S - C_{44}H), \quad \kappa_{xy} = \frac{1}{D_4}(B_{44}M - C_{44}S) \tag{5.70}$$

in which $D_4 = I_{44}^{(0)}I_{44}^{(2)} - \left(I_{44}^{(1)}\right)^2$.

A further transformation is used similar to that for Eqs. (5.58) and (5.59). Taking the coordinate of the reference plane as

$$e = \frac{I_{44}^{(1)}}{I_{44}^{(0)}} \tag{5.71}$$

we get $C_{44} = 0$, and Eqs. (5.70) reduce to

$$\gamma_{xy}^0 = \frac{S}{B_{44}}, \quad \kappa_{xy} = \frac{H}{D_{44}} \tag{5.72}$$

Using the new notations $B_{44} = B_{xy}$ and $D_{44} = D_{xy}$ and applying Eqs. (5.69) and (5.71), we arrive at

$$B_{xy} = I_{44}^{(0)}, \quad D_{xy} = I_{44}^{(2)} - \frac{\left(I_{44}^{(1)}\right)^2}{I_{44}^{(0)}} \tag{5.73}$$

where B_{xy} is the actual in-plane shear stiffness of the laminate, whereas D_{xy} needs some comments. The second equation of Eqs. (5.72) yields

$$H = D_{xy}\kappa_{xy} \tag{5.74}$$

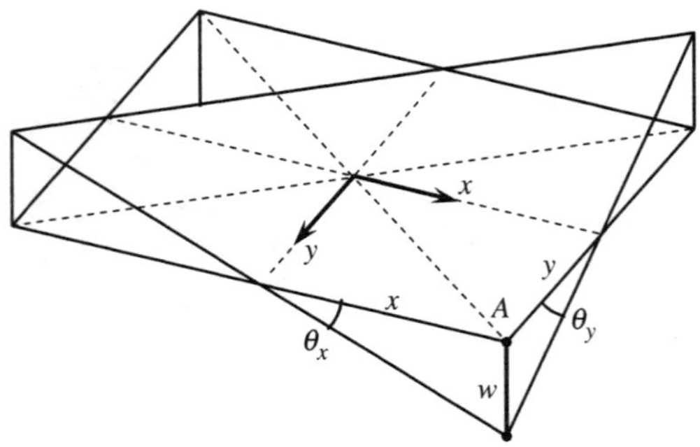

Fig. 5.17. Deformation of the element under torsion.

where κ_{xy} is given in notations to Eqs. (5.3), i.e.,

$$\kappa_{xy} = \frac{\partial \theta_x}{\partial y} + \frac{\partial \theta_y}{\partial x} \tag{5.75}$$

The deformed state of the laminated element (see Fig. 5.16) loaded with twisting moments only is shown in Fig. 5.17. Consider the deflection of point A with coordinates x and y. It follows from Fig. 5.17 that $w = x\theta_x$ or $w = y\theta_y$. Introduce the gradient of the torsional angle

$$\theta' = \frac{\partial \theta_x}{\partial y} = \frac{\partial \theta_y}{\partial x}$$

Since θ' does not depend on x and y, $\theta_x = y\theta'$, $\theta_y = x\theta'$ and $w = xy\theta'$. Using Eq. (5.75), we have $\kappa_{xy} = 2\theta$. Then, Eq. (5.74) yields

$$H = D_t^p \theta' \tag{5.76}$$

where

$$D_t^p = 2D_{xy} \tag{5.77}$$

is the plate torsional stiffness specifying the stiffness of the element which is loaded with twisting moments applied to all four edges of the element as shown in Fig. 5.16.

However, in practice we usually need the torsional stiffness of the element loaded with twisting moments applied to only two opposite edges of the element, whereas the two other edges are free. Since such loading induces not only twisting moments (see Fig. 5.4) but also transverse shear forces V (see Fig. 5.6), we must first determine the actual transverse (through-the-thickness) stiffnesses of a laminate.

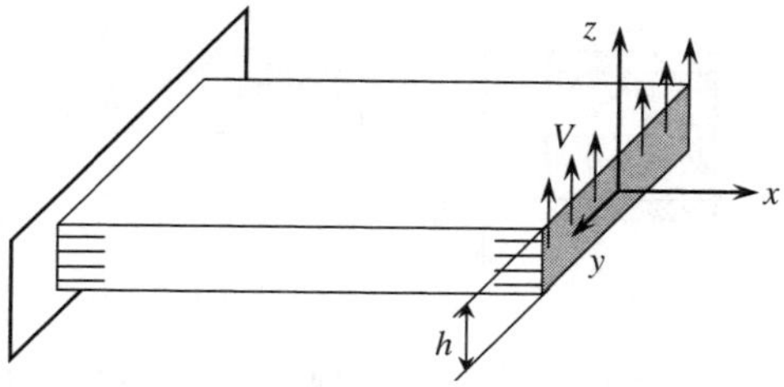

Fig. 5.18. Laminated element loaded with transverse shear forces.

Consider an orthotropic laminated element loaded with transverse shear forces $V_x = V$ uniformly distributed over the element edge as in Fig. 5.18. From Eqs. (5.20), we have two possible constitutive equations, i.e.,

$$V = S_{55}\gamma_x, \quad V = \overline{S}_{55}\gamma_x \tag{5.78}$$

in which, in accordance with Eqs. (5.46),

$$S_{55} = \sum_{i=1}^{k} A_{55}^{(i)} h_i, \quad \overline{S}_{55} = \frac{h^2}{\sum_{i=1}^{k} \frac{h_i}{A_{55}^{(i)}}} \tag{5.79}$$

For the orthotropic material, $A_{55}^{(i)} = G_{xz}^{(i)}$, where $G_{xz}^{(i)}$ is the transverse shear modulus of the ith layer. Thus, Eqs. (5.79) take the form

$$S_{55} = \sum_{i=1}^{k} G_{xz}^{(i)} h_i, \quad \overline{S}_{55} = \frac{h^2}{\sum_{i=1}^{k} \frac{h_i}{G_{xz}^{(i)}}} \tag{5.80}$$

As shown in Section 5.1, S_{55} gives the upper bound and $\overline{S}_{55}$ gives the lower bound of the actual transverse shear stiffness of the laminate. For a laminate consisting of identical layers, i.e., for the case $G_{xz}^{(i)} = G_{xz}$ for all the layers, both equations of Eqs. (5.80) give the same result $S_{55} = \overline{S}_{55} = G_{xz}h$. However, in some cases, following from Eqs. (5.80) the results can be dramatically different, whereas for engineering applications we must have instead of Eqs. (5.78) a unique constitutive equation, i.e.,

$$V = S_x\gamma_x \tag{5.81}$$

and the question arises whether S_{55} or $\overline{S}_{55}$ should be taken as S_x in this equation. Since for a homogeneous material there is no difference between S_{55} and $\overline{S}_{55}$, we can expect that this difference shows itself in the laminates consisting of layers with different transverse shear moduli.

Consider, for example, sandwich structures composed of high-stiffness thin facing layers (facings) and low-stiffness light foam core (Fig. 5.19a). The facings are made

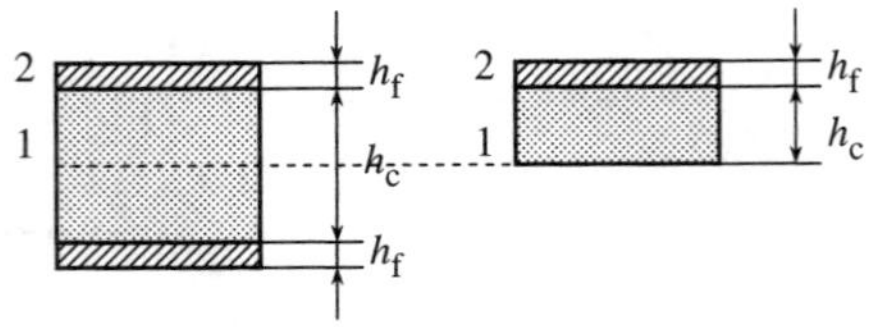

Fig. 5.19. Three-layered (sandwich) and two-layered laminates.

of aluminum alloy with modulus $E_f = 70\,\text{GPa}$ and shear modulus $G_f = 26.9\,\text{GPa}$. The foam core has $E_c = 0.077\,\text{GPa}$ and $G_c = 0.0385\,\text{GPa}$. The geometric and stiffness parameters of two sandwich beams studied experimentally (Aleksandrov et al., 1960) are presented in Table 5.1. The beams with length $l = 280\,\text{mm}$ have been tested under transverse bending. The coefficient S_a in the table corresponds to the actual shear stiffness found from experimental results. Actually, experimental study allows us to determine the shear parameter (Vasiliev, 1993)

$$k_G = \frac{D}{S_a l^2} \tag{5.82}$$

which is presented in the third column of the table and depends on the bending stiffness, D, and the beam length, l. Since the sandwich structure is symmetric, we can use Eq. (5.67) for D_x in which $2k = 2$ (the core is divided into two identical layers as in Fig. 5.19a)

$$A_{11}^{(1)} = E_c, \quad A_{11}^{(2)} = E_f,$$

$$h_1 = \frac{h_c}{2}, \quad h_2 = h_f, \quad z_0 = 0, \quad z_1 = \frac{h_c}{2}, \quad z_2 = \frac{h_c}{2} + h_f$$

The final expression is

$$D_x = \frac{2}{3}\left[\frac{1}{8}E_c h_c^3 + E_f h_f\left(\frac{3}{4}h_c^2 + \frac{3}{2}h_c h_f + h_f^2\right)\right]$$

The results of the calculation are listed in the last column of Table 5.1. The shear stiffness coefficients S_{55} and $\overline{S}_{55}$ can be found from Eqs. (5.80) which for the structure in Fig. 5.19a

Table 5.1
Parameters of sandwich structures.

h_f (mm)	h_c (mm)	k_G	Shear stiffness (GPa × mm)			Bending stiffness (GPa × mm^3)
			S_a	$\overline{S}_{55}$	S_{55}	
2.4	18.8	0.444	1.09	1.14	130	37 960
1.0	17.0	0.184	0.79	0.82	54.5	11 380

take the form

$$S_{55} = h_c G_c + 2h_f G_f$$

$$\bar{S}_{55} = \frac{(h_c + 2h_f)^2}{\dfrac{h_c}{G_c} + \dfrac{2h_f}{G_f}}$$

The results of the calculation are presented in Table 5.1. As can be seen, coefficients $\bar{S}_{55}$ are in good agreement with the corresponding experimental data, whereas coefficients S_{55} are higher by an order of magnitude. Note, that $\bar{S}_{55}$, providing the lower boundary for the exact shear stiffness, is higher than the actual stiffness S_a. The reason for this effect has been discussed in Section 5.1. Coefficient $\bar{S}_{55}$ specifies the lower boundary for the theoretical exact stiffness corresponding to the applied model of the laminate, but not for the actual stiffness following from experiment. For example, the actual shear stiffness of the sandwich beams described above can be affected by the compliance of adhesive layers which bond the facings and the core and are not allowed for in the laminate model.

So, it can be concluded that the shear stiffness coefficient $\bar{S}_{55}$ specified by the corresponding equation of Eqs. (5.79) can be used to describe the transverse shear stiffness of composite laminates. However, there are special structures for which coefficient S_{55} provides a better approximation of shear stiffness than coefficient $\bar{S}_{55}$. Consider, for example, a two-layered structure shown in Fig. 5.19b and composed of a high-stiffness facing and a low-stiffness core. Assume, as for the sandwich structure considered above, that $G_f = 26.9\,\text{GPa}$ and $G_c = 0.0385\,\text{GPa}$, so that $G_f/G_c = 699$, and take $h_c = 9.9\,\text{mm}$, and $h_f = 2.4\,\text{mm}$. It is obvious that the core, having such a low shear modulus, does not work, and the transverse shear stiffness of the laminate is governed by the facing layer. For this layer only, we get

$$S_{55} = \bar{S}_{55} = G_f h_f = 64.6\,\text{GPa} \cdot \text{mm}$$

whereas for the laminate, Eqs. (5.80) yield

$$S_{55} = h_c G_c + h_f G_f = 65\,\text{GPa} \cdot \text{mm}$$

$$\bar{S}_{55} = \frac{(h_c + h_f)^2}{\dfrac{h_c}{G_c} + \dfrac{h_f}{G_f}} = 0.59\,\text{GPa} \cdot \text{mm}$$

As can be seen, coefficient $\bar{S}_{55}$ is very far from the value that would be expected. However, structures of type for which the stiffness coefficient S_{55} is more appropriate than the coefficient $\bar{S}_{55}$ are not typical in composite technology and, being used, they usually do not require the calculation of transverse shear stiffnesses. For laminated composites it can be recommended to use the coefficient $\bar{S}_{55}$ (Chen and Tsai, 1996). Thus, the transverse shear stiffness coefficient in Eq. (5.81) can be taken in the following form

$$S_x = \frac{h^2}{\sum_{i=1}^{k} \frac{h_i}{G_{xz}^{(i)}}} \tag{5.83}$$

For shear in the *yz*-plane (see Fig. 5.18), we get a similar expression, i.e.,

$$S_y = \frac{h^2}{\sum_{i=1}^{k} \frac{h_i}{G_{yz}^{(i)}}} \tag{5.84}$$

In engineering analysis of laminated composites, transverse shear stiffnesses are mainly used to study the problems of transverse bending of composite beams and plates. Note, that the so-called classical theory of laminated beams and plates ignores the transverse shear deformation of the laminate. Consider the constitutive equations for the shear forces and write them in the following form

$$\gamma_x = \frac{V_x}{S_x}, \quad \gamma_y = \frac{V_y}{S_y}$$

Taking $S_x \to \infty$ and $S_y \to \infty$, we get $\gamma_x = 0$ and $\gamma_y = 0$. Applying Eqs. (5.14) for γ_x and γ_y, we can express the rotation angles in terms of the deflection as

$$\theta_x = -\frac{\partial w}{\partial x}, \quad \theta_y = -\frac{\partial w}{\partial y}$$

Then, the expressions for curvatures entering Eqs. (5.3) take the form

$$\kappa_x = -\frac{\partial^2 w}{\partial x^2}, \quad \kappa_y = -\frac{\partial^2 w}{\partial y^2}, \quad \kappa_{xy} = -2\frac{\partial^2 w}{\partial x \partial y}$$

For actual laminates, the transverse shear stiffness coefficients are not infinitely high, but nevertheless, the classical theories ignoring the corresponding deformation are widely used in the analysis of composite structures. To evaluate the possibility of neglecting transverse shear deformation, we can use parameter k_G specified by Eq. (5.82) and compare it with unity. The effect of the transverse shear deformation is demonstrated in Table 5.2 for the problem of transverse bending of simply supported sandwich beams with various parameters k_G listed in the table. The right hand column of the table shows the ratio of the maximum deflections of the beam, w, found with allowance for transverse shear deformation (w_G) and corresponding to the classical beam theory (w_∞). As can be seen, for beams number 4 and 5, having parameter k_G which is negligible in comparison with unity, the shear deformation practically does not affect the beams deflections.

Returning to the problem of torsion, we consider an orthotropic laminated strip with width b loaded with a torque M_t as in Fig. 5.20. In contrast to the laminate shown

Table 5.2
The effect of transverse shear deformation on the deflection of sandwich beams.

Beam number	$k_G = \dfrac{D}{Sl^2}$	$\dfrac{w_G}{w_\infty}$
1	0.444	5.386
2	0.184	2.805
3	0.015	1.152
4	0.0015	1.014
5	0.0004	1.002

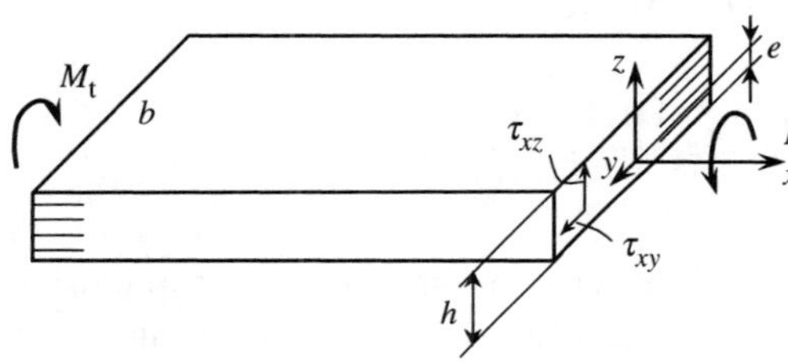

Fig. 5.20. Torsion of a laminated strip.

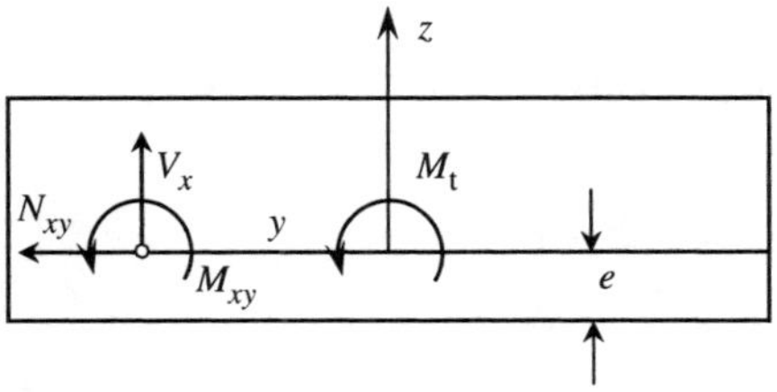

Fig. 5.21. Forces and moments acting in the strip cross section.

in Fig. 5.6, the strip in Fig. 5.20 is loaded only at the transverse edges, whereas the longitudinal edges $y = \pm b/2$ are free. The shear stresses τ_{xz} and τ_{yz} induced by torsion give rise to the shear forces N_{xy}, twisting moment M_{xy} and transverse shear force V_x shown in Fig. 5.21. Applying the corresponding constitutive equations, Eqs. (5.44) and (5.81), we get

$$N_{xy} = B_{44}\gamma_{xy}^0 + C_{44}\kappa_{xy}, \quad M_{xy} = C_{44}\gamma_{xy}^0 + D_{44}\kappa_{xy} \tag{5.85}$$

$$V_x = S_x\gamma_x \tag{5.86}$$

where the stiffness coefficients B, C, D, and S are specified by Eqs. (5.69) and (5.82). Pre-assign the coordinate of the reference plane e in accordance with Eq. (5.71).

Then, $C_{44} = 0$ and Eqs. (5.85) reduce to

$$N_{xy} = B_{xy}\gamma_{xy}^0 \tag{5.87}$$

$$M_{xy} = D_{xy}\kappa_{xy} \tag{5.88}$$

where B_{xy} and D_{xy} are given by Eqs. (5.73). Since the strip is loaded with a torque M_t only (see Fig. 5.20), $N_{xy} = 0$, and as follows from Eq. (5.87), $\gamma_{xy}^0 = 0$. So, we have only two constitutive equations, i.e., Eqs. (5.86) and (5.88) for V_x and M_{xy} which are expressed in terms of the transverse shear strain γ_x and the twisting deformation κ_{xy}. Applying Eqs. (5.14) and (5.75), we have

$$\gamma_x = \theta_x + \frac{\partial w}{\partial x}, \quad \kappa_{xy} = \frac{\partial \theta_x}{\partial y} + \frac{\partial \theta_y}{\partial x} \tag{5.89}$$

Consider the deformation of the strip. Assume that the strip cross section rotates around the longitudinal axis x through an angle θ which depends only on x (Fig. 5.22). Then, as follows from Fig. 5.22,

$$w = -y\theta, \quad \theta_y = \theta$$

Substitution into Eqs. (5.89) yields

$$\gamma_x = \theta_x - \theta' y, \quad \kappa_{xy} = \frac{\partial \theta_x}{\partial y} + \theta'$$

where $\theta' = \mathrm{d}\theta/\mathrm{d}x$. Using the first of these equations to transform the second one, we get

$$\kappa_{xy} = \frac{\partial \gamma_x}{\partial y} + 2\theta'$$

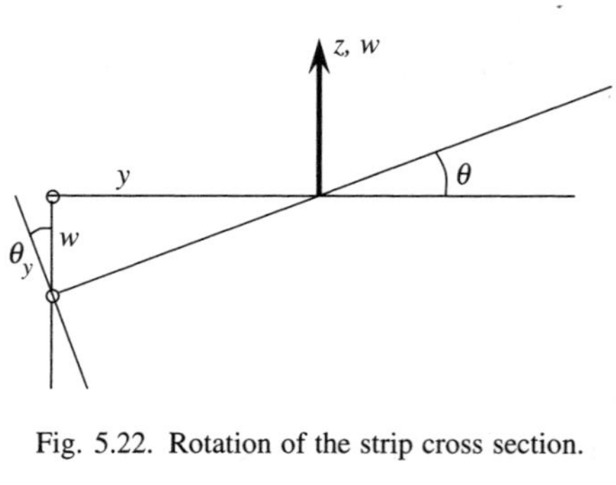

Fig. 5.22. Rotation of the strip cross section.

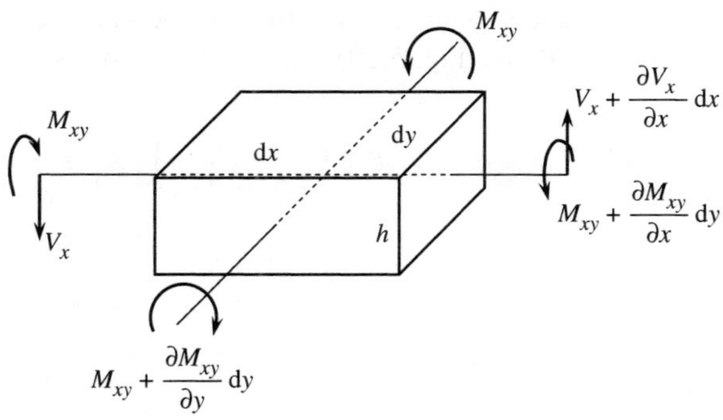

Fig. 5.23. Forces and moments acting on the strip element.

Thus, the constitutive equations, Eqs. (5.86) and (5.88) take the following final form

$$V_x = S_x\gamma_x, \quad M_{xy} = D_{xy}\left(\frac{\partial\gamma_x}{\partial y} + 2\theta'\right) \tag{5.90}$$

Consider the equilibrium of the strip element shown in Fig. 5.23. The equilibrium equations in this case are

$$\frac{\partial V_x}{\partial x} = 0, \quad \frac{\partial M_{xy}}{\partial x} = 0 \tag{5.91}$$

$$\frac{\partial M_{xy}}{\partial y} - V_x = 0 \tag{5.92}$$

The first two equations, Eqs. (5.91) show that $V_x = V_x(y)$ and $M_{xy} = M_{xy}(y)$. Then, as follows from Eq. (5.90) for V_x, $\gamma_x = \gamma_x(y)$. Substituting M_{xy} and V_x from Eqs. (5.90) into Eq. (5.92) and taking into account that θ' does not depend on y, we arrive at the following ordinary differential equation for γ_x

$$\frac{\mathrm{d}^2\gamma_x}{\mathrm{d}y^2} - k^2\gamma_x = 0$$

in which $k^2 = S_x/D_{xy}$. The general solution of this equation is

$$\gamma_x = C_1 \sinh ky + C_2 \cosh ky$$

Substitution in Eq. (5.90) for M_{xy} yields

$$M_{xy} = D_{xy}[2\theta' + k(C_1 \cosh ky + C_2 \sinh ky)]$$

The integration constants C_1 and C_2 can be found from the boundary conditions according to which M_{xy} $(y = \pm b/2) = 0$ (see Fig. 5.20). The final solution is

$$V_x = -\frac{2S_x \sinh ky}{k \cosh \lambda}\theta', \quad M_{xy} = 2D_{xy}\theta'\left(1 - \frac{\cosh ky}{\cosh \lambda}\right) \tag{5.93}$$

in which

$$\lambda = \frac{1}{2}kb = \frac{b}{2}\sqrt{\frac{S_x}{D_{xy}}} \tag{5.94}$$

Consider Fig. 5.21 and express the applied torque M_t in terms of internal forces and moments V_x and M_{xy} as

$$M_t = \int_{-b/2}^{b/2} (M_{xy} - V_x y)\mathrm{d}y$$

Substituting M_{xy} and V_x from Eqs. (5.93), we arrive at

$$M_t = D_t\theta'$$

where

$$D_t = 4D_{xy}b\left(1 - \frac{1}{\lambda}\tanh \lambda\right) \tag{5.95}$$

is the torsional stiffness of the strip. For a homogeneous orthotropic laminate discussed in Section 5.2,

$$D_{xy} = \frac{1}{12}G_{xy}h^3, \quad S_x = G_{xz}h$$

and Eq. (5.95) reduces to

$$D_t = \frac{1}{3}bh^3G_{xy}\left(1 - \frac{1}{\lambda}\tanh \lambda\right) \tag{5.96}$$

where

$$\lambda = \frac{b}{h}\sqrt{\frac{3G_{xz}}{G_{xy}}}$$

The stiffness coefficient in Eq. (5.96) is in good agreement with the exact elasticity theory solutions (Vasiliev, 1993). Particularly, for $b/h \geq 3$ the difference between D_t given by

Eq. (5.96) and the exact result is less than 2%. For a wide strip with relatively large b, the parameter λ in Eq. (5.94) is also large, and Eq. (5.95) can be approximately reduced to

$$D_{\mathrm{t}} = 4D_{xy}b \tag{5.97}$$

Dividing D_{t} by b, we can find the stiffness of the laminate with a unit width, i.e.,

$$D_{\mathrm{t}}^{b} = 4D_{xy} \tag{5.98}$$

This is a beam torsional stiffness which is twice as high as the plate stiffness specified by Eq. (5.77). The difference between Eqs. (5.77) and (5.98) is natural because Eq. (5.77) corresponds to torsion with the moments acting on all four edges of the element (see Fig. 5.16), whereas Eq. (5.98) describes torsion with only two moments applied at the transverse edges (see Fig. 5.20).

Thus, the laminate membrane, bending, transverse shear, and torsional stiffness coefficients are specified by Eqs. (5.65), (5.66), (5.82), (5.83), and (5.95).

5.6. Quasi-homogeneous laminates

Some typical layers considered in Chapter 4 were actually quasi-homogeneous laminates (see Sections 4.4 and 4.5), but being composed of a number of identical plies, they were treated as homogeneous layers. The accuracy of this assumption is evaluated below.

5.6.1. Laminate composed of identical homogeneous layers

Consider a laminate composed of layers with different thicknesses but the same stiffnesses, i.e., $A_{mn}^{(i)} = A_{mn}$ for all $i = 1, 2, 3, \ldots, k$. Then, Eqs. (5.29) and (5.32) yield

$$I_{mn}^{(r)} = \frac{A_{mn}}{r+1}h^{r+1}, \quad \overline{I}_{mn}^{(0)} = \overline{A}_{mn}h$$

This result coincides with Eqs. (5.33), which means that a laminate consisting of layers with the same mechanical properties is a homogeneous laminate (layer) as studied in Section 5.2.

5.6.2. Laminate composed of inhomogeneous orthotropic layers

Let the laminate have the following structure $[0^\circ/90^\circ]_p$, where $p = 1, 2, 3, \ldots$ specifies the number of elementary cross-ply couples of 0 and 90° plies. In Section 4.4, this laminate was treated as a homogeneous layer with material stiffness coefficients specified by Eqs. (4.114). Taking $\overline{h}_0 = \overline{h}_{90} = 0.5$ in these equations, we have

$$A_{11} = A_{22} = \frac{1}{2}\left(\overline{E}_1 + \overline{E}_2\right), \quad A_{12} = \overline{E}_1 \nu_{12}, \quad A_{44} = G_{12} \tag{5.99}$$

In accordance with Eqs. (5.36), the stiffness coefficients of this layer should be

$$B^0_{mn} = A_{mn}h, \quad C^0_{mn} = 0, \quad D^0_{mn} = \frac{1}{12}A_{mn}h^3 \tag{5.100}$$

To calculate the actual stiffnesses of the laminate, we should put $h_i = \delta$, $t_i = i\delta$, $k = 2p$, $e = h/2$, and $h = 2p\delta$ (see Fig. 5.10), where δ is the thickness of a unidirectional ply. Then, Eqs. (5.28) and (5.42) yield

$$\begin{aligned} &B_{mn} = I^{(0)}_{mn}, \quad C_{mn} = I^{(1)}_{mn} - p\delta I^{(0)}_{mn}, \\ &D_{mn} = I^{(2)}_{mn} - 2p\delta I^{(1)}_{mn} + p^2\delta^2 I^{(0)}_{mn} \end{aligned} \tag{5.101}$$

Here,

$$\begin{aligned} &I^{(0)}_{11} = I^{(0)}_{22} = p\delta\overline{E}_1(1+\alpha) = \frac{h}{2}\overline{E}_1(1+\alpha), \quad I^{(0)}_{12} = 2p\delta\overline{E}_1\nu_{12} = \overline{E}_1\nu_{12}h, \\ &I^{(0)}_{44} = 2p\delta G_{12} = G_{12}h, \quad I^{(1)}_{11} = \frac{\delta^2}{2}\overline{E}_1\sum_{j=1}^{p}[4j(1+\alpha) - (3+\alpha)], \\ &I^{(1)}_{22} = \frac{\delta^2}{2}\overline{E}_1\sum_{j=1}^{p}[4j(1+\alpha) - (3\alpha+1)], \quad I^{(1)}_{12} = \frac{1}{2}\overline{E}_1\nu_{12}h^2, \\ &I^{(1)}_{44} = \frac{1}{2}G_{12}h^2, \\ &I^{(2)}_{11} = \frac{\delta^3}{3}\overline{E}_1\sum_{j=1}^{p}\left[12j^2(1+\alpha) - 6j(3+\alpha) + 7 + \alpha\right], \\ &I^{(2)}_{22} = \frac{\delta^3}{3}\overline{E}_1\sum_{j=1}^{p}\left[12j^2(1+\alpha) - 6j(3\alpha+1) + 7\alpha + 1\right], \\ &I^{(2)}_{12} = \frac{1}{3}\overline{E}_1\nu_{12}h^3, \quad I^{(2)}_{44} = \frac{1}{3}G_{12}h^3 \end{aligned} \tag{5.102}$$

where $\alpha = E_2/E_1$.

Matching Eqs. (5.99), (5.100), (5.101), and (5.102), we can see that $B_{mn} = B^0_{mn}$, i.e., membrane stiffnesses are the same for both models of the laminate. The coupling and bending stiffnesses are also the same for $mn = 12, 44$. There is no difference between the models for $\alpha = 1$ because the laminate reduces in this case to a homogeneous layer.

Summing up the series in Eqs. (5.102) and using Eqs. (5.101), we arrive at

$$C_{11} = -C_{22} = \frac{1}{2}\overline{E}_1\delta^2 p(\alpha - 1), \quad C_{12} = C_{44} = 0,$$
$$D_{11} = D_{22} = \frac{1}{3}\overline{E}_1\delta^3 p^3(1+\alpha), \quad D_{12} = D_{12}^0, \quad D_{44} = D_{44}^0 \tag{5.103}$$

Taking into account that in accordance with Eqs. (5.100) and accepted notations

$$D_{11}^0 = D_{22}^0 = \frac{1}{3}\overline{E}_1\delta^3 p^3(1+\alpha)$$

we can conclude that the only difference between the homogeneous and the laminated models is associated with the coupling coefficients C_{11} and C_{22} which are equal to zero for the homogeneous model and are specified by Eqs. (5.103) for the laminated one. Since $p\delta = h/2$, we can write these coefficients in the form

$$C_{11} = -C_{22} = \frac{1}{4}\overline{E}_1 h\delta(1+\alpha)$$

showing that $C_{mn} \to 0$ for $\delta \to 0$.

5.6.3. *Laminate composed of angle-ply layers*

Consider a laminate with the following structure $[+\phi/-\phi]_p$, where p is the number of layers each consisting of $+\phi$ and $-\phi$ unidirectional plies. Constitutive equations Eqs. (5.5) for this laminate are

$$N_x = B_{11}\varepsilon_x^0 + B_{12}\varepsilon_y^0 + C_{14}\kappa_{xy}$$
$$N_y = B_{21}\varepsilon_x^0 + B_{22}\varepsilon_y^0 + C_{24}\kappa_{xy}$$
$$N_{xy} = B_{44}\gamma_{xy}^0 + C_{41}\kappa_x + C_{42}\kappa_y$$
$$M_x = C_{14}\gamma_{xy}^0 + D_{11}\kappa_x + D_{12}\kappa_y \tag{5.104}$$
$$M_y = C_{24}\gamma_{xy}^0 + D_{21}\kappa_x + D_{22}\kappa_y$$
$$M_{xy} = C_{41}\varepsilon_x^0 + C_{42}\varepsilon_y^0 + D_{44}\kappa_{xy}$$

in which

$$B_{mn} = A_{mn}h, \quad C_{mn} = -\frac{1}{2}A_{mn}h\delta, \quad D_{mn} = \frac{1}{12}A_{mn}h^3$$

where, h is the laminate thickness, δ the ply thickness, and A_{mn} are material stiffness coefficients specified by Eqs. (4.72). As can be seen, the laminate is anisotropic because $+\phi$ and $-\phi$ plies are located in different planes. The homogeneous model of the laminate ignores this fact and yields $C_{14} = C_{24} = 0$. Calculations show that these coefficients, although not actually equal to zero, have virtually no influence practically on the laminate behavior for $h/\delta \geq 20$.

Laminates in which any ply or layer with orientation angle $+\phi$ is accompanied by the same ply or layer but with angle $-\phi$ are referred to as balanced laminates. Being composed of only angle-ply layers, these laminates have no shear–extension coupling ($B_{14} = B_{24} = 0$), bending–stretching and shear–twisting coupling ($C_{11} = C_{12} = C_{22} = C_{44} = 0$). As follows from Eqs. (5.104), only stretching–twisting and bending–shear coupling can exist in balanced laminates. These laminates can include also 0 and 90° layers, but membrane–bending coupling can appear in such laminates.

5.7. Quasi-isotropic laminates

The layers of a laminate can be arranged in such a way that the laminate will behave as an isotropic layer under in-plane loading. Actually, the laminate is not isotropic (that is why it is called a quasi-isotropic laminate) because under transverse (normal to the laminate plane) loading and under interlaminar shear its behavior is different from that of an isotropic (e.g., metal) layer.

To derive the conditions that should be met by the structure of a quasi-isotropic laminate, consider in-plane loading with stresses σ_x, σ_y, and τ_{xy} that are shown in Fig. 5.1 and induce only in-plane strains ε_x^0, ε_y^0, and γ_{xy}^0. Taking $\kappa_x = \kappa_y = \kappa_{xy} = 0$ in Eqs. (5.5) and introducing average (through the laminate thickness h) stresses as

$$\sigma_x = N_x/h, \quad \sigma_y = N_y/h, \quad \tau_{xy} = N_{xy}/h$$

we can write the first three equations of Eqs. (5.5) in the following form

$$\begin{aligned}
\sigma_x &= \overline{B}_{11}\varepsilon_x^0 + \overline{B}_{12}\varepsilon_y^0 + \overline{B}_{14}\gamma_{xy}^0 \\
\sigma_y &= \overline{B}_{21}\varepsilon_x^0 + \overline{B}_{22}\varepsilon_y^0 + \overline{B}_{24}\gamma_{xy}^0 \\
\tau_{xy} &= \overline{B}_{41}\varepsilon_x^0 + \overline{B}_{42}\varepsilon_y^0 + \overline{B}_{44}\gamma_{xy}^0
\end{aligned} \tag{5.105}$$

in which, in accordance with Eqs. (5.28) and (5.42)

$$\overline{B}_{mn} = \sum_{i=1}^{k} A_{mn}^{(i)}\overline{h}_i, \quad \overline{h}_i = h_i/h \tag{5.106}$$

where, $\overline{h}_i$ is the thickness of the ith layer normalized to the laminate thickness and A_{mn} are the stiffness coefficients specified by Eqs. (4.72). For an isotropic layer, the constitutive equations analogous to Eqs. (5.105) are

$$\sigma_x = \overline{E}\left(\varepsilon_x^0 + \nu\varepsilon_y^0\right), \quad \sigma_y = \overline{E}\left(\varepsilon_y^0 + \nu\varepsilon_x^0\right), \quad \tau_{xy} = G\gamma_{xy}^0 \tag{5.107}$$

where

$$\overline{E} = \frac{E}{1-\nu^2}, \quad G = \frac{E}{2(1+\nu)} = \frac{1}{2}(1-\nu)\overline{E} \tag{5.108}$$

Comparing Eqs. (5.105) and (5.107), we can see that the shear–stretching coefficients of the laminate, i.e., $\overline{B}_{14} = \overline{B}_{41}$ and $\overline{B}_{24} = \overline{B}_{42}$, should be equal to zero. As follows from Eqs. (4.72) and Section 5.6.3, this means that the laminate should be balanced, i.e., it should be composed of 0°, $\pm\phi_i$ (or ϕ_i and $\pi - \phi_i$), and 90° layers only. Since the laminate stiffness in the x- and the y-directions must be the same, we require that $\overline{B}_{11} = \overline{B}_{22}$. Using Eqs. (4.72), taking $\overline{h}_i = \overline{h}$ for all i, and performing the appropriate transformation, we arrive at the following condition

$$\sum_{i=1}^{k} \cos 2\phi_i = 0$$

As can be checked by direct substitutions, for $k = 1$ this equation is satisfied if $\phi_1 = 45°$ and for $k = 2$ if $\phi_1 = 0$ and $\phi_2 = 90°$. Naturally, such one- and two-layered materials cannot be isotropic even in one plane. So, consider the case $k \geq 3$, for which the solution has the form

$$\phi_i = (i-1)\frac{\pi}{k}, \quad i = 1, 2, 3, \ldots, k \tag{5.109}$$

Using the sums that are valid for angles specified by Eq. (5.109), i.e.,

$$\sum_{i=1}^{k} \sin^2\phi_i = \sum_{i=1}^{k} \cos^2\phi_i = \frac{k}{2}$$

$$\sum_{i=1}^{k} \sin^4\phi_i = \sum_{i=1}^{k} \cos^4\phi_i = \frac{3k}{8}$$

$$\sum_{i=1}^{k} \sin^2\phi_i \cos^2\phi_i = \frac{k}{8}$$

and calculating stiffness coefficients from Eqs. (5.106) and (4.72), we get

$$\overline{B}_{11} = \overline{B}_{22} = \frac{1}{8}\left[3\left(\overline{E}_1 + \overline{E}_2\right) + 2\left(\overline{E}_1 \nu_{12} + 2G_{12}\right)\right]$$

$$\overline{B}_{12} = \frac{1}{8}\left[\overline{E}_1 + \overline{E}_2 + 2\left(3\overline{E}_1 \nu_{12} - 2G_{12}\right)\right]$$

$$\overline{B}_{44} = \frac{1}{8}\left[\overline{E}_1 + \overline{E}_2 - 2\left(\overline{E}_1 \nu_{12} - 2G_{12}\right)\right]$$

These stiffnesses provide constitutive equations in the form of Eqs. (5.107) and satisfy the conditions in Eqs. (5.108) which can be written as

$$\overline{B}_{11} = \overline{B}_{22} = \frac{E}{1-\nu^2}, \quad \overline{B}_{44} = G$$

if

$$E = \frac{\left(\overline{E}_1 + \overline{E}_2 + 2\overline{E}_1 \nu_{12}\right)\left(\overline{E}_1 + \overline{E}_2 - 2\overline{E}_1 \nu_{12} + 4G_{12}\right)}{3\left(\overline{E}_1 + \overline{E}_2\right) + 2\left(\overline{E}_1 \nu_{12} + 2G_{12}\right)}$$

$$\nu = \frac{\overline{E}_1 + \overline{E}_2 + 2\left(3\overline{E}_1 \nu_{12} - 2G_{12}\right)}{3\left(\overline{E}_1 + \overline{E}_2\right) + 2\left(\overline{E}_1 \nu_{12} + 2G_{12}\right)}, \quad G = \frac{E}{2(1+\nu)} \tag{5.110}$$

Possible solutions to Eqs. (5.109) providing quasi-isotropic properties of the laminates with different number of layers are listed in Table 5.3 for $k \leq 6$.

All quasi-isotropic laminates, having different structures determined by Eq. (5.109) for a given number of layers, k, possess the same apparent modulus and Poisson's ratio specified by Eqs. (5.110). For typical advanced composites with the properties listed in Table 3.5, these characteristics are presented in Table 5.4.

As follows from Tables 5.4 and 1.1, the specific stiffness of quasi-isotropic composites with carbon and boron fibers exceeds the corresponding characteristic of traditional isotropic structural materials – steel, aluminum, and titanium.

Table 5.3
Angles providing quasi-isotropic properties of the laminates.

Number of layers, k	Orientation angle of the ith layer					
	ϕ_1°	ϕ_2°	ϕ_3°	ϕ_4°	ϕ_5°	ϕ_6°
3	0	60	120	–	–	–
4	0	45	90	135	–	–
5	0	36	72	108	144	–
6	0	30	60	90	120	150

Table 5.4
Modulus of elasticity and Poisson's ratio of quasi-isotropic laminates made of typical advanced composites.

Property	Glass–epoxy	Carbon–epoxy	Aramid–epoxy	Boron–epoxy	Boron–Al
Modulus, E (GPa)	27.0	54.8	34.8	80.3	183.1
Poisson's ratio, ν	0.34	0.31	0.33	0.33	0.28
Specific modulus, $k_E \times 10^3$ (m)	1290	3530	2640	3820	6910

5.8. Antisymmetric laminates

In antisymmetric laminates, symmetrically located layers have mutually reversed orientations. For example, whereas laminates $[0°/90°/90°/0°]$ and $[+\phi/-\phi/-\phi/+\phi]$ are symmetric, laminates $[0°/90°/0°/90°]$ or $[0°/0°/90°/90°]$ and $[+\phi/-\phi/+\phi/-\phi]$ are antisymmetric. In contrast to symmetric laminates which have maximum bending and zero coupling stiffness coefficients, antisymmetric laminates demonstrate pronounced coupling that can be important for some special applications (e.g., robotic parts undergoing complicated deformation under simple loading, rotor blades that twist under centrifugal forces, airplane wings twisting under bending etc.).

The simplest antisymmetric laminate is a cross-ply layer consisting of two plies with angles 0 and 90°, and the same thickness $h/2$ (see Fig. 5.24). Taking $e = h/2$ and using Eqs. (5.28) and (5.41), we arrive at the following stiffness coefficients entering Eqs. (5.44)

$$B_{11} = B_{22} = \frac{h}{2}\left(\overline{E}_1 + \overline{E}_2\right), \quad B_{12} = \overline{E}_1\nu_{12}h, \quad B_{44} = G_{12}h,$$

$$C_{11} = -C_{22} = \frac{h^2}{8}\left(\overline{E}_2 - \overline{E}_1\right), \quad C_{12} = 0, \quad C_{44} = 0,$$

$$D_{11} = D_{22} = \frac{h^3}{24}\left(\overline{E}_1 + \overline{E}_2\right), \quad D_{12} = \frac{h^3}{12}\overline{E}_1\nu_{12}, \quad D_{44} = \frac{h^3}{12}G_{12}$$

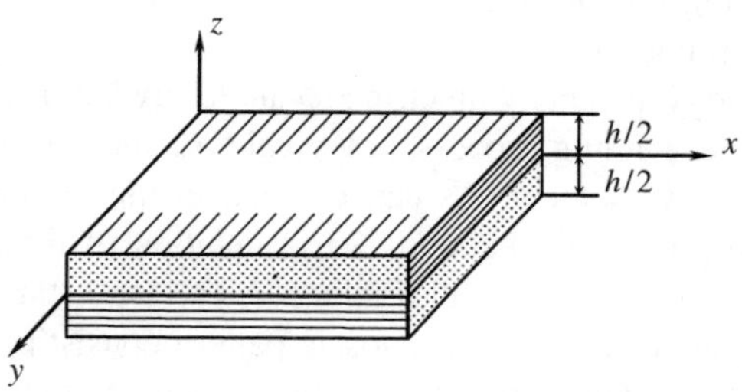

Fig. 5.24. An antisymmetric cross-ply laminate.

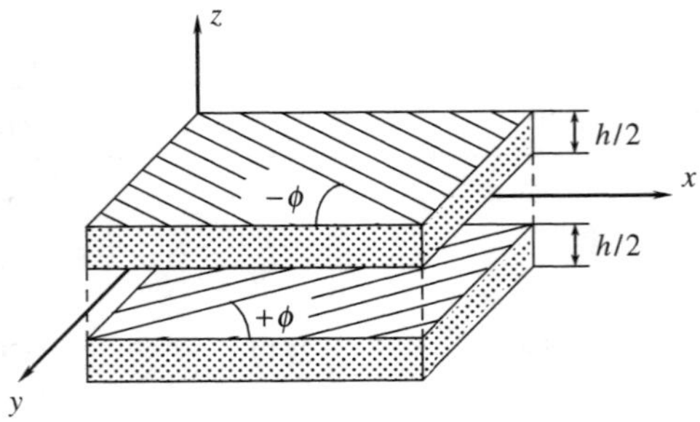

Fig. 5.25. Unbonded view of an antisymmetric angle-ply laminate.

Comparing these results with Eqs. (5.99) and (5.100), corresponding to a quasi-homogeneous cross-ply laminate, we can see that the antisymmetric cross-ply laminate has the same membrane and bending stiffnesses but nonzero coupling coefficients C_{11} and C_{22}. This fact shows, in accordance with Eqs. (5.44), that in-plane tension or compression of this laminate induces bending.

As another typical example of an antisymmetric laminate, consider an angle-ply structure consisting of two plies with the same thickness $h/2$ and orientation angles $+\phi$ and $-\phi$, respectively (see Fig. 5.25). The plies (or layers) are characterized with the following stiffness coefficients

$$A_{11}^{(1)} = A_{11}^{(2)} = A_{11}, \quad A_{12}^{(1)} = A_{12}^{(2)} = A_{12}, \quad A_{22}^{(1)} = A_{22}^{(2)} = A_{22},$$

$$A_{14}^{(1)} = -A_{14}^{(2)} = A_{14}, \quad A_{24}^{(1)} = -A_{24}^{(2)} = A_{24}, \quad A_{44}^{(1)} = A_{44}^{(2)} = A_{44}$$

where coefficients A_{mn} are specified by Eqs. (4.72). Taking again $e = h/2$, we arrive at constitutive equations in Eqs. (5.104) in which

$$B_{mn} = A_{mn}h, \quad C_{mn} = -\frac{h^2}{4}A_{mn}, \quad D_{mn} = \frac{h^3}{12}A_{mn} \tag{5.111}$$

Comparing these coefficients with those entering Eqs. (5.104) and corresponding to a quasi-homogeneous angle-ply laminate, we can conclude that the antisymmetric laminate has much larger coupling coefficients C_{14} and C_{24}, and thus a much more pronounced extension–twisting coupling effect.

In composite technology, an antisymmetric $\pm\phi$ angle-ply laminate is usually fabricated by a continuous filament winding process. A typical structure made by filament winding is shown in Fig. 4.59 of Chapter 4. As can be seen in this figure, the angle-ply layer is composed from two plies with $+\phi$ and $-\phi$ orientation of the fibers and these plies are interlaced in the process of filament winding. As a result, the structure of the layer is characterized by the distinctive regular mosaic pattern consisting of triangular-shaped, repeating in chess-board fashion, two-ply segments (T-segments) with alternating $\pm\phi$ and $\mp\phi$ reinforcement. The T-segments are arranged in regular geometric pattern around the

(a) (b) (c) (d)

Fig. 5.26. Filament-wound cylinders with various numbers n_T of T-segments: $n_T = 2$ (a), 4 (b), 8 (c), and 16 (d).

circumference and along the axis forming the so-called cross-over circles (see Fig. 4.59). Depending on the parameters of the winding process, various numbers n_T of T-segments located along the circumference can be obtained. For a cylindrical shell, the structures corresponding to $n_T = 2, 4, 8$, and 16 are shown in Fig. 5.26.

Each T-segment consists of two plies with either $[+\phi/-\phi]$ or $[-\phi/+\phi]$ structure and the plies are not interlaced within the T-segment area. If, for instance, a T-segment consists of the top triangular-shaped ply, reinforced with fibers oriented at angle $+\phi$, and the bottom one reinforced with an angle $-\phi$, then the neighboring adjacent T-segments have an inverse structure: their top plies are reinforced at angle $-\phi$, and the bottom ones are reinforced at $+\phi$.

The traditional approach used to analyze the laminates under consideration is based on the model discussed in Section 4.5 according to which the laminate is treated as a homogeneous orthotropic layer with stiffness coefficients specified by Eqs. (4.72) and (4.147). The constitutive equations are taken in accordance with Eqs. (5.44), i.e.,

$$N_x = B_{11}\varepsilon_x^0 + B_{12}\varepsilon_y^0, \quad N_y = B_{21}\varepsilon_x^0 + B_{22}\varepsilon_y^0, \quad N_{xy} = B_{44}\gamma_{xy}^0,$$

$$M_x = D_{11}\kappa_x + D_{12}\kappa_y, \quad M_y = D_{21}\kappa_x + D_{22}\kappa_y, \quad M_{xy} = D_{44}\kappa_{xy}$$

where

$$B_{mn} = A_{mn}h, \quad D_{mn} = \frac{h^3}{12}A_{mn}$$

and A_{mn} are specified by Eqs. (4.72). The approach based on these constitutive equations corresponds to an infinite number of T-segments, i.e., to $n_T \to \infty$.

Considering a T-segment as an antisymmetric laminate, we must apply a more general version of Eqs. (5.44) including the coupling stiffness coefficients, i.e.,

$$N_x = B_{11}\varepsilon_x^0 + B_{12}\varepsilon_y^0 + C_{14}\kappa_{xy}$$

$$N_y = B_{21}\varepsilon_x^0 + B_{22}\varepsilon_y^0 + C_{24}\kappa_{xy}$$

$$N_{xy} = B_{44}\gamma_{xy}^0 + C_{41}\kappa_x + C_{42}\kappa_y$$

$$M_x = C_{14}\gamma_{xy}^0 + D_{11}\kappa_x + D_{12}\kappa_y$$

$$M_y = C_{24}\gamma_{xy}^0 + D_{21}\kappa_x + D_{22}\kappa_y$$

$$M_{xy} = C_{41}\varepsilon_x^0 + C_{42}\varepsilon_y^0 + D_{44}\kappa_{xy}$$

where the stiffness coefficients are specified by Eqs. (5.111). It is important that whereas for the laminate with $[+\phi/-\phi]$ structure shown in Fig. 5.25 the coupling stiffness coefficient is negative, for the adjacent T-segment having $[-\phi/+\phi]$, this coefficient is positive. This difference results in the specific behavior of the two different laminate structures of T-segments that exhibit antisymmetric opposite anisotropic stretching–twisting and bending–shear coupling effects alternating along the circumference and axis of rotation of the shell. Due to the general alternating pattern of the T-segments (chess-board structure) and their interactions within a layer, the anisotropic effects are balancing each other, inducing at the same time, additional stresses in the plies.

To study the effect of the filament-wound mosaic pattern, the stress analysis of cylindrical shells has been performed (Morozov, 2006). The shells under consideration consist of one filament-wound $\pm\phi$ angle-ply layer and loaded with internal pressure. The solid modeling (Solid Edge) and finite-element analysis (MSC NASTRAN) techniques have been employed to model the shells with different mosaic pattern structures. Each shell is partitioned into triangular-shaped T-segments according to the particular filament-wound pattern. Correspondingly, the finite elements are also combined into the respective alternating groups. The material structure of the finite elements for each of these groups is defined as either $[+\phi/-\phi]$ or $[-\phi/+\phi]$ laminate. The cylindrical shells under consideration are reinforced with a winding angle $\phi = \pm 60°$ and loaded with internal pressure of 1 MPa. The mechanical properties of the unidirectional glass–epoxy composite ply correspond to Table 3.5. The ends of the shells are clamped and the distance between the ends (length of the cylinder) is fixed and equal to 140 mm. The diameter of the cylinder is 60 mm and total thickness of the wall is $h = 1.4$ mm (with the thickness of the unidirectional ply 0.7 mm). The stress analysis was performed for four types of shells.

The first cylinder is modeled with homogeneous orthotropic angle-ply layer and analyzed using finite-element models available within the MSC NASTRAN software. This model corresponds to $n_T \to \infty$. The other three cylinders have 2, 4, and 8 triangular-shaped segments around the circumference ($n_T = 2, 4, 8$) and are analyzed using the FE modeling of the shells with allowance for their mosaic structure. The finite-element

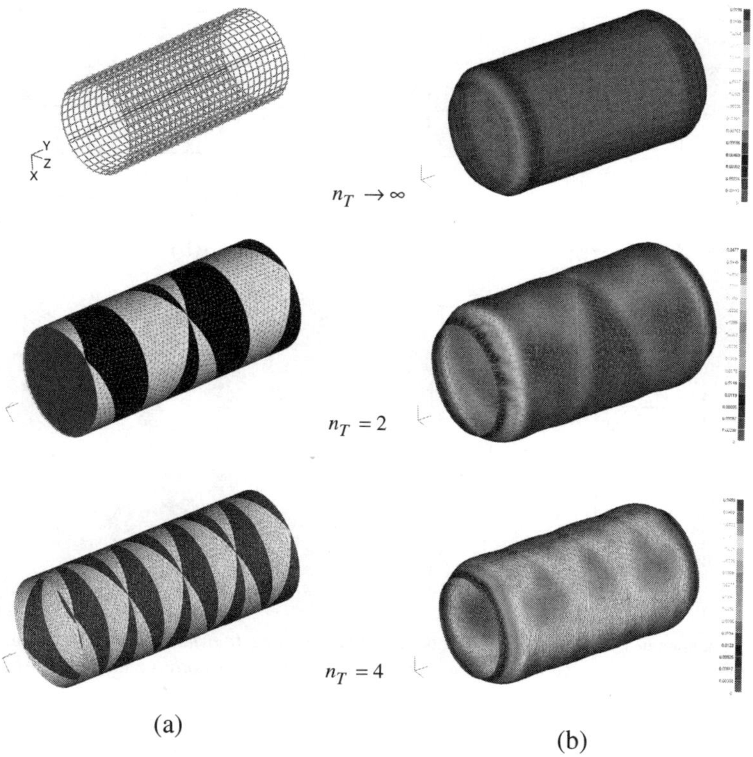

Fig. 5.27. Finite-element models (a) and deformed shapes (b) of the cylinders with $n_T \to \infty$, $n_T = 2$, and $n_T = 4$.

models and the deformed shapes for $n_T \to \infty$, $n_T = 2$, and $n_T = 4$ are shown in Fig. 5.27.

As can be seen, the deformation of the shells distinctively reflects the corresponding filament-wound mosaic texture. The calculated maximum values of stresses along and across fibers, σ_1, σ_2, and shear stresses, τ_{12}, acting in the plies are presented in Table 5.5. It can be noted from this table that the maximum stresses strongly depend on the laminate structure. The traditional model ($n_T \to \infty$) significantly underestimated the stresses.

Table 5.5
Maximum stresses in the plies of the shells with various filament-wound structures.

Structural parameter, n_T	σ_1 (MPa)	σ_2 (MPa)	τ_{12} (MPa)
∞	24.9	3.79	1.98
2	**40.99**	17.7	4.82
4	**33.2**	20.3	5.33
8	**27.30**	18.2	4.94

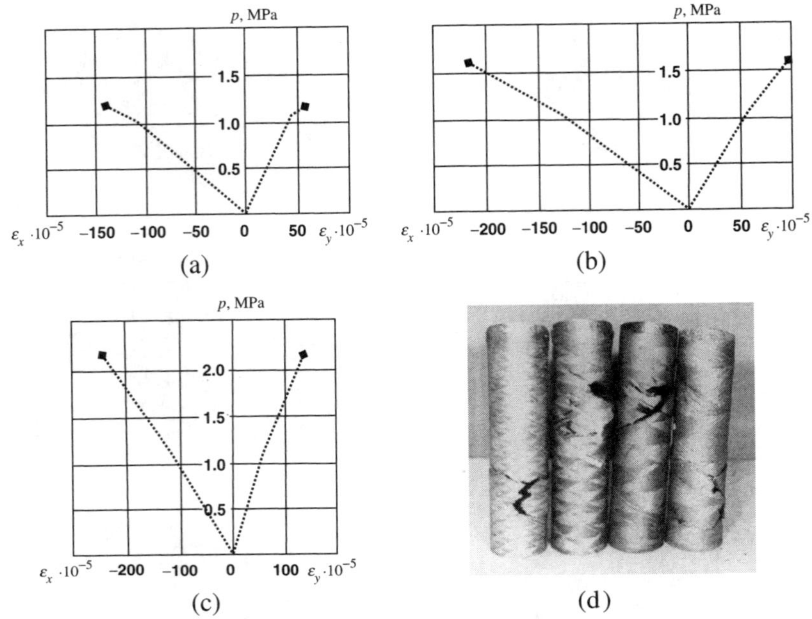

Fig. 5.28. Dependencies of the axial (ε_x) and the circumferential (ε_y) strains on internal pressure (p) for cylindrical shells with $n_T = 2$ (a), $n_T = 4$ (b), $n_T = 16$ (c) and the corresponding failure modes (d).

With an increase in the structural parameter n_T, the stresses acting along the fibers reduce and approach the value following from the traditional laminate model.

Thus, it can be expected that the higher the parameter n_T, the higher the strength of $\pm\phi$ angle-ply filament-wound structures. This prediction is confirmed by the test results presented in Fig. 5.28 (Vorobey et al., 1992). Carbon–phenolic cylindrical shells with the geometrical parameters given above have been loaded with internal pressure up to the failure. As follows from Fig. 5.28, the increase of parameter n_T from 2 (Fig. 5.28a) to 16 (Fig. 5.28c) results in a significant increase in the burst pressure.

In conclusion, it should be noted that the effect under discussion shows itself mainly in $\pm\phi$ angle-ply structures consisting of two symmetric plies. For laminated structures consisting of a system of $\pm\phi$ angle-ply layers, the coupling stiffness coefficient which causes the specific behavior discussed above is given in notations to Eqs. (5.104) and has the form

$$C_{mn} = -\frac{1}{2}A_{mn}h\delta \tag{5.112}$$

in which h is the laminate thickness and δ is the thickness of the ply. Since δ is relatively small, the coefficient C_{mn} in Eq. (5.112) is smaller than the corresponding coefficient in Eqs. (5.111), and the coupling effect caused by this coefficient is less pronounced.

5.9. Sandwich structures

Sandwich structures are three-layered laminates consisting of thin facings and a lightweight honeycomb or foam core as in Figs. 5.29 and 5.30. Since the in-plane stiffnesses of the facings are much higher than those of the core, whereas their transverse shear compliance is much lower than the same parameter of the core, the stiffness coefficients of sandwich structures are usually calculated presuming that the in-plane stiffnesses of the core are equal to zero. The transverse shear stiffnesses of the facings are assumed to be infinitely high. For the laminate shown in Fig. 5.31 this means that

$$A_{mn}^{(2)} = 0, \quad mn = 11, 12, 14, 24, 44,$$

$$A_{mn}^{(1,\,2)} \to \infty, \quad mn = 55, 56, 66$$

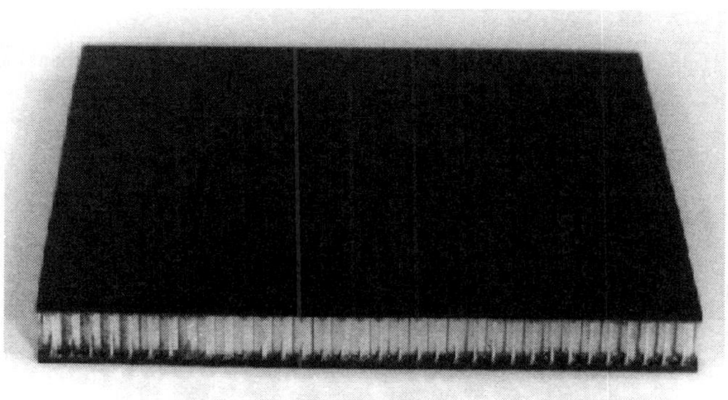

Fig. 5.29. Composite sandwich panel with honeycomb core.

Fig. 5.30. Composite sandwich rings with foam core.

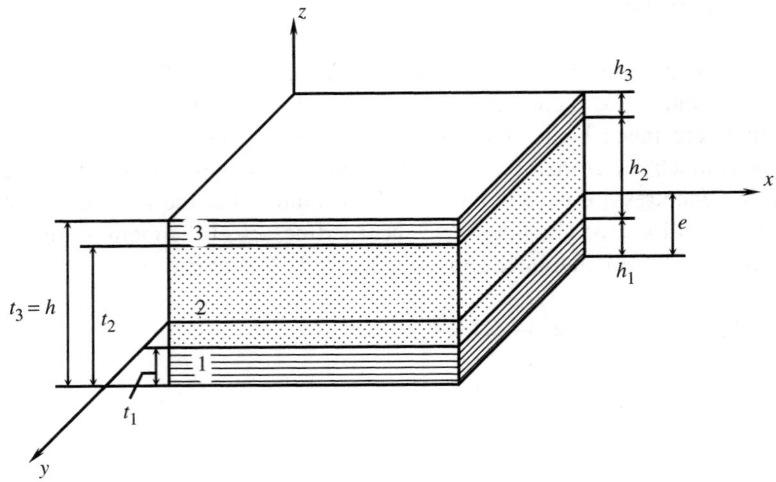

Fig. 5.31. Sandwich laminate with two laminated facings (1 and 3) and foam core (2).

As a result, the coefficients in Eqs. (5.29) become

$$I_{mn}^{(r)} = \frac{1}{r+1}\left[A_{mn}^{(1)} t_1^{r+1} + A_{mn}^{(3)} \left(t_3^{r+1} - t_2^{r+1}\right)\right] \tag{5.113}$$

where $mn = 11, 12, 22, 14, 24, 44$. Membrane, bending, and coupling stiffness coefficients can be determined using Eqs. (5.28). Transverse shear stiffnesses, Eqs. (5.83) and (5.84), for an orthotropic core, can be presented in the form

$$S_x = \frac{h^2}{h_2} G_{xz}, \quad S_y = \frac{h^2}{h_2} G_{yz}$$

where G_{xz} and G_{yz} are the shear moduli of the core.

5.10. Coordinate of the reference plane

Stiffness coefficients specified by Eqs. (5.28) include the coordinate of the reference plane e (see Fig. 5.1) that, being properly pre-assigned, allows us to simplify the constitutive equations for the laminate. As was shown in Sections 5.2 and 5.4, taking the middle plane as the reference plane, i.e., putting $e = h/2$ for homogeneous and symmetric laminate, we have $C_{mn} = 0$, and the constitutive equations take their simplest form with zero valued membrane–bending coupling terms.

Now, a natural question as to whether it is possible to reduce Eqs. (5.5) to this form in the general case arises. Taking $C_{mn} = 0$ in Eqs. (5.28), we have

$$e = \frac{I_{mn}^{(1)}}{I_{mn}^{(0)}} \tag{5.114}$$

It is important that the reference plane should be one and the same for all $mn = 11$, 12, 22, 14, 24, 44, and these six equations should give the same value of e. In the general case, this is not possible, so a universal reference plane providing $C_{mn} = 0$ cannot exist.

However, there are some other (in addition to homogeneous and symmetric structures) particular laminates for which this condition can be met. For example, consider a laminate composed of isotropic layers (see Sections 4.1 and 5.2). For such laminates,

$$A_{11}^{(i)} = A_{22}^{(i)} = \frac{E_i}{1 - \nu_i^2}, \quad A_{12}^{(i)} = \frac{E_i \nu_i}{1 - \nu_i^2}, \quad A_{44}^{(i)} = \frac{E_i}{2(1 + \nu_i)}$$

and in accordance with Eqs. (5.42)

$$I_{11}^{(0)} = I_{22}^{(0)} = \sum_{i=1}^{k} \frac{E_i h_i}{1 - \nu_i^2}, \quad I_{12}^{(0)} = \sum_{i=1}^{k} \frac{E_i \nu_i h_i}{1 - \nu_i^2}, \quad I_{44}^{(0)} = \sum_{i=1}^{k} \frac{E_i h_i}{2(1 + \nu_i)},$$

$$I_{11}^{(1)} = I_{22}^{(1)} = \frac{1}{2} \sum_{i=1}^{k} \frac{E_i h_i}{1 - \nu_i^2}(t_i + t_{i-1}), \quad I_{12}^{(1)} = \frac{1}{2} \sum_{i=1}^{k} \frac{E_i \nu_i h_i}{1 - \nu_i^2}(t_i + t_{i-1}),$$

$$I_{44}^{(1)} = \frac{1}{2} \sum_{i=1}^{k} \frac{E_i h_i}{2(1 + \nu_i)}(t_i + t_{i-1})$$

As can be seen, these parameters, when substituted into Eq. (5.114), do not provide one and the same value of e. However, if Poisson's ratio is the same for all the layers, i.e., $\nu_i = \nu$ (i = 1, 2, 3,..., k), we get

$$e = \frac{\sum_{i=1}^{k} E_i h_i (t_i + t_{i-1})}{2 \sum_{i=1}^{k} E_i h_i}$$

For practical analysis, this result is often used even if the Poisson's ratios of the layers are different. In these cases, it is assumed that all the layers can be approximately characterized with some average value of Poisson's ratio, i.e.,

$$\nu = \frac{1}{h} \sum_{i=1}^{k} \nu_i h_i$$

As another example, consider the sandwich structure described in Section 5.9. In the general case, we again fail to find the desired reference plane. However, if we assume that the facings are made of one and the same material (only the thicknesses are different), Eqs. (5.113) and (5.114) yield

$$e = \frac{h_1^2 + h_3(h_3 + 2h_1 + 2h_2)}{2(h_1 + h_3)}$$

Returning to the general case, we should emphasize that the reference plane providing $C_{mn} = 0$ for all the mn values does not exist in this case only if the laminate structure is given. If the stacking-sequence of the layers is not pre-assigned and there are sufficient number of layers, they can be arranged in such a way that $C_{mn} = 0$. Indeed, consider a laminate in Fig. 5.32 and suppose that its structure is, in general, not symmetric, i.e., $z_i' \neq z_i$ and $k' \neq k$. Using plane $z = 0$ as the reference plane, we can write the membrane–bending coupling coefficients as

$$C_{mn} = \frac{1}{2}\sum_{i=1}^{k/2} A_{mn}^{(i)} h_i (z_i + z_{i-1}) - \frac{1}{2}\sum_{i'=1}^{k'/2} A_{mn}^{(i')} h_i' \left(z_i' + z_{i-1}'\right)$$

where, $z_i \geq 0$ and $z_i' \geq 0$. Introduce a new layer coordinate $\overline{z}_i = (z_i + z_{i-1})/2$, which is the distance between the reference plane of the laminate and the middle plane of the ith layer. Then, the condition $C_{mn} = 0$ yields

$$\sum_{i=1}^{k/2} A_{mn}^{(i)} h_i \overline{z}_i = \sum_{i'=1}^{k'/2} A_{mn}^{(i')} h_i' \overline{z}_i'$$

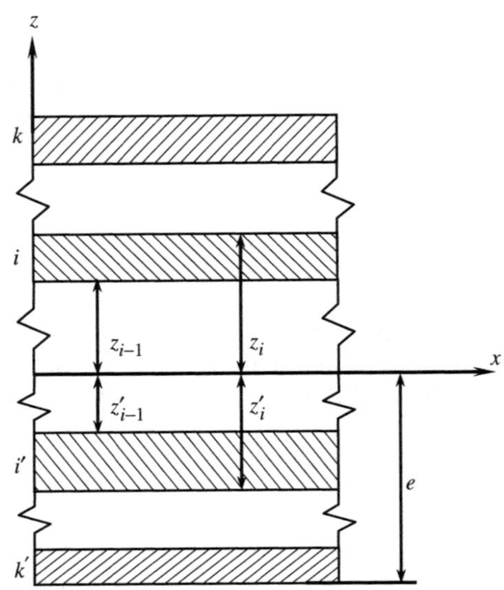

Fig. 5.32. Layer coordinates with respect to the reference plane.

Now assume that we have a group of identical layers or plies with the same stiffness coefficients A_{mn} and thicknesses. For example, the laminate could include a 1.5 mm thick 0° unidirectional layer which consists of 10 plies (the thickness of an elementary ply is 0.15 mm). Arranging these plies above $(\overline{z}_i)$ and below $(\overline{z}'_i)$ the reference plane in such a way that

$$\sum_{j=1}^{10}\left(\overline{z}_j - \overline{z}'_j\right) = 0 \tag{5.115}$$

we have no coupling for this group of plies. Doing the same with the other layers, we arrive at a laminate with no coupling. Naturally, some additional conditions following from the fact that the laminate is a continuous structure should be satisfied. However even with these conditions, Eq. (5.115) can be met with several systems of ply coordinates, and symmetric arrangement of the plies $(\overline{z}_j = \overline{z}'_j)$ is only one of these systems. The general analysis of the problem under discussion has been presented by Verchery (1999).

Return to laminates with pre-assigned stacking-sequences for the layers. It follows from Eq. (5.114), we can always make one of the coupling stiffness coefficients equal to zero, e.g., taking $e = e_{st}$ where

$$e_{st} = \frac{I_{st}^{(1)}}{I_{st}^{(0)}} \tag{5.116}$$

we get $C_{st} = 0$ (the rest of coupling coefficients are not zero).

Another way to simplify the equations for stiffnesses is to take $e = 0$, i.e., to take the surface of the laminate as the reference plane. In this case, Eqs. (5.28) take the form

$$B_{mn} = I_{mn}^{(0)}, \quad C_{mn} = I_{mn}^{(1)}, \quad D_{mn} = I_{mn}^{(2)}$$

In practical analysis, the constitutive equations for laminates with arbitrary structure are often approximately simplified using the method of reduced or minimum bending stiffnesses described, e.g., by Ashton (1969), Karmishin (1974), and Whitney (1987). To introduce this method, consider the corresponding equation of Eqs. (5.28) for bending stiffnesses, i.e.,

$$D_{mn} = I_{mn}^{(2)} - 2eI_{mn}^{(1)} + e^2 I_{mn}^{(0)} \tag{5.117}$$

and find the coordinate e delivering the minimum value of D_{mn}. Using the minimum conditions

$$\frac{\mathrm{d}}{\mathrm{d}e}D_{mn} = 0, \quad \frac{\mathrm{d}^2}{\mathrm{d}e^2}D_{mn} > 0$$

we have

$$e = e_{mn} = \frac{I_{mn}^{(1)}}{I_{mn}^{(0)}} \tag{5.118}$$

This result coincides with Eq. (5.116) and yields $C_{mn} = 0$. Thus, calculating $I_{mn}^{(1)}$ and $I_{mn}^{(0)}$, we use for each $mn = 11, 12, 22, 14, 24, 44$ the corresponding value e_{mn} specified by Eq. (5.118). Substitution yields

$$D_{mn}^r = I_{mn}^{(2)} - \frac{\left(I_{mn}^{(1)}\right)^2}{I_{mn}^{(0)}}, \quad C_{mn}^r = 0 \tag{5.119}$$

and the constitutive equations, Eqs. (5.5) become uncoupled. Naturally, this approach is only approximate because the reference plane coordinate should be the same for all stiffnesses, but it is not in the method under discussion. It follows from the foregoing derivation that the coefficients D_{mn}^r specified by Eqs. (5.119) do not exceed the actual values of bending stiffnesses, i.e., $D_{mn}^r \leq D_{mn}$. So, the method of reduced bending stiffnesses leads to underestimation of the laminate bending stiffness. In conclusion, it should be noted that this method is not formally grounded and can yield both good and poor approximation of the laminate behavior, depending on the laminate structure.

5.11. Stresses in laminates

The constitutive equations derived in the previous sections of this chapter relate forces and moments acting on the laminate to the corresponding generalized strains. For composite structures, forces and moments should satisfy equilibrium equations, whereas strains are expressed in terms of displacements. As a result, a complete set of equations is formed allowing us to find forces, moments, strains, and displacements corresponding to a given system of loads acting on the structure. Since the subject of structural mechanics is beyond the scope of this book and is discussed elsewhere (Vasiliev, 1993), we assume that this problem has already been solved, i.e., we know either generalized strains ε, γ, and κ entering Eqs. (5.5) or forces and moments N and M. If this is the case, we can use Eqs. (5.5) to find ε, γ, and κ. Now, to complete the analysis, we need to determine the stress acting in each layer of the laminate.

To do this, we should first find strains in any ith layer using Eqs. (5.3) which yield

$$\varepsilon_x^{(i)} = \varepsilon_x^0 + z_i\kappa_x, \quad \varepsilon_y^{(i)} = \varepsilon_y^0 + z_i\kappa_y, \quad \gamma_{xy}^{(i)} = \gamma_{xy}^0 + z_i\kappa_{xy} \tag{5.120}$$

where z_i is the layer normal coordinate changing over the thickness of the ith layer. If the ith layer is orthotropic with principal material axes coinciding with axes x and y, (e.g., made of fabric), Hooke's law provides the stresses we need, i.e.,

$$\sigma_x^{(i)} = \overline{E}_x^{(i)} \left(\varepsilon_x^{(i)} + \nu_{xy}^{(i)} \varepsilon_y^{(i)} \right), \quad \sigma_y^{(i)} = \overline{E}_y^{(i)} \left(\varepsilon_y^{(i)} + \nu_{yx}^{(i)} \varepsilon_x^{(i)} \right), \quad \tau_{xy}^{(i)} = G_{xy}^{(i)} \gamma_{xy}^{(i)} \tag{5.121}$$

where $\overline{E}_{x,y}^{(i)} = E_{x,y}^{(i)} / \left(1 - \nu_{xy}^{(i)} \nu_{yx}^{(i)}\right)$ and $E_x^{(i)}, E_y^{(i)}, G_{xy}^{(i)}, \nu_{xy}^{(i)}, \nu_{yx}^{(i)}$ are the elastic constants of the layer referred to the principal material axes. For an isotropic layer (e.g., metal or polymeric), we should take in Eqs. (5.121), $E_x^{(i)} = E_y^{(i)} = E_i$, $\nu_{xy}^{(i)} = \nu_{yx}^{(i)} = \nu_i$, $G_{xy}^i = G_i = E_i/2(1+\nu_i)$.

Consider a layer composed of unidirectional plies with orientation angle ϕ_i. Using Eqs. (4.69), we can express strains in the principal material coordinates as

$$\begin{aligned}
\varepsilon_1^{(i)} &= \varepsilon_x^{(i)} \cos^2 \phi_i + \varepsilon_y^{(i)} \sin^2 \phi_i + \gamma_{xy}^{(i)} \sin \phi_i \cos \phi_i \\
\varepsilon_2^{(i)} &= \varepsilon_x^{(i)} \sin^2 \phi_i + \varepsilon_y^{(i)} \cos^2 \phi_i - \gamma_{xy}^{(i)} \sin \phi_i \cos \phi_i \\
\gamma_{12}^{(i)} &= 2\left(\varepsilon_y^{(i)} - \varepsilon_x^{(i)}\right) \sin \phi_i \cos \phi_i + \gamma_{xy}^{(i)} \cos 2\phi_i
\end{aligned} \tag{5.122}$$

and find the corresponding stresses, i.e.,

$$\sigma_1^{(i)} = \overline{E}_1^{(i)} \left(\varepsilon_1^{(i)} + \nu_{12}^{(i)} \varepsilon_2^{(i)} \right), \quad \sigma_2^{(i)} = \overline{E}_2^{(i)} \left(\varepsilon_2^{(i)} + \nu_{21}^{(i)} \varepsilon_1^{(i)} \right), \quad \tau_{12}^{(i)} = G_{12}^{(i)} \gamma_{12}^{(i)} \tag{5.123}$$

where $\overline{E}_{1,2}^{(i)} = E_{1,2}^{(i)} / \left(1 - \nu_{12}^{(i)} \nu_{21}^{(i)}\right)$ and $E_1^{(i)}, E_2^{(i)}, G_{12}^{(i)}, \nu_{12}^{(i)}, \nu_{21}^{(i)}$ are the elastic constants of a unidirectional ply.

Thus, Eqs. (5.120)–(5.123) allow us to find in-plane stresses acting in each layer or in an elementary composite ply.

Compatible deformation of the layers is provided by interlaminar stresses τ_{xz}, τ_{yz}, and σ_z. To find these stresses, we need to use the three-dimensional equilibrium equations, Eqs. (2.5), which yield

$$\frac{\partial \tau_{xz}}{\partial z} = -\left(\frac{\partial \sigma_x}{\partial x} + \frac{\partial \tau_{xy}}{\partial y} \right), \quad \frac{\partial \tau_{yz}}{\partial z} = -\left(\frac{\partial \sigma_y}{\partial y} + \frac{\partial \tau_{xy}}{\partial x} \right), \quad \frac{\partial \sigma_z}{\partial z} = -\left(\frac{\partial \tau_{xz}}{\partial x} + \frac{\partial \tau_{xz}}{\partial y} \right) \tag{5.124}$$

Substituting stresses σ_x, σ_y, and τ_{xy} from Eqs. (5.4) and integrating Eqs. (5.124) with due regard to the forces that can act on the laminate surfaces, we can calculate the transverse shear and normal stresses τ_{xz}, τ_{yz}, and σ_z.

5.12. Example

As an example, consider the two-layered cylinder shown in Fig. 5.33 which consists of a $\pm 36°$ angle-ply layer with total thickness $h_1 = 0.62$ mm and 90° unidirectional layer with thickness $h_2 = 0.60$ mm. The 200 mm diameter cylinder is made by filament winding from glass–epoxy composite with the following mechanical properties: $E_1 = 44$ GPa, $E_2 = 9.4$ GPa, $G_{12} = 4$ GPa, $\nu_{21} = 0.26$. Consider two loading cases – axial compression with force P and torsion with torque T as in Fig. 5.33.

The cylinder is orthotropic, and to study the problem, we need to apply Eqs. (5.44) with some simplifications specific for this problem. First, we assume that applied loads do not induce interlaminar shear and we can take $\gamma_x = 0$ and $\gamma_y = 0$ in Eqs. (5.83) and (5.84). Hence, $V_x = 0$ and $V_y = 0$. In this case, deformations κ_x, κ_y, and κ_{xy} in Eqs. (5.3) become the changes of curvatures of the laminate. Since the loads shown in Fig. 5.33 deform the cylinder into another cylinder inducing only its axial shortening, change of radius, and rotation of the cross sections, there is no bending in the axial direction (see Fig. 5.3c) or out-of-plane twisting (see Fig. 5.3d) of the laminate. So, we can take $\kappa_x = 0$ and $\kappa_{xy} = 0$ and write constitutive equations, Eqs. (5.44), in the following form

$$
\begin{aligned}
N_x &= B_{11}\varepsilon_x^0 + B_{12}\varepsilon_y^0 + C_{12}\kappa_y \\
N_y &= B_{21}\varepsilon_x^0 + B_{22}\varepsilon_y^0 + C_{22}\kappa_y \\
N_{xy} &= B_{44}\gamma_{xy}^0 \\
M_x &= C_{11}\varepsilon_x^0 + C_{12}\varepsilon_y^0 + D_{12}\kappa_y \\
M_y &= C_{21}\varepsilon_x^0 + C_{22}\varepsilon_y^0 + D_{22}\kappa_y \\
M_{xy} &= C_{44}\gamma_{xy}^0
\end{aligned}
\tag{5.125}
$$

To determine the change of the circumferential curvature κ_y, we should take into account that the length of the cross-sectional contour being equal to $2\pi R$ before deformation becomes equal to $2\pi R\left(1 + \varepsilon_y^0\right)$ after deformation. Thus, the curvature

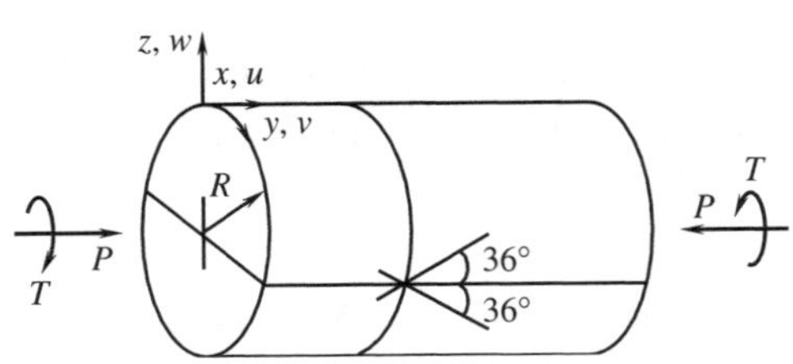

Fig. 5.33. Experimental cylinder.

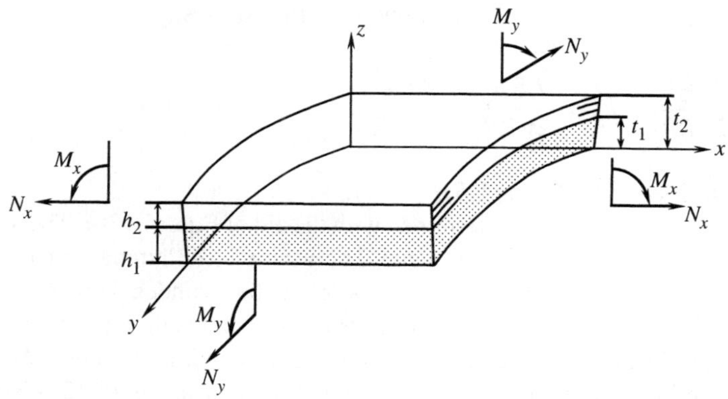

Fig. 5.34. Forces and moments acting on an element of the cylinder under axial compression.

change is

$$\kappa_y = \frac{1}{R\left(1+\varepsilon_y^0\right)} - \frac{1}{R} \approx -\frac{\varepsilon_y^0}{R} \tag{5.126}$$

The final result is obtained with the assumption that the strain is small ($\varepsilon_y^0 \ll 1$).

Consider the case of axial compression. The free body diagram for the laminate element shown in Fig. 5.34 yields (see Fig. 5.33)

$$N_x = -\frac{P}{2\pi R}, \quad N_y = 0$$

As a result, the constitutive equations of Eqs. (5.125) that we need to use for the analysis of this case become

$$B_{11}\varepsilon_x^0 + \overline{B}_{12}\varepsilon_y^0 = -\frac{P}{2\pi R}, \quad B_{21}\varepsilon_x^0 + \overline{B}_{22}\varepsilon_y^0 = 0 \tag{5.127}$$

$$M_x = C_{11}\varepsilon_x^0 + \overline{C}_{12}\varepsilon_y^0, \quad M_y = C_{21}\varepsilon_x^0 + \overline{C}_{22}\varepsilon_y^0 \tag{5.128}$$

in which

$$\begin{aligned} \overline{B}_{12} &= B_{12} - \frac{C_{12}}{R}, \quad \overline{B}_{22} = B_{22} - \frac{C_{22}}{R}, \\ \overline{C}_{12} &= C_{12} - \frac{D_{12}}{R}, \quad \overline{C}_{22} = C_{22} - \frac{D_{22}}{R} \end{aligned} \tag{5.129}$$

The first two equations, Eqs. (5.127), allow us to find strains, i.e.,

$$\varepsilon_x^0 = -\frac{P\overline{B}_{22}}{2\pi RB}, \quad \varepsilon_y^0 = \frac{PB_{21}}{2\pi RB} \tag{5.130}$$

where $B = B_{11}\overline{B}_{22} - \overline{B}_{12}B_{21}$ and $B_{21} = B_{12}$.

The bending moments can be determined with the aid of Eqs. (5.128). The axial moment, M_x, has a reactive nature in this problem. The asymmetric laminate in Fig. 5.34 tends to bend in the xz-plane under axial compression of the cylinder. However, the cylinder meridian remains straight at a distance from its ends. As a result, a reactive axial bending moment appears in the laminate. The circumferential bending moment, M_y, associated with the change in curvature of the cross-sectional contour in Eq. (5.126) is very small.

For numerical analysis, we first use Eqs. (4.72) to calculate stiffness coefficients for the angle-ply layer, i.e.,

$$A_{11}^{(1)} = 25\,\text{GPa}, \quad A_{12}^{(1)} = 10\,\text{GPa}, \quad A_{22}^{(1)} = 14.1\,\text{GPa}, \quad A_{44}^{(1)} = 11.5\,\text{GPa} \tag{5.131}$$

and for the hoop layer

$$A_{11}^{(2)} = 9.5\,\text{GPa}, \quad A_{12}^{(2)} = 2.5\,\text{GPa}, \quad A_{22}^{(2)} = 44.7\,\text{GPa}, \quad A_{44}^{(2)} = 4\,\text{GPa} \tag{5.132}$$

Then, we apply Eqs. (5.41) to find the I-coefficients that are necessary for the cases (axial compression and torsion) under study:

$$I_{11}^{(0)} = 21.2\,\text{GPa mm}, \quad I_{12}^{(0)} = 7.7\,\text{GPa mm}, \quad I_{22}^{(0)} = 35.6\,\text{GPa mm},$$

$$I_{44}^{(0)} = 9.5\,\text{GPa mm}; \quad I_{11}^{(1)} = 10.1\,\text{GPa mm}^2, \quad I_{12}^{(1)} = 3.3\,\text{GPa mm}^2,$$

$$I_{22}^{(1)} = 27.4\,\text{GPa mm}^2, \quad I_{44}^{(1)} = 4.4\,\text{GPa mm}^2; \quad I_{11}^{(2)} = 21.7\,\text{GPa mm}^3,$$

$$I_{12}^{(2)} = 5.9\,\text{GPa mm}^3, \quad I_{22}^{(2)} = 94\,\text{GPa mm}^3$$

To determine the stiffness coefficients of the laminate, we should pre-assign the coordinate of the reference surface (a cylindrical surface for the cylinder). Let us put $e = 0$ for simplicity, i.e., we take the inner surface of the cylinder as the reference surface (see Fig. 5.34). Then, Eqs. (5.28) yield

$$B_{11} = I_{11}^{(0)} = 21.2\,\text{GPa mm}, \quad B_{12} = I_{12}^{(0)} = 7.7\,\text{GPa mm},$$

$$B_{22} = I_{22}^{(0)} = 35.6\,\text{GPa mm}; \quad C_{11} = I_{11}^{(1)} = 10.1\,\text{GPa mm}^2,$$

$$C_{12} = I_{12}^{(1)} = 3.3\,\text{GPa mm}^2, \quad C_{22} = I_{22}^{(1)} = 27.4\,\text{GPa mm}^2;$$

$$D_{12} = I_{12}^{(2)} = 5.9\,\text{GPa mm}^3, \quad D_{22} = I_{22}^{(2)} = 94\,\text{GPa mm}^3$$

and in accordance with Eqs. (5.129) for $R = 100\,\text{mm}$,

$$\overline{B}_{12} = 7.7\,\text{GPa mm}, \quad \overline{B}_{22} = 35.3\,\text{GPa mm},$$

$$\overline{C}_{12} = 3.2\,\text{GPa mm}^2, \quad \overline{C}_{22} = 26.5\,\text{GPa mm}^2$$

Calculation with the aid of Eqs. (5.130) gives

$$\varepsilon_x^0 = -8.1 \cdot 10^{-5} P, \quad \varepsilon_y^0 = 1.8 \cdot 10^{-5} P$$

where P should be substituted in kN. Comparison of the obtained results with experimental data for the cylinder in Fig. 5.35 is presented in Fig. 5.36.

To determine the stresses, we first use Eqs. (5.120) which, in conjunction with Eq. (5.126) yield

$$\varepsilon_x^{(1)} = \varepsilon_x^{(2)} = \varepsilon_x^0, \quad \varepsilon_y^{(1)} = \varepsilon_y^0\left(1 - \frac{z_1}{R}\right), \quad \varepsilon_y^{(2)} = \varepsilon_y^0\left(1 - \frac{z_2}{R}\right) \tag{5.133}$$

where $0 \le z_1 \le h_1$ and $h_1 \le z_2 \le h_1 + h_2$. Since $(h_1 + h_2)/R = 0.0122$ for the cylinder under study, we can neglect z_1/R and z_2/R in comparison with unity and write

$$\varepsilon_y^{(1)} = \varepsilon_y^{(2)} = \varepsilon_y^0 \tag{5.134}$$

Fig. 5.35. Experimental composite cylinder in test fixtures.

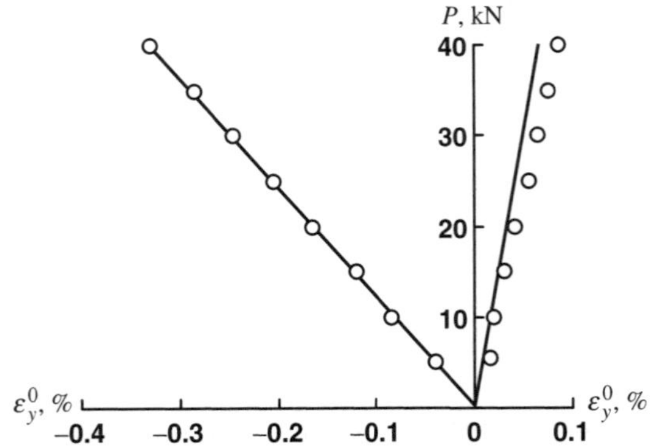

Fig. 5.36. Dependence of axial (ε_x^0) and circumferential (ε_y^0) strains of a composite cylinder on the axial force: —— analysis; O experiment.

Applying Eqs. (5.122) to calculate the strains in the plies' principal material coordinates and Eqs. (5.123) to find the stresses, we get

- in the angle-ply layer,

$$\sigma_1^{(1)} = -0.26\frac{P}{Rh}, \quad \sigma_2^{(1)} = -0.028\frac{P}{Rh}, \quad \tau_{12}^{(1)} = 0.023\frac{P}{Rh}$$

- in the hoop layer,

$$\sigma_1^{(2)} = 0.073\frac{P}{Rh}, \quad \sigma_2^{(2)} = -0.089\frac{P}{Rh}, \quad \tau_{12}^{(2)} = 0$$

where $h = h_1 + h_2$ is the total thickness of the laminate. To calculate the interlaminar stresses acting between the angle-ply and the hoop layers, we apply Eqs. (5.124). Using Eqs. (5.4) and taking Eqs. (5.133) and (5.134) into account, we first find the stresses in the layers referred to the global coordinate frame x, y, z, i.e.,

$$\sigma_x^{(i)} = A_{11}^{(i)}\varepsilon_x^0 + A_{12}^{(i)}\varepsilon_y^0, \quad \sigma_y^{(i)} = A_{21}^{(i)}\varepsilon_x^0 + A_{22}^{(i)}\varepsilon_y^0, \quad \tau_{xy}^{(i)} = 0 \tag{5.135}$$

where $i = 1, 2$ and $A_{mn}^{(i)}$ are given by Eqs. (5.131) and (5.132). Since these stresses do not depend on x and y, the first two equations in Eqs. (5.124) yield

$$\frac{\partial \tau_{xz}}{\partial z} = 0, \quad \frac{\partial \tau_{yz}}{\partial z} = 0$$

This means that both interlaminar shear stresses do not depend on z. However, on the inner and on the outer surfaces of the cylinder the shear stresses are equal to zero, so $\tau_{xz} = 0$ and $\tau_{yz} = 0$. The fact that $\tau_{yz} = 0$ is natural. Both layers are orthotropic and do not tend to twist under axial compression of the cylinder. Concerning $\tau_{xz} = 0$, a question arises as to how compatibility of the axial deformations of the layers with different stiffnesses can be provided without interlaminar shear stresses. The answer follows from the model used above to describe the stress state of the cylinder. According to this model, the transverse shear deformation γ_x is zero. Actually, this condition can be met if part of the axial force applied to the layer is proportional to the layer stiffness, i.e., as

$$P_1 = -2\pi\sigma_x^{(1)} h_1 = 2\pi h_1 \left(A_{11}^{(1)} \varepsilon_x^0 + A_{12}^{(1)} \varepsilon_y^0 \right)$$
$$P_2 = -2\pi\sigma_x^{(2)} h_2 = 2\pi h_2 \left(A_{11}^{(2)} \varepsilon_x^0 + A_{12}^{(2)} \varepsilon_y^0 \right) \tag{5.136}$$

Substituting strains from Eqs. (5.130), we can conclude that within the accuracy of a small parameter h/R (which was neglected in comparison with unity when we calculated stresses) $P_1 + P_2 = -P$, and that the axial strains are the same even if the layers are not bonded together. In the middle part of a long cylinder, the axial forces are automatically distributed between the layers in accordance with Eqs. (5.136). However, in the vicinity of the cylinder ends, this distribution depends on the loading conditions. The corresponding boundary problem will be discussed further in this section.

The third equation in Eqs. (5.124) formally yields $\sigma_z = 0$. However, this result is not correct because the equation corresponds to a plane laminate and is not valid for the cylinder. In cylindrical coordinates, the corresponding equation has the following form (see e.g., Vasiliev, 1993)

$$\frac{\partial}{\partial z}\left[\left(1+\frac{z}{R}\right)\sigma_z\right] = -\left[\left(1+\frac{z}{R}\right)\frac{\partial \tau_{xz}}{\partial x} + \frac{\partial \tau_{yz}}{\partial y} - \frac{\sigma_y}{R}\right]$$

Taking $\tau_{xz} = 0$ and $\tau_{yz} = 0$, substituting σ_y from Eqs. (5.135), and integrating, we obtain

$$\sigma_z = \frac{R}{R+z}\left[\frac{1}{R}\int_0^z \left(A_{21}\varepsilon_x^0 + A_{22}\varepsilon_y^0\right)\mathrm{d}z + C\right] \tag{5.137}$$

where, A_{mn} $(mn = 21,\ 22)$ are the step-wise functions of z, i.e.,

$$A_{mn} = A_{mn}^{(1)} \quad \text{for } 0 \le z \le h_1$$
$$A_{mn} = A_{mn}^{(2)} \quad \text{for } h_1 \le z \le h = h_1 + h_2$$

and C is the constant of integration. Since no pressure is applied to the inner surface of the cylinder, $\sigma_z(z=0) = 0$ and $C = 0$. Substitution of the stiffness coefficients, Eqs. (5.131), (5.132), and strains, Eqs. (5.130), into Eq. (5.137) yields

$$\sigma_z^{(1)} = -0.068\frac{P}{Rh}\cdot\frac{z}{R+z}$$
$$\sigma_z^{(2)} = \sigma_z^{(1)}(z=h_1) + 0.07\frac{P}{Rh}\cdot\frac{z-h_1}{R+z} \tag{5.138}$$

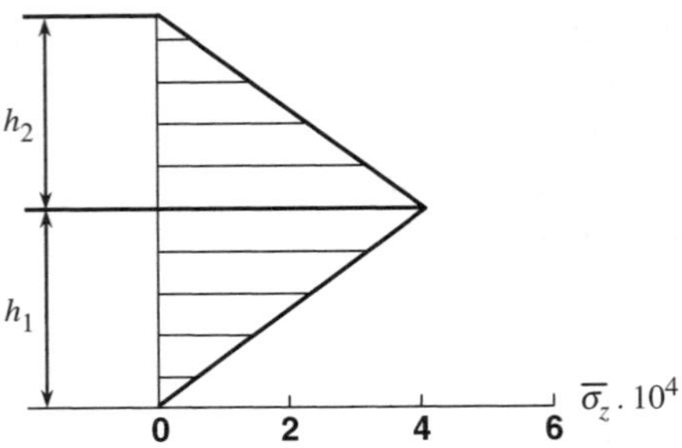

Fig. 5.37. Distribution of the normalized radial stress $\overline{\sigma}_z = \sigma_z Rh/P$ over the laminate thickness.

On the outer surface of the cylinder, $z = h$ and $\sigma_z^{(2)} = 0$ which is natural because this surface is free of any loading. The distribution of σ_z over the laminate thickness is shown in Fig. 5.37. As can be seen, interaction of the layers under axial compression of the cylinder results in radial compression that occurs between the layers.

We now return to transverse shear stress τ_{xz} and try to determine the transverse stresses taking into account the transverse shear deformation of the laminate. To do this, we should first specify the character of loading, e.g., suppose that axial force T in Fig. 5.33 is uniformly distributed over the cross-sectional contour of the angle-ply layer middle surface as in Fig. 5.38. As a result, we can take $T = 2\pi RN$ (since the cylinder is very thin, we neglect the radius change over its thickness).

To study this problem, we should supplement constitutive equations, Eqs. (5.125), with the missing equation for transverse shear, Eq. (5.83) and add the terms including the

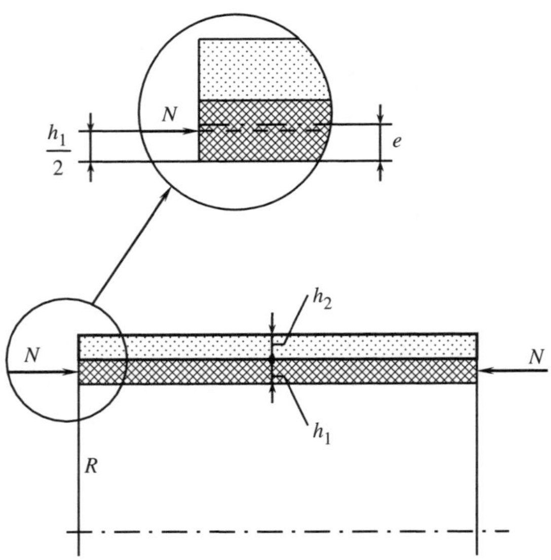

Fig. 5.38. Application of the axial forces.

change of the meridian curvature κ_x, which is no longer zero. As a result, we arrive at the following constitutive equations

$$N_x = B_{11}\varepsilon_x^0 + \overline{B}_{12}\varepsilon_y^0 + C_{11}\kappa_x \tag{5.139}$$

$$N_y = B_{21}\varepsilon_x^0 + \overline{B}_{22}\varepsilon_y^0 + C_{21}\kappa_x \tag{5.140}$$

$$M_x = C_{11}\varepsilon_x^0 + \overline{C}_{12}\varepsilon_y^0 + D_{11}\kappa_x \tag{5.141}$$

$$M_y = C_{21}\varepsilon_x^0 + \overline{C}_{22}\varepsilon_y^0 + D_{21}\kappa_x \tag{5.142}$$

$$V_x = S_x\gamma_x \tag{5.143}$$

Forces and moments in the left-hand sides of these equations are linked by equilibrium equations that can be written as (see Fig. 5.39)

$$N_x' = 0, \quad M_x' - V_x = 0, \quad V_x' - \frac{N_y}{R} = 0 \tag{5.144}$$

in which $(\ )' = \mathrm{d}(\ )/\mathrm{d}x$. The generalized strains entering Eqs. (5.139)–(5.143) are related to displacements by formulas given as notations to Eqs. (5.3) and (5.14), i.e.,

$$\varepsilon_x^0 = u', \quad \kappa_x = \theta_x', \quad \theta_x = \gamma_x - w' \tag{5.145}$$

Here, u is the axial displacement and w is the radial displacement (deflection) of the points belonging to the reference surface (see Fig. 5.33), whereas θ_x is the angle of rotation of the normal to this surface in the xz-plane and γ_x is the transverse shear deformation in this plane. The foregoing strain–displacement equations are the same as those for flat laminates. The cylindrical shape of the structure under study shows itself in the expression for circumferential strain ε_y^0. Since the radius of the cylinder after deformation becomes equal to $(R + w)$, we get

$$\varepsilon_y^0 = \frac{2\pi(R + w) - 2\pi R}{2\pi R} = \frac{w}{R} \tag{5.146}$$

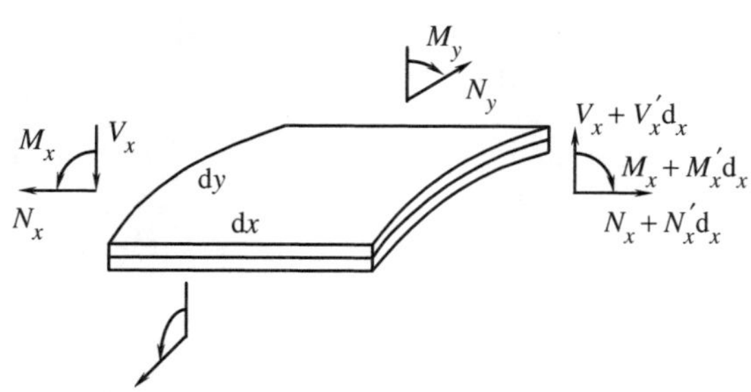

Fig. 5.39. Forces and moments acting on the cylindrical element.

To proceed with the derivation, we introduce the coordinate of the laminate reference surface, e, which gives $C_{11} = 0$, i.e., in accordance with Eq. (5.116), $e = I_{11}^{(1)}/I_{11}^{(0)}$. For the laminate under study, $e = 0.48$ mm, i.e., the reference surface is located within the internal angle-ply layer. Then, Eqs. (5.139)–(5.141), and (5.143), upon substitution of strains from Eqs. (5.145) and (5.146) can be written as

$$N_x = B_{11}u' + \overline{B}_{12}\frac{w}{R} \tag{5.147}$$

$$N_y = B_{21}u' + \overline{B}_{22}\frac{w}{R} + C_{21}\theta_x' \tag{5.148}$$

$$M_x = \overline{C}_{12}\frac{w}{R} + D_{11}\theta_x' \tag{5.149}$$

$$V_x = S_x(\theta_x + w') \tag{5.150}$$

where stiffness coefficients $B_{11}, \overline{B}_{12}, B_{21} = B_{12}, C_{21} = C_{12}, \overline{C}_{12}$ are presented above and

$$D_{11} = I_{11}^{(2)} - \frac{\left(I_{11}^{(1)}\right)^2}{I_{11}^{(0)}}, \qquad S_x = \frac{h^2}{\dfrac{h_1}{A_{55}^{(1)}} + \dfrac{h_2}{A_{55}^{(2)}}} \tag{5.151}$$

For the unidirectional ply, we take transverse shear moduli $G_{13} = G_{12} = 4$ GPa and $G_{23} = 3$ GPa. Using Eqs. (4.72), we get

$$A_{55}^{(1)} = G_{13}\cos^2\phi + G_{23}\sin^2\phi = 3.7\,\text{GPa and } A_{55}^{(2)} = 3\,\text{GPa}$$

Now, calculation with the aid of Eqs. (5.151) yields $D_{11} = 16.9\,\text{GPa mm}^3$ and $S_x = 4.05$ GPa mm.

The equilibrium equations, Eqs. (5.144), in conjunction with the constitutive equations, Eqs. (5.147)–(5.150) compose a set of seven ordinary differential equations including the same number of unknown functions – N_x, N_y, M_x, V_x, u, w, and θ_x. Thus, the set is complete and can be reduced to one governing equation for deflection w.

To do this, we integrate the first equilibrium equation in Eqs. (5.144) which shows that N_x = constant. Since at the cylinder ends $N_x = -N$, this result is valid for the whole cylinder. Using Eqs. (5.145) and (5.147), we obtain

$$\varepsilon_x^0 = u' = -\frac{1}{B_{11}}\left(N + \overline{B}_{12}\frac{w}{R}\right) \tag{5.152}$$

Substitution in Eq. (5.148) yields

$$N_y = \frac{B}{B_{11}R}w - \frac{B_{21}}{B_{11}}N + C_{21}\theta_x' \tag{5.153}$$

where $B = B_{11}\overline{B}_{22} - \overline{B}_{11}B_{21}$. We can express θ_x from Eq. (5.150) and, after differentiation, change V_x' for N_y with the aid of the last equilibrium equation in Eqs. (5.144). Substituting N_y from Eq. (5.153), we arrive at

$$\theta_x' = \frac{1}{C}\left[\frac{1}{S_xR}\left(\frac{B}{B_{11}R}w - \frac{B_{21}}{B_{11}}N\right) - w''\right] \tag{5.154}$$

where $C = 1-(C_{21}/(S_xR))$. Using Eqs. (5.149) and (5.154), we can express the bending moment in terms of deflection, i.e.,

$$M_x = \frac{D_{11}}{C}\left[\frac{1}{S_xR}\left(\frac{B}{B_{11}R}w - \frac{B_{21}}{B_{11}}N\right) - w''\right] + \overline{C}_{12}\frac{w}{R} \tag{5.155}$$

The governing equation follows now from the second equilibrium equation in Eqs. (5.144) if we differentiate it, substitute M_x'' from Eq. (5.155), express V_x' in terms of θ_x' and w'' using Eq. (5.150) and substitute θ_x' from Eq. (5.154). The final equation is as follows

$$w^{IV} - 2\alpha^2 w'' + \beta^4 w = p \tag{5.156}$$

in which

$$\alpha^2 = \frac{1}{2R}\left(\frac{C_{21}}{D_{11}} + \frac{B}{B_{11}S_xR}\right), \quad \beta^4 = \frac{B}{D_{11}B_{11}R^2},$$

$$p = \frac{B_{21}N}{D_{11}B_{11}R} = \frac{B_{21}P}{2\pi R^2 D_{11}B_{11}}$$

For the cylinder under study, $\alpha^2 = 14/R^2$ and $\beta^2 = 139/R^2$. Since $\beta > \alpha$, the solution of Eq. (5.156) can be written in the following form

$$w = \sum_{n=1}^{4} C_nF_n(x) + w_p \tag{5.157}$$

in which C_n are constants of integration and

$$F_1 = e^{-rx}\cos tx, \quad F_2 = e^{-rx}\sin tx, \quad F_3 = e^{rx}\cos tx, \quad F_4 = e^{rx}\sin tx$$

$$r = \sqrt{\frac{1}{2}(\beta^2 - \alpha^2)}, \quad t = \sqrt{\frac{1}{2}(\beta^2 + \alpha^2)}, \quad w_p = \frac{B_{21}P}{2\pi B}$$

To analyze the local effects in the vicinity of the cylinder end, e.g., $x = 0$ (the stress state of the cylinder at a distance from its ends is presented above), we should take $C_3 = 0$ and $C_4 = 0$ in Eq. (5.157) which reduces to

$$w = C_1F_1(x) + C_2F_2(x) + w_p \tag{5.158}$$

To differentiate the functions entering this solution, the following relationships can be used

$$F_1' = -(tF_2 + rF_1), \quad F_2' = tF_1 - rF_2,$$

$$F_1'' = \left(r^2 - t^2\right)F_1 + 2rtF_2, \quad F_2'' = \left(r^2 - t^2\right)F_2 - 2rtF_1,$$

$$F_1''' = -r\left(r^2 - 3t^2\right)F_1 + t\left(t^2 - 3r^2\right)F_2,$$

$$F_2''' = -r\left(r^2 - 3t^2\right)F_2 - t\left(t^2 - 3r^2\right)F_1$$

The constants of integration C_1 and C_2 entering Eq. (5.158) can be determined from the boundary conditions at $x = 0$. As follows from Figs. 5.38 and 5.39

$$M_x(x = 0) = N\left(e - \frac{h_1}{2}\right), \quad V_x(x = 0) = 0$$

in which M_x is specified by Eqs. (5.155) and (5.158), whereas V_x can be found from the second equilibrium equation in Eqs. (5.144).

For the cylinder under study, the final expressions for the strains and the rotation angle are

$$\begin{aligned}
\varepsilon_x^0 &= -\frac{P\overline{B}_{22}}{2\pi RB}\left[1 + e^{-rx}(0.11\sin tx - 0.052\cos tx)\right] \\
\varepsilon_y^0 &= \frac{PB_{21}}{2\pi RB}\left[1 + e^{-rx}(0.51\sin tx - 0.24\cos tx)\right] \\
\theta_x &= \frac{PB_{21}}{2\pi RB}e^{-rx}(6.3\cos tx - 2.3\sin tx),
\end{aligned} \tag{5.159}$$

in which $r = 7.9/R$ and $t = 8.75/R$. Thus, the solution in Eqs. (5.130) is supplemented with a boundary-layer solution that vanishes at a distance from the cylinder end.

To determine the transverse shear stress τ_{xz}, we integrate the first equation in Eqs. (5.124) subject to the condition $\tau_{xz}(z = 0) = 0$. As a result, the shear stress acting in the angle-ply layer is specified by the following expression

$$\tau_{xz}^{(1)} = -\int_0^z \frac{\mathrm{d}\sigma_x^{(1)}}{\mathrm{d}x}\mathrm{d}z$$

in which

$$\sigma_x^{(1)} = A_{11}^{(1)}\varepsilon_x^{(1)} + A_{12}^{(1)}\varepsilon_y^{(1)},$$

$$\varepsilon_x^{(1)} = \varepsilon_x^0 + z\kappa_x, \quad \varepsilon_y^{(1)} = \varepsilon_y^0 + z\kappa_y,$$

$$\kappa_x = \theta_x', \quad \kappa_y = -\frac{\varepsilon_y^0}{R}$$

Substitution of Eqs. (5.159) and rearranging yields

$$\tau_{xz}^{(1)} = \frac{P\overline{B}_{22}}{2\pi R^2 B} e^{-rx} \left[(23.75 \cos tx - 9.75 \sin tx) z + (6.3 \cos tx + 24.9 \sin tx) z^2 \right] \tag{5.160}$$

The transverse normal stress can be found from the following equation similar to Eq. (5.137)

$$\sigma_z = \frac{R}{R+z} \int_0^z \left[\frac{1}{R}\left(A_{21}\varepsilon_x^0 + A_{22}\varepsilon_y^0 \right) - \left(1 + \frac{z}{R}\right) \frac{\partial \tau_{xz}}{\partial x} \right] dz$$

For a thin cylinder, we can neglect z/R in comparison with unity. Using Eqs. (5.159) and (5.160) for the angle-ply layer, we have

$$\sigma_z^{(1)} = -0.068 \frac{P}{R^2 h} \left\{ z + e^{-rx} \left[(0.18 \cos tx - 0.0725 \sin tx) z \right. \right.$$
$$\left. \left. - (0.12 \cos tx + 0.059 \sin tx) z^2 + (0.05 \cos tx - 0.076 \sin tx) z^3 \right] \right\}$$

As can be seen, the first equation in Eqs. (5.138) follows from this solution if $x \to \infty$. The distribution of shear stress $\tau_{xz}^{(1)}(z = h_1)$ and normal stress $\sigma_z^{(1)}(z = h_1)$ acting at the interface between the angle-ply and the hoop layer of the cylinder along its length is shown in Fig. 5.40.

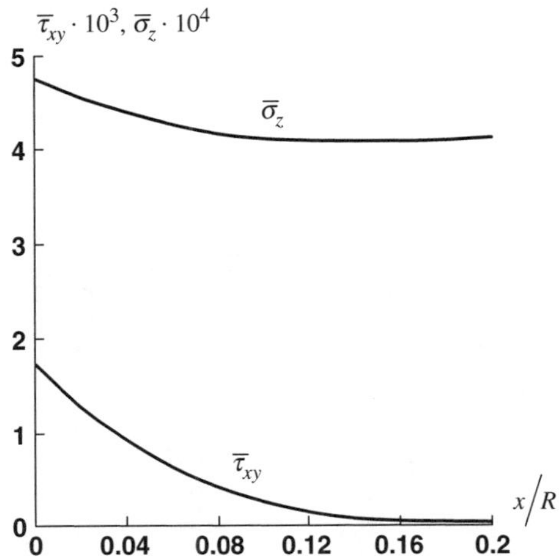

Fig. 5.40. Distribution of normalized transverse shear stress $\overline{\tau}_{xz} = \tau_{xz}^{(1)} Rh/P$ and normal stress $\overline{\sigma}_z = \sigma_z^{(1)} Rh/P$ acting on the layers interface $(z = h_1)$ along the cylinder axis.

Consider now the problem of torsion (see Fig. 5.33). The constitutive equations in Eqs. (5.125) that we need to use for this problem are

$$N_{xy} = B_{44}\gamma_{xy}^{0}, \quad M_{xy} = C_{44}\gamma_{xy} \tag{5.161}$$

Taking the coordinate of the reference surface in accordance with Eq. (5.116), i.e.,

$$e = \frac{I_{44}^{(1)}}{I_{44}^{(0)}} \tag{5.162}$$

we get $C_{44} = 0$ and $M_{44} = 0$. For the cylinder under study, $e = 0.46\,\text{mm}$, i.e., the reference surface is within the angle-ply layer. The free-body diagram for the cylinder loaded with torque T, (see Figs. 5.33 and 5.41) yields

$$N_{xy} = \frac{T}{2\pi R^2}$$

Thus,

$$\gamma_{xy}^{0} = \frac{T}{2\pi R^2 B_{44}} \tag{5.163}$$

For the experimental cylinder, shown in Fig. 5.35, normal strains were measured in the directions making ±45° angles with the cylinder meridian. To find these strains, we can use Eqs. (5.122) with $\phi_i = \pm 45°$, i.e.,

$$\varepsilon_{45}^{\pm} = \pm\frac{1}{2}\gamma_{xy}^{0}$$

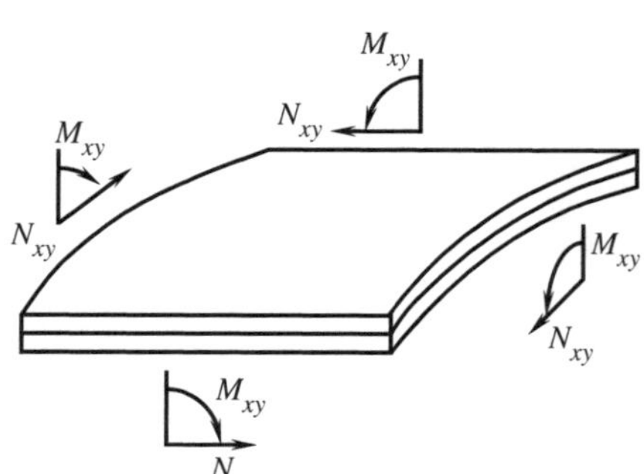

Fig. 5.41. Forces and moments acting on an element of the cylinder under torsion.

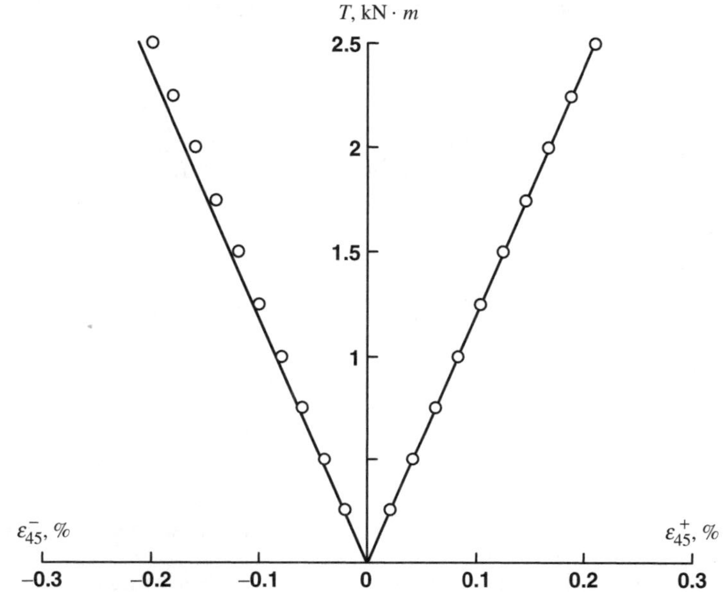

Fig. 5.42. Dependence of $\varepsilon_{45}^{\pm}$ on the torque T for a composite cylinder: —— analysis; ○ experiment.

For the cylinder under study with $B_{44} = I_{44}^{(0)} = 9.5\,\text{GPa mm}$ and $R = 100\,\text{mm}$, we get

$$\varepsilon_{45}^{\pm} = \pm\frac{T}{4\pi R^2 B_{44}} = \pm 0.84 \cdot 10^{-6} T$$

where T is measured in Nm. A comparison of the calculated results with experimental data is shown in Fig. 5.42.

To find the stresses acting in the plies, we should first use Eqs. (5.120) which for the case under study yield

$$\varepsilon_x^{(i)} = \varepsilon_y^{(i)} = 0, \quad \gamma_{xy}^{(i)} = \gamma_{xy}^0 \quad (i = 1, 2)$$

Then, Eqs. (5.122) enable us to determine the strains

- in $\pm\phi$ plies of the angle-ply layer,

$$\varepsilon_1^{\pm} = \pm\gamma_{xy}^0 \sin\phi\cos\phi, \quad \varepsilon_2^{\pm} = \mp\gamma_{xy}^0 \sin\phi\cos\phi, \quad \gamma_{12}^{\pm} = \gamma_{xy}^0 \cos 2\phi;$$

- in unidirectional plies of a hoop layer ($\phi = 90°$),

$$\varepsilon_1^{90} = \varepsilon_2^{90} = 0, \qquad \gamma_{12}^{90} = \gamma_{xy}^0$$

Finally, the stresses can be obtained with the aid of Eqs. (5.123). For the cylinder under study, we get:

- in the angle-ply layer,

$$\sigma_1^{\pm} = \pm 0.41 \frac{T}{R^2 h}, \quad \sigma_2^{\pm} = \mp 0.068 \frac{T}{R^2 h}, \quad \tau_{12}^{\pm} = 0.025 \frac{T}{R^2 h};$$

- in the hoop layer,

$$\sigma_1^{90} = \sigma_2^{90} = 0, \quad \tau_{12}^{90} = 0.082 \frac{T}{R^2 h}$$

where $h = 1.22$ mm is the total thickness of the laminate.

5.13. References

Aleksandrov, A.Ya., Brukker, L.E., Kurshin, L.M. and Prusakov, A.P. (1960). *Analysis of Sandwich Plates*. Mashinostroenie, Moscow (in Russian).

Ashton, J.E. (1969). Approximate solutions for unsymmetrically laminated plates. *Journal of Composite Materials*, **3**, 189–191.

Chen, H.-J. and Tsai, S.W. (1996). Three-dimensional effective moduli of symmetric laminates. *Journal of Composite Materials*, **30**(8).

Karmishin, A.V. (1974). Equations for nonhomogeneous thin-walled elements based on minimum stiffnesses. *Applied Mechanics, (Prikladnaya Mekhanika)*, **10**(6), 34–42 (in Russian).

Morozov, E.V. (2006). The effect of filament-winding mosaic patterns on the strength of thin-walled composite shells. *Composite Structures*, **76**, 123–129.

Vasiliev, V.V. (1993). *Mechanics of Composite Structures*. Taylor & Francis, Washington.

Verchery, G. (1999). Designing with anisotropy. Part 1: Methods and general results for laminates. In *Proc. 12th Int. Conf. on Composite Materials (ICCM-12)*, Paris, France, 5–9 July 1999, ICCM12/TCA (CD-ROM), 11 p.

Vorobey, V.V., Morozov, E.V. and Tatarnikov, O.V. (1992). *Analysis of Thermostressed Composite Structures*. Mashinostroenie, Moscow (in Russian).

Whitney, J.M. (1987). *Structural Analysis of Laminated Anisotropic Plates*. Technomic Publishing Co., Inc., Lancaster, PA, USA.

Chapter 6

FAILURE CRITERIA AND STRENGTH OF LAMINATES

Consider a laminate consisting of orthotropic layers or plies whose principal material axes 1, 2, and 3, in general, do not coincide with the global coordinates of the laminate (x, y, z) and suppose that this layer or ply is in a state of plane stress as in Fig. 6.1. It should be emphasized that, in contrast to a laminate that can be anisotropic and demonstrate coupling effects, the layer under consideration is orthotropic and is referred to its principal material axes. Using the procedure that is described in Section 5.11, we find stresses σ_1, σ_2, and τ_{12} corresponding to a given system of loads acting on the laminate. The problem that we approach now is to evaluate the laminate load-carrying capacity, i.e., to calculate the loads that cause failure of the individual layers and of the laminate as a whole. For the layer, this problem can be readily solved if we have a failure or strength criterion

$$F(\sigma_1, \sigma_2, \tau_{12}) = 1 \tag{6.1}$$

specifying the combination of stresses that causes layer fracture. In other words, the layer works while $F < 1$, fails if $F = 1$, and does not exist as a load-carrying structural element if $F > 1$. In the relevant stress space, i.e., σ_1, σ_2, and τ_{12}, Eq. (6.1) specifies the so-called failure surface (or failure envelope) shown in Fig. 6.2. Each point in this space corresponds to a particular stress state, and if the point is inside the surface, the layer withstands the corresponding combination of stresses without failure.

Thus, the problem of strength analysis is reduced to the construction of a failure criterion in its analytical, Eq. (6.1), or graphical (Fig. 6.2) form. Up to the present time, numerous variants of these forms have been proposed for traditional and composite structural materials (Gol'denblat and Kopnov, 1968; Wu, 1974; Rowlands, 1975; Tsai and Hahn, 1975; Vicario and Toland, 1975; etc.) and these have been described by the authors of many textbooks in composite materials. Omitting the history and comparative analysis of particular criteria that can be found elsewhere, we discuss here mainly the practical aspects of the problem.

6.1. Failure criteria for an elementary composite layer or ply

There exist, in general, two approaches to construct the failure surface, the first of which can be referred to as the microphenomenological approach. The term 'phenomenological'

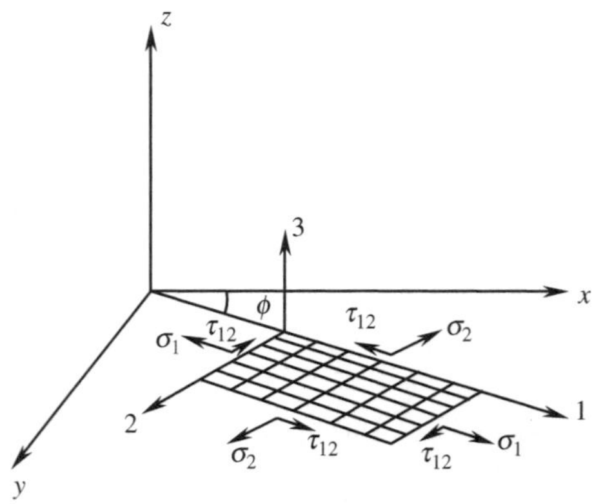

Fig. 6.1. An orthotropic layer or ply in a plane-stressed state.

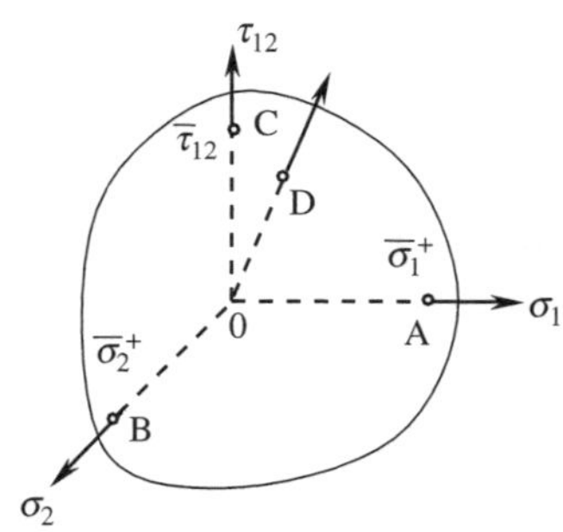

Fig. 6.2. Failure surface in the stress space.

means that the actual physical mechanisms of failure at the microscopic material level are not considered and that we deal with stresses and strains, i.e., with conventional and not actually observed state variables introduced in Chapter 2. In the micro-approach, we evaluate the layer strength using microstresses acting in the fibers and in the matrix and failure criteria proposed for homogeneous materials. Being developed up to a certain extent (see, e.g., Skudra et al., 1989), this approach requires the minimum number of experimental material characteristics, i.e., only those determining the strengths of fibers and matrices. As a result, coordinates of all the points of the failure surface in Fig. 6.2 including points A, B, and C corresponding to uniaxial and pure shear loading are found by calculation. To do this, we should simulate the layer or the ply with a suitable microstructural model (see, e.g., Section 3.3), apply a pre-assigned system of average stresses σ_1, σ_2, and τ_{12}, (e.g., corresponding to vector 0D in Fig. 6.2), find the stresses acting in the material components, specify the failure mode that can be associated with the fibers or with the matrix, and determine the ultimate combination of average stresses corresponding, e.g., to point D in Fig. 6.2. Thus, the whole failure surface can be constructed.

However, the uncertainty and approximate character of the existing micromechanical models discussed in Section 3.3 results in relatively poor accuracy using this method which, being in principle rather promising, has not found wide practical application at the present time.

The second basic approach which can be referred to as the macrophenomenological one deals with the average stresses σ_1, σ_2, and τ_{12} shown in Fig. 6.1 and ignores the ply microstructure. For a plane stress state in an orthotropic ply, this approach requires at least five experimental results specifying material strength under:

- longitudinal tension, $\overline{\sigma}_1^+$ (point A in Fig. 6.2),
- longitudinal compression, $\overline{\sigma}_1^-$,
- transverse tension, $\overline{\sigma}_2^+$ (point B in Fig. 6.2),
- transverse compression, $\overline{\sigma}_2^-$, and
- in-plane shear, $\overline{\tau}_{12}$ (point C in Fig. 6.2).

Obviously, these data are not enough to construct the complete failure surface, and two possible ways leading to two types of failure criteria can be used.

The first type referred to as structural failure criteria involves some assumptions concerning the possible failure modes that can help us to specify the shape of the failure surface. According to the second type, which provides a failure surface of an approximate form, experiments simulating a set of complicated stress states (such that two or all three stresses σ_1, σ_2, and τ_{12} are induced simultaneously) are undertaken. As a result, a system of points, like point D in Fig. 6.2, is determined and approximated with some suitable surface.

The experimental data that are necessary to construct the failure surface are usually obtained by testing thin-walled tubular specimens such as those shown in Figs. 6.3 and 6.4. These specimens are loaded with internal or external pressure p, tensile or compressive axial forces P, and end torques T, providing a known combination of axial stress, σ_x, circumferential stress, σ_y, and shear stress τ_{xy} that can be calculated as

$$\sigma_x = \frac{P}{2\pi Rh}, \quad \sigma_y = \frac{pR}{h}, \quad \tau_{xy} = \frac{T}{2\pi R^2 h}$$

Here, R is the cylinder radius and h is its thickness. For the tubular specimens shown in Fig. 6.4, which were made from unidirectional carbon–epoxy composite by circumferential winding, $\sigma_x = \sigma_2$, $\sigma_y = \sigma_1$, and $\tau_{xy} = \tau_{12}$ (see Fig. 6.1).

We shall now consider typical structural and approximation strength criteria developed for typical composite layers and plies.

6.1.1. *Maximum stress and strain criteria*

These criteria belong to a structural type and are based on the assumption that there can exist three possible modes of failure caused by stresses σ_1, σ_2, and τ_{12} or strains $\varepsilon_1, \varepsilon_2$, and γ_{12}, when they reach the corresponding ultimate values.

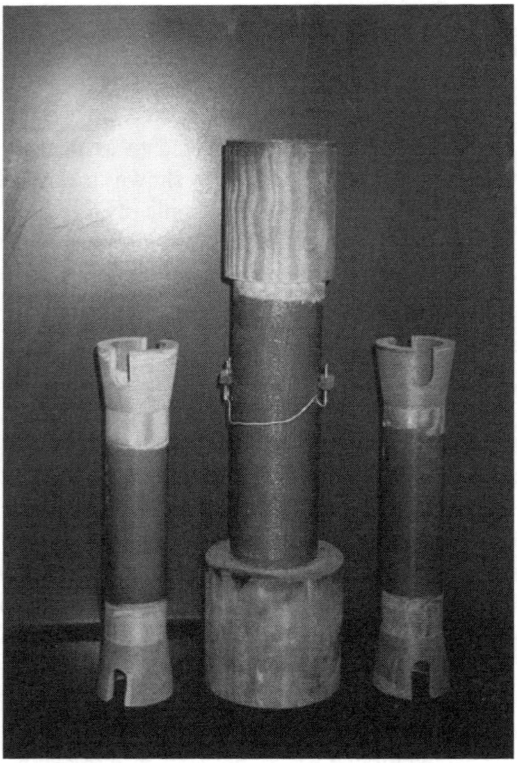

Fig. 6.3. Glass fabric–epoxy test tubular specimens.

The maximum stress criterion can be presented in the form of the following inequalities

$$
\begin{aligned}
&\sigma_1 \le \overline{\sigma}_1^+, \quad \sigma_2 \le \overline{\sigma}_2^+ \quad \text{if} \quad \sigma_1 > 0 \quad \sigma_2 > 0 \\
&|\sigma_1| \le \overline{\sigma}_1^-, \quad |\sigma_2| \le \overline{\sigma}_2^- \quad \text{if} \quad \sigma_1 < 0 \quad \sigma_2 < 0 \\
&|\tau_{12}| \le \overline{\tau}_{12}
\end{aligned}
\tag{6.2}
$$

It should be noted here and subsequently that all the ultimate stresses $\overline{\sigma}$ and $\overline{\tau}$ including compressive strength values are taken as positive quantities. The failure surface corresponding to the criterion in Eqs. (6.2) is shown in Fig. 6.5. As can be seen, according to this criterion failure is associated with independently acting stresses, and any possible stress interaction is ignored.

It can be expected that the maximum stress criterion describes adequately the behavior of those materials in which stresses σ_1, σ_2, and τ_{12} are taken by different structural elements. A typical example of such a material is the fabric composite layer discussed in Section 4.6. Indeed, warp and filling yarns (see Fig. 4.83) working independently provide material strength under tension and compression in two orthogonal directions (1 and 2), whereas

Fig. 6.4. Carbon–epoxy test tubular specimens made by circumferential winding (the central cylinder failed under axial compression and the right one under torsion).

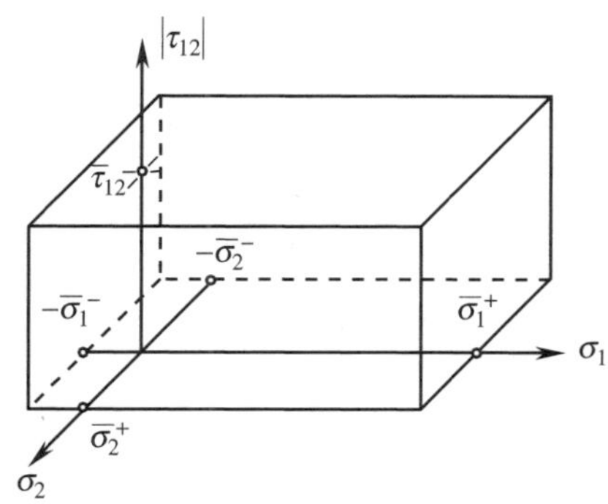

Fig. 6.5. Failure surface corresponding to maximum stress criterion.

the polymeric matrix controls the layer strength under in-plane shear. A typical failure envelope in the plane (σ_1, σ_2) for a glass–epoxy fabric composite is shown in Fig. 6.6 (the experimental data are from G. Prokhorov and N. Volkov). The corresponding results in the plane (σ_1, τ_{12}), but for a different glass fabric experimentally studied by Annin and Baev (1979), are presented in Fig. 6.7. It follows from Figs. 6.6 and 6.7, that the maximum stress criterion provides a satisfactory prediction of strength for fabric composites within the accuracy determined by the scatter of experimental results. As has been already noted, this criterion ignores the interaction of stresses. However, this interaction takes place

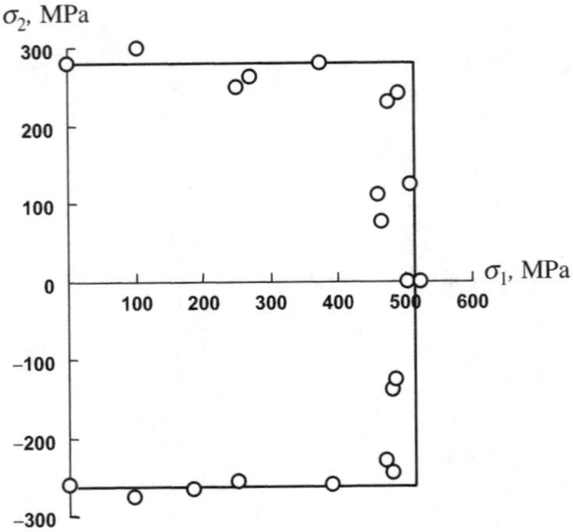

Fig. 6.6. Failure envelope for glass–epoxy fabric composite in plane (σ_1, σ_2). (———) maximum stress criterion, Eqs. (6.2); (○) experimental data.

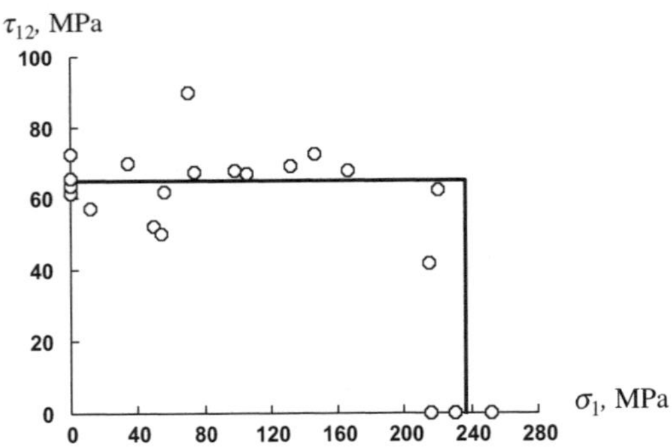

Fig. 6.7. Failure envelope for glass–epoxy fabric composite in plane (σ_1, τ_{12}). (———) maximum stress criterion, Eqs. (6.2); (○) experimental data.

in fabric composites which are loaded with compression in two orthogonal directions, because compression of the filling yarns increases the strength in the warp direction and vice versa. The corresponding experimental results from Belyankin et al. (1971) are shown in Fig. 6.8. As can be seen, there is a considerable discrepancy between the experimental data and the maximum stress criterion shown with solid lines. However, even in such cases this criterion is sometimes used to design composite structures, because it is simple and conservative, i.e., it underestimates material strength, thus increasing the safety factor for the structure under design. There exist fabric composites for which the interaction of normal stresses is exhibited in tension as well. An example of such a material is presented in Fig. 6.9 (experimental data from Gol'denblat and Kopnov (1968)). Naturally, the maximum stress criterion (solid lines in Fig. 6.9) should not be used in this case because it overestimates the material strength, and the structure can fail under loads that are lower than those predicted by this criterion.

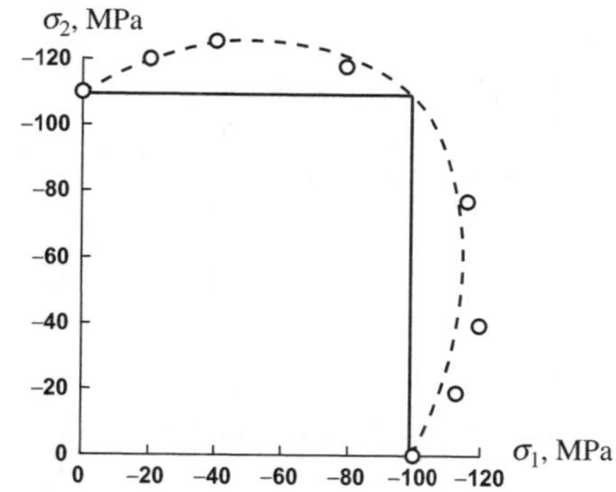

Fig. 6.8. Failure envelope for glass–phenolic fabric composite loaded with compression in plane (σ_1, σ_2). (———) maximum stress criterion, Eqs. (6.2); (- - -) polynomial criterion, Eqs. (6.16); (○) experimental data.

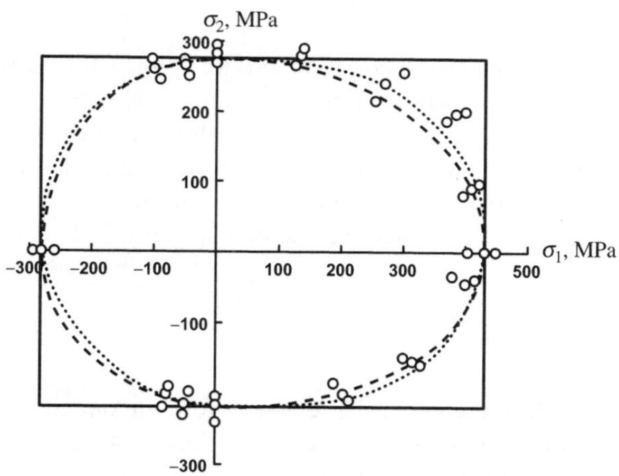

Fig. 6.9. Failure envelope for glass–epoxy fabric composite in plane (σ_1, σ_2). (———) maximum stress criterion, Eqs. (6.2); (- - - -) approximation criterion, Eqs. (6.11) and (6.12); (·········) approximation criterion, Eqs. (6.15); (○) experimental data.

The foregoing discussion concerns fabric composites. Now consider a unidirectional ply and try to apply the maximum stress criterion in this situation. First of all, because the longitudinal strength of the ply is controlled by the fibers whose strength is much higher than that of the matrix, it is natural to neglect the interaction of stress σ_1 on one side and stresses σ_2 and τ_{12}, on the other side. In other words, we can apply the maximum stress criterion to predict material strength under tension or compression in the fiber direction and, hence, use the first part of Eqs. (6.2), i.e.,

$$\begin{aligned} &\sigma_1 \leq \overline{\sigma}_1^+ \quad \text{if} \quad \sigma_1 > 0 \\ &|\sigma_1| \leq \overline{\sigma}_1^- \quad \text{if} \quad \sigma_1 < 0 \end{aligned} \tag{6.3}$$

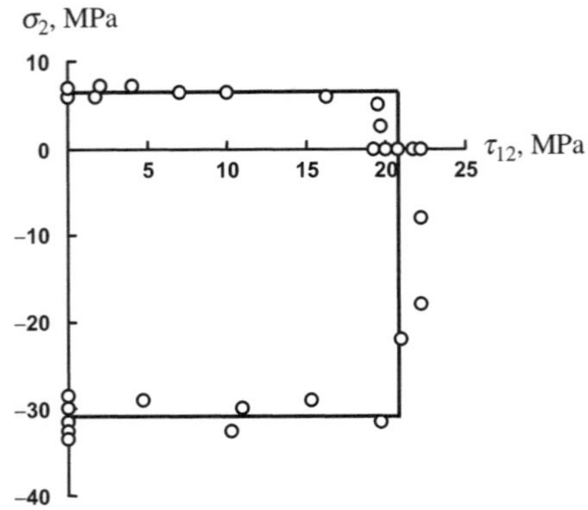

Fig. 6.10. Failure envelope for carbon–carbon unidirectional composite in plane (σ_2, τ_{12}). (———) maximum stress criterion, Eqs. (6.2); (○) experimental data.

Actually, there exist unidirectional composites with a very brittle matrix (carbon or ceramic) for which the other conditions in Eqs. (6.2) can also be applied. As an example, Fig. 6.10 displays the failure envelope for a carbon–carbon unidirectional material (experimental data from Vorobey et al., 1992). However, for the majority of unidirectional composites, the interaction of transverse normal and shear stresses is essential and should be taken into account. This means that we should apply Eq. (6.1) but can simplify it as follows

$$F(\sigma_2, \tau_{12}) = 1 \tag{6.4}$$

The simplest way to induce a combined stress state for a unidirectional ply is to use the off-axis tension or compression test as discussed in Section 4.3.1. Applying stress σ_x as in Figs. 4.22 and 4.23, we have stresses σ_1, σ_2, and τ_{12} specified by Eq. (4.78). Then, Eqs. (6.2) yield the following ultimate stresses:
For $\sigma_x > 0$,

$$\overline{\sigma}_x = \frac{\overline{\sigma}_1^+}{\cos^2\phi}, \quad \overline{\sigma}_x = \frac{\overline{\sigma}_2^+}{\sin^2\phi}, \quad \overline{\sigma}_x = \frac{\overline{\tau}_{12}}{\sin\phi\cos\phi} \tag{6.5}$$

For $\sigma_x < 0$,

$$\overline{\sigma}_x = \frac{\overline{\sigma}_1^-}{\cos^2\phi}, \quad \overline{\sigma}_x = \frac{\overline{\sigma}_2^-}{\sin^2\phi}, \quad \overline{\sigma}_x = \frac{\overline{\tau}_{12}}{\sin\phi\cos\phi} \tag{6.6}$$

The actual ultimate stress is the minimum $\overline{\sigma}_x$ value of the three values provided by Eqs. (6.5) for tension or Eqs. (6.6) for compression. The experimental data of S.W. Tsai taken from (Jones, 1999) and corresponding to a glass–epoxy unidirectional composite are presented in Fig. 6.11. As can be seen, the maximum stress criterion (solid lines)

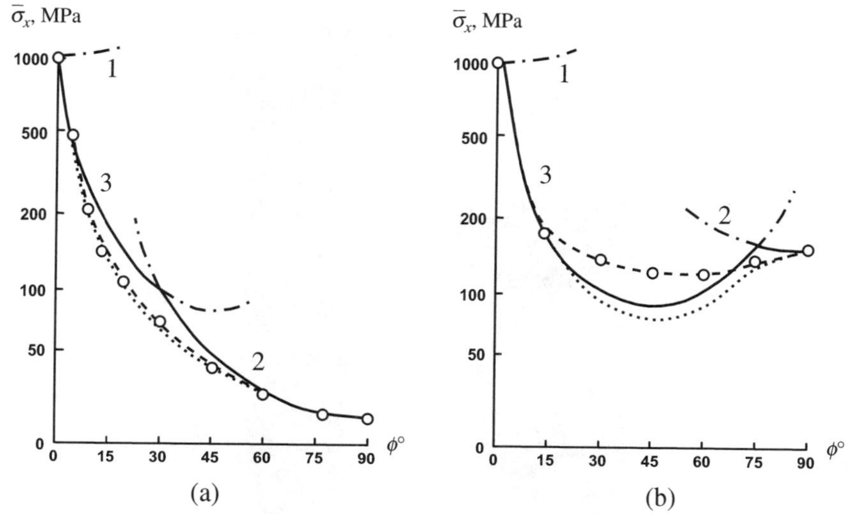

Fig. 6.11. Dependence of the stress on the fiber orientation angle for off-axes tension (a) and compression (b) of glass–epoxy unidirectional composite. (——— · –) maximum stress criterion, Eqs. (6.2); (·········) approximation criterion, Eqs. (6.3) and (6.17); (– – – – –) approximation criterion, Eqs. (6.3) and (6.18).

demonstrates fair agreement with experimental results for angles close to 0 and 90° only. An important feature of this criterion belonging to a structural type is its ability to predict the failure mode. Curves 1, 2, and 3 in Fig. 6.11 correspond to the first, the second, and the third equations of Eqs. (6.5) and (6.6). It follows from Fig. 6.11a, that fiber failure occurs only for $\phi = 0°$. For $0° < \phi < 30°$, material failure is associated with in-plane shear, whereas for $30° < \phi \leq 90°$, it is caused by the transverse normal stress σ_2.

The maximum strain failure criterion is similar to the maximum stress criterion discussed above, but is formulated in terms of strains, i.e.,

$$
\begin{aligned}
&\varepsilon \leq \bar{\varepsilon}_1^+, \quad \varepsilon_2 \leq \bar{\varepsilon}_2^+ \quad \text{if} \quad \varepsilon_1 > 0 \quad \varepsilon_2 > 0 \\
&|\varepsilon_1| \leq \bar{\varepsilon}_1^-, \quad |\varepsilon_2| \leq \bar{\varepsilon}_2^- \quad \text{if} \quad \varepsilon_1 < 0 \quad \varepsilon_2 < 0 \\
&|\gamma_{12}| \leq \bar{\gamma}_{12}
\end{aligned}
\tag{6.7}
$$

where

$$
\varepsilon_1 = \frac{\sigma_1}{E_1} - \nu_{12}\frac{\sigma_2}{E_2}, \quad \varepsilon_2 = \frac{\sigma_2}{E_2} - \nu_{21}\frac{\sigma_1}{E_1}, \quad \gamma_{12} = \frac{\tau_{12}}{G_{12}} \tag{6.8}
$$

The maximum strain criterion ignores the strain interaction but allows for the stress interaction due to Poisson's effect. This criterion provides results that are generally closely similar to those following from the maximum stress criterion.

There exists a unique stress state which can only be studied using the maximum strain criterion. This is longitudinal compression of a unidirectional ply as discussed earlier

Fig. 6.12. Failure modes of a unidirectional glass–epoxy composite under longitudinal compression.

in Section 3.4.4. Under this type of loading, only longitudinal stress σ_1 is induced, whereas $\sigma_2 = 0$ and $\tau_{12} = 0$. Nevertheless, fracture is accompanied with cracks parallel to the fibers (see Fig. 6.12 showing tests performed by Katarzhnov (1982)). These cracks are caused by transverse tensile strain ε_2 induced by Poisson's effect. The corresponding strength condition follows from Eqs. (6.7) and (6.8) and can be written as

$$|\sigma_1| \leq \overline{\varepsilon}_2^+ \frac{E_1}{\nu_{21}}$$

It should be emphasized that the test shown in Fig. 6.12 can be misleading because transverse deformation of the ply is not restricted in this test, whereas it is normally restricted in actual laminated composite structural elements. Indeed, a long cylinder with material structure $[0^\circ_{11}]$ being tested under compression yields a material strength $\overline{\sigma}_1^- = 300\,\text{MPa}$ whereas the same cylinder with material structure $[0^\circ_{10}/90^\circ]$ gives $\overline{\sigma}_1^- = 505\,\text{MPa}$ (Katarzhnov, 1982). Thus, if we change one longitudinal ply for a circumferential ply that practically does not bear any of the load in compression along the cylinder axis, but restricts its circumferential deformation, we increase the material strength in compression by 68.3%. Correspondingly, the failure mode becomes quite different (see Fig. 6.13).

(a) (b)

Fig. 6.13. Failure mode of a glass–epoxy tubular specimen with 10 longitudinal plies and one outside circumferential ply: (a) inside view; (b) outside view.

6.1.2. Approximation strength criteria

In contrast to structural strength criteria, approximation criteria do not indicate the mode of failure and are constructed by approximation of available experimental results with some appropriate function depending on stresses σ_1, σ_2, and τ_{12}. The simplest and the most widely used criterion is a second-order polynomial approximation, typical forms of which are presented in Fig. 6.14. In the stress space shown in Fig. 6.2, the polynomial criterion corresponding to Fig. 6.14a can be written as

$$F(\sigma_1, \sigma_2, \tau_{12}) = R_{11}\sigma_1^2 + R_{22}\sigma_2^2 + S_{12}\tau_{12}^2 = 1 \tag{6.9}$$

To determine the coefficients R and S, we need to perform three tests providing material strength under uniaxial loading in 1 and 2 directions and in shear. Then, applying the following conditions

$$\begin{aligned} &F(\sigma_1 = \overline{\sigma}_1, \sigma_2 = 0, \tau_{12} = 0) = 1 \\ &F(\sigma_1 = 0, \sigma_2 = \overline{\sigma}_2, \tau_{12} = 0) = 1 \\ &F(\sigma_1 = 0, \sigma_2 = 0, \tau_{12} = \overline{\tau}_{12}) = 1 \end{aligned} \tag{6.10}$$

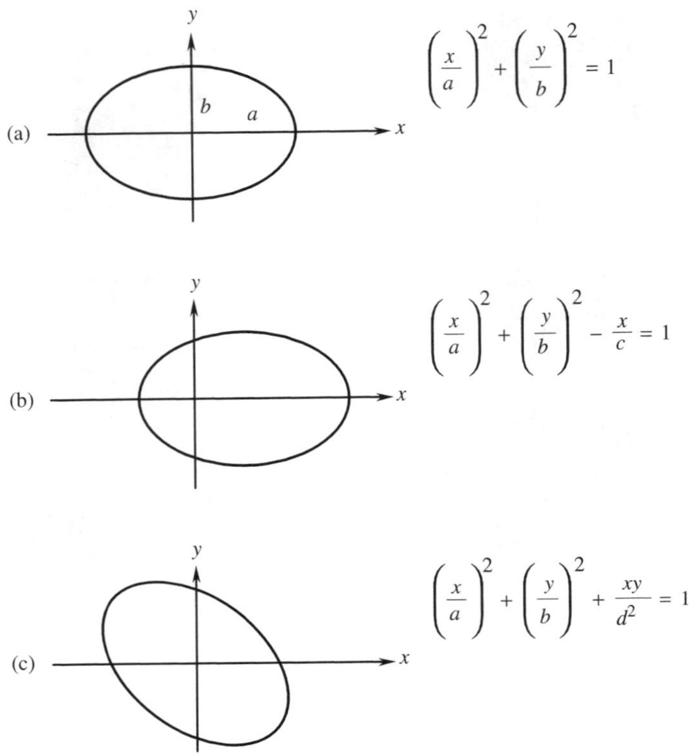

Fig. 6.14. Typical shapes of the curves corresponding to the second-order polynomials.

we can find R and S and write Eq. (6.9) in its final form

$$\left(\frac{\sigma_1}{\overline{\sigma}_1}\right)^2 + \left(\frac{\sigma_2}{\overline{\sigma}_2}\right)^2 + \left(\frac{\tau_{12}}{\overline{\tau}_{12}}\right)^2 = 1 \tag{6.11}$$

It appears as though this criterion yields the same strength estimate in tension and compression. However, it can be readily made specific for tension or compression. It is important to realize that when evaluating a material's strength we usually know the stresses acting in this material. Thus, we can take in Eq. (6.10)

$$\begin{aligned} &\overline{\sigma}_1 = \overline{\sigma}_1^+ \quad \text{if} \quad \sigma_1 > 0 \quad \text{or} \quad \overline{\sigma}_1 = \overline{\sigma}_1^- \quad \text{if} \quad \sigma_1 < 0 \\ &\overline{\sigma}_2 = \overline{\sigma}_2^+ \quad \text{if} \quad \sigma_2 > 0 \quad \text{or} \quad \overline{\sigma}_2 = \overline{\sigma}_2^- \quad \text{if} \quad \sigma_2 < 0 \end{aligned} \tag{6.12}$$

thus describing the cases of tension and compression. The failure criterion given by Eqs. (6.11) and (6.12) is demonstrated in Fig. 6.9 with application to a fabric composite loaded with stresses σ_1 and σ_2 ($\tau_{12} = 0$). Naturally, this criterion is specified by different equations for different quadrants in Fig. 6.9.

For some problems, e.g., for the problem of design, for which we usually do not know the signs of stresses, we may need to use a universal form of the polynomial criterion valid both for tension and compression. In this case, we should apply an approximation of the type shown in Fig. 6.14b and generalize Eq. (6.9) as

$$F(\sigma_1, \sigma_2, \tau_{12}) = R_1\sigma_1 + R_2\sigma_2 + R_{11}\sigma_1^2 + R_{22}\sigma_2^2 + S_{12}\tau_{12}^2 = 1 \tag{6.13}$$

Using criteria similar to Eqs. (6.10), i.e.,

$$\begin{aligned}
&F(\sigma_1 = \overline{\sigma}_1^+, \sigma_2 = 0, \tau_{12} = 0) = 1 \quad \text{if} \quad \sigma_1 > 0 \\
&F(\sigma_1 = -\overline{\sigma}_1^-, \sigma_2 = 0, \tau_{12} = 0) = 1 \quad \text{if} \quad \sigma_1 < 0 \\
&F(\sigma_1 = 0, \sigma_2 = \overline{\sigma}_2^+, \tau_{12} = 0) = 1 \quad \text{if} \quad \sigma_2 > 0 \\
&F(\sigma_1 = 0, \sigma_2 = -\overline{\sigma}_2^-, \tau_{12} = 0) = 1 \quad \text{if} \quad \sigma_2 < 0 \\
&F(\sigma_1 = 0, \sigma_2 = 0, \tau_{12} = \overline{\tau}_{12}) = 1
\end{aligned} \tag{6.14}$$

we arrive at

$$\sigma_1\left(\frac{1}{\overline{\sigma}_1^+} - \frac{1}{\overline{\sigma}_1^-}\right) + \sigma_2\left(\frac{1}{\overline{\sigma}_2^+} - \frac{1}{\overline{\sigma}_2^-}\right) + \frac{\sigma_1^2}{\overline{\sigma}_1^+\overline{\sigma}_1^-} + \frac{\sigma_2^2}{\overline{\sigma}_2^+\overline{\sigma}_2^-} + \left(\frac{\tau_{12}}{\overline{\tau}_{12}}\right)^2 = 1 \tag{6.15}$$

Comparison of this criterion with the criteria discussed above and with experimental results is presented in Fig. 6.9. As can be seen, the criteria specified by Eqs. (6.11), (6.12), and (6.15) provide close results which are in fair agreement with the experimental data for all the stress states except, possibly, biaxial compression for which there are practically no experimental results shown in Fig. 6.9. Such results are presented in Fig. 6.8 and allow us to conclude that the failure envelope can be approximated in this case by a polynomial of the type shown in Fig. 6.14c, i.e.,

$$F(\sigma_1, \sigma_2, \tau_{12}) = R_{11}\sigma_1^2 + R_{12}\sigma_1\sigma_2 + R_{22}\sigma_2^2 + S_{12}\tau_{12}^2 = 1$$

The coefficients R_{11}, R_{22}, and S_{12} can be found as earlier from Eqs. (6.10), and we need to use an additional strength condition to determine the coupling coefficient, R_{12}. A reasonable form of this condition is $F(\sigma_1 = -\overline{\sigma}_1^-, \sigma_2 = -\overline{\sigma}_2^-, \tau_{12} = 0) = 1$. This means that whereas for $|\sigma_1| < \overline{\sigma}_1^-$ and $|\sigma_2| < \overline{\sigma}_2^-$ the interaction of stresses increases material strength under compression, the combination of compressive failure stresses $|\sigma_1| = \sigma_1^-$ and $|\sigma_2| = \sigma_2^-$ results in material failure. Then

$$\left(\frac{\sigma_1}{\overline{\sigma}_1^-}\right)^2 - \frac{\sigma_1\sigma_2}{\overline{\sigma}_1^-\overline{\sigma}_2^-} + \left(\frac{\sigma_2}{\overline{\sigma}_2^-}\right)^2 + \left(\frac{\tau_{12}}{\overline{\tau}_{12}}\right)^2 = 1 \tag{6.16}$$

Comparison of this criterion with experimental data is presented in Fig. 6.8.

Now consider unidirectional composites and return to Fig. 6.11. As can be seen, the maximum stress criterion (solid lines), ignoring the interaction of stresses σ_2 and τ_{12} demonstrates rather poor agreement with experimental data. The simplest approximation criterion, Eqs. (6.11) and (6.12), takes, for the case under study, the form

$$F(\sigma_2, \tau_{12}) = \left(\frac{\sigma_2}{\bar{\sigma}_2}\right)^2 + \left(\frac{\tau_{12}}{\bar{\tau}_{12}}\right)^2 = 1 \tag{6.17}$$

and the corresponding failure envelope is shown in Fig. 6.11 with dotted lines. Although providing fair agreement with experimental results for tension (Fig. 6.11a), this criterion fails to predict material strength under compression (Fig. 6.11b). Moreover, for this case, the approximation criterion yields worse results than those demonstrated by the maximum stress criterion. There are simple physical reasons for this discrepancy. In contrast to the maximum stress criterion, Eq. (6.17) allows for stress interaction, but in such a way that the transverse stress σ_2 reduces the material strength under shear. However, this holds true only if the transverse stress is tensile. As can be seen in Fig. 6.15, in which the experimental results taken from Barbero's (1998) book are presented, a compressive stress σ_2 increases the ultimate value of shear stress τ_{12}. As a result, the simplest polynomial criterion in Eq. (6.17), being, as it has been already noted, quite adequate for $\sigma_2 > 0$, significantly underestimates material strength for $\sigma_2 < 0$ (solid line in Fig. 6.15). As also follows from Fig. 6.15, a reasonable approximation to the experimental results can be achieved if we use a curve of the type shown in Fig. 6.14b, (but moved to the left with respect to the y-axis), i.e., if we apply for this case the criterion presented by Eq. (6.15) which can be written as

$$F(\sigma_2, \tau_{12}) = \sigma_2\left(\frac{1}{\bar{\sigma}_2^+} - \frac{1}{\bar{\sigma}_2^-}\right) + \frac{\sigma_2^2}{\bar{\sigma}_2^+\bar{\sigma}_2^-} + \left(\frac{\tau_{12}}{\bar{\tau}_{12}}\right)^2 = 1 \tag{6.18}$$

The corresponding approximations are shown in Figs. 6.11 and 6.15 with dashed lines.

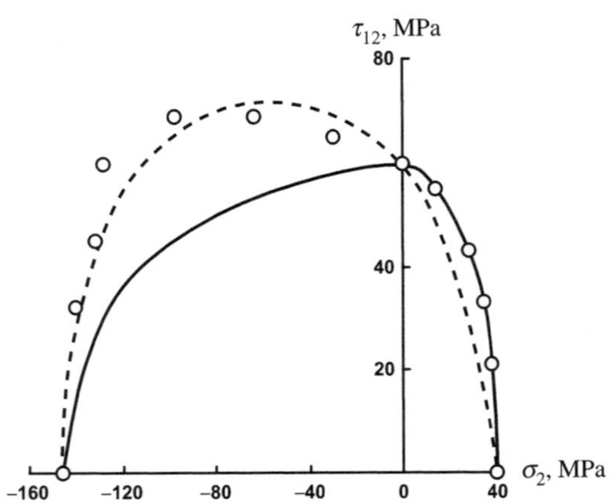

Fig. 6.15. Failure envelope for glass–epoxy unidirectional composite in plane (σ_2, τ_{12}). (———) approximation criterion, Eqs. (6.12) and (6.17); (- - - -) approximation criterion, Eqs. (6.18); (○) experimental data.

In conclusion, it should be noted that there exist more complicated polynomial strength criteria than those considered above, e.g., the fourth-order criterion of Ashkenazi (1966) and cubic criterion proposed by Tennyson et al. (1980).

6.1.3. *Tensor strength criteria*

The polynomial approximation strength criteria discussed in Section 6.1.2 have been introduced as some formal approximations of the experimental data in the principal material coordinates. When written in some other coordinate frame, these criteria become much more complicated. Consider for example an orthotropic material shown in Fig. 6.16 and referred to the principal material axes 1 and 2 and to some axes 1′ and 2′ which make an angle $\phi = 45°$ with the principal axes. For the principal material axes 1 and 2, apply a generalized form of the criterion in Eq. (6.13), i.e.,

$$F(\sigma_1, \sigma_2, \tau_{12}) = R_1\sigma_1 + R_2\sigma_2 + R_{11}\sigma_1^2 + R_{12}\sigma_1\sigma_2 + R_{22}\sigma_2^2 + S_{12}\tau_{12}^2 = 1 \quad (6.19)$$

Using the strength conditions in Eqs. (6.14) to determine the coefficients R and S, we arrive at

$$F(\sigma_1, \sigma_2, \tau_{12}) = \left(\frac{1}{\bar{\sigma}_1^+} - \frac{1}{\bar{\sigma}_1^-}\right)\sigma_1 + \left(\frac{1}{\bar{\sigma}_2^+} - \frac{1}{\bar{\sigma}_2^-}\right)\sigma_2 + \frac{\sigma_1^2}{\bar{\sigma}_1^+\bar{\sigma}_1^-} + R_{12}\sigma_1\sigma_2 + \frac{\sigma_2^2}{\bar{\sigma}_2^+\bar{\sigma}_2^-} + \left(\frac{\tau_{12}}{\bar{\tau}_{12}}\right)^2 = 1 \quad (6.20)$$

This criterion is similar to the criterion in Eq. (6.15), but it includes the coefficient R_{12} which cannot be found from simple tests using Eqs. (6.14). Treating Eq. (6.20) as the approximation strength criterion, we can apply some additional testing or additional assumptions similar to those used to derive Eq. (6.16) and determine the coefficient R_{12}.

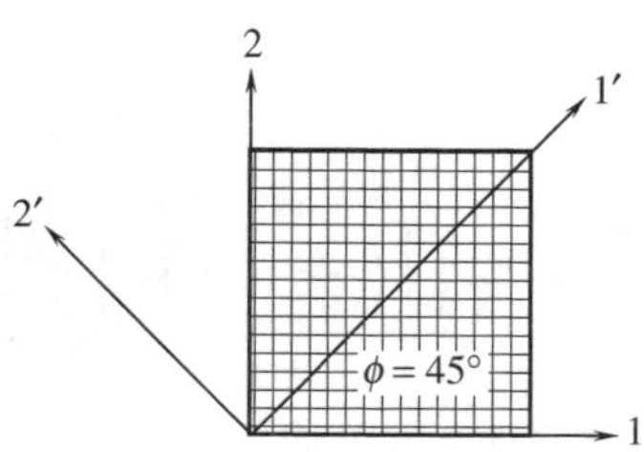

Fig. 6.16. An orthotropic material referred to coordinates (1, 2) and (1′, 2′).

We can also simplify the problem and take $R_{12} = 0$ arriving at Eq. (6.15), i.e.,

$$F(\sigma_1, \sigma_2, \tau_{12}) = \left(\frac{1}{\overline{\sigma}_1^+} - \frac{1}{\overline{\sigma}_1^-}\right)\sigma_1 + \left(\frac{1}{\overline{\sigma}_2^+} - \frac{1}{\overline{\sigma}_2^-}\right)\sigma_2 + \frac{\sigma_1^2}{\overline{\sigma}_1^+\overline{\sigma}_1^-} + \frac{\sigma_2^2}{\overline{\sigma}_2^+\overline{\sigma}_2^-} + \left(\frac{\tau_{12}}{\overline{\tau}_{12}}\right)^2 = 1 \quad (6.21)$$

which is in good agreement with experimental results (see Fig. 6.9). To simplify the analysis, assume that the material strength in tension and compression is the same for both principal directions 1 and 2, i.e.,

$$\overline{\sigma}_1^+ = \overline{\sigma}_1^- = \overline{\sigma}_2^+ = \overline{\sigma}_2^- = \overline{\sigma}_0, \quad \overline{\tau}_{12} = \overline{\tau}_0 \quad (6.22)$$

Then, Eq. (6.21) reduces to

$$F(\sigma_1, \sigma_2, \tau_{12}) = \frac{(\sigma_1^2 + \sigma_2^2)}{\overline{\sigma}_0^2} + \left(\frac{\tau_{12}}{\overline{\tau}_0}\right)^2 = 1 \quad (6.23)$$

Now, let us write Eq. (6.23) in coordinates $1'$ and $2'$ (see Fig. 6.16). To transform the stresses σ_1, σ_2, and τ_{12} to the stresses σ_1^{45}, σ_2^{45}, and τ_{12}^{45} corresponding to coordinates $1'$ and $2'$, we can use Eqs. (4.68). Taking $\phi = 45°$, $\sigma_x = \sigma_1$, $\sigma_y = \sigma_2$, $\tau_{xy} = \tau_{12}$ and $\sigma_1 = \sigma_1^{45}$, $\sigma_2 = \sigma_2^{45}$, $\tau_{12} = \tau_{12}^{45}$, we get

$$\begin{aligned} \sigma_1 &= \frac{1}{2}\left(\sigma_1^{45} + \sigma_2^{45}\right) - \tau_{12}^{45} \\ \sigma_2 &= \frac{1}{2}\left(\sigma_1^{45} + \sigma_2^{45}\right) + \tau_{12}^{45} \\ \tau_{12} &= \frac{1}{2}\left(\sigma_1^{45} - \sigma_2^{45}\right) \end{aligned} \quad (6.24)$$

Substitution in Eq. (6.23) yields

$$F\left(\sigma_1^{45}, \sigma_2^{45}, \tau_{12}^{45}\right) = \frac{1}{4}\left(\frac{2}{\overline{\sigma}_0^2} + \frac{1}{\overline{\tau}_0^2}\right)\left[\left(\sigma_1^{45}\right)^2 + \left(\sigma_2^{45}\right)^2\right] + \frac{1}{2}\left(\frac{2}{\overline{\sigma}_0^2} - \frac{1}{\overline{\tau}_0^2}\right)\sigma_1^{45}\sigma_2^{45} + \frac{2}{\overline{\sigma}_0^2}\left(\tau_{12}^{45}\right)^2 = 1 \quad (6.25)$$

For tension in the directions of axes 1′ and 2′ in Fig. 6.16 and for shear in plane 1′2′, we can write Eq. (6.25) in the following forms similar to Eqs. (6.10)

$$
\begin{aligned}
&F\left(\sigma_1^{45}=\overline{\sigma}_{45},\sigma_2^{45}=0,\tau_{12}^{45}=0\right)=1\\
&F\left(\sigma_1^{45}=0,\sigma_2^{45}=\overline{\sigma}_{45},\tau_{12}^{45}=0\right)=1\\
&F\left(\sigma_1^{45}=0,\sigma_2^{45}=0,\tau_{12}^{45}=\overline{\tau}_{45}\right)=1
\end{aligned}
\tag{6.26}
$$

Here, $\overline{\sigma}_{45}$ and $\overline{\tau}_{45}$ determine material strength in coordinates 1′ and 2′ (see Fig. 6.16). Then, Eq. (6.25) can be reduced to

$$
F\left(\sigma_1^{45},\sigma_2^{45},\tau_{12}^{45}\right)=\frac{1}{\overline{\sigma}_{45}^2}\left[\left(\sigma_1^{45}\right)^2+\left(\sigma_2^{45}\right)^2\right]+\left(\frac{1}{\overline{\tau}_{45}^2}-\frac{2}{\overline{\sigma}_{45}^2}\right)\sigma_1^{45}\sigma_2^{45}+\left(\frac{\tau_{12}^{45}}{\overline{\tau}_{45}}\right)=1
\tag{6.27}
$$

where $\overline{\sigma}_{45}$ and $\overline{\tau}_{45}$ are given by

$$
\frac{1}{\overline{\sigma}_{45}^2}=\frac{1}{4}\left(\frac{2}{\overline{\sigma}_0^2}+\frac{1}{\overline{\tau}_0^2}\right),\quad \overline{\tau}_{45}^2=\frac{1}{2}\overline{\sigma}_0^2
$$

Comparing Eq. (6.27) with Eq. (6.23), we can see that Eq. (6.27), in contrast to Eq. (6.23), includes a term with the product of stresses σ_1^{45} and σ_2^{45}. So, the strength criterion under study changes its form with a transformation of the coordinate frame (from 1 and 2 to 1′ and 2′ in Fig. 6.16) which means that the approximation polynomial strength criterion in Eq. (6.23) and, hence, the original criterion in Eq. (6.21) is not invariant with respect to the rotation of the coordinate frame.

Consider the class of invariant strength criteria which are formulated in a tensor-polynomial form as linear combinations of mixed invariants of the stress tensor σ_{ij} and the strength tensors of different ranks S_{ij}, S_{ijkl}, etc., i.e.,

$$
\sum_{i,k} S_{ik}\sigma_{ik}+\sum_{i,k,m,n} S_{ikmn}\sigma_{ik}\sigma_{mn}+\cdots=1
\tag{6.28}
$$

Using the standard transformation for tensor components we can readily write this equation for an arbitrary coordinate frame. However, the fact that the strength components form a tensor induces some conditions that should be imposed on these components and not necessarily correlate with experimental data.

To be specific, consider a second-order tensor criterion. Introducing contracted notations for tensor components and restricting ourselves to the consideration of orthotropic materials referred to the principal material coordinates 1 and 2 (see Fig. 6.16), we can present Eq. (6.22) as

$$F(\sigma_1, \sigma_2, \tau_{12}) = R_1^0\sigma_1 + R_2^0\sigma_2 + R_{11}^0\sigma_1^2 + 2R_{12}^0\sigma_1\sigma_2 + R_{22}^0\sigma_2^2 + 4S_{12}^0\tau_{12}^2 = 1 \tag{6.29}$$

which corresponds to Eq. (6.28) if we put

$$\sigma_{11} = \sigma_1, \quad \sigma_{12} = \tau_{12}, \quad \sigma_{22} = \sigma_2 \quad \text{and} \quad S_{11} = R_1, \quad S_{22} = R_2, \quad S_{1111} = R_{11}^0,$$

$$S_{1122} = S_{2211} = R_{12}^0, \quad S_{2222} = R_{22}^0, \quad S_{1212} = S_{2121} = S_{1221} = S_{2112} = S_{12}^0$$

The superscript '0' indicates that the components of the strength tensors are referred to the principal material coordinates. Applying the strength conditions in Eqs. (6.14), we can reduce Eq. (6.29) to the following form

$$F(\sigma_1, \sigma_2, \tau_{12}) = \sigma_1\left(\frac{1}{\overline{\sigma}_1^+} - \frac{1}{\overline{\sigma}_1^-}\right) + \sigma_2\left(\frac{1}{\overline{\sigma}_2^+} - \frac{1}{\overline{\sigma}_2^-}\right)$$
$$+ \frac{\sigma_1^2}{\overline{\sigma}_1^+\overline{\sigma}_1^-} + 2R_{12}^0\sigma_1\sigma_2 + \frac{\sigma_2^2}{\overline{\sigma}_2^+\overline{\sigma}_2^-} + \left(\frac{\tau_{12}}{\overline{\tau}_{12}}\right)^2 = 1 \tag{6.30}$$

This equation looks similar to Eq. (6.20), but there is a principal difference between them. Whereas Eq. (6.20) is only an approximation to the experimental results, and we can take any suitable value of coefficient R_{12} (in particular, we put $R_{12} = 0$), the criterion in Eq. (6.30) has an invariant tensor form, and coefficient R_{12}^0 should be determined using this property of the criterion.

Following Gol'denblat and Kopnov (1968) consider two cases of pure shear in coordinates $1'$ and $2'$ shown in Fig. 6.17 and assume that $\tau_{45}^+ = \overline{\tau}_{45}^+$ and $\tau_{45}^- = \overline{\tau}_{45}^-$, where the overbar denotes, as earlier, the ultimate value of the corresponding stress. In the general case, $\overline{\tau}_{45}^+ \neq \overline{\tau}_{45}^-$. Indeed, for a unidirectional composite, stress τ_{45}^+ induces tension in

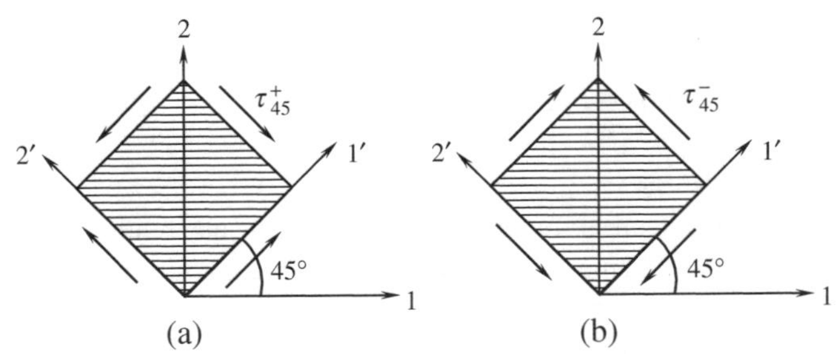

Fig. 6.17. Pure shear in coordinates $(1', 2')$ rotated by 45° with respect to the principal material coordinates (1, 2).

the fibers, whereas τ_{45}^{-} causes compression of the fibers, and the corresponding ultimate values can be different. Using the results presented in Section 2.4, we can conclude that for the loading case shown in Fig. 6.17a, $\sigma_1 = \tau_{45}^{+}$, $\sigma_2 = -\tau_{45}^{+}$, and $\tau_{12} = 0$, whereas for the case in Fig. 6.17b, $\sigma_1 = -\tau_{45}^{-}$, $\sigma_2 = \tau_{45}^{-}$, and $\tau_{12} = 0$. Applying the strength criterion in Eq. (6.30) for these loading cases, we arrive at

$$F(\sigma_1 = \overline{\tau}_{45}^{+}, \sigma_2 = -\overline{\tau}_{45}^{+}, \tau_{12} = 0)$$

$$= \overline{\tau}_{45}^{+}\left(\frac{1}{\overline{\sigma}_1^{+}} - \frac{1}{\overline{\sigma}_1^{-}} - \frac{1}{\overline{\sigma}_2^{+}} + \frac{1}{\overline{\sigma}_2^{-}}\right) + \left(\overline{\tau}_{45}^{+}\right)^2\left(\frac{1}{\overline{\sigma}_1^{+}\overline{\sigma}_1^{-}} + \frac{1}{\overline{\sigma}_2^{+}\overline{\sigma}_2^{-}} - 2R_{12}^{0}\right) = 1$$

$$F(\sigma_1 = -\overline{\tau}_{45}^{-}, \sigma_2 = \overline{\tau}_{45}^{-}, \tau_{12} = 0)$$

$$= \overline{\tau}_{45}^{-}\left(\frac{1}{\overline{\sigma}_1^{-}} - \frac{1}{\overline{\sigma}_1^{+}} + \frac{1}{\overline{\sigma}_2^{+}} - \frac{1}{\overline{\sigma}_2^{-}}\right) + \left(\overline{\tau}_{45}^{-}\right)^2\left(\frac{1}{\overline{\sigma}_1^{+}\overline{\sigma}_1^{-}} + \frac{1}{\overline{\sigma}_2^{+}\overline{\sigma}_2^{-}} - 2R_{12}^{0}\right) = 1$$

In general, these two equations give different solutions for R_{12}^{0}. A unique solution exists if the following compatibility condition is valid

$$\frac{1}{\overline{\sigma}_1^{+}} - \frac{1}{\overline{\sigma}_1^{-}} - \frac{1}{\overline{\sigma}_2^{+}} + \frac{1}{\overline{\sigma}_2^{-}} = \frac{1}{\overline{\tau}_{45}^{+}} - \frac{1}{\overline{\tau}_{45}^{-}} \tag{6.31}$$

If the actual material strength characteristics do not satisfy this equation, the strength criteria in Eq. (6.30) cannot be applied to this material. If they do, the coefficient R_{12}^{0} can be found as

$$R_{12}^{0} = \frac{1}{2}\left(\frac{1}{\overline{\sigma}_1^{+}\overline{\sigma}_1^{-}} + \frac{1}{\overline{\sigma}_2^{+}\overline{\sigma}_2^{-}} - \frac{1}{\overline{\tau}_{45}^{+}\overline{\tau}_{45}^{-}}\right) \tag{6.32}$$

For further analysis, consider for the sake of brevity a special orthotropic material shown in Fig. 6.16 for which, in accordance with Eqs. (6.22), $\overline{\sigma}_1^{+} = \overline{\sigma}_1^{-} = \overline{\sigma}_2^{+} = \overline{\sigma}_2^{-} = \overline{\sigma}_0$, $\overline{\tau}_{45}^{+} = \overline{\tau}_{45}^{-} = \overline{\tau}_{45}$, and $\overline{\tau}_{12} = \overline{\tau}_0$. As can be seen, Eq. (6.31) is satisfied in this case, and the strength criterion, Eq. (6.30), referred to the principal material coordinates (1, 2) in Fig. 6.16 takes the form

$$\frac{1}{\overline{\sigma}_0^2}\left(\sigma_1^2 + \sigma_2^2\right) + 2R_{12}^{0}\sigma_1\sigma_2 + \left(\frac{\tau_{12}}{\overline{\tau}_0}\right)^2 = 1 \tag{6.33}$$

where, in accordance with Eq. (6.32),

$$R_{12}^{0} = \frac{1}{\overline{\sigma}_0^2} - \frac{1}{2\overline{\tau}_{45}^2} \tag{6.34}$$

Substituting Eq. (6.34) into Eq. (6.33), we arrive at the final form of the criterion under consideration

$$F(\sigma_1, \sigma_2, \tau_{12}) = \frac{\left(\sigma_1^2 + \sigma_2^2\right)}{\bar{\sigma}_0^2} + \left(\frac{2}{\bar{\sigma}_0^2} - \frac{1}{\bar{\tau}_{45}^2}\right)\sigma_1\sigma_2 + \left(\frac{\tau_{12}}{\bar{\tau}_0}\right)^2 = 1 \tag{6.35}$$

Now, presenting Eq. (6.32) in the following matrix form

$$\{\sigma\}^{\mathrm{T}}\left[R^0\right]\{\sigma\} = 1 \tag{6.36}$$

where

$$\{\sigma\} = \begin{Bmatrix} \sigma_1 \\ \sigma_2 \\ \tau_{12} \end{Bmatrix}, \quad \left[R^0\right] = \begin{bmatrix} R_{11}^0 & R_{12}^0 & 0 \\ R_{12}^0 & R_{11}^0 & 0 \\ 0 & 0 & 4S_{12}^0 \end{bmatrix}$$

$$R_{11}^0 = \frac{1}{\bar{\sigma}_0^2}, \quad R_{12}^0 = \frac{1}{\bar{\sigma}_0^2} - \frac{1}{2\bar{\tau}_{45}^2}, \quad S_{12}^0 = \frac{1}{4\bar{\tau}_0^2} \tag{6.37}$$

Superscript 'T' means transposition converting the column vector $\{\sigma\}$ into the row vector $\{\sigma\}^{\mathrm{T}}$.

Let us transform stresses referred to axes (1, 2) into stresses corresponding to axes ($1'$ and $2'$) shown in Fig. 6.16. Such a transformation can be performed with the aid of Eqs. (6.24). The matrix form of this transformation is

$$\{\sigma\} = [T]\left\{\sigma^{45}\right\}, \tag{6.38}$$

where

$$[T] = \begin{bmatrix} \frac{1}{2} & \frac{1}{2} & -1 \\ \frac{1}{2} & \frac{1}{2} & 1 \\ \frac{1}{2} & -\frac{1}{2} & 0 \end{bmatrix}$$

Substitution of the stresses in Eq. (6.38) into Eq. (6.36) yields

$$\left\{\sigma^{45}\right\}^{\mathrm{T}}[T]^{\mathrm{T}}\left[R^0\right][T]\left\{\sigma^{45}\right\} = 1$$

This equation, being rewritten as

$$\left\{\sigma^{45}\right\}^{\mathrm{T}}\left[R^{45}\right]\left\{\sigma^{45}\right\} = 1 \tag{6.39}$$

specifies the strength criterion for the same material but referred to coordinates $(1', 2')$. The strength matrix has the following form

$$\left[R^{45}\right] = [T]^{\mathrm{T}}\left[R^{0}\right][T] = \begin{bmatrix} R_{11}^{45} & R_{12}^{45} & 0 \\ R_{12}^{45} & R_{11}^{45} & 0 \\ 0 & 0 & 4S_{12}^{45} \end{bmatrix}$$

where

$$\begin{aligned} R_{11}^{45} &= \frac{1}{\overline{\sigma}_0^2} + \frac{1}{4}\left(\frac{1}{\overline{\tau}_0^2} - \frac{1}{\overline{\tau}_{45}^2}\right) \\ R_{12}^{45} &= \frac{1}{\overline{\sigma}_0^2} - \frac{1}{4}\left(\frac{1}{\overline{\tau}_0^2} + \frac{1}{\overline{\tau}_{45}^2}\right) \\ S_{12}^{45} &= \frac{1}{4\overline{\tau}_{45}^2} \end{aligned} \tag{6.40}$$

The explicit form of Eq. (6.39) is

$$\begin{aligned} &\left[\frac{1}{\overline{\sigma}_0^2} + \frac{1}{4}\left(\frac{1}{\overline{\tau}_0^2} - \frac{1}{\overline{\tau}_{45}^2}\right)\right]\left[\left(\sigma_1^{45}\right)^2 + \left(\sigma_2^{45}\right)^2\right] \\ &\quad + 2\left[\frac{1}{\overline{\sigma}_0^2} - \frac{1}{4}\left(\frac{1}{\overline{\tau}_0^2} + \frac{1}{\overline{\tau}_{45}^2}\right)\right]\sigma_1^{45}\sigma_2^{45} + \left(\frac{\tau_{12}^{45}}{\overline{\tau}_{45}}\right)^2 = 1 \end{aligned} \tag{6.41}$$

Now apply the strength conditions in Eqs. (6.26) to give

$$\frac{1}{\overline{\sigma}_{45}^2} = \frac{1}{\overline{\sigma}_0^2} + \frac{1}{4}\left(\frac{1}{\overline{\tau}_0^2} - \frac{1}{\overline{\tau}_{45}^2}\right) \tag{6.42}$$

Then, the strength criterion in Eq. (6.41) can be presented as

$$\begin{aligned} F\left(\sigma_1^{45}, \sigma_2^{45}, \tau_{12}^{45}\right) &= \frac{1}{\overline{\sigma}_{45}^2}\left[\left(\sigma_1^{45}\right)^2 + \left(\sigma_2^{45}\right)^2\right] \\ &\quad + \left(\frac{2}{\overline{\sigma}_{45}^2} - \frac{1}{\overline{\tau}_0^2}\right)\sigma_1^{45}\sigma_2^{45} + \left(\frac{\tau_{12}^{45}}{\overline{\tau}_{45}}\right)^2 = 1 \end{aligned} \tag{6.43}$$

Thus, we have two formulations of the strength criterion under consideration which are specified by Eq. (6.35) for coordinates 1 and 2 and by Eq. (6.43) for coordinates $1'$ and $2'$

(see Fig. 6.16). As can be seen, Eqs. (6.35) and (6.43) have similar forms and follow from each other if we change the stresses in accordance with the following rule

$$\sigma_1 \leftrightarrow \sigma_1^{45}, \quad \sigma_2 \leftrightarrow \sigma_2^{45}, \quad \tau_{12} \leftrightarrow \tau_{12}^{45}, \quad \overline{\sigma}_0 \leftrightarrow \overline{\sigma}_{45}, \quad \overline{\tau}_0 \leftrightarrow \overline{\tau}_{45}$$

However, such correlation is possible under the condition imposed by Eq. (6.42) which can be presented in the form

$$I_s = \frac{1}{\overline{\sigma}_0^2} + \frac{1}{4\overline{\tau}_0^2} = \frac{1}{\overline{\sigma}_{45}^2} + \frac{1}{4\overline{\tau}_{45}^2} \tag{6.44}$$

This result means that I_s is the invariant of the strength tensor, i.e., that its value does not depend on the coordinate frame for which the strength characteristics entering Eq. (6.44) have been found.

If the actual material characteristics do not satisfy Eq. (6.44), the tensor strength criterion cannot be applied to this material. However, if this equation is consistent with experimental data, the tensor criterion offers considerable possibilities to study material strength. Indeed, restricting ourselves to two terms presented in Eq. (6.28) let us write this equation in coordinates $(1', 2')$ shown in Fig. 6.16 and suppose that $\phi \neq 45°$. Then

$$\sum_{i,k} S_{ik}^{\phi}\sigma_{ik}^{\phi} + \sum_{i,k,m,n} S_{ikmn}^{\phi}\sigma_{ik}^{\phi}\sigma_{mn}^{\phi} = 1 \tag{6.45}$$

Here, S_{ik}^{ϕ} and S_{ikmn}^{ϕ} are the components of the second and the fourth rank strength tensors which are transformed in accordance with tensor calculus as

$$S_{ik}^{\phi} = \sum_{p,q} l_{ip}l_{kq}S_{pq}^{0}$$
$$S_{ikmn}^{\phi} = \sum_{p,q,r,s} l_{ip}l_{kq}l_{mr}l_{ns}S_{pqrs}^{0} \tag{6.46}$$

Here, l are directional cosines of axes $1'$ and $2'$ on the plane referred to coordinates 1 and 2 (see Fig. 6.16), i.e., $l_{11} = \cos\phi$, $l_{12} = \sin\phi$, $l_{21} = -\sin\phi$, and $l_{22} = \cos\phi$. Substitution of Eqs. (6.46) in Eq. (6.45) yields the strength criterion in coordinates $(1', 2')$ but written in terms of strength components corresponding to coordinates (1, 2), i.e.,

$$\sum_{i,k}\sum_{p,q} l_{ip}l_{kq}S_{pq}^{0}\sigma_{ik}^{\phi} + \sum_{i,k,m,n}\sum_{p,q,r,s} l_{ip}l_{kq}l_{mr}l_{ns}S_{pqrs}^{0}\sigma_{ik}^{\phi}\sigma_{mn}^{\phi} = 1 \tag{6.47}$$

Apply Eq. (6.47) to the special orthotropic material studied above (see Fig. 6.16) and for which, in accordance with Eq. (6.22),

$$
\begin{aligned}
&S_{pq} = 0, \quad S_{1111} = S_{2222} = R_{11}^0 = R_{22}^0 = \frac{1}{\bar{\sigma}_0^2} \\
&S_{1122} = S_{2211} = R_{12}^0 = \frac{1}{\bar{\sigma}_0^2} - \frac{1}{2\bar{\tau}_{45}^2} \\
&S_{1212} = S_{2121} = S_{1221} = S_{2112} = S_{12}^0 = \frac{1}{4\bar{\tau}_0^2}
\end{aligned}
\tag{6.48}
$$

Following Gol'denblat and Kopnov (1968), consider the material strength under tension in the $1'$-direction and in shear in plane $(1', 2')$. Taking first $\sigma_{11}^{\phi} = \bar{\sigma}_{\phi}, \sigma_{22}^{\phi} = 0, \tau_{12}^{\phi} = 0$ and then $\tau_{12}^{\phi} = \bar{\tau}_{\phi}, \sigma_{11}^{\phi} = 0, \sigma_{22}^{\phi} = 0$, we get from Eq. (6.47)

$$
\bar{\sigma}_{\phi}^2 = \frac{1}{\sum\limits_{p,q,r,s} l_{1p}l_{1q}l_{1r}l_{1s}S_{pqrs}^0}, \qquad \bar{\tau}_{\phi}^2 = \frac{1}{\sum\limits_{p,q,r,s} l_{1p}l_{2q}l_{1r}l_{2s}S_{pqrs}^0}
$$

or in explicit form

$$
\begin{aligned}
\bar{\sigma}_{\phi}^2 &= \left[R_{11}^0\left(\cos^4\phi + \sin^4\phi\right) + 2\left(S_{12}^0 + 2R_{12}^0\right)\sin^2\phi\cos^2\phi\right]^{-1} \\
\bar{\tau}_{\phi}^2 &= 4\left[2\left(R_{11}^0 - R_{12}^0\right)\sin^2\phi\cos^2\phi + S_{12}^0\cos^2 2\phi\right]^{-1}
\end{aligned}
\tag{6.49}
$$

These equations allow us to calculate the material strength in any coordinate frame whose axes make angle ϕ with the corresponding principal material axes. Taking into account Eqs. (6.44) and (6.48), we can derive the following relationship from Eqs. (6.49)

$$
\frac{1}{\bar{\sigma}_{\phi}^2} + \frac{1}{4\bar{\tau}_{\phi}^2} = \frac{1}{\bar{\sigma}_0^2} + \frac{1}{4\bar{\tau}_0^2} = I_s \tag{6.50}
$$

So, I_s is indeed the invariant of the strength tensor whose value for a given material does not depend on ϕ.

Thus, tensor-polynomial strength criteria provide universal equations that can be readily written in any coordinate frame, but on the other hand, material mechanical characteristics, particularly material strength in different directions, should follow the rules of tensor transformation, i.e., composed invariants (such as I_s) must be the same for all coordinate frames.

6.1.4. Interlaminar strength

The failure of composite laminates can also be associated with interlaminar fracture caused by transverse normal and shear stresses σ_3 and τ_{13}, τ_{23} or σ_z and τ_{xz}, τ_{yz}

(see Fig. 4.18). Since $\sigma_3 = \sigma_z$ and shear stresses in coordinates (1, 2, 3) are linked with stresses in coordinates (x, y, z) by simple relationships in Eqs. (4.67) and (4.68), the strength criterion is formulated here in terms of stresses σ_z, τ_{xz}, τ_{yz} which can be found directly from Eqs. (5.124). Since the laminate strength in tension and compression across the layers is different, we can use the polynomial criterion similar to Eq. (6.15). For the stress state under study, we get

$$\sigma_z\left(\frac{1}{\overline{\sigma}_3^+} - \frac{1}{\overline{\sigma}_3^-}\right) + \left(\frac{\tau_r}{\overline{\tau}_i}\right)^2 = 1 \tag{6.51}$$

where

$$\tau_r = \sqrt{\tau_{13}^2 + \tau_{23}^2} = \sqrt{\tau_{xz}^2 + \tau_{yz}^2}$$

is the resultant transverse shear stress, and $\overline{\tau}_i$ determines the interlaminar shear strength of the material.

In thin-walled structures, the transverse normal stress is usually small and can be neglected in comparison with the shear stress. Then, Eq. (6.51) can be simplified and written as

$$\tau_r = \overline{\tau}_i \tag{6.52}$$

As an example, Fig. 6.18 displays the dependence of the normalized maximum deflection w/R on the force P for a fiberglass–epoxy cross-ply cylindrical shell of radius R loaded with a radial concentrated force P (Vasiliev, 1970). The shell failure was caused by delamination. The shadowed interval shows the possible values of the ultimate force

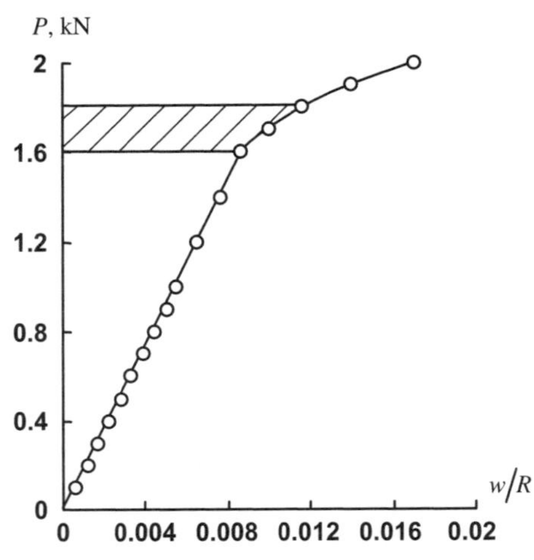

Fig. 6.18. Experimental dependence of the normalized maximum deflection of a fiberglass–epoxy cylindrical shell on the radial concentrated force.

calculated with the aid of Eq. (6.52) (this value is not unique because of the scatter in interlaminar shear strength).

6.2. Practical recommendations

As follows from the foregoing analysis, for practical strength evaluation of fabric composites, we can use either the maximum stress criterion, Eqs. (6.2) or second-order polynomial criterion in Eq. (6.15) in conjunction with Eq. (6.16) for the case of biaxial compression. For unidirectional composites with polymeric matrices, we can apply Eqs. (6.3) and (6.4) in which function F is specified by Eq. (6.18). It should be emphasized that experimental data usually have rather high scatter, and the accuracy of more complicated and rigorous strength criteria can be more apparent than real.

Comparing tensor-polynomial and approximation strength criteria, we can conclude the following. The tensor criteria should be used if our purpose is to develop a theory of material strength, because a consistent physical theory must be covariant, i.e., the constraints that are imposed on material properties within the framework of this theory should not depend on a particular coordinate frame. For practical applications, the approximation criteria are more suitable, but in the forms they are presented here they should be used only for orthotropic unidirectional plies or fabric layers in coordinates whose axes coincide with the fibers' directions.

To evaluate the laminate strength, we should first determine the stresses acting in the plies or layers (see Section 5.11), identify the layer that is expected to fail first and apply one of the foregoing strength criteria. The fracture of the first ply or layer may not necessarily result in failure of the whole laminate. Then, simulating the failed element with a suitable model (see, e.g., Section 4.4.2), the strength analysis is repeated and continued up to failure of the last ply or layer.

In principle, failure criteria can be constructed for the whole laminate as a quasi-homogeneous material. This is not realistic for design problems, since it would be necessary to compare the solutions for numerous laminate structures which cannot practically be tested. However, this approach can be used successfully for structures that are well developed and in mass production. For example, the segments of two structures of composite drive shafts – one made of fabric and the other of unidirectional composite, are shown in Fig. 6.19. Testing these segments in tension, compression, and torsion, we can plot the strength envelope on the plane (M, T), where M is the bending moment and T is the torque, and evaluate the shaft strength for different combinations of M and T with high accuracy and reliability.

6.3. Examples

For the first example, consider a problem of torsion of a thin-walled cylindrical drive shaft (see Fig. 6.20) made by winding a glass–epoxy fabric tape at angles $\pm 45°$. The material properties are $E_1 = 23.5\,\text{GPa}$, $E_2 = 18.5\,\text{GPa}$, $G_{12} = 7.2\,\text{GPa}$, $\nu_{12} = 0.16$,

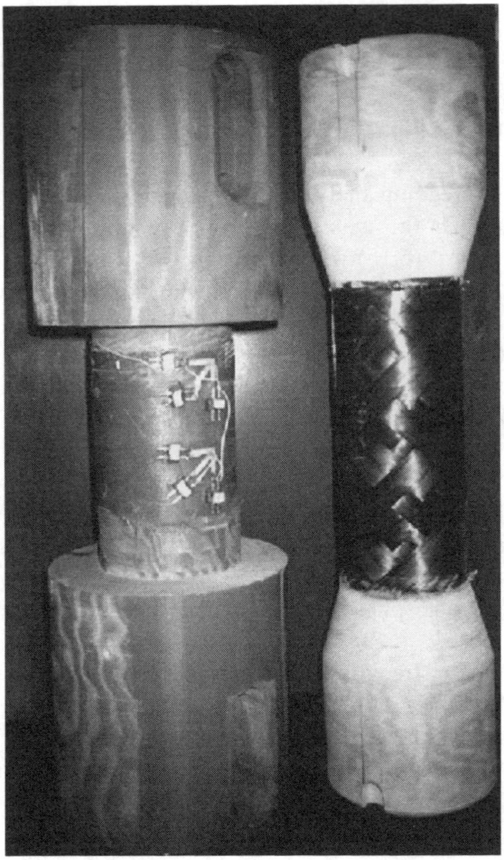

Fig. 6.19. Segments of composite drive shafts with test fixtures. Courtesy of CRISM.

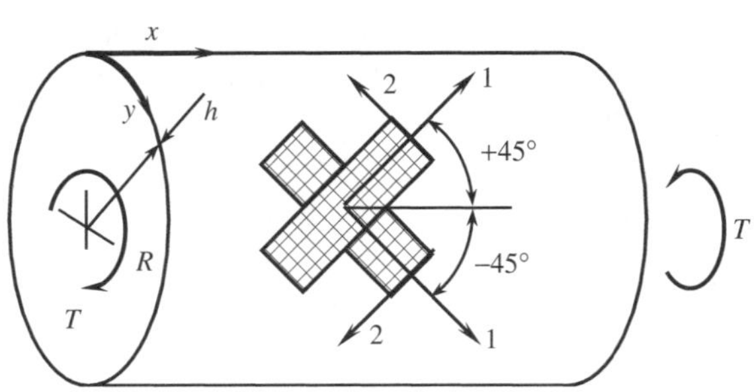

Fig. 6.20. Torsion of a drive shaft.

$\nu_{21} = 0.2$, $\overline{\sigma}_1^+ = 510\,\text{MPa}$, $\overline{\sigma}_1^- = 460\,\text{MPa}$, $\overline{\sigma}_2^+ = 280\,\text{MPa}$, $\overline{\sigma}_2^- = 260\,\text{MPa}$, $\overline{\tau}_{12} = 85\,\text{MPa}$. The shear strain induced by torque T is (Vasiliev, 1993)

$$\gamma_{xy} = \frac{T}{2\pi R^2 B_{44}}$$

Here, T is the torque, $R = 0.05\,\text{m}$ is the shaft radius, and B_{44} is the shear stiffness of the wall. According to Eqs. (5.39), $B_{44} = A_{44}h$, where $h = 5\,\text{mm}$ is the wall thickness, and A_{44} is specified by Eqs. (4.72) and can be presented as ($\phi = 45°$)

$$A_{44} = \frac{1}{4(1-\nu_{12}\nu_{21})}(E_1 + E_2 - 2E_1\nu_{12})$$

Using Eqs. (5.122), we can determine strains in the principal material coordinates 1 and 2 of $\pm 45°$ layers (see Fig. 6.20)

$$\varepsilon_1^{\pm} = \pm\frac{1}{2}\gamma_{xy}, \quad \varepsilon_2^{\pm} = \mp\frac{1}{2}\gamma_{xy}, \quad \gamma_{12}^{\pm} = 0$$

Applying Eqs. (5.123) and the foregoing results, we can express stresses in terms of T as

$$\sigma_1^{\pm} = \pm\frac{TE_1(1-\nu_{12})}{\pi R^2 h(E_1 + E_2 - 2E_1\nu_{12})}$$

$$\sigma_2^{\pm} = \mp\frac{TE_2(1-\nu_{21})}{\pi R^2 h(E_1 + E_2 - 2E_1\nu_{12})}$$

$$\tau_{12}^{\pm} = 0$$

The task is to determine the ultimate torque, $\overline{T}_u$.

First, use the maximum stress criterion, Eqs. (6.2), which gives the following four values of the ultimate torque corresponding to tensile or compressive failure of $\pm 45°$ layers

$$\sigma_1^+ = \overline{\sigma}_1^+, \quad T_u = 34\,\text{kNm}$$

$$\sigma_1^- = \overline{\sigma}_1^-, \quad T_u = 30.7\,\text{kNm}$$

$$\sigma_2^+ = \overline{\sigma}_2^+, \quad T_u = 25.5\,\text{kNm}$$

$$\sigma_2^- = \overline{\sigma}_2^-, \quad T_u = 23.7\,\text{kNm}$$

The actual ultimate torque is the lowest of these values, i.e., $\overline{T}_u = 23.7\,\text{kNm}$.

Now apply the polynomial criterion in Eq. (6.15), which has the form

$$\sigma_1^{\pm}\left(\frac{1}{\overline{\sigma}_1^{+}}-\frac{1}{\overline{\sigma}_1^{-}}\right)+\sigma_2^{\pm}\left(\frac{1}{\overline{\sigma}_2^{+}}-\frac{1}{\overline{\sigma}_2^{-}}\right)+\frac{\left(\sigma_1^{\pm}\right)^2}{\overline{\sigma}_1^{+}\overline{\sigma}_1^{-}}+\frac{\left(\sigma_2^{\pm}\right)^2}{\overline{\sigma}_2^{+}\overline{\sigma}_2^{-}}=1$$

For +45 and −45° layers, we get, respectively $T_u = 21.7\,\text{kNm}$ and $T_u = 17.6\,\text{kNm}$. Thus, $T_u = 17.6\,\text{kNm}$.

As a second example, consider the cylindrical shell described in Section 5.12 (see Fig. 5.30) and loaded with internal pressure p. Axial, N_x, and circumferential, N_y, stress resultants can be found as

$$N_x = \frac{1}{2}pR, \quad N_y = pR$$

where $R = 100\,\text{mm}$ is the shell radius. Applying constitutive equations, Eqs. (5.125), and neglecting the change in the cylinder curvature ($\kappa_y = 0$), we arrive at the following equations for strains

$$B_{11}\varepsilon_x^0 + B_{12}\varepsilon_y^0 = \frac{1}{2}pR, \quad B_{12}\varepsilon_x^0 + B_{22}\varepsilon_y^0 = pR \tag{6.53}$$

Using Eqs. (5.122) and (5.123) to determine strains and stresses in the principal material coordinates of the plies, we have

$$\begin{aligned}
\sigma_1^{(i)} &= \frac{pR}{2B}\overline{E}_1\Big[(B_{22}-2B_{12})\left(\cos^2\phi_i+\nu_{12}\sin^2\phi_i\right)\\
&\qquad +(2B_{11}-B_{12})\left(\sin^2\phi_i+\nu_{12}\cos^2\phi_i\right)\Big]\\
\sigma_2^{(i)} &= \frac{pR}{2B}\overline{E}_2\Big[(B_{22}-2B_{12})\left(\sin^2\phi_i+\nu_{21}\cos^2\phi_i\right)\\
&\qquad +(2B_{11}-B_{12})\left(\cos^2\phi_i+\nu_{21}\sin^2\phi_i\right)\Big]\\
\tau_{12}^{(i)} &= \frac{pR}{2B}G_{12}\left(2B_{11}+B_{12}-B_{22}\right)\sin 2\phi_i
\end{aligned} \tag{6.54}$$

Here, $B = B_{11}B_{22} - B_{12}^2$, and the membrane stiffnesses B_{mn} for the shell under study are presented in Section 5.12. Subscript 'i' in Eqs. (6.54) indicates the helical plies for which $i = 1, \phi_1 = \phi = 36°$ and circumferential plies for which $i = 2$ and $\phi_2 = 90°$.

The task that we consider is to find the ultimate pressure $\overline{p}_u$. For this purpose, we use the strength criteria in Eqs. (6.3), (6.4), and (6.17), and the following material properties $\overline{\sigma}_1^+ = 1300\,\text{MPa}$, $\overline{\sigma}_2^+ = 27\,\text{MPa}$, $\overline{\tau}_{12} = 45\,\text{MPa}$.

Calculation with the aid of Eqs. (6.54) yields

$$\sigma_1^{(1)} = 83.9p, \quad \sigma_2^{(1)} = 24.2p, \quad \tau_{12}^{(1)} = 1.9p,$$

$$\sigma_1^{(2)} = 112p, \quad \sigma_2^{(2)} = 19.5p, \quad \tau_{12}^{(2)} = 0$$

Applying Eqs. (6.3) to evaluate the plies' strength along the fibers, we get

$$\sigma_1^{(1)} = \overline{\sigma}_1^+, \quad p_{\rm u} = 14.9\,{\rm MPa}$$

$$\sigma_1^{(2)} = \overline{\sigma}_1^+, \quad p_{\rm u} = 11.2\,{\rm MPa}$$

The failure of the matrix can be identified using Eq. (6.17), i.e.,

$$\left(\frac{\sigma_2^{(1)}}{\overline{\sigma}_2^+}\right)^2 + \left(\frac{\tau_{12}^{(1)}}{\overline{\tau}_{12}}\right)^2 = 1, \quad p_{\rm u} = 1.1\,{\rm MPa}$$

$$\left(\frac{\sigma_2^{(2)}}{\overline{\sigma}_2^+}\right)^2 + \left(\frac{\tau_{12}^{(2)}}{\overline{\tau}_{12}}\right)^2 = 1, \quad p_{\rm u} = 1.4\,{\rm MPa}$$

Thus, we can conclude that failure occurs initially in the matrix of helical plies and takes place at an applied pressure $p_{\rm u}^{(1)} = 1.1$ MPa. This pressure destroys only the matrix of the helical plies, whereas the fibers are not damaged and continue to work. According to the model of a unidirectional layer with failed matrix discussed in Section 4.4.2, we should take $E_2 = 0$, $G_{12} = 0$, and $\nu_{12} = 0$ in the helical layer. Then, the stiffness coefficients, Eqs. (4.72) for this layer, become

$$A_{11}^{(1)} = E_1 \cos^4\phi, \quad A_{12}^{(1)} = E_1 \sin^2\phi\cos^2\phi, \quad A_{22}^{(1)} = E_1 \sin^4\phi \tag{6.55}$$

Calculating again the membrane stiffnesses B_{mn} (see Section 5.12) and using Eqs. (6.53), we get for $p \geq p_{\rm u}^{(1)}$

$$\sigma_1^{(1)} = 92.1p, \quad \sigma_2^{(1)} = 24.2p_{\rm u}^{(1)} = 26.6\,{\rm MPa}, \quad \tau_{12}^{(1)} = 1.9p_{\rm u}^{(1)} = 2.1\,{\rm MPa},$$

$$\sigma_1^{(2)} = 134.6p, \quad \sigma_2^{(2)} = 22.6p, \quad \tau_{12}^{(2)} = 0$$

For a pressure $p \geq p_{\rm u}^{(1)}$, three modes of failure are possible. The pressure causing failure of the helical plies under longitudinal stress $\sigma_1^{(1)}$ can be calculated from the following equation

$$\sigma_1^{(1)} = 83.9p_{\rm u}^{(1)} + 92.1\left(p_{\rm u} - p_{\rm u}^{(1)}\right) = \overline{\sigma}_1^+$$

which yields $p_u = 14.2\,\text{MPa}$. The analogous value for the circumferential ply is determined by the following condition

$$\sigma_1^{(2)} = 112p_u^{(1)} + 134.6\left(p_u - p_u^{(1)}\right) = \overline{\sigma}_1^+$$

which gives $p_u = 9.84\,\text{MPa}$. Finally, the matrix of the circumferential layer can fail under tension across the fibers. Since $\tau_{12}^{(2)} = 0$, we put

$$\sigma_2^{(2)} = 19.5p_u^{(1)} + 22.6\left(p_u - p_u^{(1)}\right) = \overline{\sigma}_2^+$$

and find $p_u = 1.4\,\text{MPa}$

Thus, the second failure stage takes place at $p_u^{(2)} = 1.4\,\text{MPa}$ and is associated with cracks in the matrix of the circumferential layer (see Fig. 4.36).

For $p \geq p_u^{(2)}$, we should put $E_2 = 0$, $G_{12} = 0$, and $\nu_{12} = 0$ in the circumferential layer whose stiffness coefficients become

$$A_{11}^{(2)} = 0, \quad A_{12}^{(2)} = 0, \quad A_{22}^{(2)} = E_1 \tag{6.56}$$

The membrane stiffnesses of the structure now correspond to the monotropic model of a composite unidirectional ply (see Section (3.3)) and can be calculated as

$$B_{mn} = A_{mn}^{(1)}h_1 + A_{mn}^{(2)}h_2$$

where A_{mn} are specified by Eqs. (6.55) and (6.56), and $h_1 = 0.62\,\text{mm}$ and $h_2 = 0.6\,\text{mm}$ are the thicknesses of the helical and the circumferential layers. Using again Eqs. (6.54), we have for $p \geq p_u^{(2)}$

$$\sigma_1^{(1)} = 137.7p, \quad \sigma_1^{(2)} = 122.7p$$

The cylinder's failure can now be caused by fracture of either helical fibers or circumferential fibers. The corresponding values of the ultimate pressure can be found from the following equations

$$\sigma_1^{(1)} = 83.9p_u^{(1)} + 92.1\left(p_u^{(2)} - p_u^{(1)}\right) + 137.7\left(p_u - p_u^{(2)}\right) = \overline{\sigma}_1^+$$

$$\sigma_1^{(2)} = 112p_u^{(1)} + 134.6\left(p_u^{(2)} - p_u^{(1)}\right) + 122.7\left(p_u - p_u^{(2)}\right) = \overline{\sigma}_1^+$$

in which $p_u^{(1)} = 1.1\,\text{MPa}$, and $p_u^{(2)} = 1.4\,\text{MPa}$.

The first of these equations yields $p_u = 10\,\text{MPa}$, whereas the second gives $p_u = 10.7\,\text{MPa}$.

Thus, failure of the structure under study occurs at $\overline{p}_u = 10\,\text{MPa}$ as a result of fiber fracture in the helical layer.

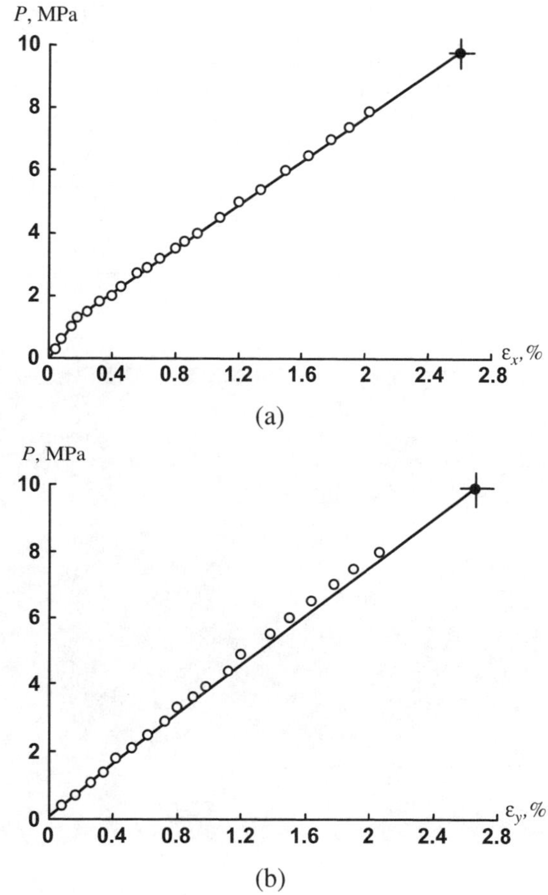

Fig. 6.21. Dependence of the axial (a) and the circumferential (b) strains on internal pressure. (———) analysis; (○) experimental data.

The dependencies of strains, which can be calculated using Eqs. (6.53), and the appropriate values of B_{mn} are shown in Fig. 6.21 (solid lines). As can be seen, the theoretical prediction is in fair agreement with the experimental results. The same conclusion can be drawn for the burst pressure which is listed in Table 6.1 for two types of filament-wound fiberglass pressure vessels. A typical example of the failure mode for the vessels presented in Table 6.1 is shown in Fig. 6.22.

6.4. Allowable stresses for laminates consisting of unidirectional plies

It follows from Section 6.3 (see also Section 4.4.2) that a unidirectional fibrous composite ply can experience two modes of failure associated with

- fiber failure, and
- cracks in the matrix.

Table 6.1
Burst pressure for the filament-wound fiberglass pressure vessels.

Diameter of the vessel (mm)	Layer thickness (mm)		Calculated burst pressure (MPa)	Number of tested vessels	Experimental burst pressure	
	h_1	h_2			Mean value (MPa)	Variation coefficient (%)
200	0.62	0.60	10	5	9.9	6.8
200	0.92	0.93	15	5	13.9	3.3

Fig. 6.22. The failure mode of a composite pressure vessel.

The first mode can be identified using the strength criterion in Eqs. (6.3), i.e.,

$$\begin{aligned} &\sigma_1^{(i)} \leq \overline{\sigma}_1^+ \quad \text{if} \quad \sigma_1^{(i)} > 0 \\ &\left|\sigma_1^{(i)}\right| \leq \overline{\sigma}_1^- \quad \text{if} \quad \sigma_1^{(i)} < 0 \end{aligned} \tag{6.57}$$

in which $\overline{\sigma}_1^+$ and $\overline{\sigma}_1^-$ are the ultimate stresses of the ply under tension and compression along the fibers, and i is the ply number. For the second failure mode, we have the strength criterion in Eq. (6.18), i.e.,

$$\sigma_2^{(i)}\left(\frac{1}{\overline{\sigma}_2^+}-\frac{1}{\overline{\sigma}_2^-}\right)+\frac{\left(\sigma_2^{(i)}\right)^2}{\overline{\sigma}_2^+\overline{\sigma}_2^-}+\left(\frac{\tau_{12}^{(i)}}{\overline{\tau}_{12}}\right)^2=1 \tag{6.58}$$

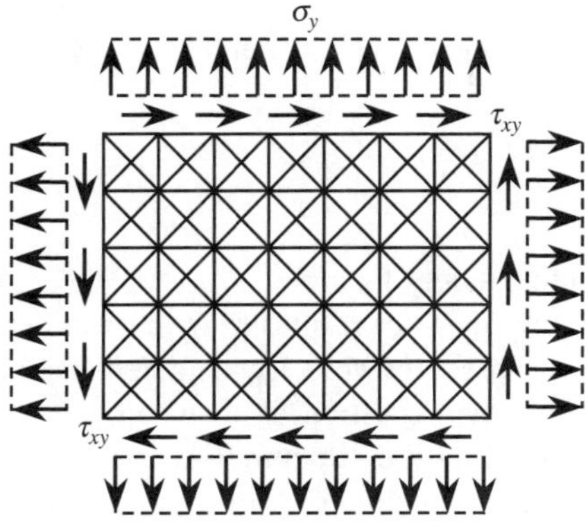

Fig. 6.23. A laminate loaded with normal and shear stresses.

in which $\overline{\sigma}_2^+$ and $\overline{\sigma}_2^-$ are the ultimate stresses of the ply under tension and compression across the fibers, and $\overline{\tau}_{12}$ is the ultimate in-plane shear stress.

Consider a laminate loaded with normal, σ_x and σ_y, and shear, τ_{xy}, stresses as in Fig. 6.23. Assume that the stresses are increased in proportion to some loading parameter p. Applying the strength criteria in Eqs. (6.57) and (6.58), we can find two values of this parameter, i.e., $p = p_{\mathrm{f}}$ which corresponds to fiber failure in at least one of the plies and $p = p_{\mathrm{m}}$ for which the loading causes matrix failure in one or more plies. Since the parameters p_{f} and p_{m} usually do not coincide with each other for modern composites, a question concerning the allowable level of stresses acting in the laminate naturally arises. Obviously, the stresses under which the fibers fail must not be treated as allowable stresses. Moreover, the allowable value p_{a} of the loading parameter must be less than p_{f} by a certain safety factor s_{f}, i.e.,

$$p_{\mathrm{a}}^{\mathrm{f}} = p_{\mathrm{f}}/s_{\mathrm{f}} \tag{6.59}$$

However, for matrix failure, the answer is not evident, and at least two different situations may take place.

First, the failure of the matrix can result in failure of the laminate. As an example, we can take a $\pm\phi$ angle-ply layer discussed in Section 4.5 whose moduli in the x- and y-directions are specified by Eqs. (4.147), i.e.,

$$E_x = A_{11} - \frac{A_{12}^2}{A_{22}}, \quad E_y = A_{22} - \frac{A_{12}^2}{A_{11}}$$

Ignoring the load-carrying capacity of the failed matrix, i.e., taking $E_2 = 0, G_{12} = 0$, and $\nu_{12} = \nu_{21} = 0$ in Eqs. (4.72) to get

$$A_{11} = E_1 \cos^4\phi, \quad A_{12} = E_1 \sin^2\phi \cos^2\phi, \quad A_{22} = E_1 \cos^4\phi$$

we arrive at $E_x = 0$ and $E_y = 0$ which means that the layer under consideration cannot work without the matrix. For such a mode of failure, we should take the allowable loading parameter as

$$p_a^m = p_m/s_m \tag{6.60}$$

where s_m is the corresponding safety factor.

Second, the matrix fracture does not result in laminate failure. As an example of such a structure, we can take the pressure vessel considered in Section 6.3 (see Figs. 6.21 and 6.22). Now we have another question as to whether the cracks in the matrix are allowed even if they do not affect the structure's strength. The answer to this question depends on the operational requirements for the structure. For example, suppose that the pressure vessel in Fig. 6.22 is a model of a filament-wound solid propellant rocket motor case which works only once and for a short period of time. Then, it is appropriate to ignore the cracks appearing in the matrix and take the allowable stresses in accordance with Eq. (6.59). We can also suppose that the vessel may be a model of a pressurized passenger cabin in a commercial airplane for which no cracks in the material are allowed in flight. Then, in principle, we must take the allowable stresses in accordance with Eq. (6.60). However, it follows from the examples considered in Sections 4.4.2 and 6.3, that for modern composites the loading parameter p_m is reached at the initial stage of the loading process. As a result, the allowable loading parameter, p_a^m in Eq. (6.60), is so small that modern composite materials cannot demonstrate their high strength governed by the fibers and cannot compete with metal alloys. A more realistic approach allows the cracks in the matrix to occur but only if p_m is higher than the operational loading parameter p_o. Using Eq. (6.60), we can presume that

$$p_o = p_a^m = p_m/s_m$$

The ultimate loading parameter $\overline{p}$ is associated with fiber failure, so that $\overline{p} = p_f$. Thus, the actual safety factor for the structure becomes

$$s = \frac{\overline{p}}{p_o} = \frac{p_f}{p_m} s_m \tag{6.61}$$

and depends on the ratio p_f/p_m.

To be specific, consider a four-layered [0°/45°/–45°/90°] quasi-isotropic carbon–epoxy laminate shown in Fig. 6.23 which is widely used in aircraft composite structures. The mechanical properties of quasi-isotropic laminates are discussed in Section 5.7. The constitutive equations for such laminates have the form typical for isotropic materials, i.e.,

$$\varepsilon_x = \frac{1}{E}(\sigma_x - \nu\sigma_y), \quad \varepsilon_y = \frac{1}{E}(\sigma_y - \nu\sigma_x), \quad \gamma_{xy} = \frac{\tau_{xy}}{G} \tag{6.62}$$

where E, ν, and G are given by Eqs. (5.110). For a typical carbon–epoxy fibrous composite, Table 5.4 yields $E = 54.8$ GPa, $\nu = 0.31$, $G = 20.9$ GPa. The strains in the plies' principal coordinates (see Fig. 4.18) can be found using Eqs. (4.69) from which we have for the 0° layer,

$$\varepsilon_1 = \varepsilon_x, \quad \varepsilon_2 = \varepsilon_y, \quad \gamma_{12} = \gamma_{xy}$$

for the ±45° layer,

$$\varepsilon_1 = \frac{1}{2}(\varepsilon_x + \varepsilon_y \pm \gamma_{xy}), \quad \varepsilon_2 = \frac{1}{2}(\varepsilon_x + \varepsilon_y \mp \gamma_{xy}), \quad \gamma_{12} = \pm(\varepsilon_y - \varepsilon_x)$$

for the 90° layer,

$$\varepsilon_1 = \varepsilon_y, \quad \varepsilon_2 = \varepsilon_x, \quad \gamma_{12} = \gamma_{xy}$$

The stresses in the plies are

$$\sigma_1 = \overline{E}_1(\varepsilon_1 + \nu_{12}\varepsilon_2), \quad \sigma_2 = \overline{E}_2(\varepsilon_2 + \nu_{21}\varepsilon_1), \quad \tau_{12} = G_{12}\gamma_{12} \tag{6.63}$$

where $\overline{E}_{1,2} = E_{1,2}/(1 - \nu_{12}\nu_{21})$, $\nu_{12}E_1 = \nu_{21}E_2$, and the elastic constants E_1, E_2, ν_{21}, and G_{12}, are given in Table 3.5. For given combinations of the acting stresses σ_x, σ_y, and τ_{xy} (see Fig. 6.23), the strains ε_x, ε_y, and γ_{xy}, found with the aid of Eqs. (6.62), are transformed to the ply strains $\varepsilon_1, \varepsilon_2$, and γ_{12}, and then substituted into Eqs. (6.63) for the stresses σ_1, σ_2, and τ_{12}. These stresses are substituted into the strength criteria in Eqs. (6.57) and (6.58), the first of which gives the combination of stresses σ_x, σ_y, and τ_{xy} causing failure of the fibers, whereas the second one enables us to determine the stresses inducing matrix failure.

Consider the biaxial loading with stresses σ_x and σ_y as shown in Fig. 6.23. The corresponding failure envelopes are presented in Fig. 6.24. The solid lines determine the domain within which the fibers do not fail, whereas within the area bound by the dashed lines no cracks in the matrix appear. Consider, for example, the cylindrical pressure vessel discussed in Section 6.3 for which

$$\sigma_x = \frac{pR}{2h}, \quad \sigma_y = \frac{pR}{h} \tag{6.64}$$

are the axial and circumferential stresses expressed in terms of the vessel radius and thickness, R and h, and the applied internal pressure p. Let us take $R/h = 100$. Then, $\sigma_x = 50p$ and $\sigma_y = 100p$, whereby $\sigma_y/\sigma_x = 2$. This combination of stresses is shown with the line 0BA in Fig. 6.24. Point A corresponds to failure of the fibers in the circumferential (90°) layer and gives the ultimate loading parameter (which is the pressure in this case) $\overline{p} = p_f = 8$ MPa. Point B corresponds to matrix failure in the axial (0°) layer and to the loading parameter $p_m = 2.67$ MPa. Taking $s_m = 1.5$, which is a typical value of the safety factor preventing material damage under the operational pressure, we get $s = 4.5$ from Eq. (6.61). In case we do not need such a high safety factor, we should either allow the cracks in the matrix to appear under the operational pressure, or to change the carbon–epoxy composite to some other material. It should be noted that the significant

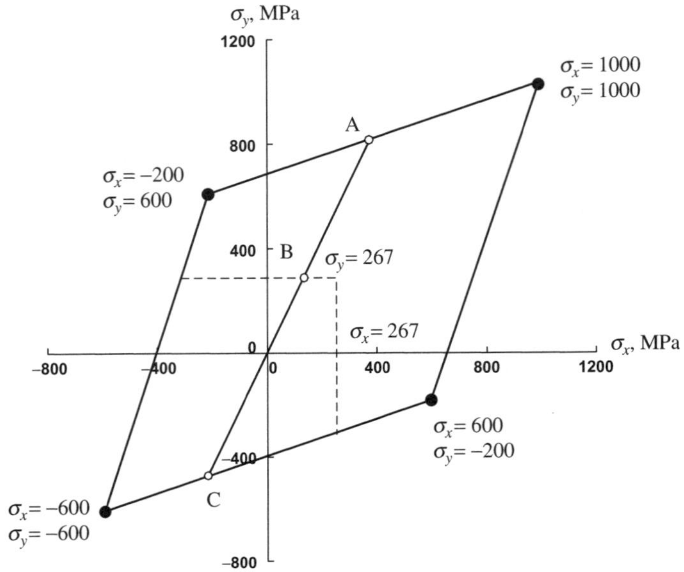

Fig. 6.24. Failure envelopes for biaxial loading corresponding to the failure criteria in Eqs. (6.57) (———), and (6.58) (- - - - - -).

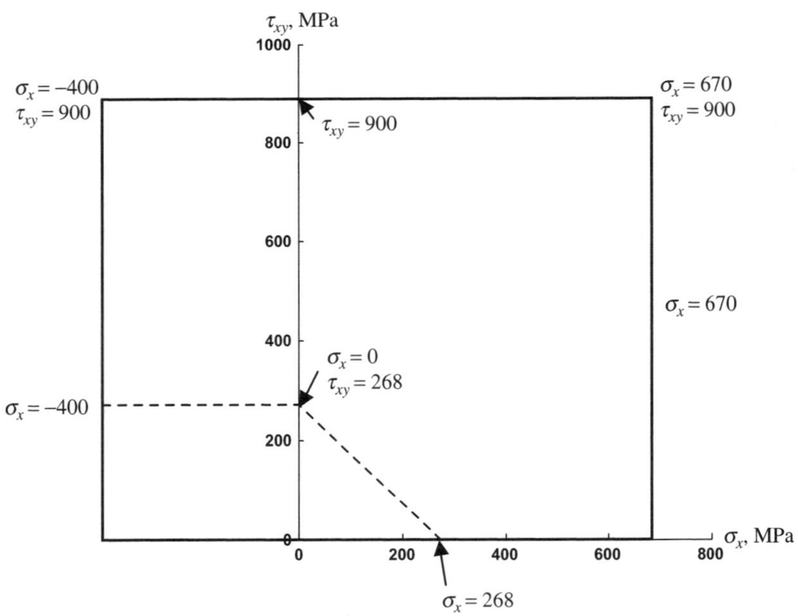

Fig. 6.25. Failure envelopes for uniaxial tension and compression combined with shear corresponding to the failure criteria in Eqs. (6.57) (———), and (6.58) (- - - - - -).

difference between the loading parameters p_f and p_m is typical mainly for tension. For compressive loads, the fibers usually fail before the matrix (see point C in Fig. 6.24).

Similar results for uniaxial tension with stresses σ_x combined with shear stresses τ_{xy} (see Fig. 6.23) are presented in Fig. 6.25. As can be seen, shear can induce the same effect as tension.

The problem of matrix failure discussed above significantly reduces the application of modern fibrous composites to structures subjected to long-term cyclic loading. It should be noted that Figs. 6.24 and 6.25 correspond to static loading at room temperature. Temperature, moisture, and fatigue can considerably reduce the areas bounded by the dashed lines in Figs. 6.24 and 6.25 (see Sections 7.1.2 and 7.3.3). Some methods developed to solve the problem of matrix failure are discussed in Sections 4.4.3 and 4.4.4.

6.5. References

Annin, B.D. and Baev, L.V. (1979). Criteria of composite material strength. In *Proceedings of the First USA–USSR Symposium on Fracture of Composite Materials*, Riga, USSR, Sept. 1978 (G.C. Sih and V.P. Tamuzh eds.). Sijthoff & Noordhoff, The Netherlands, pp. 241–254.

Ashkenazi, E.K. (1966). *Strength of Anisotropic and Synthetic Materials*. Lesnaya Promyshlennost, Moscow (in Russian).

Barbero, E.J. (1998). *Introduction to Composite Materials Design*. West Virginia University, USA.

Belyankin, F.P., Yatsenko, V.F. and Margolin, G.G. (1971). *Strength and Deformability of Fiberglass Plastics under Biaxial Compression*. Naukova Dumka, Kiev (in Russian).

Gol'denblat, I.I. and Kopnov, V.A. (1968). *Criteria of Strength and Plasticity for Structural Materials*. Mashinostroenie, Moscow (in Russian).

Jones, R.M. (1999). *Mechanics of Composite Materials*, 2nd edn. Taylor and Francis, Philadelphia, PA.

Katarzhnov, Yu.I. (1982). *Experimental Study of Load Carrying Capacity of Hollow Circular and Box Composite Beams under Compression and Torsion*. PhD thesis, Riga (in Russian).

Rowlands, R.E. (1975). Flow and failure of biaxially loaded composites: experimental – theoretical correlation. In *AMD – Vol. 13, Inelastic Behavior of Composite Materials*. ASME Winter Annual Meeting, Houston, TX (C.T. Herakovich ed.). ASME, New York, pp. 97–125.

Skudra, A.M., Bulavs, F.Ya., Gurvich, M.R. and Kruklinsh, A.A. (1989). *Elements of Structural Mechanics of Composite Truss Systems*. Zinatne, Riga (in Russian).

Tennyson, R.C., Nanyaro, A.P. and Wharram, G.E. (1980). Application of the cubic polynomial strength criterion to the failure analysis of composite materials. *Journal of Composite Materials*, **14**(suppl), 28–41.

Tsai, S.W. and Hahn, H.T. (1975). Failure analysis of composite materials. In *AMD – Vol. 13, Inelastic Behavior of Composite Materials*. ASME Winter Annual Meeting, Houston, TX (C.T. Herakovich ed.). ASME, New York, pp. 73–96.

Vasiliev, V.V. (1970). Effect of a local load on an orthotropic glass-reinforced plastic shell. *Polymer Mechanics/Mechanics of Composite Materials*, **6**(1), 80–85.

Vasiliev, V.V. (1993). *Mechanics of Composite Structures*. Taylor & Francis, Washington.

Vicario, A.A. Jr. and Toland, R.H. (1975). Failure criteria and failure analysis of composite structural components. In *Composite Materials* (L.J. Broutman and R.H. Krock eds.), *Vol. 7, Structural Design and Analysis. Part I* (C.C. Chamis ed.). Academic Press, New York, pp. 51–97.

Vorobey, V.V., Morozov, E.V. and Tatarnikov, O.V. (1992). *Analysis of Thermostressed Composite Structures*. Mashinostroenie, Moscow (in Russian).

Wu, E.M. (1974). Phenomenological anisotropic failure criterion. In *Composite Materials* (L.J. Broutman and R.H. Krock eds.), *Vol. 2, Mechanics of Composite Materials* (G.P. Sendecky ed.). Academic Press, New York, pp. 353–431.

Chapter 7

ENVIRONMENTAL, SPECIAL LOADING, AND MANUFACTURING EFFECTS

The properties of composite materials, as well as those of all structural materials, are affected by environmental and operational conditions. Moreover, for polymeric composites, this influence is more pronounced than for conventional metal alloys, because polymers are more sensitive to temperature, moisture, and time, than are metals. There exists also a specific feature of composites associated with the fact that they do not exist apart from composite structures and are formed while these structures are fabricated. As a result, the material characteristics depend on the type and parameters of the manufacturing process, e.g., unidirectional composites made by pultrusion, hand lay-up, and filament winding can demonstrate different properties.

This section of the book is concerned with the effect of environmental, loading, and manufacturing factors on the mechanical properties and behavior of composites.

7.1. Temperature effects

Temperature is the most important of environmental factors affecting the behavior of composite materials. First of all, polymeric composites are rather sensitive to temperature and have relatively low thermal conductivity. This combination of properties allows us, on one hand, to use these materials in structures subjected to short-term heating, and on the other hand, requires the analysis of these structures to be performed with due regard to temperature effects. Secondly, there exist composite materials, e.g., carbon–carbon and ceramic composites, that are specifically developed for operation under intense heating and materials such as mineral-fiber composites that are used to form heatproof layers and coatings. Thirdly, the fabrication of composite structures is usually accompanied with more or less intensive heating (e.g., for curing or carbonization), and the subsequent cooling induces thermal stresses and strains, to calculate which we need to utilize the equations of thermal conductivity and thermoelasticity, as discussed below.

7.1.1. Thermal conductivity

Heat flow through a unit area of a surface with normal n is related to the temperature gradient in the n-direction according to Fourier's law as

$$q = -\lambda \frac{\partial T}{\partial n} \tag{7.1}$$

where λ is the thermal conductivity of the material. The temperature distribution along the n-axis is governed by the following equation

$$\frac{\partial}{\partial n}\left(\lambda \frac{\partial T}{\partial n}\right) = c\rho \frac{\partial T}{\partial t} \tag{7.2}$$

in which c and ρ are the specific heat and density of the material, and t is time. For a steady (time-independent) temperature distribution, $\partial T/\partial t = 0$, and Eq. (7.2) yields

$$T = C_1 \int \frac{dn}{\lambda} + C_2 \tag{7.3}$$

Consider a laminated structure referred to coordinates x, z as shown in Fig. 7.1. To determine the temperature distribution along the x axis only, we should take into account that λ does not depend on x, and assume that T does not depend on z. Using conditions $T(x = 0) = T_0$ and $T(x = l) = T_l$ to find the constants C_1 and C_2 in Eq. (7.3), in which $n = x$, we get

$$T = T_0 + \frac{x}{l}(T_l - T_0)$$

Introduce the apparent thermal conductivity of the laminate in the x direction, λ_x, and write Eq. (7.1) for the laminate as

$$q_x = -\lambda_x \frac{T_l - T_0}{l}$$

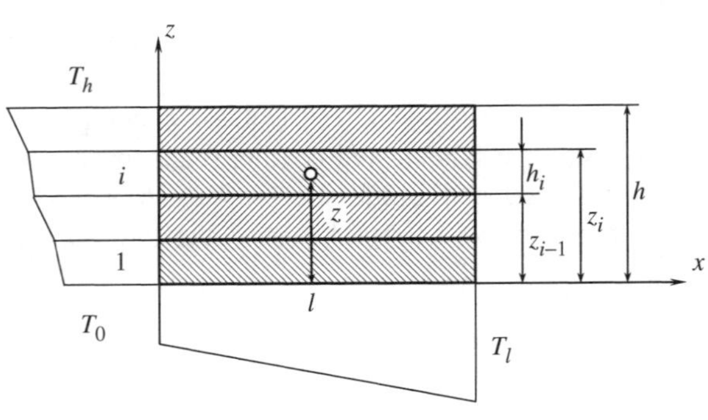

Fig. 7.1. Temperature distribution in a laminate.

The same equation can be written for the ith layer, i.e.,

$$q_i = -\lambda_i \frac{T_l - T_0}{l}$$

The total heat flow through the laminate in the x direction is

$$q_x h = \sum_{i=1}^{k} q_i h_i$$

Combining the foregoing results, we arrive at

$$\lambda_x = \sum_{i=1}^{k} \lambda_i \overline{h}_i \tag{7.4}$$

where $\overline{h}_i = h_i / h$.

Consider the heat transfer in the z direction and introduce the apparent thermal conductivity λ_z in accordance with the following form of Eq. (7.1)

$$q_z = -\lambda_z \frac{T_h - T_0}{h} \tag{7.5}$$

Taking $n = z$ and $\lambda = \lambda_i$ for $z_{i-1} \leq z \leq z_i$ in Eq. (7.3), using step-wise integration and the conditions $T(z = 0) = T_0$, $T(z = h) = T_h$ to find constants C_1, and C_2, we obtain for the ith layer

$$T_i = T_0 + \frac{T_h - T_0}{\sum_{i=1}^{k} \frac{h_i}{\lambda_i}} \left(\frac{z - z_{i-1}}{\lambda_i} + \sum_{j=1}^{i-1} \frac{h_j}{\lambda_j} \right) \tag{7.6}$$

The heat flow through the ith layer follows from Eqs. (7.1) and (7.6), i.e.,

$$q_i = -\lambda_i \frac{\partial T_i}{\partial z} = -\frac{T_h - T_0}{\sum_{i=1}^{k} \frac{h_i}{\lambda_i}}$$

Obviously, $q_i = q_z$ (see Fig. 7.1), and with due regard to Eq. (7.5)

$$\frac{1}{\lambda_z} = \sum_{i=1}^{k} \frac{\overline{h}_i}{\lambda_i} \tag{7.7}$$

where, as earlier, $\overline{h}_i = h_i / h$.

The results obtained, Eqs. (7.4) and (7.7), can be used to determine the thermal conductivity of a unidirectional composite ply. Indeed, comparing Fig. 7.1 with Fig. 3.34

showing the structure of the first-order ply model, we can write the following equations specifying thermal conductivity of a unidirectional ply along and across the fibers

$$\lambda_1 = \lambda_{1f} v_f + \lambda_m v_m$$
$$\frac{1}{\lambda_2} = \frac{v_f}{\lambda_{2f}} + \frac{v_m}{\lambda_m} \tag{7.8}$$

Here, λ_{1f} and λ_{2f} are the thermal conductivities of the fiber in the longitudinal and transverse directions (for some fibers, they are different), λ_m is the corresponding characteristic of the matrix, and v_f, $v_m = 1 - v_f$ are the fiber and matrix volume fractions, respectively. The conductivity coefficients in Eqs. (7.8) are analogous to elastic constants specified by Eqs. (3.76) and (3.78), and the discussion presented in Section 3.3 is valid for Eqs. (7.8) as well. In particular, it should be noted that application of higher-order microstructural models has practically no effect on λ_1, but substantially improves λ_2 determined by Eqs. (7.8). Typical properties for unidirectional and fabric composites are listed in Table 7.1.

Consider heat transfer in an orthotropic ply or layer in coordinate frame x, y whose axes x and y make angle ϕ with the principal material coordinates x_1 and x_2 as in Fig. 7.2. Heat flows in coordinates x, y and x_1, x_2 are linked by the following equations

$$q_x = q_1 \cos\phi - q_2 \sin\phi, \quad q_y = q_1 \sin\phi + q_2 \cos\phi \tag{7.9}$$

Here, in accordance with Eq. (7.1)

$$q_1 = -\lambda_1 \frac{\partial T}{\partial x_1}, \quad q_2 = -\lambda_2 \frac{\partial T}{\partial x_2}$$

Table 7.1
Typical thermal conductivity and expansion coefficients of composite materials.

Property	Glass–epoxy	Carbon–epoxy	Aramid–epoxy	Boron–epoxy	Glass fabric–epoxy	Aramid fabric–epoxy
Longitudinal conductivity λ_1 (W/mK)	0.6	1	0.17	0.5	0.35	0.13
Transverse conductivity λ_2 (W/mK)	0.4	0.6	0.1	0.3	0.35	0.13
Longitudinal CTE 10^6 α_1 (1/°C)	7.4	−0.3	−3.6	4.1	8	0.8
Transverse CTE 10^6 α_2 (1/°C)	22.4	34	60	19.2	8	0.8

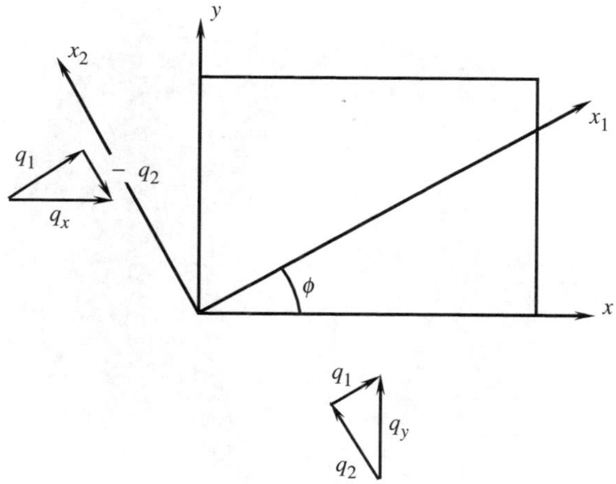

Fig. 7.2. Heat flows in coordinates x, y and x_1, x_2.

Changing variables x_1, x_2 to x, y with the aid of the following transformation relationships

$$x = x_1 \cos\phi - x_2 \sin\phi, \quad y = x_1 \sin\phi + x_2 \cos\phi$$

and substituting q_1 and q_2 into Eqs. (7.9), we arrive at

$$q_x = -\lambda_x \frac{\partial T}{\partial x} - \lambda_{xy} \frac{\partial T}{\partial y}, \quad q_y = -\lambda_y \frac{\partial T}{\partial y} + \lambda_{xy} \frac{\partial T}{\partial x}$$

where

$$\begin{aligned} \lambda_x &= \lambda_1 \cos^2\phi + \lambda_2 \sin^2\phi \\ \lambda_y &= \lambda_1 \sin^2\phi + \lambda_2 \cos^2\phi \\ \lambda_{xy} &= (\lambda_2 - \lambda_1) \sin\phi \cos\phi \end{aligned} \tag{7.10}$$

can be treated as the ply thermal conductivities in coordinates x, y. Since the ply is anisotropic in these coordinates, the heat flow in the, for example x direction, induces a temperature gradient not only in the x direction, but in the y direction as well. Using Eq. (7.4), we can now determine the in-plane thermal conductivities of the laminate as

$$\Lambda_x = \sum_{i=1}^{k} \lambda_x^{(i)} \overline{h}_i, \quad \Lambda_y = \sum_{i=1}^{k} \lambda_y^{(i)} \overline{h}_i, \quad \Lambda_{xy} = \sum_{i=1}^{k} \lambda_{xy}^{(i)} \overline{h}_i \tag{7.11}$$

where $\lambda_{x,y}^{(i)}$ are specified by Eqs. (7.10) in which $\lambda_{1,2} = \lambda_{1,2}^{(i)}$ and $\phi = \phi_i$. For $\pm\phi$ angle-ply laminates that are orthotropic, $\Lambda_{xy} = 0$.

Fig. 7.3. A composite section of a space telescope. Courtesy of CRISM.

As an example, consider the composite body of a space telescope, the section of which is shown in Fig. 7.3. The cylinder having diameter $D = 1$ m and total thickness $h = 13.52$ mm consists of four layers, i.e.,

- $\pm\phi_s$ angle-ply carbon–epoxy external skin with the following parameters:
  $\phi_s = 20°$, $h_s^e = 3.5$ mm, $E_1^e = 120$ GPa,
  $E_2^e = 11$ GPa, $G_{12}^e = 5.5$ GPa, $\nu_{21}^e = 0.27$,
  $\lambda_1^e = 1$ W/m K, $\lambda_2^e = 0.6$ W/m K,
  $\alpha_1^e = -0.3 \cdot 10^{-6}$ 1/°C, $\alpha_2^e = 34 \cdot 10^{-6}$ 1/°C,
- carbon–epoxy lattice layer (see Fig. 4.93) formed by a system of $\pm\phi_r$ helical ribs with
  $\phi_r = 26°$, $h_r = 9$ mm, $\delta_r = 4$ mm, $a_r = 52$ mm, $E_r = 80$ GPa,
  $\lambda_r = 0.9$ W/m K, $\alpha_r = -1 \cdot 10^{-6}$ 1/°C,
- internal skin made of aramid fabric with
  $h_s^i = 1$ mm, $E_x^i = E_y^i = 34$ GPa, $G_{xy}^i = 5.6$ GPa,
  $\nu_{xy}^i = \nu_{yx}^i = 0.15$, $\lambda_x^i = \lambda_y^i = 0.13$ W/m K,
  $\alpha_x^i = \alpha_y^i = 0.8 \cdot 10^{-6}$ 1/°C (x and y are the axial and circumferential coordinates of the cylinder),
- internal layer of aluminum foil with $h_f = 0.02$ mm, $E_f = 70$ GPa, $\nu_f = 0.3$,
  $\lambda_f = 210$ W/m K, $\alpha_f = 22.3 \cdot 10^{-6}$ 1/°C

The apparent thermal conductivity of the cylinder wall can be found with the aid of Eqs. (7.10), (7.11), and the continuum model of the lattice layer described in Section 4.7 as

$$\Lambda_x = \frac{1}{h}\left[\left(\lambda_1^e \cos^2\phi_e + \lambda_2^e \sin^2\phi_e\right) h_s^e + \frac{2}{a_r} h_r \delta_r \lambda_r \cos^2\phi_r + \lambda_x^i h_s^i + \lambda_f h_f\right]$$

Calculation yields $\Lambda_x = 0.64\ \mathrm{W/mK}$. The thermal resistance of a unit length of this structure is

$$r_x = \frac{1}{\Lambda_x \pi D h} = 36.8 \frac{\mathrm{K}}{\mathrm{W\,m}}$$

7.1.2. Thermoelasticity

It is known that heating gives rise to thermal strains that, when restricted, induce thermal stresses. Assume that the temperature distribution in a composite structure is known, and consider the problem of thermoelasticity.

Consider first the thermoelastic behavior of a unidirectional composite ply studied in Section 3.3 and shown in Fig. 3.29. The generalized Hooke's law, Eqs. (3.58), allowing for temperature effects, can be written as

$$\varepsilon_{1T} = \varepsilon_1 + \varepsilon_1^T, \quad \varepsilon_{2T} = \varepsilon_2 + \varepsilon_2^T, \quad \gamma_{12T} = \gamma_{12} \tag{7.12}$$

Here, and subsequently, the subscript "T" shows the strains that correspond to the problem of thermoelasticity, whereas the superscript "T" indicates temperature terms. Elastic strains ε_1, ε_2, and γ_{12} in Eqs. (7.12) are related to stresses by Eqs. (3.58). Temperature strains, to a first approximation, can be taken as linear functions of the temperature change, i.e.,

$$\varepsilon_1^T = \alpha_1 \Delta T, \quad \varepsilon_2^T = \alpha_2 \Delta T \tag{7.13}$$

where α_1 and α_2 are the coefficients of thermal expansion (CTE) along and across the fibers, and $\Delta T = T - T_0$ is the difference between the current temperature T and some initial temperature T_0 at which thermal strains are zero. The inverse form of Eqs. (7.12) is

$$\begin{aligned}
\sigma_1 &= \overline{E}_1(\varepsilon_{1T} + \nu_{12}\varepsilon_{2T}) - \overline{E}_1\left(\varepsilon_1^T + \nu_{12}\varepsilon_2^T\right) \\
\sigma_2 &= \overline{E}_2(\varepsilon_{2T} + \nu_{21}\varepsilon_{1T}) - \overline{E}_2\left(\varepsilon_2^T + \nu_{21}\varepsilon_1^T\right) \\
\tau_{12} &= G_{12}\gamma_{12T}
\end{aligned} \tag{7.14}$$

where $\overline{E}_{1,2} = E_{1,2}/(1 - \nu_{12}\nu_{21})$.

To describe the thermoelastic behavior of a ply, apply the first-order micromechanical model shown in Fig. 3.34. Since the CTE (and elastic constants) of some fibers can be different in the longitudinal and transverse directions, generalize the first two equations of Eqs. (3.63) as

$$\begin{aligned}
\varepsilon_{1T}^{\mathrm{f,m}} &= \frac{1}{E_{\mathrm{f1,m}}}\left(\sigma_1^{\mathrm{f,m}} - \nu_{\mathrm{f2,m}}\sigma_2^{\mathrm{f,m}}\right) + \alpha_{\mathrm{f1,m}}\Delta T \\
\varepsilon_{2T}^{\mathrm{f,m}} &= \frac{1}{E_{\mathrm{f2,m}}}\left(\sigma_2^{\mathrm{f,m}} - \nu_{\mathrm{f1,m}}\sigma_1^{\mathrm{f,m}}\right) + \alpha_{\mathrm{f2,m}}\Delta T
\end{aligned} \tag{7.15}$$

Repeating the derivation of Eqs. (3.76)–(3.79), we arrive at

$$
\begin{aligned}
&E_1 = E_{f1}v_f + E_m v_m, \qquad \nu_{21} = \nu_{f2}v_f + \nu_m v_m \\
&\frac{1}{E_2} = (1-\nu_{f1}\nu_{f2})\frac{v_f}{E_{f2}} + \left(1-\nu_m^2\right)\frac{v_m}{E_m} + \frac{\nu_{21}^2}{E_1} \\
&\alpha_1 = \frac{1}{E_1}\left(E_{f1}\alpha_{f1}v_f + E_m\alpha_m v_m\right) \\
&\alpha_2 = (\alpha_{f2} + \nu_{f2}\alpha_{f1})\,v_f + (1+\nu_m)\,\alpha_m v_m - \nu_{21}\alpha_1
\end{aligned}
\tag{7.16}
$$

These equations generalize Eqs. (3.76)–(3.79) for the case of anisotropic fibers and specify the apparent CTE of a unidirectional ply.

As an example, consider the high-modulus carbon–epoxy composite tested by Rogers et al. (1977). The microstructural parameters for this material are as follows ($T = 27°$C)

$$
\begin{aligned}
&E_{f1} = 411\,\text{GPa}, \quad E_{f2} = 6.6\,\text{GPa}, \quad \nu_{f1} = 0.06 \\
&\nu_{f2} = 0.35, \quad \alpha_{f1} = -1.2\cdot 10^{-6}\,1/°\text{C}, \quad \alpha_{f2} = 27.3\cdot 10^{-6}\,1/°\text{C} \\
&E_m = 5.7\,\text{GPa}, \quad \nu_m = 0.316, \quad \alpha_m = 45\cdot 10^{-6}\,1/°\text{C}, \quad v_f = v_m = 0.5
\end{aligned}
$$

For these properties, Eqs. (7.16) yield

$$
\begin{aligned}
&E_1 = 208.3\,\text{GPa}, \quad E_2 = 6.5\,\text{GPa}, \quad \nu_{21} = 0.33 \\
&\alpha_1 = -0.57\cdot 10^{-6}\,1/°\text{C}, \quad \alpha_2 = 43.4\cdot 10^{-6}\,1/°\text{C}
\end{aligned}
$$

whereas the experimental results were

$$
\begin{aligned}
&E_1 = 208.6\,\text{GPa}, \quad E_2 = 6.3\,\text{GPa}, \quad \nu_{21} = 0.33 \\
&\alpha_1 = -0.5\cdot 10^{-6}\,1/°\text{C}, \quad \alpha_2 = 29.3\cdot 10^{-6}\,1/°\text{C}
\end{aligned}
$$

Thus, it can be concluded that the first-order microstructural model provides good results for the longitudinal material characteristics, but fails to predict α_2 with the required accuracy. The discussion and conclusions concerning this problem presented in Section 3.3 for elastic constants are valid for thermal expansion coefficients as well. For practical applications, α_1 and α_2 are normally determined by experimental methods. However, in contrast to the elasticity problem for which the knowledge of experimental elastic constants and material strength excludes consideration of the micromechanical models, for the thermoelasticity problems, these models provide us with useful information even if we know the experimental thermal expansion coefficients. Indeed, consider a unidirectional ply that is subjected to uniform heating that induces only thermal strains, i.e., $\varepsilon_{1T} = \varepsilon_1^T$, $\varepsilon_{2T} = \varepsilon_2^T$, and $\gamma_{12T} = 0$. Then, Eqs. (7.14) yield $\sigma_1 = 0$, $\sigma_2 = 0$, and $\tau_{12} = 0$. For homogeneous materials, this means that no stresses occur under uniform heating. However, this is not the case for a composite ply. Generalizing Eqs. (3.74) that specify longitudinal stresses in the fibers and in the matrix, we obtain

$$\sigma_1^{\mathrm{f}} = E_{\mathrm{f}1}(\alpha_1 - \alpha_{\mathrm{f}1})\Delta T, \quad \sigma_1^{\mathrm{m}} = E_{\mathrm{m}}(\alpha_1 - \alpha_{\mathrm{m}})\Delta T$$

where α_1 and α_2 are specified by Eqs. (7.16). Thus, because the thermal expansion coefficients of the fibers and the matrix are different from those of the material, there exist microstructural thermal stresses in the composite structural elements. These stresses are self-balanced.

Indeed,

$$\sigma_1 = \sigma_1^{\mathrm{f}} v_{\mathrm{f}} + \sigma_1^{\mathrm{m}} v_{\mathrm{m}} = 0$$

Consider an orthotropic layer referred to coordinate axes x, y making angle ϕ with the principal material coordinate axes (see Fig. 7.2). Using Eqs. (7.14) instead of Eqs. (4.56) and repeating the derivation of Eqs. (4.71), we arrive at

$$\begin{aligned}
\sigma_x &= A_{11}\varepsilon_{xT} + A_{12}\varepsilon_{yT} + A_{14}\gamma_{xyT} - A_{11}^T \\
\sigma_y &= A_{21}\varepsilon_{xT} + A_{22}\varepsilon_{yT} + A_{24}\gamma_{xyT} - A_{22}^T \\
\tau_{xy} &= A_{41}\varepsilon_{xT} + A_{42}\varepsilon_{yT} + A_{44}\gamma_{xyT} - A_{12}^T
\end{aligned} \tag{7.17}$$

where A_{mn} are specified by Eqs. (4.72), and the thermal terms are

$$\begin{aligned}
A_{11}^T &= \overline{E}_1\varepsilon_{12}^T \cos^2\phi + \overline{E}_2\varepsilon_{21}^T \sin^2\phi \\
A_{22}^T &= \overline{E}_1\varepsilon_{12}^T \sin^2\phi + \overline{E}_2\varepsilon_{21}^T \cos^2\phi \\
A_{12}^T &= \left(\overline{E}_1\varepsilon_{12}^T - \overline{E}_2\varepsilon_{21}^T\right)\sin\phi\cos\phi
\end{aligned} \tag{7.18}$$

Here,

$$\varepsilon_{12}^T = \varepsilon_1^T + \nu_{12}\varepsilon_2^T, \quad \varepsilon_{21}^T = \varepsilon_2^T + \nu_{21}\varepsilon_1^T$$

and ε_1^T, ε_2^T are determined by Eqs. (7.13). The inverse form of Eqs. (7.17) is

$$\varepsilon_{xT} = \varepsilon_x + \varepsilon_x^T, \quad \varepsilon_{yT} = \varepsilon_y + \varepsilon_y^T, \quad \gamma_{xyT} = \gamma_{xy} + \gamma_{xy}^T \tag{7.19}$$

Here, ε_x, ε_y, and γ_{xy} are expressed in terms of stresses σ_x, σ_y, and τ_{xy} by Eqs. (4.75), whereas the thermal strains are

$$\varepsilon_x^T = \varepsilon_1^T \cos^2\phi + \varepsilon_2^T \sin^2\phi$$
$$\varepsilon_y^T = \varepsilon_1^T \sin^2\phi + \varepsilon_2^T \cos^2\phi$$
$$\gamma_{xy}^T = \left(\varepsilon_1^T - \varepsilon_2^T\right) \sin 2\phi$$

Introducing thermal expansion coefficients in the xy coordinate frame with the following equations

$$\varepsilon_x^T = \alpha_x \Delta T, \quad \varepsilon_y^T = \alpha_y \Delta T, \quad \gamma_{xy}^T = \alpha_{xy} \Delta T \tag{7.20}$$

and using Eqs. (7.13), we obtain

$$\begin{aligned} \alpha_x &= \alpha_1 \cos^2\phi + \alpha_2 \sin^2\phi \\ \alpha_y &= \alpha_1 \sin^2\phi + \alpha_2 \cos^2\phi \\ \alpha_{xy} &= (\alpha_1 - \alpha_2) \sin 2\phi \end{aligned} \tag{7.21}$$

It follows from Eqs. (7.19) that, in an anisotropic layer, uniform heating induces not only normal strains, but also a shear thermal strain. As can be seen in Fig. 7.4, Eqs. (7.21)

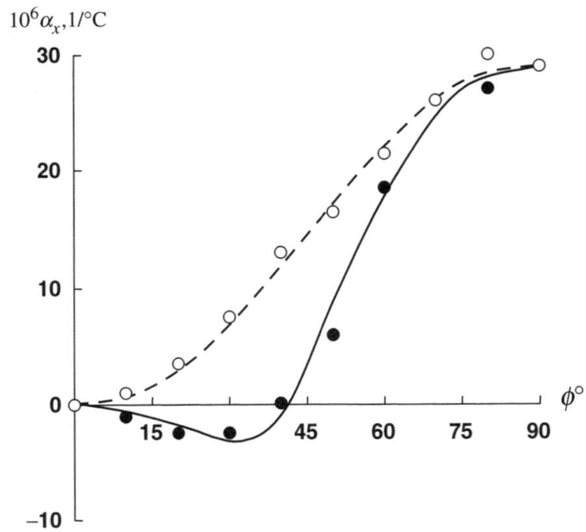

Fig. 7.4. Calculated (lines) and experimental (circles) dependencies of thermal expansion coefficients on the ply orientation angle for unidirectional thermoplastic carbon composite (– – –, ○) and a $\pm\phi$ angle-ply layer (———, ●).

provide fair agreement with the experimental results of Barnes et al. (1989) for composites with carbon fibers and thermoplastic matrix (dashed line and light circles).

Consider a symmetric $\pm\phi$ angle-ply layer (see Section 4.5.1). This layer is orthotropic, and the corresponding constitutive equations of thermoelastisity have the form of Eqs. (7.17) in which $A_{14} = A_{41} = 0$, $A_{24} = A_{42} = 0$, and $A_{12}^T = 0$. The inverse form of these equations is

$$\varepsilon_{xT} = \varepsilon_x + \varepsilon_x^T, \quad \varepsilon_{yT} = \varepsilon_y + \varepsilon_y^T, \quad \gamma_{xyT} = \gamma_{xy}$$

where ε_x, ε_y, and γ_{xy} are expressed in terms of stresses by Eqs. (4.146), whereas the thermal strains are

$$\varepsilon_x^T = \frac{A_{11}^T A_{22} - A_{22}^T A_{12}}{A_{11}A_{22} - A_{12}^2}, \quad \varepsilon_y^T = \frac{A_{22}^T A_{11} - A_{11}^T A_{12}}{A_{11}A_{22} - A_{12}^2}$$

Using Eqs. (4.147), (7.13), (7.18), and (7.20), we arrive at the following expressions for apparent thermal expansion coefficients

$$\alpha_x = \frac{1}{E_x}\left(a_{11}^T - \nu_{yx}a_{22}^T\right), \quad \alpha_y = \frac{1}{E_y}\left(a_{22}^T - \nu_{xy}a_{11}^T\right) \tag{7.22}$$

in which

$$a_{11}^T = \overline{E}_1(\alpha_1 + \nu_{12}\alpha_2)\cos^2\phi + \overline{E}_2(\alpha_2 + \nu_{21}\alpha_1)\sin^2\phi$$

$$a_{22}^T = \overline{E}_1(\alpha_1 + \nu_{12}\alpha_2)\sin^2\phi + \overline{E}_2(\alpha_2 + \nu_{21}\alpha_1)\cos^2\phi$$

Comparison of α_x with the experimental results of Barnes et al. (1989) for a thermoplastic carbon composite is presented in Fig. 7.4 (solid line and dots). As can be seen in this figure, there exists an interval ($0 \le \phi \le 40°$) within which the coefficient α_x of the angle-ply layer is negative. The same type of behavior is demonstrated by aramid epoxy angle-ply composites. A comparison of calculated values based on Eqs. (7.22) with the experimental results of Strife and Prevo (1979) is presented in Fig. 7.5. Looking at Figs. 7.4 and 7.5, we can hypothesize that supplementing an angle-ply laminate with plies having small thermal elongations in the x direction, we can synthesize composite materials with zero thermal expansion in this direction. Such materials are important, for example, for space telescopes (Fig. 7.3), antennas, measuring instruments, and other high-precision, thermally stable structures (Hamilton and Patterson, 1993).

Consider laminates with arbitrary structural parameters (see Chapter 5). Repeating the derivation of Eqs. (5.5) and using the thermoelasticity constitutive equations, Eqs. (7.17),

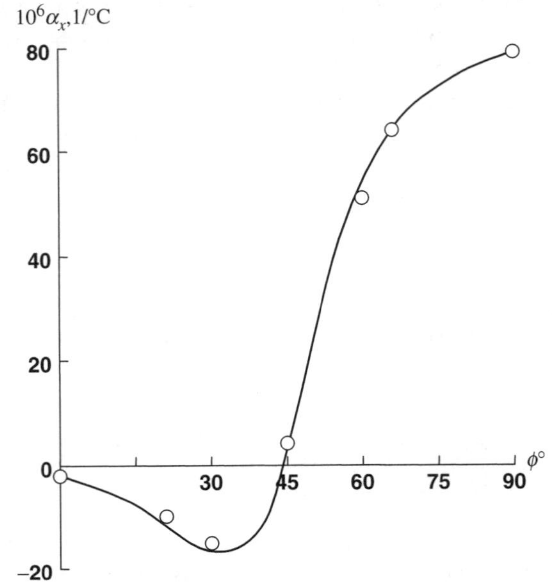

Fig. 7.5. Calculated (line) and experimental (circles) dependencies of thermal expansion coefficient on the ply orientation angle for an aramid–epoxy $\pm\phi$ angle-ply layer.

instead of Eqs. (4.71), we arrive at

$$\begin{aligned}
N_x &= B_{11}\varepsilon_{xT}^0 + B_{12}\varepsilon_{yT}^0 + B_{14}\gamma_{xyT}^0 + C_{11}\kappa_{xT} + C_{12}\kappa_{yT} + C_{14}\kappa_{xyT} - N_{11}^T \\
N_y &= B_{21}\varepsilon_{xT}^0 + B_{22}\varepsilon_{yT}^0 + B_{24}\gamma_{xyT}^0 + C_{21}\kappa_{xT} + C_{22}\kappa_{yT} + C_{24}\kappa_{xyT} - N_{22}^T \\
N_{xy} &= B_{41}\varepsilon_{xT}^0 + B_{42}\varepsilon_{yT}^0 + B_{44}\gamma_{xyT}^0 + C_{41}\kappa_{xT} + C_{42}\kappa_{yT} + C_{44}\kappa_{xyT} - N_{12}^T \\
M_x &= C_{11}\varepsilon_{xT}^0 + C_{12}\varepsilon_{yT}^0 + C_{14}\gamma_{xyT}^0 + D_{11}\kappa_{xT} + D_{12}\kappa_{yT} + D_{14}\kappa_{xyT} - M_{11}^T \\
M_y &= C_{21}\varepsilon_{xT}^0 + C_{22}\varepsilon_{yT}^0 + C_{24}\gamma_{xyT}^0 + D_{21}\kappa_{xT} + D_{22}\kappa_{yT} + D_{24}\kappa_{xyT} - M_{22}^T \\
M_{xy} &= C_{41}\varepsilon_{xT}^0 + C_{42}\varepsilon_{yT}^0 + C_{44}\gamma_{xyT}^0 + D_{41}\kappa_{xT} + D_{42}\kappa_{yT} + D_{44}\kappa_{xyT} - M_{12}^T
\end{aligned} \tag{7.23}$$

These equations should be supplemented with Eqs. (5.15) for transverse shear forces, i.e.,

$$V_x = S_{55}\gamma_{xT} + S_{56}\gamma_{yT}, \quad V_y = S_{65}\gamma_{xT} + S_{66}\gamma_{yT} \tag{7.24}$$

The temperature terms entering Eqs. (7.23) have the following form

$$N_{mn}^T = \int_{-e}^{s} A_{mn}^T \mathrm{d}z, \quad M_{mn}^T = \int_{-e}^{s} A_{mn}^T z \mathrm{d}z$$

where A_{mn}^{T} are specified by Eqs. (7.18). Performing the transformation that is used in Section 5.1 to reduce Eqs. (5.6), (5.7), and (5.8) to Eqs. (5.28) and (5.29), we get

$$N_{mn}^{\mathrm{T}} = J_{mn}^{(0)}, \quad M_{mn}^{\mathrm{T}} = J_{mn}^{(1)} - eJ_{mn}^{(0)} \tag{7.25}$$

Here (see Fig. 5.8),

$$J_{mn}^{(r)} = \int_0^h A_{mn}^{T} t^r \mathrm{d}t \tag{7.26}$$

where $r = 0,\ 1$ and $mn = 11,\ 12,\ 22$.

For a laminate, the temperature governed by Eq. (7.6) is linearly distributed over the layers' thicknesses (see Fig. 7.1). The same law can be, obviously, assumed for the temperature coefficients in Eqs. (7.18), i.e., for the ith layer in Fig. 5.10

$$A_{mn}^{Ti} = \left(A_{mn}^{T}\right)_{i-1} + \frac{1}{h_i}\left[\left(A_{mn}^{T}\right)_i - \left(A_{mn}^{T}\right)_{i-1}\right](t - t_{i-1})$$

where $\left(A_{mn}^{T}\right)_{i-1} = A_{mn}^{T}\,(t = t_{i-1})$ and $\left(A_{mn}^{T}\right)_i = A_{mn}^{T}\,(t = t_i)$. Then, Eq. (7.26) takes the form

$$J_{mn}^{(r)} = \sum_{i=1}^{k} \frac{1}{h_i}\left\{\left[\left(A_{mn}^{T}\right)_{i-1} t_i - \left(A_{mn}^{T}\right)_i t_{i-1}\right]\frac{t_i^{r+1} - t_{i-1}^{r+1}}{r+1} + \left[\left(A_{mn}^{T}\right)_i - \left(A_{mn}^{T}\right)_{i-1}\right]\frac{t_i^{r+2} - t_{i-1}^{r+2}}{r+2}\right\}$$

If the temperature variation over the thickness of the ith layer can be neglected, we can introduce some average value

$$\left(\overline{A}_{mn}^{T}\right)_i = \frac{1}{2}\left[\left(A_{mn}^{T}\right)_{i-1} + \left(A_{mn}^{T}\right)_i\right]$$

and present Eq. (7.26) as

$$J_{mn}^{(r)} = \frac{1}{r+1}\sum_{i=1}^{k}\left(\overline{A}_{mn}^{T}\right)_i\left(t_i^{r+1} - t_{i-1}^{r+1}\right) \tag{7.27}$$

Total (elastic and temperature) generalized strains ε_T, γ_T, and κ_T entering Eqs. (7.23) and (7.24) can be expressed in terms of the displacements and rotational angles of the laminate element with the aid of Eqs. (5.3) and (5.14), i.e.,

$$\varepsilon_{xT}^0 = \frac{\partial u}{\partial x}, \quad \varepsilon_{yT}^0 = \frac{\partial v}{\partial y}, \quad \gamma_{xyT}^0 = \frac{\partial u}{\partial y} + \frac{\partial v}{\partial x} \tag{7.28}$$

$$\kappa_{xT} = \frac{\partial \theta_x}{\partial x}, \quad \kappa_{yT} = \frac{\partial \theta_y}{\partial y}, \quad \kappa_{xyT} = \frac{\partial \theta_x}{\partial y} + \frac{\partial \theta_y}{\partial x} \tag{7.29}$$

$$\gamma_{xT} = \theta_x + \frac{\partial w}{\partial x}, \quad \gamma_{yT} = \theta_y + \frac{\partial w}{\partial y} \tag{7.30}$$

It follows from Eqs. (7.23) that in the general case, uniform heating of laminates induces, in contrast to homogeneous materials, not only in-plane strains but also changes to the laminate curvatures and twist. Indeed, assume that the laminate is free from edge and surface loads so that forces and moments in the left-hand sides of Eqs. (7.23) are equal to zero. Since the CTE of the layers, in the general case, are different, the thermal terms N^T and M^T in the right-hand sides of Eqs. (7.23) are not equal to zero even for a uniform temperature field, and these equations enable us to find ε_T, γ_T, and κ_T specifying the laminate in-plane and out-of-plane deformation. Moreover, using the approach described in Section 5.11, we can conclude that uniform heating of the laminate is accompanied, in the general case, by stresses acting in the layers and between the layers.

As an example, consider the four-layered structure of the space telescope described in Section 7.1.1.

First, we calculate the stiffness coefficients of the layers, i.e.,

- for the internal layer of aluminum foil,

$$A_{11}^{(1)} = A_{22}^{(1)} = \overline{E}_f = 76.92\,\text{GPa}, \quad A_{12}^{(1)} = \nu_\text{f}\overline{E}_\text{f} = 23.08\,\text{GPa}$$

- for the inner skin,

$$A_{11}^{(2)} = A_{22}^{(2)} = \overline{E}_x^i = 34.87\,\text{GPa}, \quad A_{12}^{(2)} = \nu_{xy}^i\overline{E}_x^i = 5.23\,\text{GPa}$$

- for the lattice layer,

$$A_{11}^{(3)} = 2E_r\frac{\delta_r}{a_r}\cos^4\phi_r = 14.4\,\text{GPa}$$

$$A_{22}^{(3)} = 2E_r\frac{\delta_r}{a_r}\sin^4\phi_r = 0.25\,\text{GPa}$$

$$A_{12}^{(3)} = 2E_r\frac{\delta_r}{a_r}\sin^2\phi\cos^2\phi = 1.91\,\text{GPa}$$

- for the external skin,

$$A_{11}^{(4)} = \overline{E}_1^\text{e}\cos^4\phi_\text{e} + \overline{E}_2^\text{e}\sin^4\phi_\text{e} + 2\left(\overline{E}_1^\text{e}\nu_{12}^\text{e} + 2G_{12}^\text{e}\right)\sin^2\phi_\text{e}\cos^2\phi_\text{e} = 99.05\,\text{GPa}$$

$$A_{22}^{(4)} = \overline{E}_1^{\text{e}} \sin^4 \phi_{\text{e}} + \overline{E}_2^{\text{e}} \cos^4 \phi_{\text{e}} + 2\left(\overline{E}_1^{\text{e}} \nu_{12}^{\text{e}} + 2G_{12}^{\text{e}}\right) \sin^2 \phi_{\text{e}} \cos^2 \phi_{\text{e}} = 13.39\,\text{GPa}$$

$$A_{12}^{(4)} = \overline{E}_1^{\text{e}} \nu_{12}^{\text{e}} + \left[\overline{E}_1^{\text{e}} + \overline{E}_2^{\text{e}} - 2\left(\overline{E}_1^{\text{e}} \nu_{12}^{e} + 2G_{12}^{\text{e}}\right)\right] \sin^2 \phi_{\text{e}} \cos^2 \phi_{\text{e}} = 13.96\,\text{GPa}$$

Using Eqs. (7.18), we find the thermal coefficients of the layers (the temperature is uniformly distributed over the laminate thickness)

$$\left(A_{11}^T\right)_1 = \left(A_{22}^T\right)_1 = \overline{E}_{\text{f}} \alpha_{\text{f}} \Delta T = 1715 \cdot 10^{-6} \Delta T \text{ GPa/°C}$$

$$\left(A_{11}^T\right)_2 = \left(A_{22}^T\right)_2 = \overline{E}_x^i \left(1 + \nu_{xy}^i\right) \alpha_x^i \Delta T = 32.08 \cdot 10^{-6} \Delta T \text{ GPa/°C}$$

$$\left(A_{11}^T\right)_3 = 2E_r \frac{\delta_r}{a_r} \alpha_r \cos^2 \phi_r \Delta T = 4.46 \cdot 10^{-6} \Delta T \text{ GPa/°C}$$

$$\left(A_{22}^T\right)_3 = 2E_r \frac{\delta_r}{a_r} \alpha_r \sin^2 \phi_r \Delta T = 1.06 \cdot 10^{-6} \Delta T \text{ GPa/°C}$$

$$\left(A_{11}^T\right)_4 = \left[\overline{E}_1^{\text{e}} (\alpha_1^{\text{e}} + \nu_{12}^{\text{e}} \alpha_2^{\text{e}}) \cos^2 \phi + \overline{E}_2^{\text{e}} (\alpha_2^{\text{e}} + \nu_{21}^{\text{e}} \alpha_1^{\text{e}}) \sin^2 \phi\right] \Delta T$$
$$= 132.43 \cdot 10^{-6} \Delta T \text{ GPa/°C}$$

$$\left(A_{22}^T\right)_4 = \left[\overline{E}_1^{\text{e}} (\alpha_1^{\text{e}} + \nu_{12}^{\text{e}} \alpha_2^{\text{e}}) \sin^2 \phi + \overline{E}_2^{\text{e}} (\alpha_2^{\text{e}} + \nu_{21}^{\text{e}} \alpha_1^{\text{e}}) \cos^2 \phi\right] \Delta T$$
$$= 317.61 \cdot 10^{-6} \Delta T \text{ GPa/°C}$$

Since the layers are orthotropic, $A_{12}^T = 0$ for all of them. Specifying the coordinates of the layers (see Fig. 5.10) i.e.,

$$t_0 = 0\,\text{mm}, \quad t_1 = 0.02\,\text{mm}, \quad t_2 = 1.02\,\text{mm}, \quad t_3 = 10.02\,\text{mm}, \quad t_4 = 13.52\,\text{mm}$$

and applying Eq. (7.27), we calculate the parameters $J_{mn}^{(r)}$ for the laminate

$$J_{11}^{(0)} = \left(A_{11}^T\right)_1 (t_1 - t_0) + \left(A_{11}^T\right)_2 (t_2 - t_1) + \left(A_{11}^T\right)_3 (t_3 - t_2)$$
$$+ \left(A_{11}^T\right)_4 (t_4 - t_3) = 570 \cdot 10^{-6} \Delta T \text{ GPa mm/°C}$$

$$J_{22}^{(0)} = 1190 \cdot 10^{-6} \Delta T \text{ GPa mm/°C}$$

$$J_{11}^{(1)} = \frac{1}{2}\left[\left(A_{11}^T\right)_1 \left(t_1^2 - t_0^2\right) + \left(A_{11}^T\right)_2 \left(t_2^2 - t_1^2\right)\right.$$
$$\left. + \left(A_{11}^T\right)_3 \left(t_3^2 - t_2^2\right) + \left(A_{11}^T\right)_4 \left(t_4^2 - t_3^2\right)\right] = 5690 \cdot 10^{-6} \Delta T \text{ GPa mm/°C}$$

$$J_{22}^{(1)} = 13150 \cdot 10^{-6} \Delta T \text{ GPa mm/°C}$$

To determine M_{mn}^T, we need to specify the reference surface of the laminate. Assume that this surface coincides with the middle surface, i.e., that e $= h/2 = 6.76\,\text{mm}$. Then, Eqs. (7.25) yield

$$N_{11}^T = J_{11}^{(0)} = 570 \cdot 10^{-6} \Delta T \text{ GPa mm/°C}$$

$$N_{22}^T = J_{22}^{(0)} = 1190 \cdot 10^{-6} \Delta T \text{ GPa mm/°C}$$

$$M_{11}^T = J_{11}^{(1)} - eJ_{11}^{(0)} = 1840 \cdot 10^{-6} \Delta T \text{ GPa mm/°C}$$

$$M_{22}^T = 5100 \cdot 10^{-6} \Delta T \text{ GPa mm/°C}$$

Thus, the thermal terms entering the constitutive equations of thermoplasticity, Eqs. (7.23), are specified. Using these results, we can determine the apparent coefficients of thermal expansion for the space telescope section under study (see Fig. 7.3). We can assume that, under uniform heating, the curvatures do not change in the middle part of the cylinder so that $\kappa_{xT} = 0$ and $\kappa_{yT} = 0$. Since there are no external loads, the free body diagram enables us to conclude that $N_x = 0$ and $N_y = 0$. As a result, the first two equations of Eqs. (7.23) for the structure under study become

$$B_{11}\varepsilon_{xT}^0 + B_{12}\varepsilon_{yT}^0 = N_{11}^T$$

$$B_{21}\varepsilon_{xT}^0 + B_{22}\varepsilon_{yT}^0 = N_{22}^T$$

Solving these equations for thermal strains and taking into account Eqs. (7.20), we get

$$\varepsilon_{xT}^0 = \frac{1}{B}\left(B_{22}N_{11}^T - B_{12}N_{22}^T\right) = \alpha_x \Delta T$$

$$\varepsilon_{yT}^0 = \frac{1}{B}\left(B_{11}N_{22}^T - B_{12}N_{11}^T\right) = \alpha_y \Delta T$$

where $B = B_{11}B_{22} - B_{12}^2$. For the laminate under study, calculation yields

$$\alpha_x = -0.94 \cdot 10^{-6}\ 1/°\text{C}, \quad \alpha_y = 14.7 \cdot 10^{-6}\ 1/°\text{C}$$

Return to Eqs. (7.13) and (7.20) based on the assumption that the coefficients of thermal expansion do not depend on temperature. For moderate temperatures, this is a reasonable approximation. This conclusion follows from Fig. 7.6, in which the experimental results of Sukhanov et al. (1990) (shown with solid lines) are compared with Eqs. (7.20), in which $\Delta T = T - 20°\text{C}$ (dashed lines) represent carbon–epoxy angle-ply laminates. However, for relatively high temperatures, some deviation from linear behavior can be observed. In this case, Eqs. (7.13) and (7.20) for thermal strains can be generalized as

$$\varepsilon^T = \int_{T_0}^{T} \alpha(T)\mathrm{d}T$$

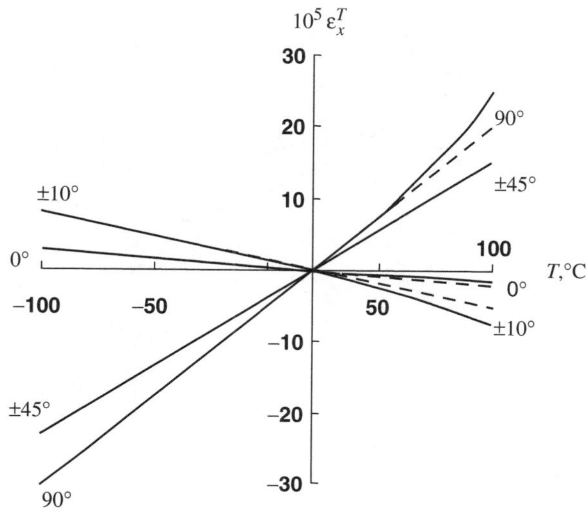

Fig. 7.6. Experimental dependencies of thermal strains on temperature (solid lines) for $\pm\phi$ angle-ply carbon–epoxy composite and the corresponding linear approximations (dashed lines).

Temperature variations can also result in a change in material mechanical properties. As follows from Fig. 7.7, in which the circles correspond to the experimental data of Ha and Springer (1987), elevated temperatures result in either higher or lower reduction of material strength and stiffness characteristics, depending on whether the corresponding material characteristic is controlled mainly by the fibers or by the matrix. The curves presented in Fig. 7.7 correspond to a carbon–epoxy composite, but they are typical for polymeric unidirectional composites. The longitudinal modulus and tensile strength, being controlled by the fibers, are less sensitive to temperature than longitudinal compressive strength, and transverse and shear characteristics. Analogous results for a more temperature-sensitive thermoplastic composite studied by Soutis and Turkmen (1993) are presented in Fig. 7.8. Metal matrix composites demonstrate much higher thermal resistance, whereas ceramic and carbon–carbon composites have been specially developed to withstand high temperatures. For example, carbon–carbon fabric composite under heating up to 2500°C demonstrates only a 7% reduction in tensile strength and about 30% reduction in compressive strength without significant change of stiffness.

Analysis of thermoelastic deformation for materials whose stiffness characteristics depend on temperature presents substantial difficulties because thermal strains are caused not only by material thermal expansion, but also by external forces. Consider, for example, a structural element under temperature T_0 loaded with some external force P_0, and assume that the temperature is increased to a value T_1. Then, the temperature change will cause a thermal strain associated with material expansion, and the force P_0, being constant, also induces additional strain because the material stiffness at temperature T_1 is less than its stiffness at temperature T_0. To determine the final stress and strain state of the structure,

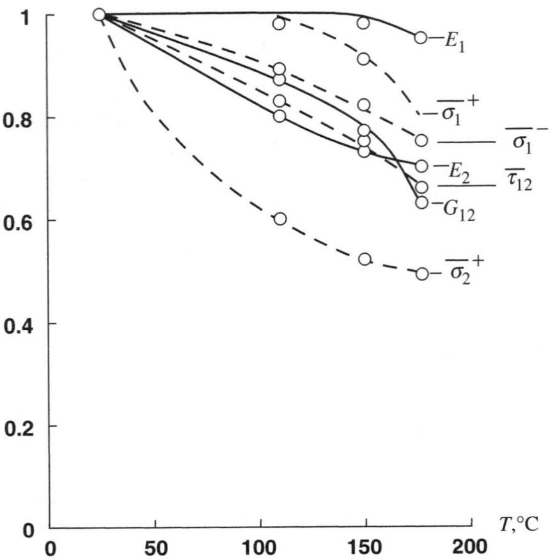

Fig. 7.7. Experimental dependencies of normalized stiffness (solid lines) and strength (dashed lines) characteristics of unidirectional carbon–epoxy composite on temperature.

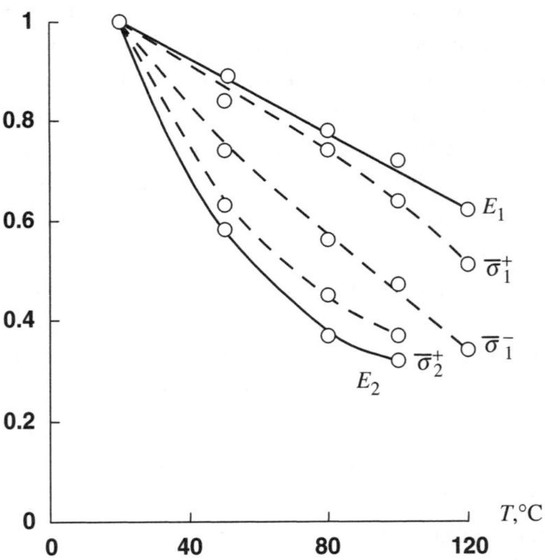

Fig. 7.8. Experimental dependencies of normalized stiffness (solid lines) and strength (dashed lines) characteristics of unidirectional glass–polypropylene composite on temperature.

we should describe the process of loading and heating using, e.g., the method of successive loading (and heating) presented in Section 4.1.2.

7.2. Hygrothermal effects and aging

Effects that are similar to temperature variations, i.e., expansion and degradation of properties, can also be caused by moisture. Moisture absorption is governed by Fick's law, which is analogous to Fourier's law, Eq. (7.1), for thermal conductivity, i.e.,

$$q_W = -D\frac{\partial W}{\partial n} \tag{7.31}$$

in which q_W is the diffusion flow through a unit area of surface with normal n, D is the diffusivity of the material whose moisture absorption is being considered, and W is the relative mass moisture concentration in the material, i.e.,

$$W = \frac{\Delta m}{m} \tag{7.32}$$

where Δm is the increase in the mass of a unit volume material element due to moisture absorption and m is the mass of the dry material element. Moisture distribution in the material is governed by the following equation, similar to Eq. (7.2) for thermal conductivity

$$\frac{\partial}{\partial n}\left(D\frac{\partial W}{\partial n}\right) = \frac{\partial W}{\partial t} \tag{7.33}$$

Consider a laminated composite material shown in Fig. 7.9 for which n coincides with the z axis. Despite the formal correspondence between Eq. (7.2) for thermal conductivity and Eq. (7.32) for moisture diffusion, there is a difference in principle between these problems. This difference is associated with the diffusivity coefficient D, which is much lower than

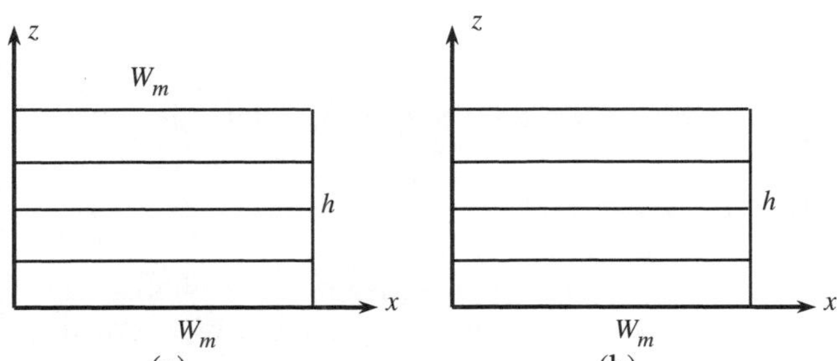

Fig. 7.9. Composite material exposed to moisture on both surfaces $z = 0$ and $z = h$ (a), and on the surface $z = 0$ only (b).

the thermal conductivity λ of the same material. As is known, there are materials, e.g., metals, with relatively high λ and practically zero D coefficients. Low D-value means that moisture diffusion is a rather slow process. As shown by Shen and Springer (1976), the temperature increase in time inside a surface-heated composite material reaches a steady (equilibrium) state temperature about 10^6 times faster than the moisture content approaching the corresponding stable state. This means that, in contrast to Section 7.1.1 in which the steady (time-independent) temperature distribution is studied, we must consider the time-dependent process of moisture diffusion. To simplify the problem, we can neglect the possible variation of the mass diffusion coefficient D over the laminate thickness, taking $D =$ constant for polymeric composites. Then, Eq. (7.33) reduces to

$$D\frac{\partial^2 W}{\partial z^2} = \frac{\partial W}{\partial t} \tag{7.34}$$

Consider the laminate in Fig. 7.9a. Introduce the maximum moisture content W_m that can exist in the material under the preassigned environmental conditions. Naturally, W_m depends on the material nature and structure, temperature, relative humidity (RH) of the gas (e.g., humid air), or on the nature of the liquid (distilled water, salted water, fuel, lubricating oil, etc.) to the action of which the material is exposed. Introduce also the normalized moisture concentration as

$$w(z, t) = \frac{W(z, t)}{W_m} \tag{7.35}$$

Obviously, for $t \to \infty$, we have $w \to 1$. Then, the function $w(z, t)$ can be presented in the form

$$w(z,\ t) = 1 - \sum_{n=1}^{\infty} w_n(z)\mathrm{e}^{-k_n t} \tag{7.36}$$

Substitution into Eq. (7.34), with due regard to Eq. (7.35), yields the following ordinary differential equation

$$w_n'' + r_n^2 w_n = 0$$

in which $r_n^2 = k_n/D$ and $(\ )' = \mathrm{d}(\ )/\mathrm{d}z$. The general solution is

$$w_n = C_{1n}\sin r_n z + C_{2n}\cos r_n z$$

The integration constants can be found from the boundary conditions on the surfaces $z = 0$ and $z = h$ (see Fig. 7.9a). Assume that on these surfaces $W = W_m$ or $w = 1$. Then, in accordance with Eq. (7.36), we get

$$w_n(0,\ t) = 0, \quad w_n(h,\ t) = 0 \tag{7.37}$$

The first of these conditions yields $C_{2n} = 0$, whereas from the second condition we have $\sin r_n h = 0$, which yields

$$r_n h = (2n - 1)\pi \quad (n = 1, 2, 3, \ldots) \tag{7.38}$$

Thus, the solution in Eq. (7.36) takes the form

$$w(z, t) = 1 - \sum_{n=1}^{\infty} C_{1n} \sin\left(\frac{2n-1}{h}\pi z\right) \exp\left[-\left(\frac{2n-1}{h}\right)^2 \pi^2 \mathrm{Dt}\right] \tag{7.39}$$

To determine C_{1n}, we must use the initial condition, according to which

$$w(0 < z < h, \ t = 0) = 0$$

Using the following Fourier series

$$1 = \frac{4}{\pi} \sum_{n=1}^{\infty} \frac{\sin(2n-1)z}{2n-1}$$

we get $C_{1n} = 4/(2n - 1)\pi$, and the solution in Eq. (7.39) can be written in its final form

$$w(z, t) = 1 - \frac{4}{\pi} \sum_{n=1}^{\infty} \frac{\sin(2n-1)\pi\bar{z}}{2n-1} \exp\left[-\left(\frac{2n-1}{h}\right)^2 \pi^2 \mathrm{Dt}\right] \tag{7.40}$$

where $\bar{z} = z/h$.

For the structure in Fig. 7.9b, the surface $z = h$ is not exposed to moisture, and hence $q_W(z = h) = 0$. So, in accordance with Eq. (7.31), the second boundary condition in Eqs. (7.37) must be changed to $w'(h, t) = 0$. Then, instead of Eq. (7.38), we must use

$$r_n h = \frac{\pi}{2}(2n - 1)$$

Comparing this result with Eq. (7.38), we can conclude that for the laminate in Fig. 7.9b, $w(z, t)$ is specified by the solution in Eq. (7.40) in which we must change h to $2h$.

The mass increase of the material with thickness h is

$$\Delta M = A \int_0^h \Delta m \mathrm{d}z$$

where A is the surface area. Using Eqs. (7.32) and (7.35), we get

$$\Delta M = A m W_m \int_0^h w \mathrm{d}z$$

Switching to a dimensionless variable $\bar{z} = z/h$ and taking the total moisture content as

$$C = \frac{\Delta M}{Amh} \tag{7.41}$$

we arrive at

$$C = W_m \int_0^1 w \, d\bar{z}$$

where w is specified by Eq. (7.40). Substitution of this equation and integration yields

$$\bar{C} = \frac{C}{W_m} = 1 - \frac{8}{\pi^2} \sum_{n=1}^{\infty} \frac{1}{(2n-1)^2} \exp\left[-\left(\frac{2n-1}{h}\right)^2 \pi^2 Dt\right] \tag{7.42}$$

For numerical analysis, consider a carbon–epoxy laminate for which $D = 10^{-3}$ mm^2/hour (Tsai, 1987) and $h = 1$ mm. The distributions of the moisture concentration over the laminate thickness are shown in Fig. 7.10 for $t = 1, 10, 50, 100, 200$, and 500 h. As can be seen, complete impregnation of 1-mm-thick material takes about 500 h. The dependence of $\bar{C}$ on t found in accordance with Eq. (7.42) is presented in Fig. 7.11.

An interesting interpretation of the curve in Fig. 7.11 can be noted if we change the variable t to $\sqrt{t}$. The resulting dependence is shown in Fig. 7.12. As can be seen, the initial

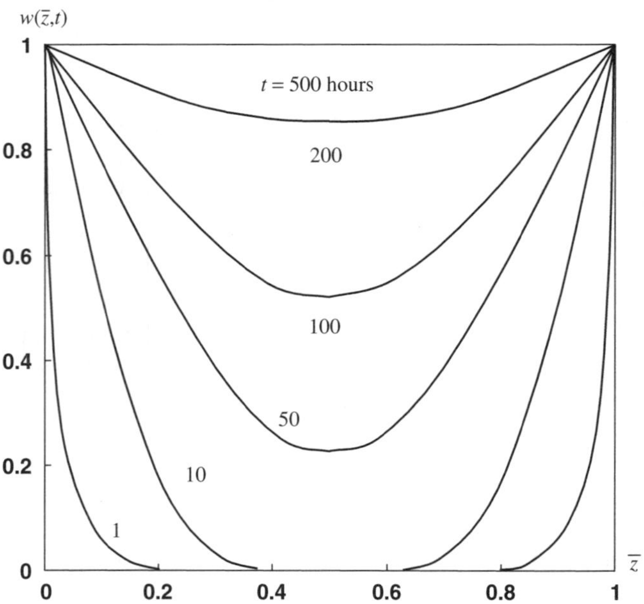

Fig. 7.10. Distribution of the normalized moisture concentration w over the thickness of 1-mm-thick carbon–epoxy composite for various exposure times t.

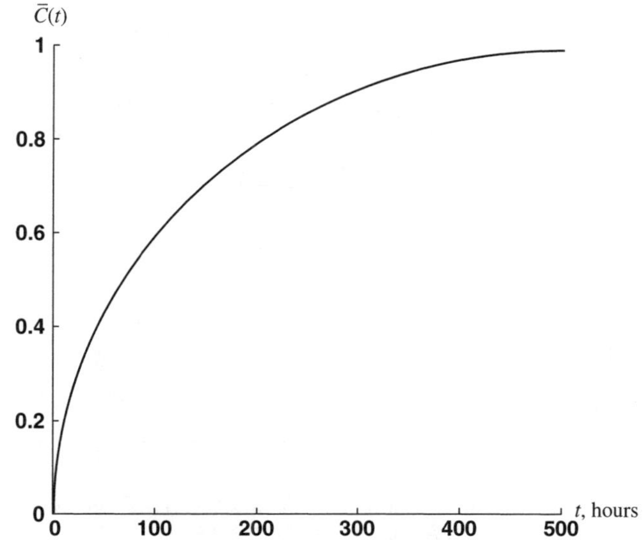

Fig. 7.11. Dependence of the normalized moisture concentration $\overline{C}$ on time t.

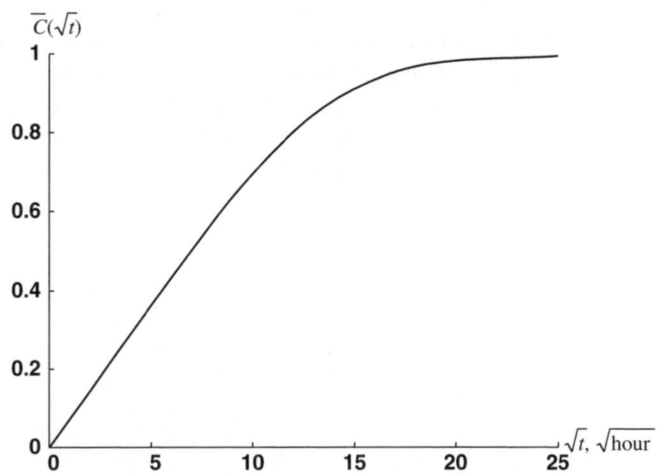

Fig. 7.12. Dependence of the normalized moisture concentration on $\sqrt{t}$.

part of the curve is close to a straight line whose slope can be used to determine the diffusion coefficient of the material matching the theoretical dependence $C(t)$ with the experimental one. Note that experimental methods usually result in rather approximate evaluation of the material diffusivity D with possible variations up to 100% (Tsai, 1987). The maximum value of the function $C(t)$ to which it tends to approach determines the maximum moisture content $C_m = W_m$.

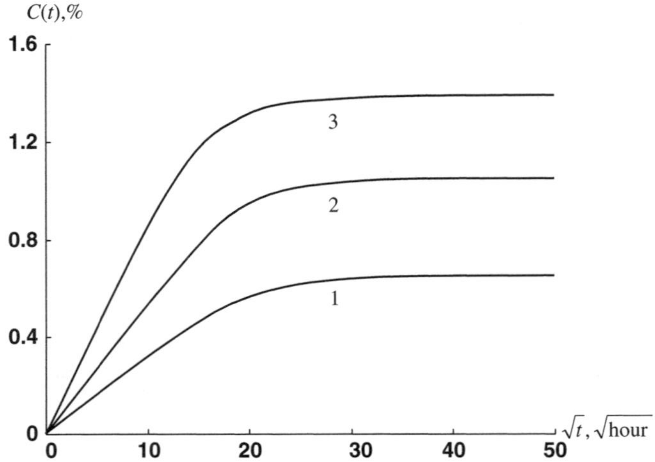

Fig. 7.13. Dependence of the moisture content on time for a carbon–epoxy composite exposed to air with 45% *RH* (1), 75% *RH* (2), 95% *RH* (3).

Thus, the material behavior under the action of moisture is specified by two experimental parameters – D and C_m – which can depend on the ambient media, its moisture content, and temperature. The experimental dependencies of C in Eq. (7.41) on t for 0.6-mm-thick carbon–epoxy composite exposed to humid air with various relative humidity (RH) levels are shown in Fig. 7.13 (Survey, 1984). As can be seen, the moisture content is approximately proportional to the air humidity. The gradients of the curves in Fig. 7.13 depend on the laminate thickness (Fig. 7.14, Survey, 1984).

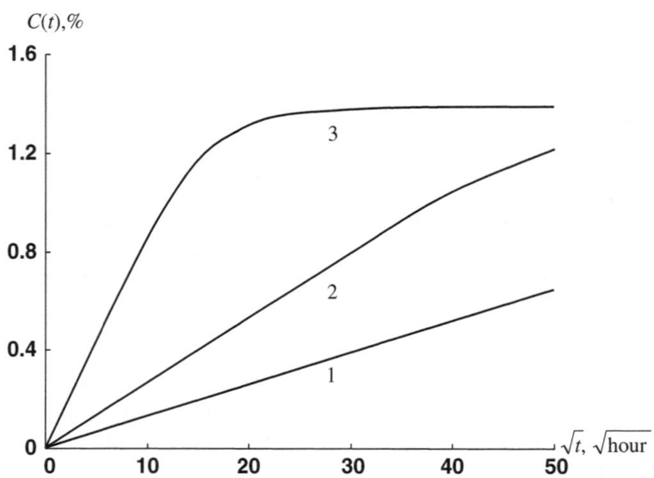

Fig. 7.14. Dependencies of the moisture content on time for a carbon–epoxy composite with thickness 3.6 mm (1), 1.2 mm (2), and 0.6 mm (3) exposed to humid air with 75% *RH*.

Among polymeric composites, the highest capacity for moisture absorption under room temperature is demonstrated by aramid composites ($7 \pm 0.25\%$ by weight) in which both the polymeric matrix and fibers are susceptible to moisture. Glass and carbon polymeric composites are characterized with moisture content $3.5 \pm 0.2\%$ and $2 \pm 0.75\%$, respectively. In real aramid–epoxy and carbon–epoxy composite structures, the moisture content is usually about 2% and 1%, respectively. The lowest susceptibility to moisture is demonstrated by boron composites. Metal matrix, ceramic, and carbon–carbon composites are not affected by moisture.

The material diffusivity coefficient D depends on temperature in accordance with the Arrhenius relationship (Tsai, 1987)

$$D(T_a) = \frac{D_0}{e^{k/T_a}}$$

in which D_0 and k are some material constants and T_a is the absolute temperature. Experimental dependencies of the moisture content on time in a 1.2-mm-thick carbon–epoxy composite exposed to humid air with 95% *RH* at various temperatures are presented in Fig. 7.15 (Survey, 1984). The most pronounced effect of temperature is observed for aramid–epoxy composites. The corresponding experimental results of Milyutin et al. (1989) are shown in Fig. 7.16.

When a material absorbs moisture, it expands, demonstrating effects that are similar to thermal effects, which can be modeled using the equations presented in Section 7.1.2, if we treat α_1, α_2 and α_x, α_y as coefficients of moisture expansion and change ΔT for C. Similar to temperature, increase in moisture reduces material strength and stiffness. For carbon–epoxy composites, this reduction is about 12%, for aramid–epoxy composites, about 25%, and for glass–epoxy materials, about 35%. After drying out, the effect of moisture usually disappears.

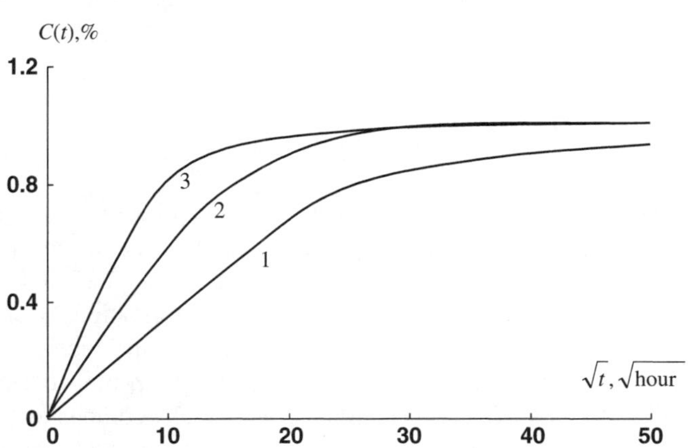

Fig. 7.15. Dependencies of the moisture content on time for 1.2-mm-thick carbon–epoxy composite exposed to humid air with 95% *RH* under temperatures 25°C (1), 50°C (2), and 80°C (3).

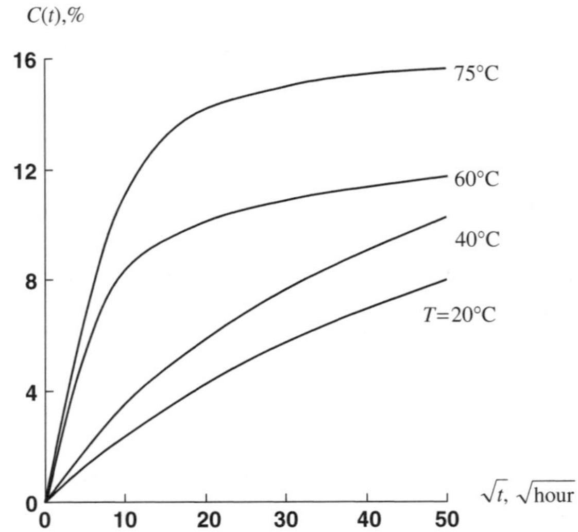

Fig. 7.16. Moisture content as a function of time and temperature for aramid–epoxy composites.

The cyclic action of temperature, moisture, or sun radiation results in material aging, i.e., in degradation of the material properties during the process of material or structure storage. For some polymeric composites, exposure to elevated temperature, which can reach 70°C, and radiation, whose intensity can be as high as 1 kW/m^2, can cause more complete curing of the resin and some increase of material strength in compression, shear, or bending. However, under long-term action of the aforementioned factors, the material strength and stiffness decrease. To evaluate the effect of aging, testing under transverse bending (see Fig. 4.98) is usually performed. The flexural strength obtained

$$\overline{\sigma}_{\mathrm{f}} = \frac{3\overline{P}l}{2bh^2}$$

allows for both fiber and matrix material degradation in the process of aging. Experimental results from G.M. Gunyaev et al. showing the dependence of the normalized flexural strength on time for advanced composites are presented in Fig. 7.17. The most dramatic is the effect of aging on the ultimate transverse tensile deformation $\overline{\varepsilon}_2$ of unidirectional composites: the low value of which results in cracking of the matrix as discussed in Sections 4.4.2 and 6.4. After accelerated aging, i.e., long-term moisture conditioning at temperature 70°C, a 0.75% moisture content in carbon–epoxy composites results in about 20% reduction of $\overline{\varepsilon}_2$, whereas a 1.15% moisture content causes about 45% reduction.

Environmental effects on composite materials are discussed in detail elsewhere (Tsai, 1987; Springer, 1981, 1984, 1988).

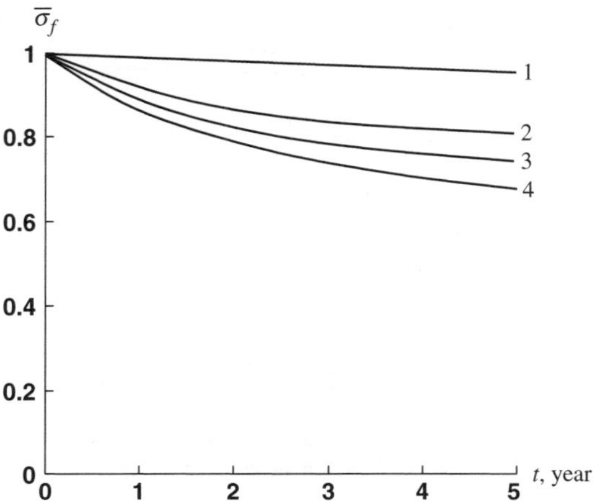

Fig. 7.17. Dependence of the normalized flexural strength on the time of aging for boron (1), carbon (2), aramid (3), and glass (4) epoxy composites.

7.3. Time and time-dependent loading effects

7.3.1. Viscoelastisity

Polymeric matrices are characterized with pronounced viscoelastic properties resulting in time-dependent behavior of polymeric composites that manifests itself in creep (see Section 1.1), stress relaxation, and dependence of the stress–strain diagram on the rate of loading. It should be emphasized that in composite materials, viscoelastic deformation of the polymeric matrix is restricted by the fibers that are usually linear elastic and do not demonstrate time-dependent behavior. The one exception to existing fibers is represented by aramid fibers that are actually polymeric themselves by their nature. The properties of metal matrix, ceramic, and carbon–carbon composites under normal conditions do not depend on time. Rheological (time-dependent) characteristics of structural materials are revealed in creep tests allowing us to plot the dependence of strain on time under constant stress. Such diagrams are shown in Fig. 7.18 for the aramid–epoxy composite described by Skudra et al. (1989). An important characteristic of the material can be established if we plot the so-called isochrone stress–strain diagrams shown in Fig. 7.19. Three curves in this figure are plotted for $t = 0$, $t = 100$, and $t = 1000$ days, and the points on these curves correspond to points 1, 2, 3 in Fig. 7.18. As can be seen, the initial parts of the isochrone diagrams are linear, which means that under moderate stress, the material under study can be classified as a linear-viscoelastic material. To characterize such a material, we need to have only one creep diagram, whereby the other curves can be plotted, increasing strains in proportion to stress. For example, the creep curve corresponding to $\sigma_1 = 450\,\text{MPa}$ in Fig. 7.18 can be obtained if we multiply strains corresponding to $\sigma_1 = 300\,\text{MPa}$ by 1.5.

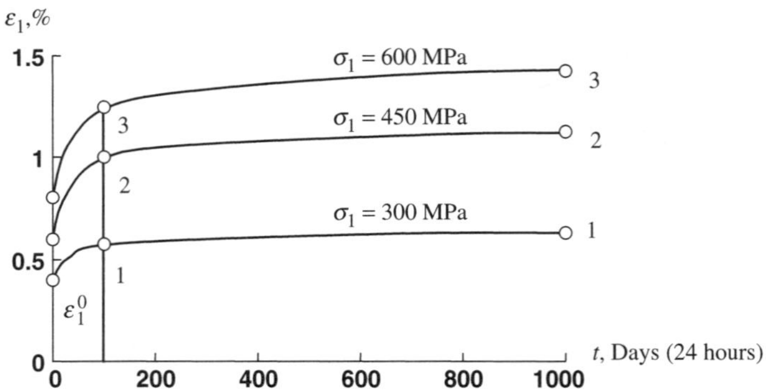

Fig. 7.18. Creep strain response of unidirectional aramid–epoxy composite under tension in longitudinal direction with three constant stresses.

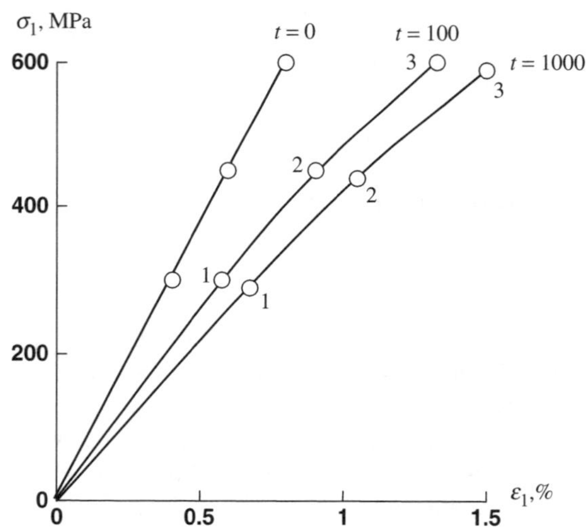

Fig. 7.19. Isochrone stress–strain diagrams corresponding to creep curves in Fig. 7.18.

Linear-viscoelastic material behavior is described with reasonable accuracy by the hereditary theory, according to which the dependence of strain on time is expressed as

$$\varepsilon(t) = \frac{1}{E}\left[\sigma(t) + \int_0^t C(t-\tau)\sigma(\tau)\mathrm{d}\tau\right] \tag{7.43}$$

Here, t is the current time, τ is some moment of time in the past ($0 \le \tau \le t$) at which stress $\sigma(\tau)$ acts, and $C(t-\tau)$ is the creep compliance (or creep kernel) depending on time passing from the moment τ to the moment t. The constitutive equation of hereditary theory, Eq. (7.43), is illustrated in Fig. 7.20. As can be seen, the total strain $\varepsilon(t)$ is composed of the elastic strain ε_e governed by the current stress $\sigma(t)$ and the viscous

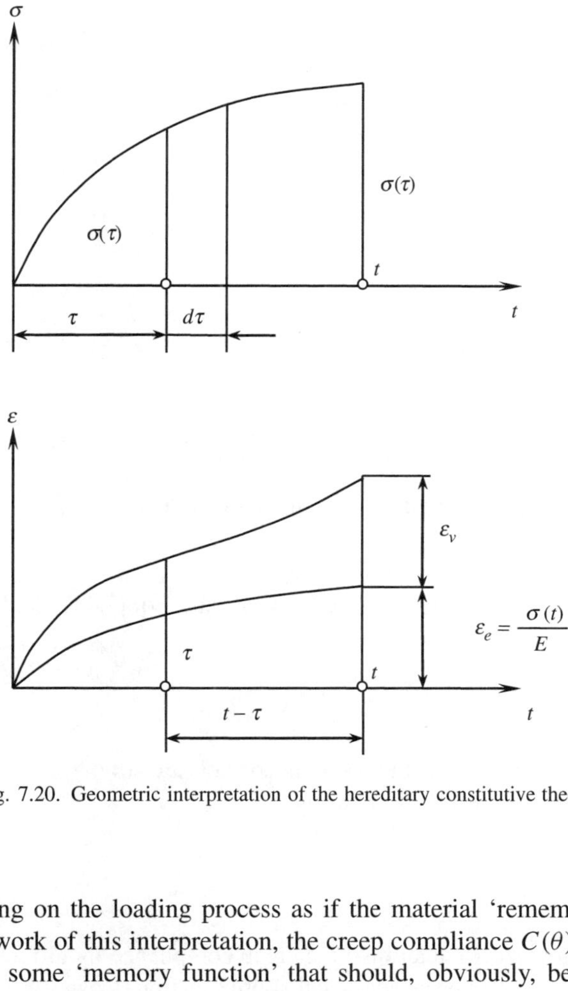

Fig. 7.20. Geometric interpretation of the hereditary constitutive theory.

strain ε_v depending on the loading process as if the material 'remembers' this process. Within the framework of this interpretation, the creep compliance $C(\theta)$, where $\theta = t - \tau$ can be treated as some 'memory function' that should, obviously, be infinitely high at $\theta = 0$ and tend to zero for $\theta \to \infty$, as in Fig. 7.21.

The inverse form of Eq. (7.43) is

$$\sigma(t) = E\left[\varepsilon(t) - \int_0^t R(t-\tau)\varepsilon(\tau)\mathrm{d}\tau\right] \tag{7.44}$$

Here, $R(t-\tau)$ is the relaxation modulus or the relaxation kernel that can be expressed, as shown below, in terms of $C(t-\tau)$.

The creep compliance is determined using experimental creep diagrams. Transforming to a new variable $\theta = t - \tau$, we can write Eq. (7.43) in the following form

$$\varepsilon(t) = \frac{1}{E}\left[\sigma(t) + \int_0^t C(\theta)\sigma(\theta - t)\mathrm{d}\theta\right] \tag{7.45}$$

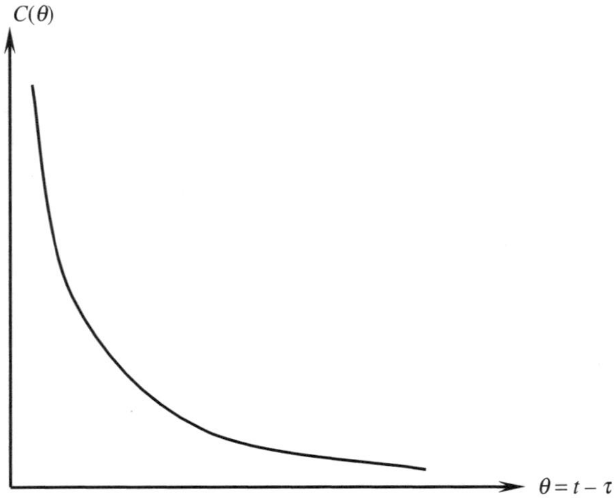

Fig. 7.21. Typical form of the creep compliance function.

For a creep test, the stress is constant, so $\sigma = \sigma^0$, and Eq. (7.44) yields

$$\varepsilon(t) = \varepsilon^0 \left[1 + \int_0^t C(\theta) \mathrm{d}\theta \right] \tag{7.46}$$

where $\varepsilon^0 = \sigma^0/E = \varepsilon(t = 0)$ is the instantaneous elastic strain (see Fig. 7.18). Differentiating this equation with respect to t, we get

$$C(t) = \frac{1}{\varepsilon^0} \frac{\mathrm{d}\varepsilon(t)}{\mathrm{d}t} x$$

This expression allows us to determine the creep compliance by differentiating the given experimental creep diagram or its analytical approximation. However, for practical analysis, $C(\theta)$ is usually determined directly from Eq. (7.46) introducing some approximation for $C(\theta)$ and matching the function obtained $\varepsilon(t)$ with the experimental creep diagram. For this purpose, Eq. (7.46) is written in the form

$$\frac{\varepsilon(t)}{\varepsilon^0} = 1 + \int_0^t C(\theta) \mathrm{d}\theta \tag{7.47}$$

Experimental creep diagrams for unidirectional glass–epoxy composite are presented in this form in Fig. 7.22 (solid lines).

The simplest form is an exponential approximation of the type

$$C(\theta) = \sum_{n=1}^{N} A_n \mathrm{e}^{-\alpha_n \theta} \tag{7.48}$$

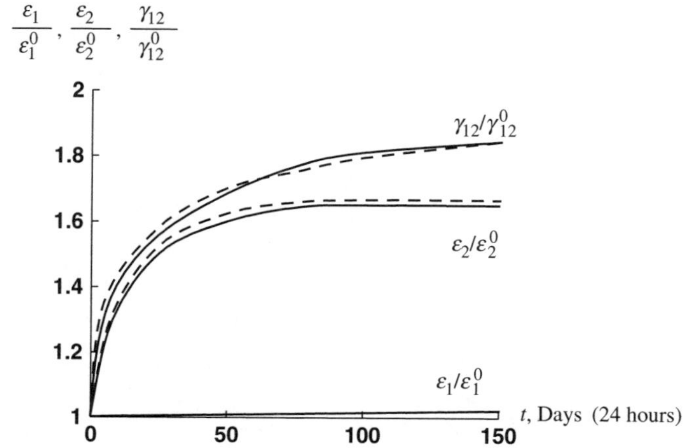

Fig. 7.22. Creep strain diagrams for unidirectional glass–epoxy composite (solid lines) under tension in longitudinal direction ($\varepsilon_1/\varepsilon_1^0$), transverse direction ($\varepsilon_2/\varepsilon_2^0$), and under in-plane shear ($\gamma_{12}/\gamma_{12}^0$) and the corresponding exponential approximations (dashed lines).

Substituting Eq. (7.48) into Eq. (7.47), we obtain

$$\frac{\varepsilon(t)}{\varepsilon^0} = 1 + \sum_{n=1}^{N} \frac{A_n}{\alpha_n}(1 - \mathrm{e}^{-\alpha_n t})$$

For the curves presented in Fig. 7.22, calculation yields

- longitudinal tension: $N = 1$, $A_1 = 0$;
- transverse tension: $N = 1$, $A_1 = 0.04$, $\alpha_1 = 0.06$ 1/day;
- in-plane shear: $N = 2$, $A_1 = 0.033$, $\alpha_1 = 0.04$ 1/day, $A_2 = 0.06$, $\alpha_2 = 0.4$ 1/day.

The corresponding approximations are shown in Fig. 7.22 with dashed lines. The main shortcoming of the exponential approximation in Eq. (7.48) is associated with the fact that, in contrast to Fig. 7.21, $C(\theta)$ has no singularity at $\theta = 0$, which means that it cannot properly describe material behavior in the vicinity of $t = 0$.

It should be emphasized that the one-term exponential approximation corresponds to a simple rheological mechanical model shown in Fig. 7.23. The model consists of two linear springs simulating material elastic behavior in accordance with Hooke's law

$$\sigma_1 = E_1\varepsilon_1, \quad \sigma_2 = E_2\varepsilon_2 \tag{7.49}$$

and one dash-pot simulating material viscous behavior obeying the Newton flow law

$$\sigma_v = \eta\frac{\mathrm{d}\varepsilon_v}{\mathrm{d}t} \tag{7.50}$$

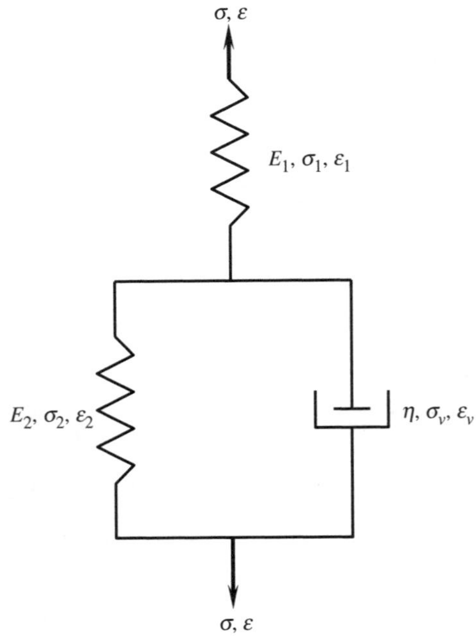

Fig. 7.23. Three-element mechanical model.

Equilibrium and compatibility conditions for the model in Fig. 7.23 are

$$\sigma = \sigma_2 + \sigma_v, \quad \sigma_1 = \sigma$$

$$\varepsilon_v = \varepsilon_2, \quad \varepsilon_1 + \varepsilon_2 = \varepsilon$$

Using the first of these equations and Eqs. (7.49)–(7.50), we get

$$\sigma = E_2\varepsilon_2 + \eta\frac{d\varepsilon_v}{dt}$$

Taking into account that

$$\varepsilon_2 = \varepsilon_v = \varepsilon - \frac{\sigma}{E}$$

we finally arrive at the following constitutive equation relating the apparent stress σ to the apparent strain ε

$$\sigma\left(1 + \frac{E_2}{E_1}\right) + \frac{\eta}{E_1}\frac{d\sigma}{dt} = E_2\varepsilon + \eta\frac{d\varepsilon}{dt} \tag{7.51}$$

This equation allows us to introduce some useful material characteristics. Indeed, consider a very fast loading, i.e., such that stress σ and strain ε can be neglected in comparison with their rates. Then, integration yields $\sigma = E_i\varepsilon$, where $E_i = E_1$ is the instantaneous modulus of the material. Now assume that the loading is so slow that stress and strain rates can be neglected. Then, Eq. (7.51) yields $\sigma = E_l\varepsilon$, where

$$E_l = \frac{E_1 E_2}{E_1 + E_2} \tag{7.52}$$

is the long-time modulus.

We can now apply the model under study to describe material creep. Taking $\sigma = \sigma_0$ and integrating Eq. (7.51) with initial condition $\varepsilon_0(0) = \sigma_0/E$, we get

$$\varepsilon = \frac{\sigma_0}{E_1}\left[1 + \frac{E_1}{E_2}\left(1 - \mathrm{e}^{-\frac{E_2}{\eta}t}\right)\right]$$

The corresponding creep diagram is shown in Fig. 7.24. As follows from this figure, $\varepsilon(t \to \infty) = \sigma_0/E_l$, where E_l is specified by Eq. (7.52). This means that there exists some limit for the creep strain, and materials that can be described with this model should possess the so-called limited creep.

Now assume that the model is loaded in such a way that the apparent strain is constant, i.e., that $\varepsilon = \varepsilon_0$. Then, the solution of Eq. (7.51) that satisfies the condition $\sigma(0) = E_1\varepsilon_0$ is

$$\sigma = \frac{E_1\varepsilon_0}{E_1 + E_2}\left(E_2 + E_1\mathrm{e}^{-t/t_r}\right), \quad t_r = \frac{\eta}{E_1 + E_2}$$

The corresponding dependence is presented in Fig. 7.25 and illustrates the process of stress relaxation. The parameter t_r is called the time of relaxation. During this time, the stress decreases by the factor of e.

Consider again Eq. (7.51) and express E_1, E_2, and η in terms of E_i, E_l, and t_r. The resulting equation is as follows

$$\sigma + t_r\frac{\mathrm{d}\sigma}{\mathrm{d}t} = E_l\varepsilon + E_i t_r\frac{\mathrm{d}\varepsilon}{\mathrm{d}t} \tag{7.53}$$

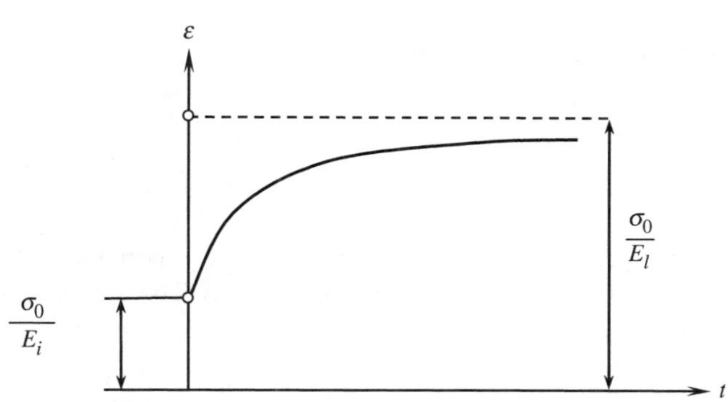

Fig. 7.24. Creep diagram corresponding to the mechanical model in Fig. 7.23.

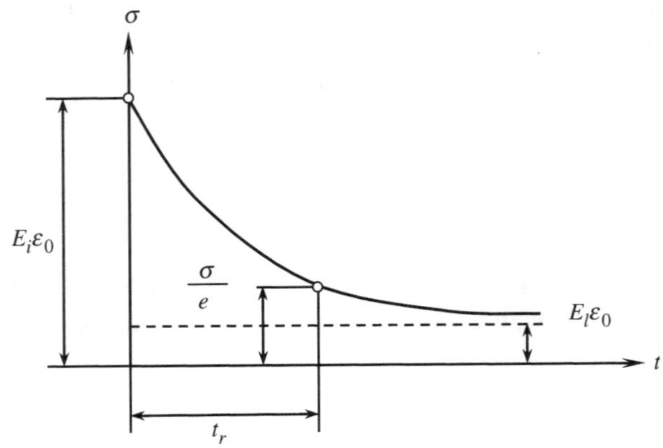

Fig. 7.25. Relaxation diagram corresponding to the mechanical model in Fig. 7.23.

This first-order differential equation can be solved for ε in the general case. Omitting rather cumbersome transformations, we arrive at the following solution

$$\varepsilon(t) = \frac{1}{E_i}\left[\sigma(t) + \frac{1}{t_r}\left(1 - \frac{E_l}{E_i}\right)\int_0^t \mathrm{e}^{-\frac{E_l}{E_i t_r}(t-\tau)}\sigma(\tau)\mathrm{d}\tau\right]$$

This result corresponds to Eq. (7.45) of the hereditary theory with one-term exponential approximation of the creep compliance in Eq. (7.48), in which $N = 1$. Taking more terms in Eq. (7.48), we get more flexibility in the approximation of experimental results with exponential functions. However, the main features of material behavior are, in principle, the same as that for the one-term approximation (see Figs. 7.23 and 7.24). In particular, there exists the long-time modulus that follows from Eq. (7.46) if we examine the limit for $t \to \infty$, i.e.,

$$\varepsilon(t) \to \frac{\sigma_0}{E_l}, \quad E_l = \frac{E}{1 + \int_0^\infty C(\theta)\mathrm{d}\theta}$$

For the exponential approximation in Eq. (7.48),

$$I = \int_0^\infty C(\theta)\mathrm{d}\theta = \sum_{n=1}^{N} \frac{A_n}{\alpha_n}$$

Since the integral I has a finite value, the exponential approximation of the creep compliance can be used only for materials with limited creep. There exist more complicated singular approximations, e.g.,

$$C(\theta) = \frac{A}{\theta^\alpha}, \quad C(\theta) = \frac{A}{\theta^\alpha}\mathrm{e}^{-\beta\theta}$$

for which $I \rightarrow \infty$ and $E_l = 0$. This means that for such materials, the creep strain can be infinitely high.

A useful interpretation of the hereditary theory constitutive equations can be constructed with the aid of the integral Laplace transformation, according to which a function $f(t)$ is associated with its Laplace transform $f^*(p)$ as

$$f^*(p) = \int_0^{\infty} f(t)\mathrm{e}^{-pt}\mathrm{d}t$$

For some functions that we need to use for the examples presented below, we have

$$\begin{aligned} &f(t) = 1, \quad f^*(p) = \frac{1}{p} \\ &f(t) = \mathrm{e}^{-\alpha t}, \quad f^*(p) = \frac{1}{\alpha + p} \end{aligned} \tag{7.54}$$

The importance of the Laplace transformation for the hereditary theory is associated with the existence of the so-called convolution theorem, according to which

$$\left[\int_0^t f_1(\theta) f_2(\theta - t)\mathrm{d}\theta\right]^* = f_1^*(p) f_2^*(p)$$

Using this theorem and applying Laplace transformation to Eq. (7.45), we get

$$\varepsilon^*(p) = \frac{1}{E}\left[\sigma^*(p) + C^*(p)\sigma^*(p)\right]$$

This result can be presented in a form similar to Hookes's law, i.e.,

$$\sigma^*(p) = E^*(p)\varepsilon^*(p) \tag{7.55}$$

where

$$E^* = \frac{E}{1 + C^*(p)}$$

Applying Laplace transformation to Eq. (7.44), we arrive at Eq. (7.55) in which

$$E^* = E[1 - R^*(p)] \tag{7.56}$$

Comparing Eqs. (7.55) and (7.56), we can relate Laplace transforms of the creep compliance to the relaxation modulus, i.e.,

$$\frac{1}{1 + C^*(p)} = 1 - R^*(p)$$

With due regard to Eq. (7.55), we can formulate the elastic–viscoelastic analogy or the correspondence principle, according to which the solution of the linear viscoelasticity problem can be obtained in terms of the corresponding Laplace transforms from the solution of the linear elasticity problem if E is replaced with E^* and all the stresses, strains, displacements, and external loads are replaced with their Laplace transforms.

For an orthotropic material in a plane stress state, e.g., for a unidirectional composite ply or layer referred to the principal material axes, Eqs. (4.55) and (7.43) can be generalized as

$$\begin{aligned}
\varepsilon_1(t) &= \frac{1}{E_1}\left[\sigma_1(t) + \int_0^t C_{11}(t-\tau)\sigma_1(\tau)\mathrm{d}\tau\right] \\
&\quad - \frac{\nu_{12}}{E_2}\left[\sigma_2(t) + \int_0^t C_{12}(t-\tau)\sigma_2(\tau)\mathrm{d}\tau\right] \\
\varepsilon_2(t) &= \frac{1}{E_2}\left[\sigma_2(t) + \int_0^t C_{22}(t-\tau)\sigma_2(\tau)\mathrm{d}\tau\right] \\
&\quad - \frac{\nu_{21}}{E_1}\left[\sigma_1(t) + \int_0^t C_{21}(t-\tau)\sigma_1(\tau)\mathrm{d}\tau\right] \\
\gamma_{12}(t) &= \frac{1}{G_{12}}\left[\tau_{12}(t) + \int_0^t K_{12}(t-\tau)\tau_{12}(\tau)\mathrm{d}\tau\right]
\end{aligned}$$

Applying Laplace transformation to these equations, we can reduce them to a form similar to Hooke's law, Eqs. (4.55), i.e.,

$$\begin{aligned}
\varepsilon_1^*(p) &= \frac{\sigma_1^*(p)}{E_1^*(p)} - \frac{\nu_{12}^*(p)}{E_2^*(p)}\sigma_2^*(p) \\
\varepsilon_2^*(p) &= \frac{\sigma_2^*(p)}{E_2^*(p)} - \frac{\nu_{21}^*(p)}{E_1^*(p)}\sigma_1^*(p) \\
\gamma_{12}^*(p) &= \frac{\tau_{12}^*(p)}{G_{12}^*(p)}
\end{aligned} \tag{7.57}$$

where

$$\begin{aligned}
E_1^*(p) &= \frac{E_1}{1+C_{11}^*(p)}, \quad E_2^*(p) = \frac{E_2}{1+C_{22}^*(p)}, \quad G_{12}^*(p) = \frac{G_{12}}{1+K_{12}^*(p)} \\
\nu_{12}^*(p) &= \frac{1+C_{12}^*(p)}{1+C_{22}^*(p)}\nu_{12}, \quad \nu_{21}^*(p) = \frac{1+C_{21}^*(p)}{1+C_{11}^*(p)}\nu_{21}
\end{aligned} \tag{7.58}$$

For the unidirectional composite ply whose typical creep diagrams are shown in Fig. 7.22, the foregoing equations can be simplified by neglecting material creep in the longitudinal direction ($C_{11} = 0$) and assuming that Poisson's effect is linear elastic and symmetric, i.e., that

$$\frac{\nu_{12}^*}{E_2^*} = \frac{\nu_{12}}{E_2}, \quad \frac{\nu_{21}^*}{E_1^*} = \frac{\nu_{21}}{E_1}$$

Then, Eqs. (7.57) take the form

$$\begin{aligned} \varepsilon_1^*(p) &= \frac{\sigma_1^*(p)}{E_1} - \frac{\nu_{12}}{E_2}\sigma_2^*(p) \\ \varepsilon_2^*(p) &= \frac{\sigma_2^*(p)}{E_2^*} - \frac{\nu_{21}}{E_1}\sigma_1^*(p) \\ \gamma_{12}^*(p) &= \frac{\tau_{12}^*(p)}{G_{12}^*(p)} \end{aligned} \tag{7.59}$$

Supplementing constitutive equations, Eqs. (7.57) or (7.59), with strain-displacement and equilibrium equations written in terms of Laplace transforms of stresses, strains, displacements, and external loads and solving the problem of elasticity, we can find Laplace transforms for all the variables. To represent the solution obtained in this way in terms of time t, we need to take the inverse Laplace transformation, and this is the most difficult stage of the problem solution. There exist exact and approximate analytical and numerical methods for performing the inverse Laplace transformation discussed, for example, by Schapery (1974). The most commonly used approach is based on approximation of the solution written in terms of the transformation parameter p with some functions for which the inverse Laplace transformation is known.

As an example, consider the problem of torsion for an orthotropic cylindrical shell similar to that shown in Fig. 6.20. The shear strain induced by torque T is specified by Eq. (5.163). Using the elastic–viscoelastic analogy, we can write the corresponding equation for the creep problem as

$$\gamma_{xy}^*(p) = \frac{T^*(p)}{2\pi R^2 B_{44}^*(p)} \tag{7.60}$$

Here, $B_{44}^*(p) = A_{44}^*(p)h$, where h is the shell thickness.

Let the shell be made of glass–epoxy composite whose mechanical properties are listed in Table 3.5 and creep diagrams are shown in Fig. 7.22. To simplify the analysis, we suppose that for the unidirectional composite under study $E_2/E_1 = 0.22$, $G_{12}/E_1 = 0.06$, and $\nu_{12} = \nu_{21} = 0$, and introduce the normalized shear strain

$$\overline{\gamma} = \gamma_{xy}\left(\frac{T}{R^2 h E_1}\right)^{-1}$$

Consider a ±45° angle-ply material discussed in Section 4.5 for which, with due regard to Eqs. (4.72), and (7.58), we can write

$$A_{44}^*(p) = \frac{1}{4}\left(E_1 + E_2^*\right) = \frac{1}{4}\left[E_1 + \frac{E_2}{1 + C_{22}^*(p)}\right]$$

Exponential approximation, Eq. (7.48), of the corresponding creep curve in Fig. 7.22 (the lower dashed line) is

$$C_{22} = A_1 \mathrm{e}^{-\alpha_1 \theta}$$

where $A_1 = 0.04$ and $\alpha_1 = 0.06$ 1/day. Using Eqs. (7.54), we arrive at the following Laplace transforms of the creep compliance and the torque which is constant

$$C_{22}^*(p) = \frac{A_1}{\alpha_1 + p}, \quad T^*(p) = \frac{T}{p}$$

The final expression for the Laplace transform of the normalized shear strain is

$$\overline{\gamma}^*(p) = \frac{2E(\alpha_1 + A_1 + p)}{\pi p(\alpha_1 + A_1 E + p)} \tag{7.61}$$

where $E = E_1/(E_1 + E_2)$

To use Eqs. (7.54) for the inverse Laplace transformation, we should decompose the right-hand part of Eq. (7.61) as

$$\overline{\gamma}^*(p) = \frac{2E}{\pi(\alpha_1 + A_1 E)}\left[\frac{\alpha_1 + A_1}{p} - \frac{A_1(1-E)}{\alpha_1 + A_1 E + p}\right]$$

Applying Eqs. (7.54), we get

$$\overline{\gamma}(t) = \frac{2E}{\pi(\alpha_1 + A_1 E)}\left[\alpha_1 + A_1 - A_1(1-E)\mathrm{e}^{-(\alpha_1 + A_1 E)t}\right]$$

This result is demonstrated in Fig. 7.26. As can be seen, there is practically no creep because the cylinder's deformation is controlled mainly by the fibers.

Quite different behavior is demonstrated by the cylinder made of 0°/90° cross-ply composite material discussed in Section 4.4. In accordance with Eqs. (4.114) and (7.58), we have

$$A_{44}^*(p) = G_{12}^*(p) = \frac{G_{12}}{1 + K_{12}^*(p)}$$

Exponential approximation, Eq. (7.48), of the shear curve in Fig. 7.22 (the upper dashed line) results in the following equation for the creep compliance

$$K_{12} = A_1 \mathrm{e}^{-\alpha_1 \theta} + A_2 \mathrm{e}^{-\alpha_2 \theta}$$

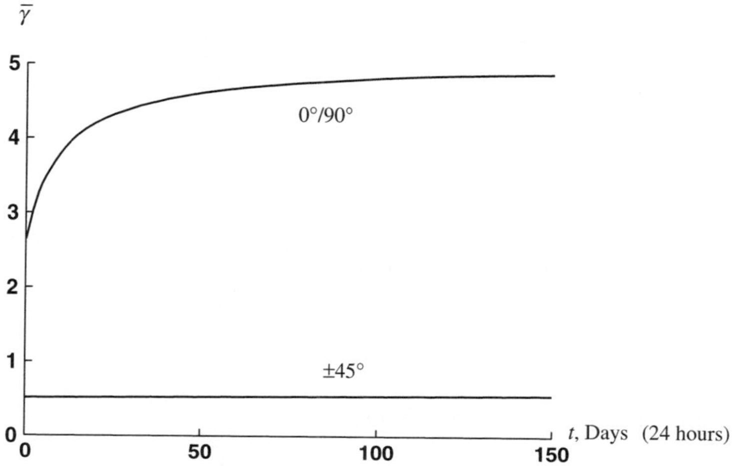

Fig. 7.26. Dependencies of the normalized shear strain on time for 0°/90° cross-ply and ±45° angle-ply glass–epoxy composite cylinders under torsion.

in which $A_1 = 0.033$, $\alpha_1 = 0.04$ 1/day, $A_2 = 0.06$, and $\alpha_2 = 0.4$ 1/day. Omitting simple transformations, we finally get

$$\bar{\gamma} = \frac{E_1}{2\pi G_{12}}\left[1 + \frac{A_1}{\alpha_1}\left(1 - e^{-\alpha_1 t}\right) + \frac{A_2}{\alpha_2}\left(1 - e^{-\alpha_2 t}\right)\right]$$

The corresponding creep diagram is shown in Fig. 7.26.

Under relatively high stresses, polymeric composites demonstrate nonlinear viscoelastic behavior. The simplest approach to study nonlinear creep problems is based on experimental isochrone stress–strain diagrams of the type shown in Fig. 7.19. Using the curves corresponding to time moments $t_1 < t_2 < t_3$, etc., we can solve a sequence of nonlinear elasticity problems for these time moments and thus determine the change of strains and stresses with time. This approach, sometimes referred to as the aging theory, is approximate and can be used to study structures loaded with forces that do not change with time, or change very slowly.

There also exist several variants of nonlinear hereditary theory described, e.g., by Rabotnov (1980). According to the most common versions, Eq. (7.43) is generalized as

$$\varepsilon(t) = \frac{1}{E}\left[\sigma(t) + \int_0^t C_1(t-\tau)\sigma(\tau)d\tau \right.$$
$$\left. + \int_0^t \int_0^t C_2(t-\tau_1, t-\tau_2)\sigma(\tau_1)\sigma(\tau_2)d\tau_1 d\tau_2 + \cdots \right]$$

or

$$\varepsilon(t) = \sum_k A_k \left[\sigma(t) + \int_0^t C(t-\tau)\sigma(\tau)\mathrm{d}\tau \right]^k$$

or

$$f[\varepsilon(t)] = \frac{1}{E}\left[\sigma(t) + \int_0^t C(t-\tau)\sigma(\tau)\mathrm{d}\tau \right]$$

or

$$\varepsilon(t) = \phi[\sigma(t)] + \int_0^t C(t-\tau)\psi[\sigma(\tau)]\mathrm{d}\tau$$

In conclusion, it should be noted that correctly designed composite structures (see the next chapter) in which the material behavior is controlled by fibers usually do not exhibit pronounced time-dependent behavior. For example, consider the filament-wound glass–epoxy pressure vessel studied in Section 6.3 (see Fig. 6.22 and the second row in Table 6.1 for parameters of the vessel). The vessel consists of $\pm 36^\circ$ helical plies and circumferential plies, and has structural parameters that are close to optimal (see Section 8.3.1). The experimental dependence of circumferential strain on time for step-wise loading with internal pressure p presented in Fig. 7.27 does not indicate any

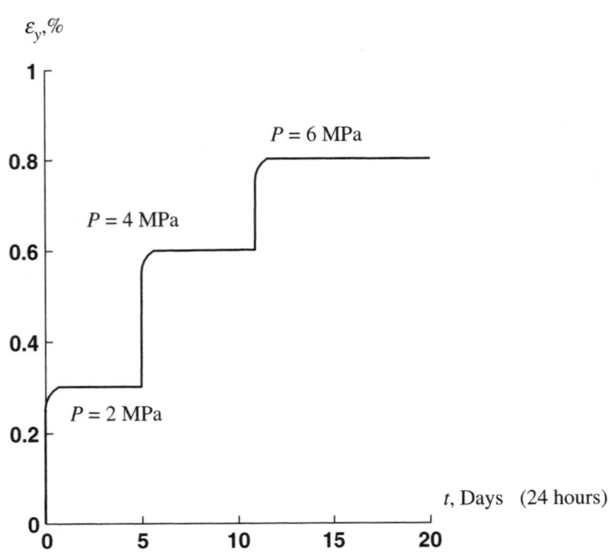

Fig. 7.27. Dependence of the circumferential strain on time for a glass–epoxy cylindrical pressure vessel loaded in steps with internal pressure p.

significant creep deformation. It should be emphasized that this conclusion is valid for normal conditions only; at elevated temperatures, composite structures can exhibit significant creep deformation.

7.3.2. Durability

Composite materials, to be applied to structures with long service life, need to be guaranteed for the corresponding period of time from failure, which is usually a result of an evolutionary process of material degradation in the service environment. To provide proper durability of the material, we need, in turn, to study its long-term behavior under load and its endurance limits. The most widely used durability criteria establishing the dependence of material strength on the time of loading are based on the concept of the accumulation of material damage induced by acting stresses and intensified by the degrading influence of service conditions such as temperature, moisture, etc. Particular criteria depend on the accepted models simulating the material damage accumulation. Although there exist microstructural approaches to the durability evaluation of composite materials (see, e.g., Skudra et al., 1989), for practical purposes, the experimental dependencies of the ultimate stresses on the time of their action are usually evaluated. In particular, these experiments allow us to conclude that fibers, which are the major load-carrying elements of composite materials, possess some residual strength $\overline{\sigma}_\infty = \overline{\sigma}(t \to \infty)$, which is about 50% to 70% of the corresponding static strength $\overline{\sigma}_0 = \overline{\sigma}(t = 0)$, depending on the fiber type. Typical dependencies of the long-term strength of composite materials on time are presented in Fig. 7.28. As can be seen, the time of loading dramatically affects material strength. However, being unloaded at any moment of time t, composite materials demonstrate practically the same static strength that they had before long-term loading.

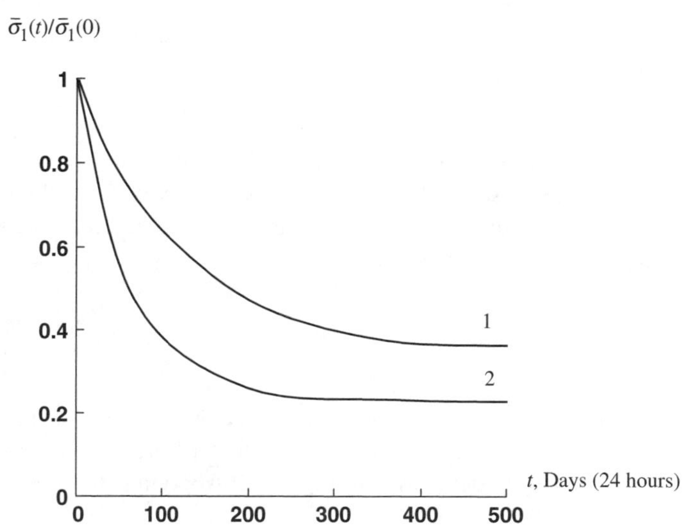

Fig. 7.28. Normalized long-term longitudinal strength of aramid–epoxy (1) and glass–epoxy (2) unidirectional composites.

Approximation of the curves shown in Fig. 7.28 can be performed using exponential functions as follows

$$\overline{\sigma}(t) = \overline{\sigma}_\infty + \sum_n A_n e^{-\lambda_n t} \tag{7.62}$$

in which $\overline{\sigma}_\infty$, A_n, and λ_n are coefficients providing the appropriate approximation. The initial static strength is

$$\overline{\sigma}(0) = \overline{\sigma}_0 = \overline{\sigma}_\infty + \sum_n A_n$$

The simplest is a one-term approximation

$$\overline{\sigma}(t) = \overline{\sigma}_\infty + (\overline{\sigma}_0 - \overline{\sigma}_\infty)e^{-\lambda t} \tag{7.63}$$

To approximate the initial part of the curve, we can put $\overline{\sigma}_\infty = 0$ and arrive at the following equation

$$\overline{\sigma}(t) = \overline{\sigma}_0 e^{-\lambda t} \tag{7.64}$$

Now assume that we can solve Eqs. (7.62), (7.63), or (7.64) for t and find the material durability $t_d(\sigma)$, i.e., the time during which the material can withstand stress σ. Consider the process of loading as a system of k stages such that the duration of each stage is t_i and the stress acting at this stage is σ_i $(i = 1, 2, 3, \ldots, k)$. Then, the whole period of time during which the material can withstand such step-wise loading can be calculated with the aid of the following equation

$$\sum_{i=1}^{k} \frac{t_i}{t_d(\sigma_i)} = 1$$

in which $t_d(\sigma_i)$ is the material durability corresponding to stress σ_i.

The strength criteria discussed in Chapter 6 can be generalized for the case of long-term loading if we change the static ultimate stresses entering these criteria for the corresponding long-term strength characteristics.

7.3.3. Cyclic loading

Consider the behavior of composite materials under the action of loads periodically changing with time. For qualitative analysis, consider first a material that can be simulated with the simple mechanical model shown in Fig. 7.23. Applying stress acting according to the following form

$$\sigma(t) = \sigma_0 \sin \omega t \tag{7.65}$$

where σ_0 is the amplitude of stress and ω is the frequency, we can solve Eq. (7.53), which describes the model under study for strain $\varepsilon(t)$. The result is

$$\varepsilon(t) = \varepsilon_0 \sin(\omega t + \theta) \tag{7.66}$$

where

$$\varepsilon_0 = \sigma_0 \sqrt{\frac{1 + t_r^2\omega^2}{E_l^2 + E_i^2 t_r^2 \omega^2}}$$
$$\tan\theta = -\frac{t_r\omega(E_i - E_l)}{E_l + E_i t_r^2\omega^2} \tag{7.67}$$

It follows from these equations that a viscoelastic material is characterized with a phase shift of strain with respect to stress. Eliminating the time variable from Eqs. (7.65) and (7.66), we arrive at the following relationship between stress and strain

$$\left(\frac{\sigma}{\sigma_0}\right)^2 + \left(\frac{\varepsilon}{\varepsilon_0}\right)^2 - 2\cos\theta\frac{\sigma\varepsilon}{\sigma_0\varepsilon_0} = \sin^2\theta$$

This is the equation of an ellipse shown in Fig. 7.29a. The absolute value of the area A, inside this ellipse (its sign depends on the direction of integration along the contour) determines the energy dissipation per single cycle of vibration, i.e.,

$$\Delta W = |A| = \pi\sigma_0\varepsilon_0 \,|\sin\theta| \tag{7.68}$$

Folowing Zinoviev and Ermakov (1994), we can introduce the dissipation factor as the ratio of energy loss in a loading cycle, ΔW, to the value of the elastic potential energy in

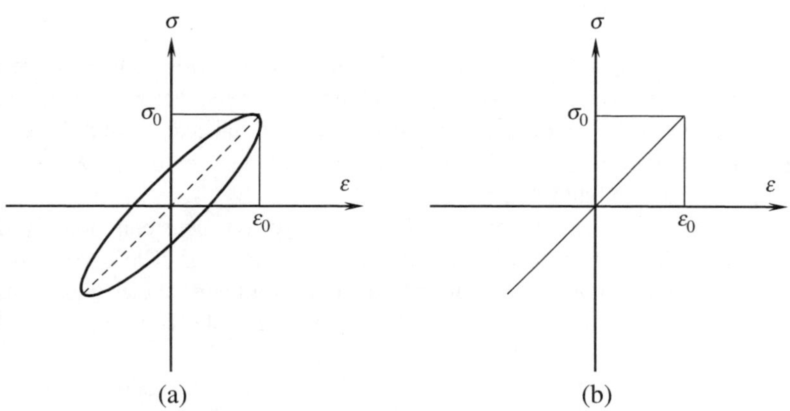

Fig. 7.29. Stress–strain diagrams for viscoelastic (a) and elastic (b) materials.

a cycle, W, as

$$\psi = \frac{\Delta W}{W}$$

where, in accordance with Fig. 7.29b, $W = (1/2)\sigma_0\varepsilon_0$. Transforming Eq. (7.68) with the aid of Eqs. (7.67), we arrive at

$$\psi = \frac{2\pi t_r \omega}{1 + t_r^2\omega^2}\left(1 - \frac{E_l}{E_i}\right)$$

As follows from this equation, ψ depends on the number of oscillations accomplished during the period of time equal to the material relaxation time, t_r, and reaches a maximum value for $t_r\omega = 1$.

As shown by Zinoviev and Ermakov (1994), for anisotropic materials, the dissipation factor depends also on the direction of loading. Particularly, for a unidirectional composite ply, referred to axes x and y and making angle ϕ with the principal material axes 1 and 2 as in Fig. 4.18, the dissipation factors are

$$\psi_x = E_x\left[\left(\frac{\psi_1}{E_1}\cos^2\phi - \frac{\psi_2}{E_2}\sin^2\phi\right)\cos 2\phi + \psi_{45}\mu_{12}\sin^2\phi\cos^2\phi\right]$$

$$\psi_y = E_y\left[\left(\frac{\psi_2}{E_2}\cos^2\phi - \frac{\psi_1}{E_1}\sin^2\phi\right)\cos 2\phi + \psi_{45}\mu_{12}\sin^2\phi\cos^2\phi\right]$$

$$\psi_{xy} = G_{xy}\left[\left(\frac{2\psi_1}{E_1} + \frac{2\psi_2}{E_2} - \psi_{45}\mu_{12}\right)\sin^2\phi\cos^2\phi + \frac{\psi_{12}}{G_{12}}\cos^2 2\phi\right]$$

where

$$\mu_{12} = \frac{1 - \nu_{12}}{E_1} + \frac{1 - \nu_{21}}{E_2} + \frac{1}{G_{12}}$$

E_x, E_y, and G_{xy} are specified by Eqs. (4.76), and ψ_1, ψ_2, ψ_{12}, and ψ_{45} are the ply dissipation factors corresponding to loading along the fibers, across the fibers, under in-plane shear, and at 45° with respect to principal material axes 1 and 2. As follows from Fig. 7.30, calculations based on the foregoing equations provide fair agreement with experimental results of Ni and Adams (1984).

Energy dissipation in conjunction with the relatively low heat conductivity of composite materials induces their self-heating during cyclic loading. The dependence of an aramid–epoxy composite material's temperature on the number of cycles under tensile and compressive loading with frequency 10^3 cycles per minute is shown in Fig. 7.31 (Tamuzh and Protasov, 1986).

Under cyclic loading, structural materials experience a fatigue fracture caused by material damage accumulation. As already noted in Section 3.2.4, the heterogeneous structure of composite materials provides relatively high resistance of these materials to crack

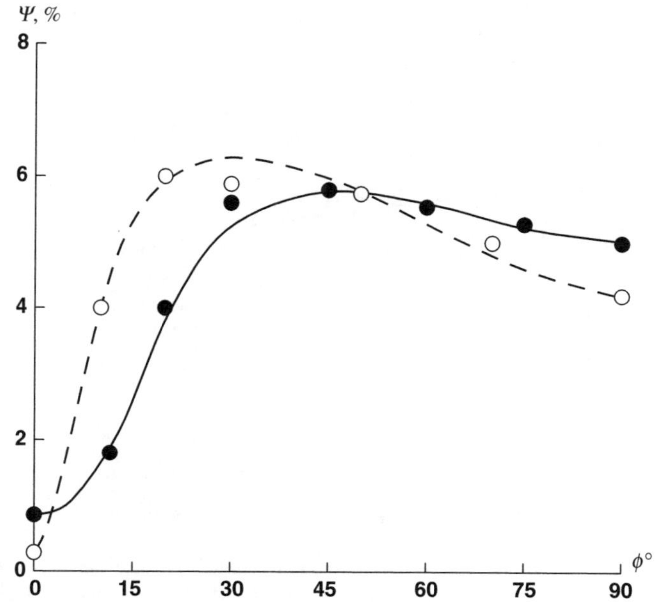

Fig. 7.30. Calculated (lines) and experimental (circles) dependencies of dissipation factor on the ply orientation for glass–epoxy (——— ●) and carbon–epoxy (- - - ○) unidirectional composites.

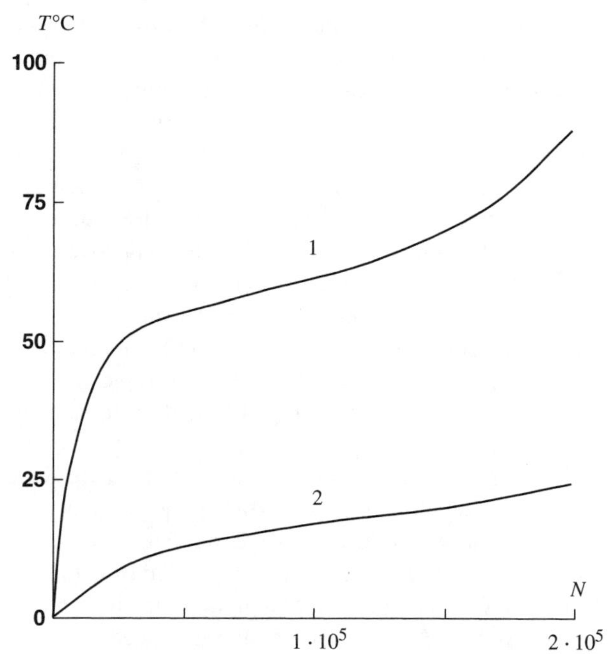

Fig. 7.31. Temperature of an aramid–epoxy composite as a function of the number of cycles under tension (1) and compression (2).

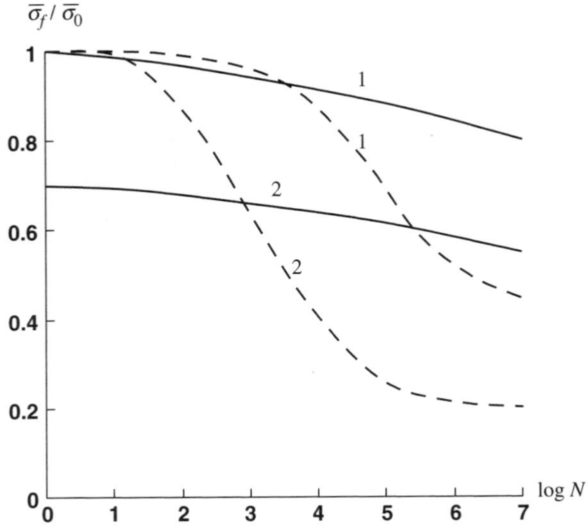

Fig. 7.32. Typical fatigue diagrams for carbon–epoxy composite (solid lines) and aluminum alloy (dashed lines) specimens without (1) and with (2) stress concentration (fatigue strength is normalized to static strength of specimens without stress concentration).

propagation, resulting in their specific behavior under cyclic loading. It follows from Fig. 7.32, showing experimental results obtained by V. F. Kutinov, that stress concentration in aluminum specimens, which has practically no effect on the material's static strength due to plasticity of aluminum, dramatically reduces its fatigue strength. Conversely, the static strength of carbon–epoxy composites, which are brittle materials, is reduced by stress concentration that has practically no effect on the slope of the fatigue curve. On average, the residual strength of carbon composites after 10^6 loading cycles is 70–80% of the material's static strength, in comparison to 30–40% for aluminum alloys. Qualitatively, this comparative evaluation is true for all fibrous composites that are widely used in structural elements subjected to intensive vibrations, such as helicopter rotor blades, airplane propellers, drive shafts, automobile leaf springs, etc.

The forgoing discussion concerns the fatigue strength of unidirectional composites loaded along the fibers. However, composites are anisotropic materials having different strength in different directions and, naturally, different response under cyclic loading. As shown in Fig. 7.33 presenting the approximations of the experimental results given by Tsai (1987), the degradation of material strength under tension across the fibers (line 2) is much higher than under tension along the fibers (line 1). Recall that the stress σ_2 induces the cracks in the matrix discussed in Sections 4.4.2 and 6.4.

A typical composite materials fatigue diagram, constructed from the experimental results of Apinis et al. (1991), is shown in Fig. 7.34. Standard fatigue diagrams usually determine the material strength for $10^3 \leq N \leq 10^6$ and are approximated as

$$\sigma_R = a - b \log N \tag{7.69}$$

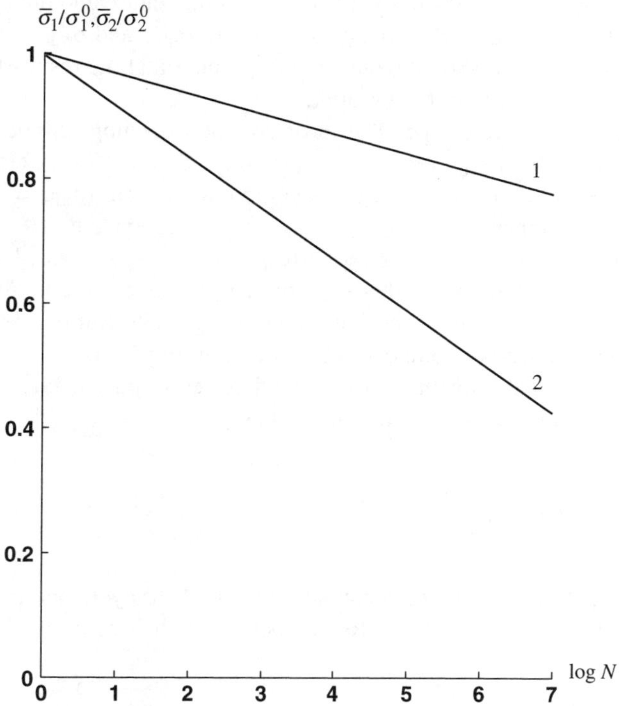

Fig. 7.33. Normalized fatigue strength of carbon–epoxy composites loaded along (1) and across (2) the fibers.

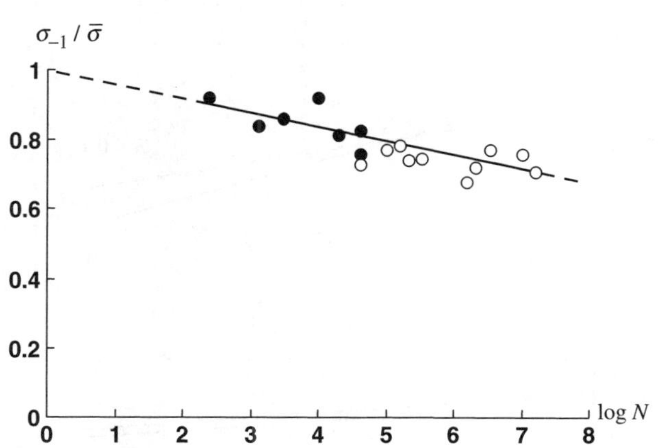

Fig. 7.34. Normalized fatigue diagram for fabric carbon–carbon composite material ($\bar{\sigma}$ static strength), ● —— ○ experimental part of the diagram for the loading frequency of 6 Hz (●) and 330 Hz (○), - - - extrapolation.

Here, N is the number of cycles to failure under stress σ_R, a and b are experimental constants depending on frequency of cyclic loading, temperature and other environmental factors, and on the stress ratio $R = \sigma_{\min}/\sigma_{\max}$, where $\sigma_{\max}$ and $\sigma_{\min}$ are the maximum and the minimum stresses. It should be taken into account that the results for fatigue tests are characterized, as a rule, with high scatter.

Factor R specifies the cycle type. The most common bending fatigue test provides a symmetric cycle for which $\sigma_{\min} = -\sigma$, $\sigma_{\max} = \sigma$, and $R = -1$. A tensile load cycle ($\sigma_{\min} = 0$, $\sigma_{\max} = \sigma$) has $R = 0$, whereas a compressive cycle ($\sigma_{\min} = -\sigma$, $\sigma_{\max} = 0$) has $R \to -\infty$. Cyclic tension with $\sigma_{\max} > \sigma_{\min} > 0$ corresponds to $0 < R < 1$, whereas cyclic compression with $0 > \sigma_{\max} > \sigma_{\min}$ corresponds to $1 < R < \infty$. Fatigue diagrams for unidirectional aramid–epoxy composite studied by Limonov and Anderson (1991) corresponding to various R-values are presented in Fig. 7.35. Similar results (Anderson et al., 1991) for carbon–epoxy composites are shown in Fig. 7.36.

Since only σ_{-1} is usually available from standard tests under cyclic bending, fatigue strengths for other load cycles are approximated as

$$\sigma_R = \sigma_{-1} + \sigma_{\mathrm{m}}\left(1 - \frac{\sigma_{-1}}{\overline{\sigma}_t}\right)$$

where $\sigma_{\mathrm{m}} = (\sigma_{\min} + \sigma_{\max})/2$ is the mean stress of the load cycle and $\overline{\sigma}_t$ is the material long-term strength (see Section 7.3.2) for the period of time equal to that of the cyclic loading.

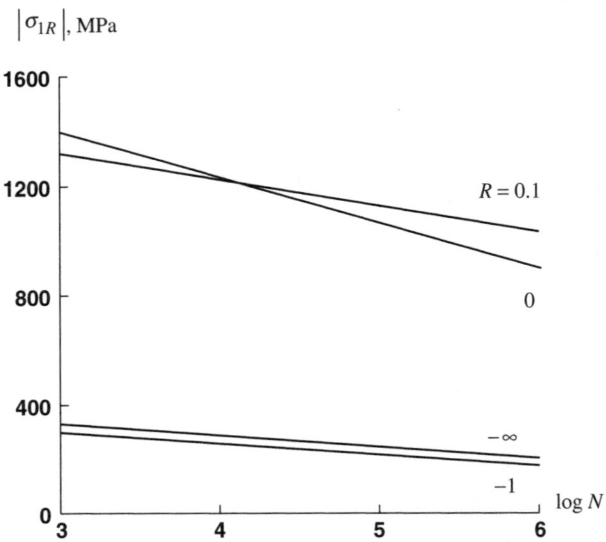

Fig. 7.35. Fatigue diagrams for unidirectional aramid–epoxy composite loaded along the fibers with various stress ratios.

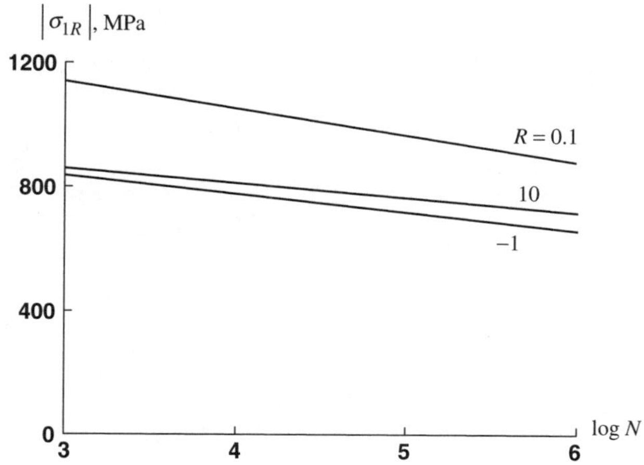

Fig. 7.36. Fatigue diagrams for a unidirectional carbon–epoxy composite loaded along the fibers with various stress ratios.

Fabric composites are more susceptible to cyclic loading than materials reinforced with straight fibers. This fact is illustrated in Fig. 7.37 showing the experimental results of Schulte et al. (1987).

The foregoing discussion deals with high-cycle fatigue. The initial interval $1 \leq N \leq 10^3$ corresponding to so-called low-cycle fatigue is usually studied separately, because the slope of the approximation in Eq. (7.69) can be different for high stresses. A typical fatigue diagram for this case is shown in Fig. 7.38 (Tamuzh and Protasov, 1986).

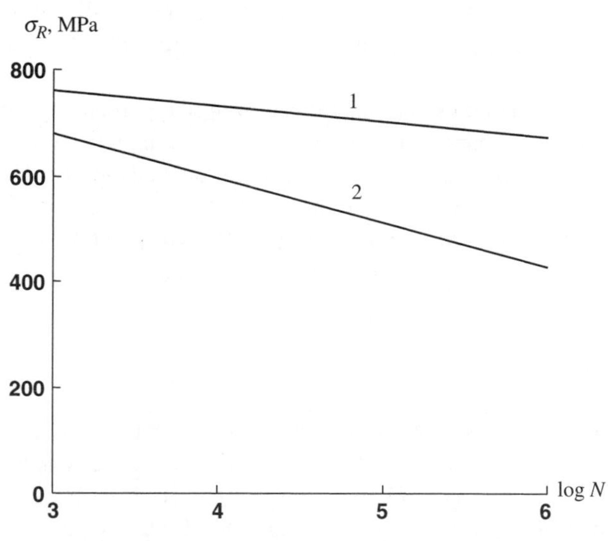

Fig. 7.37. Tensile fatigue diagrams for a cross-ply (1) and fabric (2) carbon–epoxy composites.

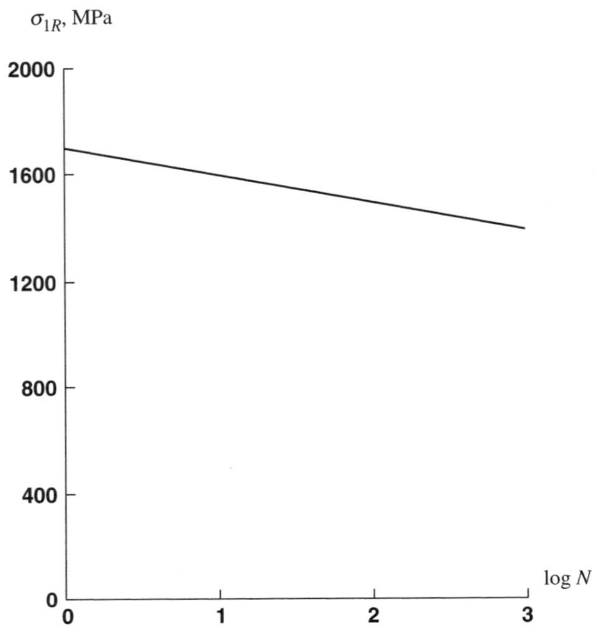

Fig. 7.38. Low-cycle fatigue diagram for unidirectional aramid–epoxy composite loaded along the fibers with $R = 0.1$.

Fatigue has also some effect on the stiffness of composite materials. This can be seen in Fig. 7.39 demonstrating a reduction in the elastic modulus for a glass–fabric–epoxy–phenolic composite under low-cycle loading (Tamuzh and Protasov, 1986). This effect should be accounted for in the application of composites to the design of structural members such as automobile leaf-springs that, being subjected to cyclic loading, are designed under stiffness constraints.

Stiffness degradation can be used as an indication of material damage to predict fatigue failure. The most sensitive characteristic of the stiffness change is the tangent modulus E_t specified by the second equation in Eqs. (1.8). The dependence of E_t on the number of cycles, $\overline{N}$, normalized to the number of cycles that cause material fatigue fracture under the preassigned stress, is presented in Fig. 7.40 corresponding to a $\pm 45^\circ$ angle-ply carbon–epoxy laminate studied by Murakami et al. (1991).

7.3.4. Impact loading

Thin-walled composite laminates possessing high in-plane strength and stiffness are rather susceptible to damage initiated by transverse impact loads that can cause fiber breakage, cracks in the matrix, delamination, and even material penetration by the impactor. Depending on the impact energy determined by the impactor mass and velocity and the properties of laminate, impact loading can result in considerable reduction

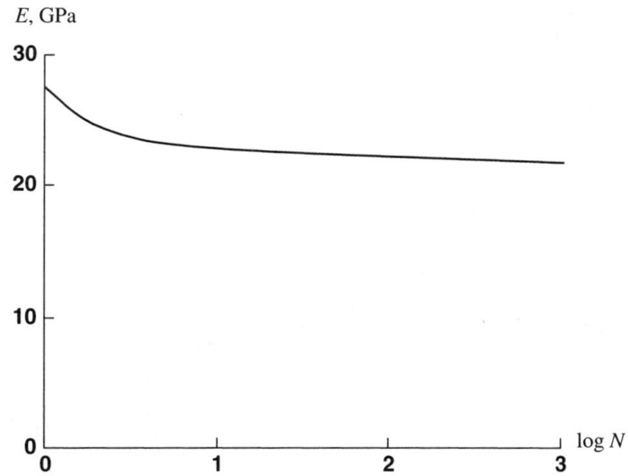

Fig. 7.39. Dependence of elastic modulus of glass fabric–epoxy–phenolic composite on the number of cycles at stress $\sigma = 0.5\bar{\sigma}$ ($\bar{\sigma}$ is the static ultimate stress).

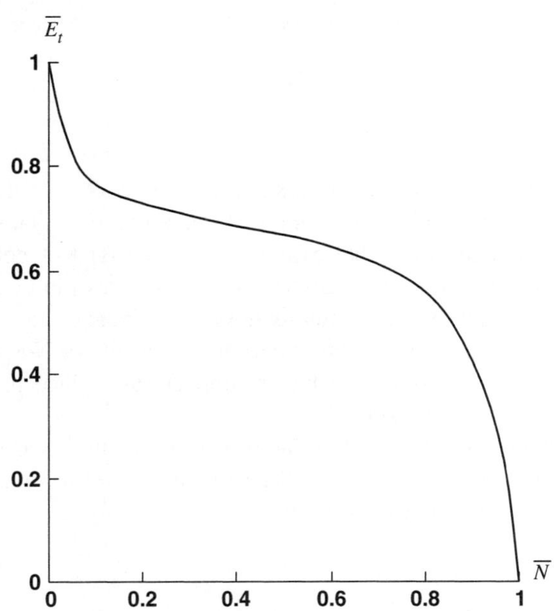

Fig. 7.40. Dependence of the tangent modulus normalized to its initial value on the number of cycles related to the ultimate number corresponding to fatigue failure under stress $\sigma_{max} = 120\,\text{MPa}$ and $R = -1$ for $\pm 45°$ angle-ply carbon–epoxy laminate.

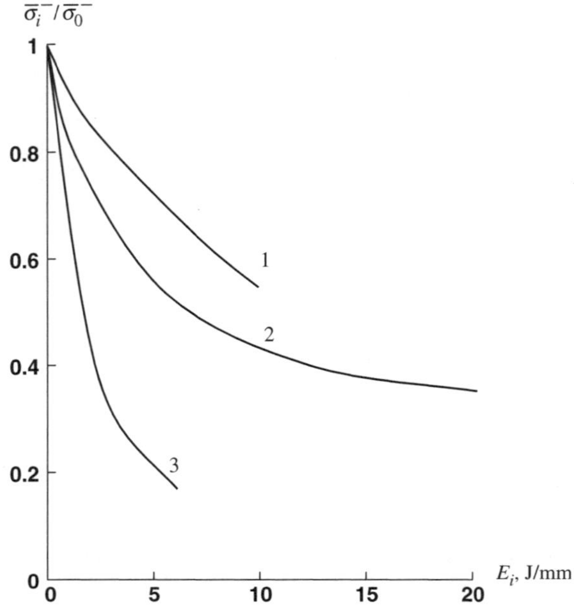

Fig. 7.41. Dependence of compression strength after impact normalized to the initial compressive strength on the impact energy related to the plate thickness for glass fabric–epoxy (1), unidirectional glass–epoxy (2), and carbon–epoxy composite plates (3).

in material strength under tension, compression, and shear. One of the most dangerous consequences of an impact loading is an internal delamination in laminates, which can sometimes be hardly noticed by visual examination. This type of defect causes a dramatic reduction in the laminate compressive strength and results in unexpected failure of thin-walled composite structures due to microbuckling of fibers or local buckling of plies. As follows from Fig. 7.41, showing the experimental results of Verpoest et al. (1989) for unidirectional and fabric composite plates, impact can reduce material strength in compression by a factor of 5 or more.

To study the mechanism of material interlaminar delamination, consider the problem of wave propagation through the thickness of the laminate shown in Fig. 7.42. The motion equation has the following well-known form

$$\frac{\partial}{\partial z}\left(E_z \frac{\partial u_z}{\partial z}\right) = \rho \frac{\partial^2 u_z}{\partial t^2} \tag{7.70}$$

Here, u_z is the displacement in the z-direction, E_z is material modulus in the same direction depending, in the general case on z, and ρ is the material density. For the laminate in Fig. 7.42, the solution of Eq. (7.70) should satisfy the following boundary and

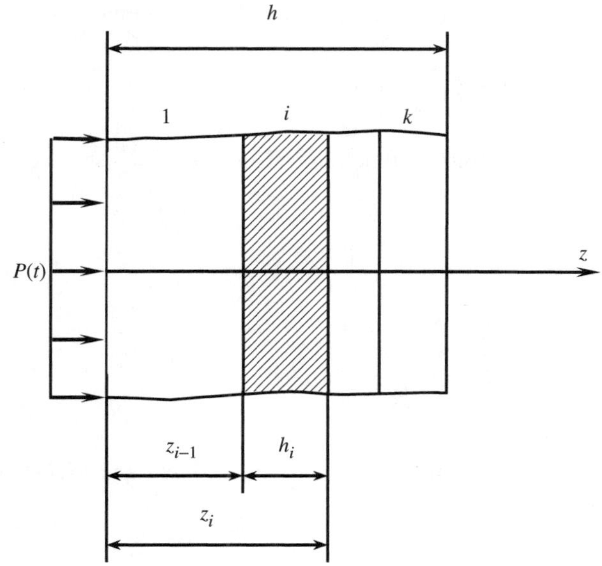

Fig. 7.42. Laminate under impact load.

initial conditions

$$\sigma_z(z = 0,\ t) = -p(t), \quad \sigma_z(z = h,\ t) = 0 \tag{7.71}$$

$$u_z(z,\ t = 0) = 0, \quad \frac{\partial u_z}{\partial t}(z > 0,\ t = 0) = 0 \tag{7.72}$$

in which

$$\sigma_z = E_z \frac{\partial u_z}{\partial z} \tag{7.73}$$

is the interlaminar normal stress.

Consider first a homogeneous layer such that E_z and ρ do not depend on z. Then, Eq. (7.70) takes the form

$$c^2 \frac{\partial^2 u_z}{\partial z^2} = \frac{\partial^2 u_z}{\partial t^2}$$

where $c^2 = E_z/\rho$. Transform this equation introducing new variables, i.e., $x_1 = z + ct$ and $x_2 = z - ct$. Performing conventional transformation and rearrangement, we arrive at

$$\frac{\partial^2 u_z}{\partial x_1 \partial x_2} = 0$$

The solution for this equation can be readily found and presented as

$$u_z = \phi_1(x_1) + \phi_2(x_2) = \phi_1(z + ct) + \phi_2(z - ct)$$

Here, ϕ_1 and ϕ_2 are some arbitrary functions. Using Eq. (7.73), we get

$$\sigma_z = E_z[f_1(x + ct) + f_2(x - ct)]$$

where

$$f_1 = \frac{\partial \phi_1}{\partial z}, \quad f_2 = \frac{\partial \phi_2}{\partial z}$$

Applying the boundary and initial conditions, Eqs. (7.71) and (7.72), we arrive at the following final result

$$\sigma_z = E[f(x + ct) - f(x - ct)] \tag{7.74}$$

in which the form of function f is governed by the shape of the applied pulse. As can be seen, the stress wave is composed of two components having opposite signs and moving in opposite directions with one and the same speed c, which is the speed of sound in the material. The first term in Eq. (7.74) corresponds to the applied pulse that propagates to the free surface $z = h$ (see Fig. 7.43, demonstrating the propagation of a rectangular pulse), whereas the second term corresponds to the pulse reflected from the free surface $z = h$. It is important that for a compressive direct pulse (which is usually the case), the reflected pulse is tensile and can cause material delamination since the strength of laminated composites under tension across the layers is very low.

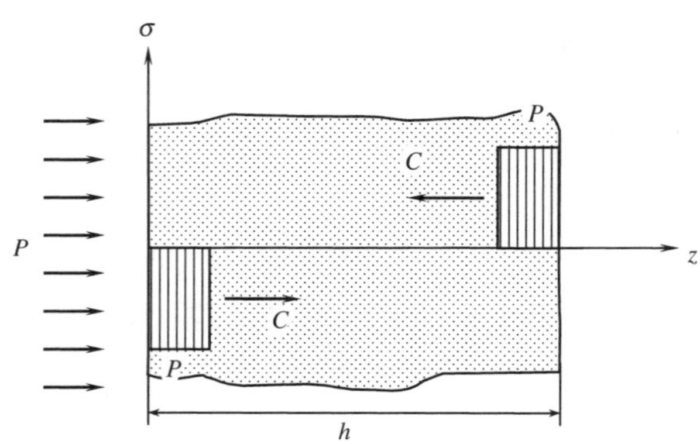

Fig. 7.43. Propagation of direct and reflected pulses through the layer thickness.

Note that the speed of sound in a homogeneous material, i.e.,

$$c = \sqrt{\frac{E_z}{\rho}} \tag{7.75}$$

is the same for the tensile and compressive waves in Fig. 7.43. This means that the elastic modulus in Eq. (7.75) must be the same for both tension and compression. For composite materials, tensile and compressive tests sometimes produce modulus values that are slightly different. Usually, the reason for such a difference is that the different specimens and experimental techniques are used for tensile and compression tests. Testing of fiberglass fabric coupons (for which the difference in the experimental values of tensile and compressive moduli is sometimes observed) involving continuous loading from compression to tension through zero load does not show any 'kink' in the stress–strain diagram at zero stress. Naturally, for heterogeneous materials, the apparent (effective) stiffness can be different for tension and compression as, for example, in materials with cracks that propagate under tension and close under compression. Sometimes stress–strain diagrams with a 'kink' at the origin are used to approximate nonlinear experimental diagrams that, actually, do not have a 'kink' at the zero stress level at all.

For laminates, such as in Fig. 7.42, the boundary conditions, Eqs. (7.71), should be supplemented with the interlaminar conditions $u_z^{(i)} = u_z^{(i-1)}$ and $\sigma_z^{(i)} = \sigma_z^{(i-1)}$. Omitting the rather cumbersome solution that can be found elsewhere (Vasiliev and Sibiryakov, 1985), we present some numerical results.

Consider the two-layered structure: the first layer of which has thickness 15 mm and is made of aramid–epoxy composite material with $E_z^{(1)} = 4.2\,\text{GPa}$, $\rho_1 = 1.4\,\text{g/cm}^3$, and the second layer is made of boron–epoxy composite material and has $E_z^{(2)} = 4.55\,\text{GPa}$, $\rho_2 = 2\,\text{g/cm}^3$, and $h_2 = 12\,\text{mm}$. The duration of a rectangular pulse of external pressure p acting on the surface of the first layer is $t_p = 5\times10^{-6}$ s. The dependence of the interlaminar ($z = 15\,\text{mm}$) stress on time is shown in Fig. 7.44. As can be seen, at $t \approx 3t_p$ the tensile interface stress exceeds the intensity of the pulse of pressure by the factor of 1.27. This stress is a result of interaction of the direct stress wave with the waves reflected from the laminate's inner, outer, and interface surfaces. Thus, in a laminate, each interface surface generates elastic waves.

For laminates consisting of more than two layers, the wave interaction becomes more complicated and, what is more important, can be controlled by the appropriate stacking sequence of layers. As an example, consider a sandwich structure shown in Fig. 7.45a. The first (loaded) layer is made of aluminum and has $h_1 = 1\,\text{mm}$, $E_z^{(1)} = 72\,\text{GPa}$, $\rho_1 = 2.7\,\text{g/cm}^3$, the second layer is a foam core with $h_2 = 10\,\text{mm}$, $E_z^{(2)} = 0.28\,\text{GPa}$, $\rho_2 = 0.25\,\text{g/cm}^3$, and the third (load-carrying) composite layer has $h_3 = 12\,\text{mm}$, $E_z^{(3)} = 10\,\text{GPa}$, $\rho_3 = 1.4\,\text{g/cm}^3$. The duration of a rectangular pulse of external pressure is 10^{-6} s. The maximum tensile stress occurs in the middle plane of the load-carrying layer (plane *a–a* in Fig. 7.45). The normal stress induced in this plane is presented in Fig. 7.46a. As can be seen, at the moment of time t equal to about 1.75×10^{-5} s, this stress is tensile and can cause delamination of the structure.

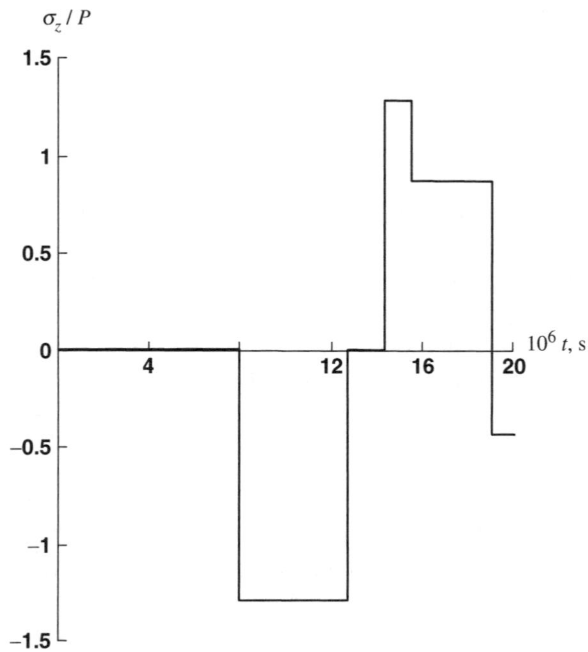

Fig. 7.44. Dependence of the interlaminar stress referred to the acting pressure on time.

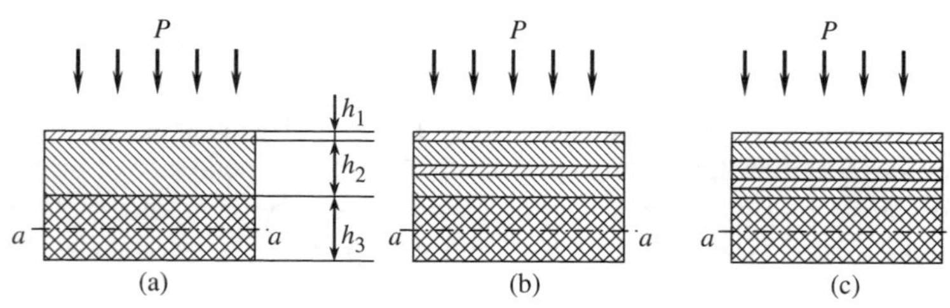

Fig. 7.45. Structure of the laminates under study.

Now introduce an additional aluminum layer in the foam core as shown in Fig. 7.45b. As follows from Fig. 7.46b, this layer suppresses the tensile stress in section *a–a*. Two intermediate aluminum layers (Fig. 7.45c) working as generators of compressive stress waves eliminate the appearance of tensile stress in this section. Naturally, the effect under discussion can be achieved for a limited period of time. However, in reality, the impact-generated tensile stress is dangerous soon after the application of the pulse. The damping capacity of real structural materials (which is not taken into account in the foregoing analysis) dramatically reduces the stress amplitude in time.

A flying projectile with relatively high kinetic energy can penetrate through the laminate. As is known, composite materials, particularly, high-strength aramid fabrics, are widely

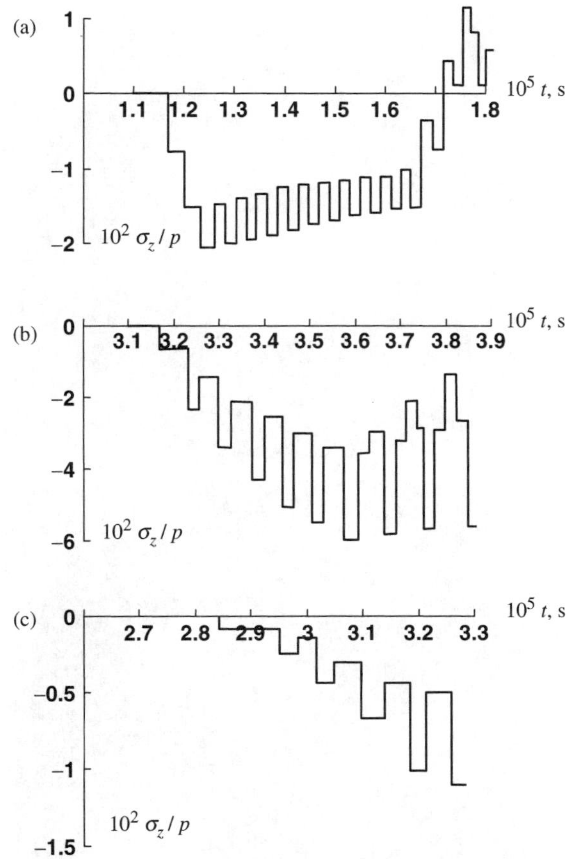

Fig. 7.46. Normal stress related to external pressure acting in section *a–a* of the laminates in Fig. 7.45 (a)–(c), respectively.

used for protection against flying objects. To demonstrate the mechanism of this protection, consider a square composite plate clamped in the steel frame shown in Fig. 7.47 and subjected to impact by a rectangular plane projectile (see Fig. 7.47) simulating the blade of the turbojet engine compressor. The plate consists of layers of thin aramid fabric impregnated with epoxy resin at a distance from the window in the frame (see Fig. 7.47) and co-cured together as shown in Fig. 7.48. The front (loaded) surface of the plate has a 1-mm-thick cover sheet made of glass fabric–epoxy composite. The results of ballistic tests are presented in Table 7.2. Front and back views of plate No. 2 are shown in Fig. 7.47, and the back view of plate No. 3 can be seen in Fig. 7.48. Since the mechanical properties of the aramid fabric used to make the plates are different in the warp and fill directions, the plates consist of couples of mutually orthogonal layers of fabric that are subsequently referred to as $0^\circ/90^\circ$ layers. All the plates listed in Table 7.2 have $n = 32$ of such couples.

(a)

(b)

Fig. 7.47. Plate no. 2 (see Table 7.2) after the impact test: (a) – front view; (b) – back view.

Fig. 7.48. Back view of plate no. 3 (see Table 7.2) after the impact test.

Table 7.2
Ballistic test of plates made of aramid fabric.

Plate no.	Projectile velocity (m/s)	Test results
1	315	No penetration
2	320	The projectile is 'caught' by the containment
3	325	Penetration

To calculate the projectile velocity below which it fails to perforate the plate (the so-called ballistic limit), we use the energy conservation law, according to which

$$\frac{1}{2}m_p\left(V_s^2 - V_r^2\right) = n(W + T) \tag{7.76}$$

where V_s is the projectile striking velocity, V_r is its residual velocity, $m_p = 0.25$ kg is the projectile mass, $n = 32$ is the number of the $0°/90°$ layers, W is the fracture work for the $0°/90°$ layers, and T is the kinetic energy of the layer. All other factors and the fiberglass cover of the plate are neglected.

The fracture work can be evaluated using the quasi-static test shown in Fig. 7.49. A couple of mutually orthogonal fabric layers is fixed along the plate contour and loaded by the projectile. The area under the force–deflection curve (solid line in Fig. 7.49) can be treated as the work of fracture which, for the fabric under study, has been found to be $W = 120\,\text{Nm}$.

To calculate T, the deformed shape of the fabric membrane has been measured. Assuming that the velocities of the membrane points are proportional to deflections f and that $\mathrm{d}f_\mathrm{m}/\mathrm{d}t = V_s$, the kinetic energy of the fabric under study (the density of the layer unit surface is $0.2\,\text{kg/m}^2$) turns out to be $T_c = 0.0006\,V_s^2$.

To find the ballistic limit, we should take $V_r = 0$ in Eq. (7.76). Substituting the foregoing results in this equation, we get $V_b = 190.5\,\text{m/s}$, which is much lower than the experimental result ($V_b = 320\,\text{m/s}$) following from Table 7.2.

Let us change the model of the process and assume that the fabric layers fail one after another rather than all of them at once, as is assumed in Eq. (7.76). The result is expected to be different because the problem under study is not linear, and the principle of superposition is not applicable. Bearing this in mind, we write Eq. (7.76) in the following incremental form

$$\frac{1}{2}m_p\left(V_{k-1}^2 - V_k^2\right) = W + T_{k-1} \tag{7.77}$$

Here, V_{k-1} and V_k are the projectile velocities before and after the failure of the kth couple of fabric layers, W is, as earlier, the fracture work consumed by the kth couple of layers, $T_{k-1} = 0.0006\,V_{k-1}^2$, and the last term in the right-hand side of Eq. (7.77) means that we account for the kinetic energy of only those fabric layers that have been already

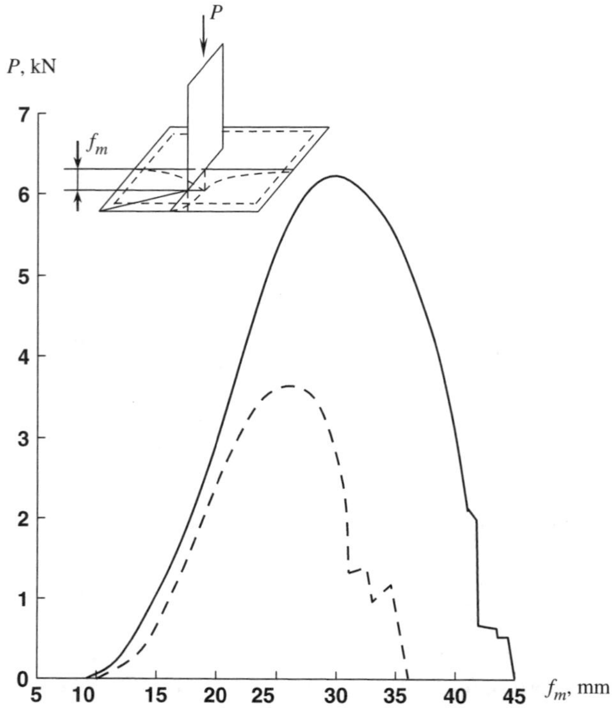

Fig. 7.49. Force–deflection diagrams for square aramid fabric membranes, ——— couple of layers with orthogonal orientations, - - - superposition of the diagrams for individually tested layers.

penetrated by the projectile. Solving Eq. (7.77) for V_k, we arrive at

$$V_k = \sqrt{[1 - 0.0048(k-1)]V_{k-1}^2 - \frac{2}{m_p}W} \tag{7.78}$$

For $k = 1$, we take $V_0 = 320$ m/s, in accordance with the experimental ballistic limit, and have $V_1 = 318.5$ m/s from Eq. (7.78). Taking $k = 2$, we repeat the calculation and find that, after the failure of the second couple of fabric layers, $V_2 = 316.2$ m/s. This process is repeated until $V_k = 0$, and the number k thus determined gives an estimate of the minimum number of $0°/90°$ layers that can stop a projectile with striking velocity $V_s = 320$ m/s. The result of the calculation is presented in Fig. 7.50, from which it follows that $k = 32$. This is exactly the same number of layers that have been used to construct the experimental plates.

Thus, it can be concluded that the high impact resistance of aramid fabrics is determined by two main factors. The first factor is the relatively high work of fracture, which is governed not only by the high strength, but also by the interaction of the fabric layers. The dashed line in Fig. 7.49 shows the fracture process constructed as a result of the superposition of experimental diagrams for individual 0° and 90° layers. The solid line

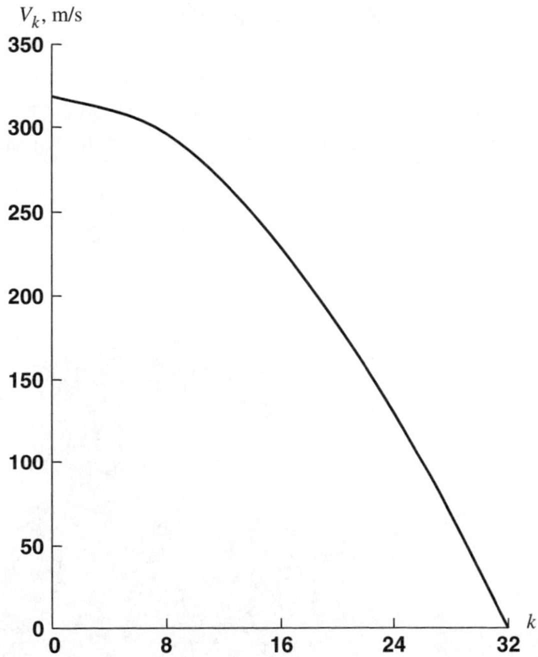

Fig. 7.50. Dependence of the residual velocity of the projectile on the number of penetrated layers.

corresponds, as noted, to 0° and 90° layers tested together (the ratio of the fabric strength under tension in the warp and the fill direction is 1.3). As can be seen, the area under the solid line is much larger that under the dashed one, which indicates the high contribution of the layers interaction to the work of fracture. If this conclusion is true, we can expect that for layers with higher anisotropy and for laminates in which the principal material axes of the adjacent layers are not orthogonal, the fracture work would be higher than for the orthotropic laminate under study. The second factor increasing the impact resistance of aramid fabrics is associated with a specific process of the failure, during which the fabric layers fail one after another, but not all at once. Plates of the same number of layers, but consisting of resin impregnated and co-cured layers that fail at once, demonstrate much lower impact resistance.

7.4. Manufacturing effects

As has been already noted, composite materials are formed in the process of fabrication of a composite structure, and their properties are strongly dependent on the type and parameters of the processing technology. This means that material specimens that are used to determine mechanical properties should be fabricated using the same manufacturing method that is expected to be applied to fabricate the structure under study.

7.4.1. Circumferential winding and tape overlap effect

To demonstrate the direct correlation that can exist between processing and material properties, consider the process of circumferential winding on a cylindrical surface as in Fig. 7.51. As a rule, the tapes are wound with some overlap w_0 shown in Fig. 7.52a. Introducing the dimensionless parameter

$$\lambda = \frac{w_0}{w} \tag{7.79}$$

Fig. 7.51. Winding of a circumferential layer. Courtesy of CRISM.

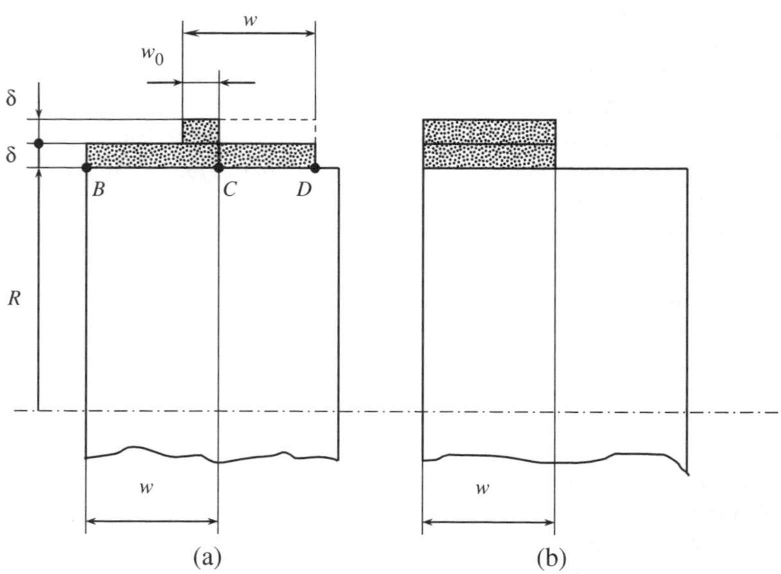

Fig. 7.52. Circumferential winding with (a) partial overlap $w_0 < w$ and (b) complete overlap $w_0 = w$.

we can conclude that for the case of complete overlap (Fig. 7.52b) we have $\lambda = 1$. The initial position of the tape placed with overlap w_0 as in Fig. 7.52a is shown in this figure with a dashed line, whereas the final position of the tapes is shown with solid lines. Assume that after the winding and curing are over, the resulting structure is a unidirectionally reinforced ring that is removed from the mandrel and loaded with internal pressure, so that the ring radius, being R before the loading, becomes R_1. Decompose the resultant force acting in the ring cross-section into two components, i.e.,

$$F = F' + F'' \tag{7.80}$$

and introduce the apparent stress acting along the fibers of the ring as

$$\sigma_1 = \frac{F}{A} \tag{7.81}$$

where $A = 2w\delta$ is the cross-sectional area of the ring made from two tapes as shown in Fig. 7.52. The force F'corresponds to part BC of the ring (Fig. 7.52a) and can be found as

$$F' = A'E_1\frac{R_1 - R}{R}$$

where $A' = (w + w_0)\delta$ is the cross-sectional area of this part of the ring and E_1 is the modulus of elasticity of the cured unidirectional composite. To calculate the force F'' that corresponds to part CD of the ring (Fig. 7.52a), we should take into account that the fibers start to take the load only when this part of the tape reaches the position indicated with dashed lines, i.e.,

$$F'' = A''E_1\frac{R_1 - (R + \delta)}{R}$$

where $A'' = (w - w_0)\delta$. With due regard to Eqs. (7.79), (7.80), and (7.81), we can write the result of the foregoing analysis in the following form

$$\sigma_1 = E_1\left[\varepsilon_1 - \frac{\delta}{2R}(1 - \lambda)\right] \tag{7.82}$$

Here, $\varepsilon_1 = (R_1 - R)/R$ is the apparent strain in the fiber direction. For complete overlap in Fig. 7.52b, $\lambda = 1$, and $\sigma_1 = E_1\varepsilon_1$. It should be noted that there exists also the so-called tape-to-tape winding for which $\lambda = 0$. This case cannot be described by Eq. (7.82) because of assumptions introduced in the derivation, and the resulting equation for this case is $\sigma_1 = E_1\varepsilon_1$.

It follows from Eq. (7.81), which is valid for winding without tension, that overlap of the tape results in reduction of material stiffness. Since the levels of loading for the fibers in the BC and CD parts of the ring (Fig. 7.52a) are different, a reduction in material strength can also be expected.

Filament winding is usually performed with some initial tension of the tape. This tension improves the material properties because it straightens the fibers and compacts the material. However, high tension may result in fiber damage and reduction in material strength. For glass and carbon fibers, the preliminary tension usually does not exceed 5% of the tape strength, whereas for aramid fibers, that are less susceptible to damage, the level of initial tension can reach 20% of the tape strength. Preliminary tension reduces the effect of the tape overlap discussed above and described by Eq. (7.82). However, this effect can show itself in a reduction in material strength, because the initial stresses which are induced by preliminary tension in the fibers can be different, and some fibers can be overloaded or underloaded by the external forces acting on the structure in operational conditions. Strength reduction of aramid–epoxy unidirectional composites with tape overlap has been observed in the experiments of Rach and Ivanovskii (1986) for winding on a 200-mm-diameter mandrel, as demonstrated in Fig. 7.53.

The absence of tape preliminary tension or low tension can cause ply waviness as shown in Fig. 7.54, which can occur in filament-wound laminates as a result of the pressure exerted by the overwrapped plies on the underwrapped plies or in flat laminates due to material shrinkage in the process of curing.

The simplest model for analysis is a regular waviness as presented in Fig. 7.54(a). To determine the apparent modulus in the x direction, we can use an expression similar to

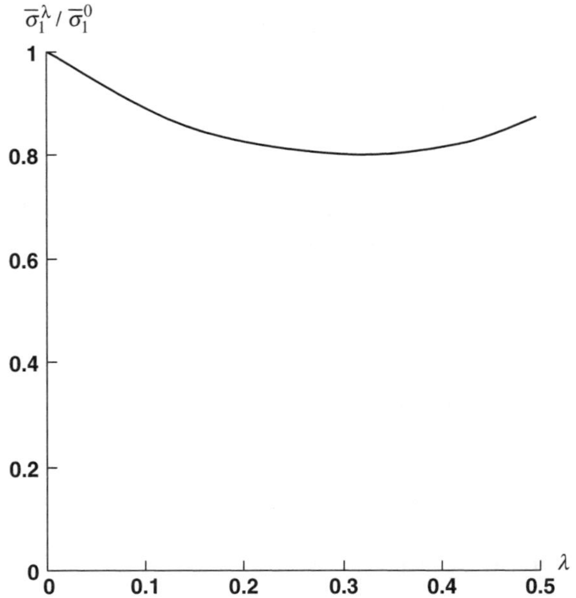

Fig. 7.53. Dependence of the normalized longitudinal strength of unidirectional aramid–epoxy composite on the tape overlap.

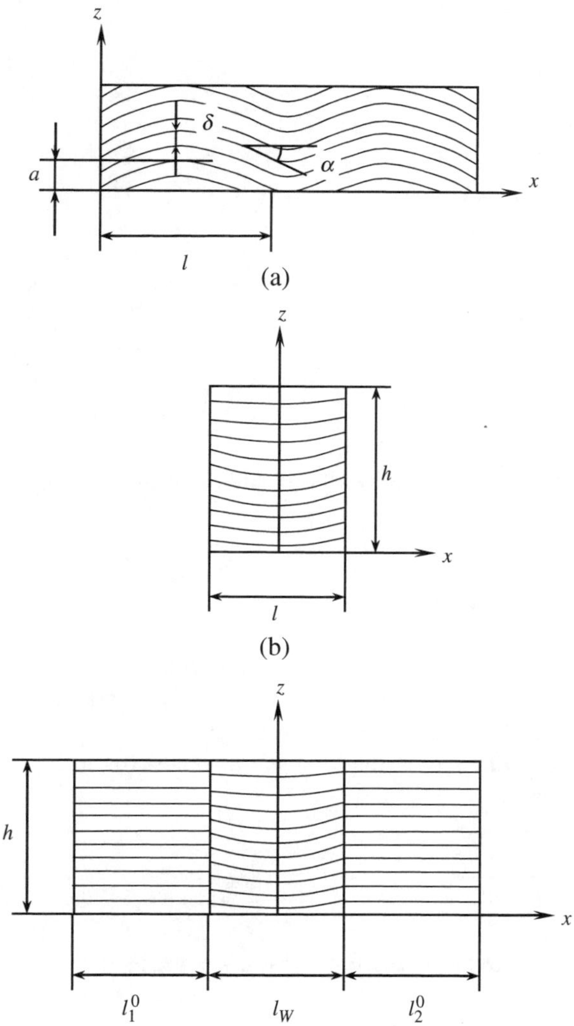

Fig. 7.54. Regular (a), through-the-thickness (b), and local (c) ply waviness.

the one presented in Eqs. (4.76), i.e.,

$$\frac{1}{E_x} = \frac{\cos^4\alpha}{E_1} + \frac{\sin^4\alpha}{E_3} + \left(\frac{1}{G_{13}} - \frac{2\nu_{31}}{E_1}\right)\sin^2\alpha\cos^2\alpha \tag{7.83}$$

Then, because the structure is periodic,

$$\frac{1}{E_x^{(r)}} = \frac{1}{l}\int_0^l \frac{\mathrm{d}x}{E_x} \tag{7.84}$$

Approximating the ply wave as

$$z = a \sin \frac{\pi x}{l}$$

where a is the amplitude, we get

$$\tan \alpha = \frac{\mathrm{d}z}{\mathrm{d}x} = f \cos \frac{\pi x}{l}$$

where $f = \pi a / l$. Substitution into Eqs. (7.83) and (7.84) and integration yields (Tarnopol'skii and Roze, 1969)

$$\frac{1}{E_x^{(r)}} = \frac{1}{2\lambda}\left[\frac{2+f^2}{E_1} + \frac{1}{E_3}\left(2\lambda - 2 - 3f^2\right) + \left(\frac{1}{G_{13}} - \frac{2\nu_{31}}{E_1}\right) f^2\right]$$

where $\lambda = (1 + f^2)^{3/2}$. Simplifying this result using the assumption that $f^2 \ll 1$, we arrive at

$$E_x^{(r)} = \frac{E_1}{1 + \frac{E_1 f^2}{2G_{13}}} \tag{7.85}$$

For glass–, carbon–, and aramid–epoxy composites with properties listed in Table 3.5, the dependencies corresponding to Eq. (7.85) are presented in comparison to the experimental results of Tarnopol'skii and Roze (1969) in Fig. 7.55.

If the ply waviness varies over the laminate thickness, as in Fig. 7.54(b), Eq. (7.85) can be generalized as

$$E_x^{(t)} = \frac{E_1}{h} \int_0^h \frac{dz}{1 + \frac{E_1}{2G_{13}} f^2(z)} \tag{7.86}$$

Finally, for only local waviness (see Fig. 7.54c), we obtain

$$\frac{1}{E_x^{(l)}} = \frac{\bar{l}_1^0}{E_1} + \frac{\bar{l}_w}{E_x^{(t)}} + \frac{\bar{l}_2^0}{E_1}$$

where

$$\bar{l}_{1,2}^0 = \frac{l_{1,2}^0}{l_1^0 + l_w + l_2^0}, \quad \bar{l}_w = \frac{l_w}{l_1^0 + l_w + l_2^0}$$

and $E_x^{(t)}$ is specified by Eq. (7.86).

Even moderate ply waviness dramatically reduces material strength under compression along the fibers, as can be seen in Fig. 7.56, which illustrates the experimental results of V. F. Kutinov for a unidirectional carbon–epoxy composite. The other strength characteristics of unidirectional composites are only slightly affected by the ply waviness.

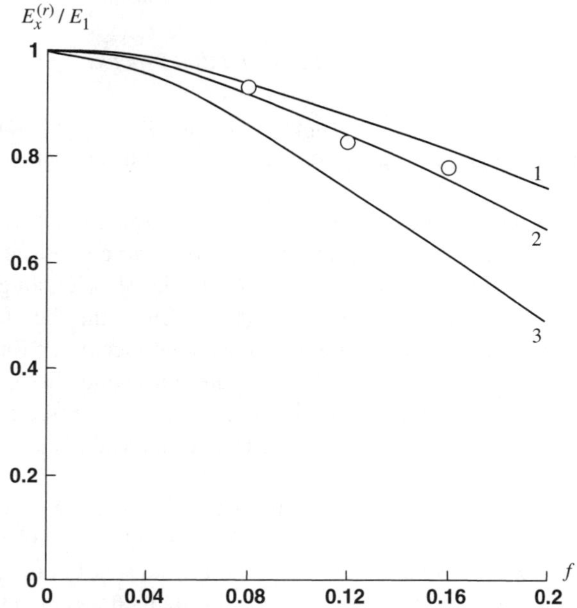

Fig. 7.55. Reduction of the normalized modulus with the ply waviness parameter, f, for (1) glass–, (2) carbon–, and (3) aramid–epoxy composites. ——— Eqs. (7.85), ○ experiment for glass–epoxy composite.

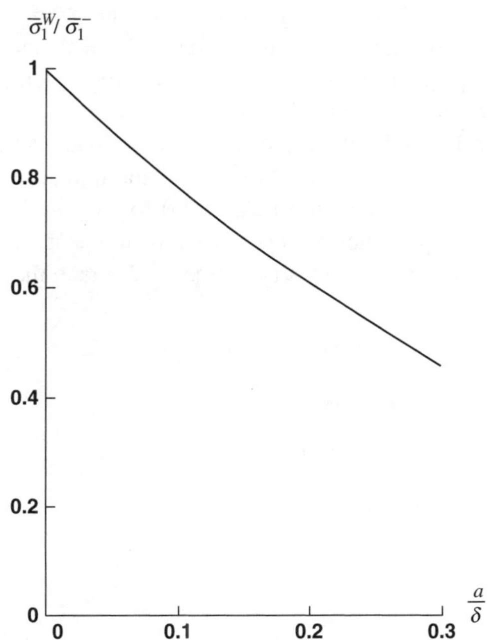

Fig. 7.56. Experimental dependence of carbon–epoxy composite longitudinal compression strength related to the corresponding strength of material without ply waviness on the ratio of the waviness amplitude to the ply thickness.

7.4.2. Warping and bending of laminates in fabrication process

There exist also some manufacturing operations that are specific for composites that cause stresses and strains appearing in composite structural elements in the process of their fabrication.

As an example, consider the problem of bending and warping of unsymmetric laminates during their fabrication. Assume that a laminated polymeric composite panel is cured at temperature T_c and cooled to room temperature T_0. Under slow cooling, the temperature change, $\Delta T = T_0 - T_c$, is the same for all the layers. Since the thus-fabricated panel is free of loading (i.e., no loads are applied to its edges or surfaces), the forces and moments in the left-hand sides of Eqs. (7.23) and (7.24) are zero, and these equations form a linear algebraic system for generalized strains ε_T, γ_T and κ_T. Integration of the strain-displacement equations, Eqs. (7.28), allows us to determine the shape of the fabricated panel.

Analysis of Eqs. (7.25) and (7.26), similar to that performed in Section 5.4, shows that for symmetric laminates $M_{mn}^T = 0$. Since $C_{mn} = 0$ for such laminates, the last three equations of Eqs. (7.23) in which $M_x = M_y = M_{xy} = 0$ form a set of homogeneous equations whose solution is $\kappa_{xT} = \kappa_{yT} = \kappa_{xyT} = 0$. This means that a flat symmetric panel does not acquire curvature in the process of cooling. Naturally, the in-plane dimensions of the panel become different from those that the panel had before cooling. The corresponding thermal strains ε_{xT}^0, ε_{yT}^0, and γ_{xyT}^0 can be found from the first three equations of Eqs. (7.23) in which $N_x = N_y = N_{xy} = 0$, but N_{11}^T, N_{22}^T, and N_{12}^T are not zero.

However, for unsymmetric laminates, in general, $M_{mn}^T \neq 0$, and these laminates experience bending and warping in the process of cooling. To demonstrate this, consider the two antisymmetric laminates studied in Section 5.8.

The first one is a two-layered orthotropic cross-ply laminate shown in Fig. 5.24. Using the stiffness coefficients calculated in Section 5.8, taking into account that for a cross-ply laminate $N_{12}^T = M_{12}^T = 0$, and applying Eqs. (7.23) for N_{xy} and M_{xy}, we get $\gamma_{xyT}^0 = 0$ and $\kappa_{xyT} = 0$. Thus, cooling of such a cross-ply laminated panel does not induce in-plane shear or twisting in it. The other four parts of Eqs. (7.23) take the form

$$\begin{aligned}
&a_{xx}\varepsilon_{xT}^0 + a_{xy}\varepsilon_{yT}^0 - c_{xx}\kappa_{xT} = n_x \\
&a_{yx}\varepsilon_{xT}^0 + a_{yy}\varepsilon_{yT}^0 + c_{yy}\kappa_{yT} = n_y \\
&-c_{xx}\varepsilon_{xT}^0 + b_{xx}\kappa_{xT} + b_{xy}\kappa_{yT} = m_x \\
&c_{yy}\varepsilon_{yT}^0 + b_{xy}\kappa_{xT} + b_{yy}\kappa_{yT} = m_y
\end{aligned} \tag{7.87}$$

where

$$a_{xx} = a_{yy} = h\overline{E}, \quad a_{xy} = a_{yx} = \overline{E}_1\nu_{12}h$$

$$c_{xx} = c_{yy} = \frac{h^2}{8}(\overline{E}_1 - \overline{E}_2), \quad b_{xx} = b_{yy} = \frac{h^3\overline{E}}{12}$$

$$b_{xy} = b_{yx} = \frac{h^3}{12}\overline{E}_1\nu_{12}, \quad \overline{E} = \frac{1}{2}(\overline{E}_1 + \overline{E}_2)$$

$$n_x = n_y = \frac{h}{2}[\overline{E}_1(\alpha_1 + \nu_{12}\alpha_2) + \overline{E}_2(\alpha_2 + \nu_{21}\alpha_1)]\Delta T$$

$$-m_x = m_y = \frac{h^2}{8}[\overline{E}_1(\alpha_1 + \nu_{12}\alpha_2) - \overline{E}_2(\alpha_2 + \nu_{21}\alpha_1)]\Delta T$$

The solutions to Eqs. (7.87) can be written as

$$\begin{aligned} \varepsilon_{xT}^0 &= \frac{n_x}{a_{xx} + a_{xy}} + \frac{c_{xx}}{a_{xx}^2 - a_{xy}^2}(a_{xx}\kappa_{xT} + a_{xy}\kappa_{yT}) \\ \varepsilon_{yT}^0 &= \frac{n_x}{a_{xx} + a_{xy}} - \frac{c_{xx}}{a_{xx}^2 - a_{xy}^2}(a_{xx}\kappa_{yT} + a_{xy}\kappa_{xT}) \end{aligned} \tag{7.88}$$

$$\kappa_{xT} = \frac{\tilde{m}_x}{e_{xx} - e_{xy}}, \quad \kappa_{yT} = -\frac{\tilde{m}_x}{e_{xx} - e_{xy}} \tag{7.89}$$

where

$$\tilde{m}_x = m_x + \frac{c_{xx}n_x}{a_{xx} + a_{xy}}$$

$$e_{xx} = b_{xx} - \frac{a_{xx}c_{xx}^2}{a_{xx}^2 - a_{xy}^2}, \quad e_{xy} = b_{xy} - \frac{a_{xy}c_{xx}^2}{a_{xx}^2 - a_{xy}^2}$$

As follows from Eqs. (7.88) and (7.89), ε and κ do not depend on x and y.

To find the in-plane displacements, we should integrate Eqs. (7.28), which have the form

$$\frac{\partial u}{\partial x} = \varepsilon_{xT}^0, \quad \frac{\partial v}{\partial y} = \varepsilon_{yT}^0, \quad \frac{\partial u}{\partial y} + \frac{\partial v}{\partial x} = 0$$

Referring the panel to coordinates x and y shown in Fig. 7.57 and assuming that $u(x = 0, y = 0) = 0$ and $v(x = 0, y = 0) = 0$, we get

$$u = \varepsilon_{xT}^0 x, \quad v = \varepsilon_{yT}^0 y \tag{7.90}$$

Now consider Eqs. (7.24) in which $V_x = V_y = 0$. Thus, $\gamma_{xT} = \gamma_{yT} = 0$, and Eqs. (7.30) yield $\theta_x = -\partial w/\partial x$, $\theta_y = -\partial w/\partial y$. The plate deflection can be found from Eqs. (7.29), which reduce to

$$\frac{\partial^2 w}{\partial x^2} = -\kappa_{xT}, \quad \frac{\partial^2 w}{\partial y^2} = -\kappa_{yT}, \quad \frac{\partial^2 w}{\partial x \partial y} = 0$$

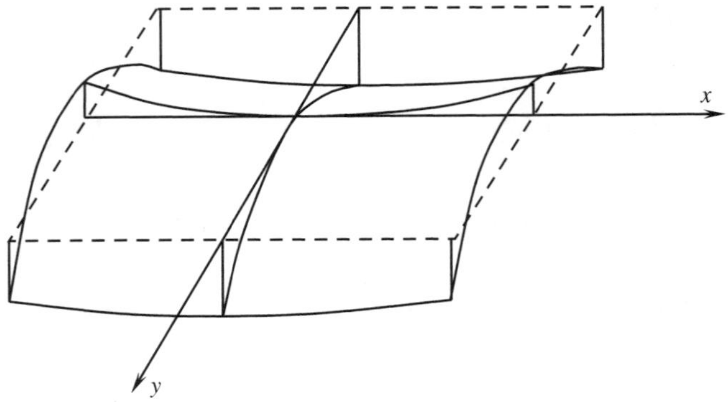

Fig. 7.57. Deformed shape of a cross-ply antisymmetric panel.

Assuming that $w(x=0,\ y=0)=0$, $\theta_x(x=0,\ y=0)=0$, and $\theta_y(x=0,\ y=0)=0$, we can write the result of the integration as

$$w=-\frac{1}{2}\left(\kappa_{xT}x^2+\kappa_{yT}y^2\right) \tag{7.91}$$

To present this solution in an explicit form, consider, for the sake of brevity, material with zero Poisson's ratios ($\nu_{12}=\nu_{21}=0$). Then, Eqs. (7.88)–(7.91) yield

$$u=\frac{\Delta Tx}{E_1+E_2}\left[E_1\alpha_1+E_2\alpha_2+6(E_1-E_2)\frac{E_1E_2(\alpha_2-\alpha_1)}{E_1^2+14E_1E_2-E_2^2}\right]$$

$$v=\frac{\Delta Ty}{E_1+E_2}\left[E_1\alpha_1+E_2\alpha_2+6(E_1-E_2)\frac{E_1E_2(\alpha_2-\alpha_1)}{E_1^2+14E_1E_2-E_2^2}\right]$$

$$w=-\frac{12\Delta T}{h}\cdot\frac{E_1E_2(\alpha_2-\alpha_1)}{E_1^2+14E_1E_2-E_2^2}\left(x^2-y^2\right)$$

The deformed shape of the panel is shown in Fig. 7.57. Note that displacements u and v correspond to the panel reference plane, which is the contact plane of the 0° and 90° layers (see Fig. 5.24).

Another typical antisymmetric structure is the two-layered angle-ply laminate shown in Fig. 5.25. Using the stiffness coefficients for this laminate calculated in Section 5.8 and Eqs. (7.25) and (7.27), we can write Eqs. (7.23) in the following form

$$A_{11}\varepsilon_{xT}^0+A_{12}\varepsilon_{yT}^0-\frac{h}{4}A_{14}\kappa_{xyT}=A_{11}^T$$

$$A_{12}\varepsilon_{xT}^0+A_{22}\varepsilon_{yT}^0-\frac{h}{4}A_{24}\kappa_{xyT}=A_{22}^T$$

$$A_{44}\gamma_{xyT}^{0} - \frac{h}{4}(A_{14}\kappa_{xT} + A_{24}\kappa_{yT}) = 0$$

$$-A_{14}\gamma_{xyT}^{0} + \frac{h}{3}(A_{11}\kappa_{xT} + A_{12}\kappa_{yT}) = 0$$

$$-A_{24}\gamma_{xyT}^{0} + \frac{h}{3}(A_{12}\kappa_{xT} + A_{22}\kappa_{yT}) = 0$$

$$A_{14}\varepsilon_{xT}^{0} + A_{24}\varepsilon_{yT}^{0} - \frac{h}{3}A_{44}\kappa_{xyT} = A_{12}^{T}$$

where

$$A_{11}^{T} = \left[\overline{E}_1(\alpha_1 + \nu_{12}\alpha_2)\cos^2\phi + \overline{E}_2(\alpha_2 + \nu_{21}\alpha_1)\sin^2\phi\right]\Delta T$$

$$A_{22}^{T} = \left[\overline{E}_1(\alpha_1 + \nu_{12}\alpha_2)\sin^2\phi + \overline{E}_2(\alpha_2 + \nu_{21}\alpha_1)\cos^2\phi\right]\Delta T$$

$$A_{12}^{T} = \left[\overline{E}_1(\alpha_1 + \nu_{12}\alpha_2) - \overline{E}_2(\alpha_2 + \nu_{21}\alpha_1)\right]\Delta T\sin\phi\cos\phi$$

The solution is

$$\varepsilon_{xT}^{0} = \frac{1}{A}\left[A_{11}^{T}A_{22} - A_{22}^{T}A_{12} + \frac{h}{4}(A_{14}A_{22} - A_{24}A_{12})\kappa_{xyT}\right]$$

$$\varepsilon_{yT}^{0} = \frac{1}{A}\left[A_{22}^{T}A_{11} - A_{11}^{T}A_{12} + \frac{h}{4}(A_{24}A_{11} - A_{14}A_{22})\kappa_{xyT}\right]$$

$$\gamma_{xyT}^{0} = 0, \quad \kappa_{xT} = 0, \quad \kappa_{yT} = 0$$

$$\kappa_{xyT} = \frac{A_{14}(A_{11}^{T}A_{22} - A_{22}^{T}A_{12}) + A_{24}\left(A_{22}^{T}A_{11} - A_{11}^{T}A_{12}\right) - A_{12}^{T}}{h\left[\frac{A}{3}A_{44} + \frac{1}{4}\left(2A_{14}A_{24}A_{12} - A_{14}^{2}A_{22} - A_{24}^{2}A_{11}\right)\right]}$$

where $A = A_{11}A_{22} - A_{12}^2$.

Thus, the panel under study experiences only in-plane deformation and twisting. Displacements u and v can be determined by Eqs. (7.28), whereas the following equations should be used to find w

$$\frac{\partial^2 w}{\partial x^2} = 0, \quad \frac{\partial^2 w}{\partial y^2} = 0, \quad \frac{\partial^2 w}{\partial x\,\partial y} = -\kappa_{xyT}$$

The result is

$$w = -\kappa_{xyT}xy$$

The deformed shape of the panel is shown in Fig. 7.58.

Depending on the laminates structures and dimensions, there exists a whole class of stable and unstable laminate configurations as studied by Hyer (1989).

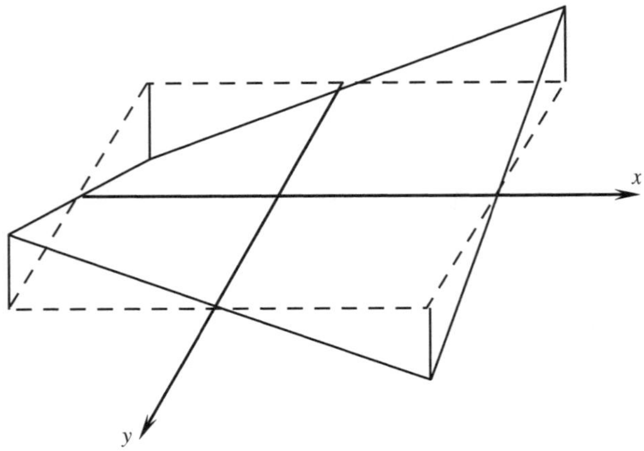

Fig. 7.58. Deformed shape of an angle-ply antisymmetric panel.

7.4.3. Shrinkage effects and residual strains

Deformation and warping of laminates appearing after the manufacturing process is completed can occur not only due to cooling of the cured composite but also as a result of material shrinkage due to release of tension in the fibers after the composite part is removed from the mandrel or chemical setting of the polymeric matrix.

To demonstrate these effects, consider a thin unidirectional layer formed from circumferential plies wound on a metallic cylindrical mandrel (see Fig. 7.59) under a tension. Since the stiffness of the mandrel is much higher than that of the layer, we can assume that, on cooling from the curing temperature T_c to room temperature T_0, the strains in the principal material coordinates of the layer are governed by the mandrel with which the cured layer is bonded, i.e.,

$$\varepsilon_1^T = \varepsilon_2^T = \alpha_0 \Delta T \tag{7.92}$$

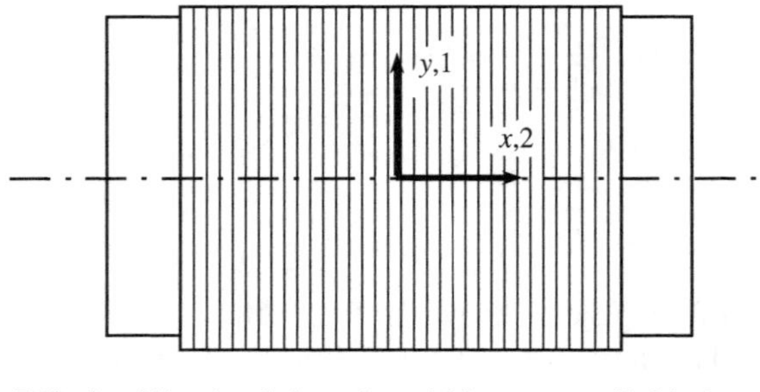

Fig. 7.59. A unidirectional circumferential layer on a cylindrical mandrel.

where α_0 is the CTE of the mandrel material, and $\Delta T = T_0 - T_c$. On the other hand, if the layer is cooled after being removed from the mandrel, its strains can be calculated as

$$\varepsilon_1 = \alpha_1 \Delta T + \varepsilon_1^0, \quad \varepsilon_2 = \alpha_2 \Delta T + \varepsilon_2^0 \tag{7.93}$$

The first terms in the right-hand sides of these equations are the free temperature strains along and across the fibers (see Fig. 7.59), whereas ε_1^0 and ε_2^0 correspond to the possible layer shrinkage in these directions.

Using Eqs. (7.92) and (7.93), we can determine the strains that appear in the layer when it is removed from the mandrel, i.e.,

$$\begin{aligned} \overline{\varepsilon}_1 &= \varepsilon_1 - \varepsilon_1^T = \varepsilon_1^0 + (\alpha_1 - \alpha_0)\Delta T \\ \overline{\varepsilon}_2 &= \varepsilon_2 - \varepsilon_2^T = \varepsilon_2^0 + (\alpha_2 - \alpha_0)\Delta T \end{aligned} \tag{7.94}$$

These strains can be readily found if we measure the layer diameter and length before and after it is removed from the mandrel. Then, the shrinkage strains can be determined as

$$\begin{aligned} \varepsilon_1^0 &= \overline{\varepsilon}_1 - (\alpha_1 - \alpha_0)\Delta T \\ \varepsilon_2^0 &= \overline{\varepsilon}_2 - (\alpha_2 - \alpha_0)\Delta T \end{aligned}$$

For a glass–epoxy composite with the following thermo-mechanical properties

$$E_1 = 37.24\,\text{GPa}, \quad E_2 = 2.37\,\text{GPa}, \quad G_{12} = 1.2\,\text{GPa}$$

$$\nu_{12} = 0.26, \quad \alpha_1 = 3.1 \times 10^{-6}\,1/^\circ\text{C}, \quad \alpha_2 = 25 \times 10^{-6}\,1/^\circ\text{C}$$

The measurements of Morozov and Popkova (1987) gave $\varepsilon_1^0 = -93.6 \times 10^{-5}$, $\varepsilon_2^0 = -64 \times 10^{-5}$. Further experiments performed for different winding tensions and mandrel materials have shown that, although the strain ε_1^0 strongly depends on these parameters, the strain ε_2^0 practically has no variation. This supports the assumption that the strain ε_2^0 is caused by chemical shrinkage of the resin and depends only on the resin's characteristics and properties.

For a cylinder in which the fibers make angle ϕ with the x-axis in Fig. 7.59, the strains induced by removal of the mandrel can be found from Eqs. (4.70), i.e.,

$$\begin{aligned} \overline{\varepsilon}_x &= \overline{\varepsilon}_1 \cos^2\phi + \overline{\varepsilon}_2 \sin^2\phi \\ \overline{\varepsilon}_y &= \overline{\varepsilon}_1 \sin^2\phi + \overline{\varepsilon}_2 \cos^2\phi \\ \overline{\gamma}_{xy} &= (\overline{\varepsilon}_1 - \overline{\varepsilon}_2)\sin 2\phi \end{aligned} \tag{7.95}$$

where $\overline{\varepsilon}_1$ and $\overline{\varepsilon}_2$ are specified by Eqs. (7.94). The dependencies of $\overline{\varepsilon}_x$, $\overline{\varepsilon}_y$, and $\overline{\gamma}_{xy}$ on ϕ, plotted with the aid of Eqs. (7.95), are shown in Fig. 7.60, together with the experimental data of Morozov and Popkova (1987). As can be seen, the composite cylinder experiences

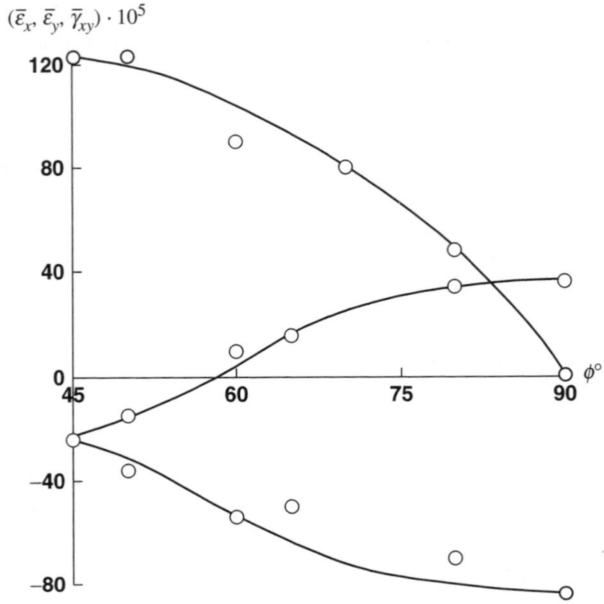

Fig. 7.60. Dependence of residual strains in a glass–epoxy filament wound cylinder on the winding angle: ——— calculation, ○ experiment.

in the general case not only a change in its length $(\bar{\varepsilon}_x)$ and diameter $\left(\bar{\varepsilon}_y\right)$, but also twist $\left(\bar{\gamma}_{xy}\right)$.

To study the $\pm\phi$ angle-ply layer, we should utilize the thermoelasticity constitutive equations, Eqs. (7.23). Neglecting the bending and coupling stiffness coefficients, we can write for the case under study

$$N_x = B_{11}\bar{\varepsilon}_x + B_{12}\bar{\varepsilon}_y - N_{1T}$$
$$N_y = B_{21}\bar{\varepsilon}_x + B_{22}\bar{\varepsilon}_y - N_{2T} \tag{7.96}$$

Applying these equations to an angle-ply composite cylinder removed from its mandrel, we should put $N_x = 0$, $N_y = 0$ because the cylinder is free of loads, and take $\varepsilon_1^T = \bar{\varepsilon}_1$, $\varepsilon_2^T = \bar{\varepsilon}_2$ in Eqs. (7.18), (7.25), and (7.26) that specify N_{1T} and N_{2T}. Then, Eqs. (7.96) yield the following expressions for the strains that appear in the angle-ply cylinder after it is removed from the mandrel

$$\bar{\varepsilon}_x = \frac{1}{B}(N_{1T}B_{22} - N_{2T}B_{12})$$

$$\bar{\varepsilon}_y = \frac{1}{B}(N_{2T}B_{11} - N_{1T}B_{12})$$

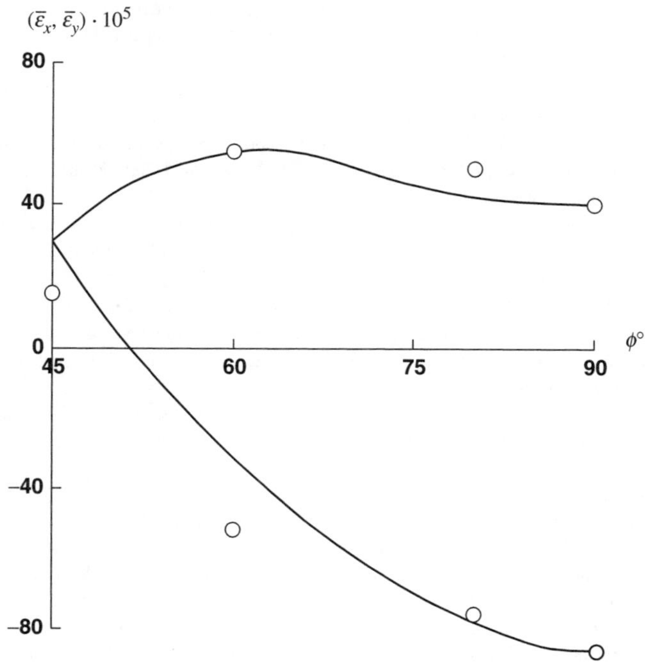

Fig. 7.61. Residual strains $\pm\phi$ angle-ply filament wound glass–epoxy cylinder, ——— calculation, ○ experiment.

where $B = B_{11}B_{22} - B_{12}^2$,

$$N_{1T} = h\left[\overline{E}_1(\bar{\varepsilon}_1 + \nu_{12}\bar{\varepsilon}_2)\cos^2\phi + \overline{E}_2(\bar{\varepsilon}_2 + \nu_{21}\bar{\varepsilon}_1)\sin^2\phi\right]$$

$$N_{2T} = h\left[\overline{E}_1(\bar{\varepsilon}_1 + \nu_{12}\bar{\varepsilon}_2)\sin^2\phi + \overline{E}_2(\bar{\varepsilon}_2 + \nu_{21}\bar{\varepsilon}_1)\cos^2\phi\right]$$

Here, $\bar{\varepsilon}_1$ and $\bar{\varepsilon}_2$ are given by Eqs. (7.94), $B_{mn} = A_{mn}h$, where A_{mn} are specified by Eqs. (4.72), and h is the cylinder thickness. The results of calculations for the experimental cylinder studied by Morozov and Popkova (1987) are presented in Fig. 7.61.

As follows from Figs. 7.60 and 7.61, the approach described above, based on constitutive equations for laminates, Eqs. (7.23), with the shrinkage characteristics of a unidirectional ply or an elementary layer determined experimentally, provides fair agreement between the predicted results and the experimental data.

7.5. References

Anderson, Ya.A., Mikelsons, M.Ya., Tamuzh, V.P. and Tarashuch I.V. (1991). Fatigue failure of laminated carbon-fiber-reinforced plastic. *Mechanics of Composite Materials*, **27**(1), 58–62.

Apinis, R.P., Mikelsons, M.Ya. and Khonichev, V.I. (1991). Fatigue resistance of carbon material in symmetric tension-compression. *Mechanics of Composite Materials*, **5**, 928–930, (in Russian).

Barnes, J.A., Simms, I.J., Farrow, G.J., Jackson, D., Wostenholm, G. and Yates, B. (1989). Thermal expansion behavior of thermoplastic composite materials. In *Proc. Am. Soc. for Composite Mater.*, Fourth Tech. Conf., Blacksburg, Virginia, Technomic Publ., pp. 717–725.

Ha, S.K. and Springer, G.S. (1987). Mechanical properties of graphite epoxy composites at elevated temperatures. In *Proc. 6th Int. Conf. on Composite Mater., (ICCM-6)*, July 1987, Vol. 4, (F.L. Matthews et al. eds.). Elsevier Applied Science, London, pp. 422–430.

Hamilton, J.G. and Patterson, J.M. (1993). Design and analysis of near-zero *CTE* laminates and application to spacecraft. In *Proc. 9th Int. Conf. on Composite Mater., (ICCM-9)*, Madrid, 12–16 July, Vol. 6, Composite Properties and Applications, Univ. of Zaragoza, Woodhead Publ. Ltd., pp. 108–119.

Hyer, M.W. (1989). Mechanics of unsymmetric laminates. *Handbook of Composites*, Vol. 2, *Structure and Design* (C.T. Herakovich and Yu.M. Tarnopol'skii eds.). Elsevier, Applied Science, pp. 86–114.

Limonov, V.A. and Anderson, Ya.A. (1991). Effect of stress ratio on the fatigue strength of organic plastics. *Mechanics of Composite Materials*, **27**(3), 276–283.

Milyutin, G.I., Bulmanis, V.N., Grakova, T.S., Popov, N.S. and Zakrzhevskii, A.M. (1989). Study and prediction of the strength characteristics of a wound epoxy organic-fiber plastic under different environmental effects. *Mechanics of Composite Materials*, **25**(2), 183–189.

Morozov, E.V. and Popkova, L.K. (1987). Combined theoretical and experimental method of determining residual stresses in wound composite shells. *Mechanics of Composite Materials*, **23**(6), 802–807.

Murakami, S., Kanagawa, Y., Ishida, T. and Tsushima, E. (1991). Inelastic deformation and fatigue damage of composite under multiaxial loading. In *Inelastic Deformation of Composite Materials* (G.J. Dvorak ed.). Springer Verlag, New York, pp. 675–694.

Ni, R.G. and Adams, R.D. (1984). The damping and dynamic moduli of symmetric laminated composite beams – theoretical and experimental results. *Mechanics of Composite Materials*, **18**(2), pp. 104–121.

Rabotnov, Yu. N. (1980). *Elements of Hereditary Solid Mechanics*. Mir Publishers, Moscow.

Rach, V.A. and Ivanovskii, V.S. (1986). On the effect of fiber length variation in filament wound structures. *Mechanics of Composite Materials*, Riga, 67–72 (in Russian).

Rogers, E.F., Phillips, L.M., Kingston-Lee, D.M. et al. (1977). The thermal expansion of carbon fiber-reinforced plastics. *Journal of Material Science*, **1**(12), 718–734.

Schapery, R.A. (1974). Viscoelastic behavior and analysis of composite materials. In *Composite Materials* (L.J. Broutman and R.H. Krock eds.), Vol. 2, *Mechanics of Composite Materials*, (G.P. Sendeckyj ed.). Academic Press Inc., New York, pp. 85–168.

Schulte, K., Reese, E. and Chou, T.-W. (1987). Fatigue behavior and damage development in woven fabric and hybrid fabric composites. In *Proc. 6th Int. Conf. on Composite Materials and 2nd European Conf. on Composite Mater., (ICCM and ECCM)*, Vol. 4, (F.L. Matthews, N.C.R. Buskel, J.M. Hodgkinson, and J. Morton eds.). Elsevier Science Ltd, London, pp. 89–99.

Shen, S.H. and Springer, G.S. (1976). Moisture absorption and desorption of composite materials. *Journal of Composite Materials*, **10**, 2–20.

Skudra, A.M., Bulavs, F.Ya., Gurvich, M.R. and Kruklinsh, A.A. (1989). *Elements of Structural Mechanics of Composite Truss Systems*. Zinatne, Riga, (in Russian).

Soutis, C. and Turkmen, D. (1993). High-temperature effects on the compressive strength of glass fiber-reinforced composites. In *Proc. 9th Int. Conf. on Composite Mater., (ICCM-9)*, Madrid, 12–16 July, Vol. 6, *Composite Properties and Applications*, Univ. of Zaragoza, Woodhead Publ. Ltd., pp. 581–588.

Springer, G.S. (ed.) (1981). *Environmental Effects on Composite Materials*. Vol. 1. Technomic Publ.

Springer, G.S. (ed.) (1984). *Environmental Effects on Composite Materials*. Vol. 2. Technomic Publ.

Springer, G.S. (ed.) (1988). *Environmental Effects on Composite Materials*. Vol. 3. Technomic Publ.

Strife, J.R. and Prevo, K.M. (1979). The thermal expansion behavior of unidirectional kevlar/epoxy composites. *Mechanics of Composite Materials*, **13**, 264–276.

Sukhanov, A.V., Lapotkin, V.A., Artemchuk, V.Ya. and Sobol', L.A. (1990). Thermal deformation of composites for dimensionally stable structures. *Mechanics of Composite Materials*, **26**(4), 432–436.

Survey (1984). *Application of Composite Materials in Aircraft Technology*. Central Aero-hydrodynamics Institute (in Russian).

Tamuzh, V.P. and Protasov V.D. (eds.). (1986). *Fracture of Composite Structures*. Zinatne, Riga (in Russian).

Tarnopol'skii, Yu.M. and Roze, A.V. (1969). *Specific Features of Analysis for Structural Elements of Reinforced Plastics*. Zinatne, Riga (in Russian).

Tsai, S.W. (1987). *Composite Design*, Think Composites Publ.

Vasiliev, V.V. and Sibiryakov, A.V. (1985). Propagation of elastic waves in a laminated strip. *Mech. Solids*, **1**, 104–109.

Verpoest, I., Li, L. and Doxsee, L. (1989). A combined micro-and macromechanical study of the impact behavior of carbon and glass fiber epoxy composites. In *Proc. 7th Int. Conf. on Composite Mater., (ICCM-7)*, November 1989, Guangzhou, China, Vol. 2, (W. Yunshu, G. Zhenlong and W. Rrenjie eds.). Pergamon Press, Oxford, pp. 316–321.

Zinoviev, P.A. and Ermakov, Yu.N. (1994). *Energy Dissipation in Composite Materials*, Technomic Publ.

Chapter 8

OPTIMAL COMPOSITE STRUCTURES

Advanced composite materials are characterized by high specific strength and stiffness and, in combination with automatic manufacturing processes, make it possible to fabricate composite structures with high levels of weight and cost efficiency. The replacement of metal alloys by composite materials, in general, reduces the structure's mass by 20–30%. However, in some special cases, the number of which progressively increases, the combination of material directional properties with design concept utilizing these properties, being supported by the advantages of modern composite technology, provides a major improvement in the structural performance. Such efficiency is demonstrated by composite structures of uniform strength in which the load is taken by uniformly stressed fibers.

To introduce composite structures of uniform strength, consider a laminated panel shown in Fig. 8.1 and loaded with in-plane forces N_x, N_y, and N_{xy} uniformly distributed along the panel edges. Let the laminate consist of k unidirectional composite layers characterized with thicknesses h_i and fiber orientation angles ϕ_i ($i = 1, 2, 3, \ldots, k$). For a plane stress state, the stacking-sequence of the layers is not important.

8.1. Optimal fibrous structures

To derive the optimality criterion specifying the best structure for the panel in Fig. 8.1, we first use the simplest monotropic model of a unidirectional composite (see Section 3.3) assuming that the forces N_x, N_y, and N_{xy} are taken by the fibers only. For the design problem, this is a reasonable approach because the transverse and shear strengths of a unidirectional composite ply (stresses $\overline{\sigma}_2$ and $\overline{\tau}_{12}$) are much lower than the ply strength in the longitudinal direction (stress $\overline{\sigma}_1$). Using Eqs. (4.68) in which we put $\sigma_2 = 0$ and $\tau_{12} = 0$, we can write the following equilibrium equations relating the applied forces to the stresses $\sigma_1^{(1)}$ in the direction of the fibers of the ith layer

$$N_x = \sum_{i=1}^{k} \sigma_x^{(i)} h_i = \sum_{i=1}^{k} \sigma_1^{(i)} h_i \cos^2 \phi_i, \quad N_y = \sum_{i=1}^{k} \sigma_y^{(i)} h_i = \sum_{i=1}^{k} \sigma_1^{(i)} h_i \sin^2 \phi_i$$
$$N_{xy} = \sum_{i=1}^{k} \tau_{xy}^{(i)} h_i = \sum_{i=1}^{k} \sigma_1^{(i)} h_i \sin \phi_i \cos \phi_i \tag{8.1}$$

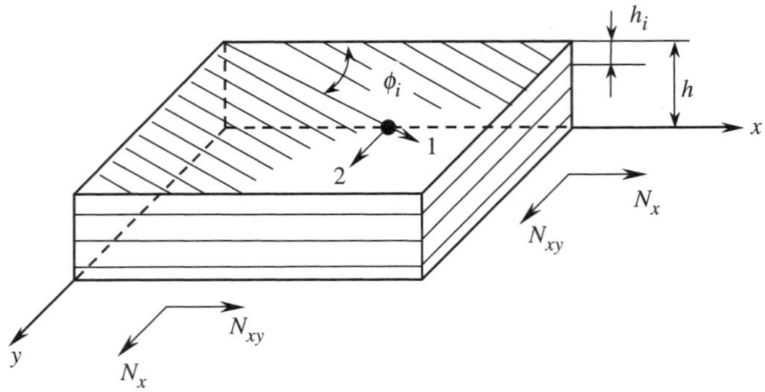

Fig. 8.1. A laminated plate in a plane state of stress.

The strain $\varepsilon_1^{(i)}$ in the fiber direction in the ith layer can be expressed in terms of strains in coordinates x, y with the aid of the first equation of Eqs. (4.69). Using the constitutive equations for the monotropic model of the ply, Eqs. (3.61), we arrive at

$$\sigma_1^{(i)} = E_1\varepsilon_1^{(i)} = E_1(\varepsilon_x \cos^2\phi_i + \varepsilon_y \sin^2\phi_i + \gamma_{xy}\sin\phi_i\cos\phi_i) \tag{8.2}$$

It is assumed that the layers are made of one and the same material.

Consider the design problem and stipulate, for example, that the best structure for the laminate is the one providing the minimum total thickness

$$h = \sum_{i=1}^{k} h_i \tag{8.3}$$

for the given combination of loads. Thus, we should minimize the laminate thickness in Eq. (8.3) subject to the constraints imposed by Eqs. (8.1) and (8.2). To solve this problem, we can use the method of Lagrange multipliers, according to which we should introduce multipliers λ and minimize the following augmented function

$$L = \sum_{i=1}^{k} h_i + \lambda_x\left(N_x - \sum_{i=1}^{k}\sigma_1^{(i)}h_i\cos^2\phi_i\right) + \lambda_y\left(N_y - \sum_{i=1}^{k}\sigma_1^{(i)}h_i\sin^2\phi_i\right)$$

$$+ \lambda_{xy}\left(N_{xy} - \sum_{i=1}^{k}\sigma_1^{(i)}h_i\sin\phi_i\cos\phi_i\right)$$

$$+ \sum_{i=1}^{k}\lambda_i\left[\sigma_i - E_1(\varepsilon_x\cos^2\phi_i + \varepsilon_y\sin^2\phi_i + \gamma_{xy}\sin\phi_i\cos\phi_i)\right]$$

with respect to the design variables h_i, ϕ_i, and multipliers λ, i.e.,

$$\frac{\partial L}{\partial h_i}=0, \quad \frac{\partial L}{\partial \phi_i}=0, \quad \frac{\partial L}{\partial \lambda_x}=\frac{\partial L}{\partial \lambda_y}=\frac{\partial L}{\partial \lambda_{xy}}=\frac{\partial L}{\partial \lambda_i}=0 \tag{8.4}$$

Minimization with respect to λ gives, obviously, the constraints in Eqs. (8.1) and (8.2), whereas the first two of Eqs. (8.4) yield

$$\sigma_1^{(i)}(\lambda_x \cos^2\phi_i+\lambda_y \sin^2\phi_i+\lambda_{xy}\sin\phi_i\cos\phi_i)=1 \tag{8.5}$$

$$h_i\sigma_1^{(i)}\left[(\lambda_y-\lambda_x)\sin 2\phi_i+\lambda_{xy}\cos 2\phi_i\right]=E_1\lambda_i\left[(\varepsilon_y-\varepsilon_x)\sin 2\phi_i+\varepsilon_{xy}\cos 2\phi_i\right] \tag{8.6}$$

The solution of Eq. (8.6) is

$$\lambda_x=E_1\varepsilon_x\frac{\lambda_i}{h_i\sigma_1^{(i)}}, \quad \lambda_y=E_1\varepsilon_y\frac{\lambda_i}{h_i\sigma_1^{(i)}}, \quad \lambda_{xy}=E_1\gamma_{xy}\frac{\lambda_i}{h_i\sigma_1^{(i)}}$$

These equations allow us to conclude that

$$\frac{\lambda_i}{h_i\sigma_1^{(i)}}=\frac{\lambda_x}{E_1\varepsilon_x}=\frac{\lambda_y}{E_1\varepsilon_y}=\frac{\lambda_{xy}}{E_1\gamma_{xy}}=\frac{1}{c^2}$$

where c is some constant. Substituting λ_x, λ_y, and λ_{xy} from these equations into Eq. (8.5) and taking into account Eq. (8.2), we have

$$\left(\sigma_1^{(i)}\right)^2=c^2 \tag{8.7}$$

This equation has two solutions: $\sigma_1^{(i)}=\pm c$.

Consider the first case, i.e., $\sigma_1^{(i)}=c$. Adding the first two equations of Eqs. (8.1) and taking into account Eq. (8.9), we have

$$h=\frac{1}{c}(N_x+N_y)$$

Obviously, the minimum value of h corresponds to $c=\overline{\sigma}_1$, where $\overline{\sigma}_1$ is the ultimate stress. Thus, the total thickness of the optimal plate is

$$h=\frac{1}{\overline{\sigma}_1}(N_x+N_y) \tag{8.8}$$

Taking now $\sigma_1^{(i)}=\overline{\sigma}_1$ in Eqs. (8.1) and eliminating $\overline{\sigma}_1$ with the aid of Eq. (8.8), we arrive at the following two optimality conditions in terms of the design variables and acting forces

$$\sum_{i=1}^{k} h_i(N_x \sin^2\phi_i - N_y \cos^2\phi_i) = 0 \tag{8.9}$$

$$\sum_{i=1}^{k} h_i[(N_x + N_y)\sin\phi_i \cos\phi_i - N_{xy}] = 0 \tag{8.10}$$

Thus, $2k$ design variables, i.e., k values of h_i and k values of ϕ_i, should satisfy these three equations, Eqs. (8.8)–(8.10). All possible optimal laminates have the same total thickness in Eq. (8.8). As follows from Eq. (8.2), the condition $\sigma_1^{(i)} = \overline{\sigma}_1$ is valid, in the general case, if $\varepsilon_x = \varepsilon_y = \varepsilon$ and $\gamma_{xy} = 0$. Applying Eqs. (4.69) to determine the strains in the principal material coordinates of the layers, we arrive at the following result $\varepsilon_1 = \varepsilon_2 = \varepsilon$ and $\gamma_{12} = 0$. This means that the optimal laminate is the structure of uniform stress and strain in which the fibers in each layer coincide with the directions of principal strains. An important feature of the optimal laminate follows from the last equation of Eqs. (4.168) which yields $\phi_i' = \phi_i$. Thus, the optimal angles do not change under loading.

Introducing the new variables

$$\overline{h}_i = \frac{h_i}{h}, \quad n_y = \frac{N_y}{N_x}, \quad n_{xy} = \frac{N_{xy}}{N_x}, \quad \lambda = \frac{1}{1 + n_y}$$

and taking into account that

$$\sum_{i=1}^{k} \overline{h}_i = 1 \tag{8.11}$$

we can transform Eqs. (8.8)–(8.10) that specify the structural parameters of the optimal laminate to the following final form

$$h = \frac{N_x}{\lambda \overline{\sigma}_1} \tag{8.12}$$

$$\sum_{i=1}^{k} \overline{h}_i \cos^2\phi_i = \lambda, \quad \sum_{i=1}^{k} \overline{h}_i \sin^2\phi_i = \lambda n_y \tag{8.13}$$

$$\sum_{i=1}^{k} \overline{h}_i \sin\phi_i \cos\phi_i = \lambda n_{xy} \tag{8.14}$$

For uniaxial tension in the x-direction, we have $n_y = n_{xy} = 0$, $\lambda = 1$. Then, Eqs. (8.13) yield $\phi_i = 0$ $(i = 1, 2, 3, \ldots, k)$ and Eq. (8.12) gives the obvious result $h = N_x/\overline{\sigma}$.

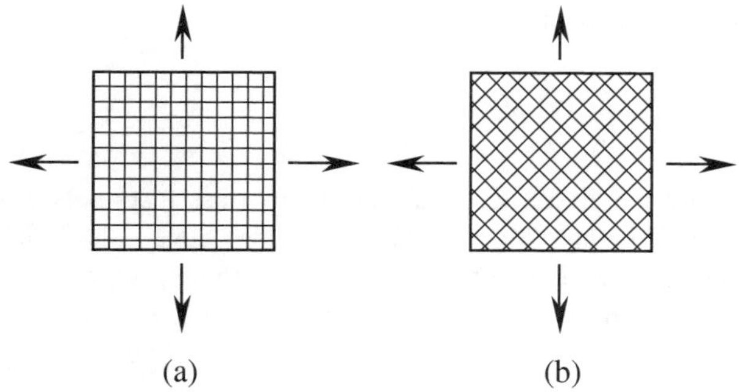

Fig. 8.2. Cross-ply (a) and ±45° angle-ply (b) optimal structures for uniform tension.

To describe tension in two orthogonal directions x and y, we should put $n_{xy} = 0$. It follows from Eq. (8.14) that the laminate structure in this case should be symmetric, i.e., each layer with angle $+\phi_i$ should be accompanied by a layer of the same thickness but with angle $-\phi_i$.

Consider, for example, uniform biaxial tension such that $N_x = N_y = N$, $N_{xy} = 0$, $n_y = 1$, $n_{xy} = 0$, $\lambda = 0.5$. For this case, Eqs. (8.12) and (8.13) yield

$$h = \frac{2N}{\overline{\sigma}_1}, \quad \sum_{i=1}^{k} \overline{h}_i \cos 2\phi_i = 0 \tag{8.15}$$

The natural structure for this case corresponds to the cross-ply laminate for which $k = 2$, $\phi_1 = 0°$, $\phi_2 = 90°$ (Fig. 8.2a). Then, the second equation of Eqs. (8.15) gives the evident result $\overline{h}_1 = \overline{h}_2$.

Consider the first equation, from which it follows that the total thickness of the optimal laminate is twice the thickness of a metal plate under the same loading conditions. This result is quite natural because, in contrast to isotropic materials, the monotropic layer can work in only one direction – along the fibers. So, we need to have the 0°-layer to take $N_x = N$ and the same, but 90°-layer to take $N_y = N$. From this we can conclude that the directional character of a composite ply's stiffness and strength is actually a material shortcoming rather than its advantage. The real advantages of composite materials are associated with their high specific strength provided by thin fibers (see Section 3.2.1), and if we had isotropic materials with such specific strength, no composites would be developed and implemented.

Return now to the second equation of Eqs. (8.15) which shows that, in addition to a cross-ply laminate, there exists an infinite number of optimal structures. For example, this equation is satisfied for a symmetric ±45° angle-ply laminate (Fig. 8.2b). Moreover, all the quasi-isotropic laminates discussed in Section 5.7 and listed in Table 5.3 satisfy the optimality conditions for uniform tension.

A loading case, which is important for actual applications, corresponds to a cylindrical pressure vessel, as considered in Section 6.3. The winding of such a vessel is shown in

Fig. 7.51. For this type of loading,

$$N_x = \frac{1}{2}pR, \quad N_y = pR, \quad N_{xy} = 0$$

where N_x and N_y are the circumferential and axial stress resultants, respectively, p the internal pressure and R is the cylinder radius. Thus, we have $n_y = 2$ and $\lambda = 1/3$. Since $N_{xy} = 0$, the structure of the laminate is symmetric with respect to the cylinder meridian, and Eqs. (8.12)–(8.14) can be reduced to

$$h = \frac{3pR}{2\overline{\sigma}_1} \tag{8.16}$$

$$\sum_{i=1}^{k} \overline{h}_i (3\cos^2\phi_i - 1) = 0 \tag{8.17}$$

Comparing Eq. (8.16) with the corresponding expression for the thickness of a metal pressure vessel, which is $h_m = pR/\overline{\sigma}$, we can see that the thickness of an optimal composite vessel is 1.5 times more than h_m. Nevertheless, because of their higher strength and lower density, composite pressure vessels are significantly lighter than metal ones. To show this, consider pressure vessels with radius $R = 100$ mm made of different materials and designed for a burst pressure $p = 20$ MPa. The results are listed in Table 8.1. As can be seen, the thickness of a glass–epoxy vessel is the same as that for the thickness of a steel vessel, because the factor 1.5 in Eq. (8.16) is compensated by the composite's strength which is 1.5 times greater than the strength of steel. However, the density of a glass–epoxy composite is much lower than the density of steel, and as a result, the mass of unit surface area of the composite vessel is only 27% of the corresponding value for a steel vessel. The most promising materials for pressure vessels are aramid and carbon composites, which have the highest specific tensile strength (see Table 8.1).

Consider Eq. (8.17) which shows that there can exist an infinite number of optimal laminates with one and the same thickness specified by Eq. (8.16).

Table 8.1
Parameters of metal and composite pressure vessels.

Parameter	Material					
	Steel	Aluminum	Titanium	Glass–epoxy	Carbon–epoxy	Aramid–epoxy
Strength, $\overline{\sigma}$, $\overline{\sigma}_1$ (MPa)	1200	500	900	1800	2000	2500
Density, ρ (g/cm^3)	7.85	2.7	4.5	2.1	1.55	1.32
Thickness of the vessel, h_m, h (mm)	1.67	4.0	2.22	1.67	1.5	1.2
Mass of the unit surface area, ρh (kg/m^2)	13.11	10.8	10.0	3.51	2.32	1.58

The simplest is a cross-ply laminate having $k = 2$, $\phi_1 = 0°$, $h_1 = h_0$, and $\phi_2 = 90°$, $h_2 = h_{90}$. For this structure, Eq. (8.17) yields $h_{90} = 2h_0$. This result seems obvious because $N_y/N_x = 2$. For symmetric $\pm\phi$ angle-ply laminate, we should take $k = 2$, $h_1 = h_2 = h_\phi/2$, $\phi_1 = +\phi$, $\phi_2 = -\phi$. Then,

$$\cos^2\phi = \frac{1}{3}, \quad \phi = \phi_0 = 54.44°$$

As a rule, helical plies are combined with circumferential plies as in Fig. 7.51. For this case, $k = 3$, $h_1 = h_2 = h_\phi/2$, $\phi_1 = -\phi_2 = \phi$, $h_3 = h_{90}$, $\phi_3 = 90°$, and Eq. (8.17) gives

$$\frac{h_{90}}{h_\phi} = 3\cos^2\phi - 1 \tag{8.18}$$

Since the thickness cannot be negative, this equation is valid for $0 \le \phi \le \phi_0$. For $\phi_0 \le \phi \le 90°$, the helical layers should be combined with an axial one, i.e., we should put $k = 3$, $h_1 = h_2 = h_\phi/2$, $\phi_1 = -\phi_2 = \phi$ and $h_3 = h_0$, $\phi_3 = 0°$. Then,

$$\frac{h_0}{h_\phi} = \frac{1}{2}(1 - 3\cos^2\phi) \tag{8.19}$$

The dependencies corresponding to Eqs. (8.18) and (8.19) are presented in Fig. 8.3. As an example, consider a filament wound pressure vessel whose parameters are listed in Table 6.1. The cylindrical part of the vessel shown in Figs. 4.14 and 6.22 consists of a $\pm 36°$ angle-ply helical layer and a circumferential layer whose thicknesses $h_1 = h_\phi$ and $h_2 = h_{90}$ are presented in Table 6.1. The ratio h_{90}/h_ϕ for two experimental vessels is 0.97 and 1.01, whereas Eq. (8.18) gives for this case $h_{90}/h_\phi = 0.96$ which shows that both vessels are close to optimal structures. Laminates reinforced with uniformly stressed fibers can exist under some restrictions imposed on the acting forces N_x, N_y, and N_{xy}. Such restrictions follow from Eqs. (8.13) and (8.14) under the conditions that $\overline{h}_i \ge 0$, $0 \le \sin^2\phi_i$, $\cos^2\phi_i \le 1$ and have the form

$$0 \le \lambda \le 1, \quad -\frac{1}{2} \le \lambda n_{xy} \le \frac{1}{2}$$

In particular, Eqs. (8.13) and (8.14) do not describe the case of pure shear for which only the shear stress resultant, N_{xy}, is not zero. This is quite natural because the strength condition $\sigma_1^{(i)} = \overline{\sigma}_1$ under which Eqs. (8.12)–(8.14) are derived is not valid for shear inducing tension and compression in angle-ply layers.

To study in-plane shear of the laminate, we should use both solutions of Eq. (8.7) and assume that for some layers, e.g., with $i = 1, 2, 3, \ldots, n-1$, $\sigma_1^{(i)} = \overline{\sigma}_1$ whereas for the other layers ($i = n, n+1, n+2, \ldots, k$), $\sigma_1^{(i)} = -\overline{\sigma}_1$. Then, Eqs. (8.1) can be reduced to

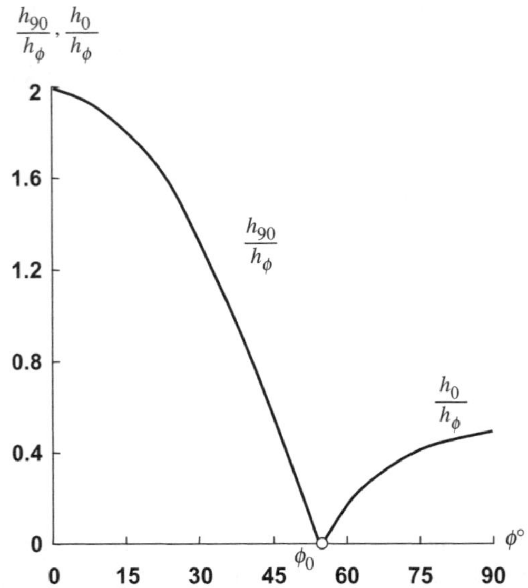

Fig. 8.3. Optimal thickness ratios for a cylindrical pressure vessel consisting of $\pm\phi$ helical plies combined with circumferential (90°) or axial (0°) plies.

the following form

$$N_x + N_y = \overline{\sigma}_1(h^+ - h^-) \tag{8.20}$$

$$N_x - N_y = \overline{\sigma}_1\left(\sum_{i=1}^{n-1} h_i^+ \cos 2\phi_i - \sum_{i=n}^{k} h_i^- \cos 2\phi_i\right) \tag{8.21}$$

$$N_{xy} = \frac{1}{2}\overline{\sigma}_1\left(\sum_{i=1}^{n-1} h_i^+ \sin 2\phi_i - \sum_{i=n}^{k} h_i^- \sin 2\phi_i\right) \tag{8.22}$$

where

$$h^+ = \sum_{i=1}^{n-1} h_i^+, \quad h^- = \sum_{i=n}^{k} h_i^-$$

are the total thicknesses of the plies with tensile and compressive stresses in the fibers, respectively.

For the case of pure shear ($N_x = N_y = 0$), Eqs. (8.20) and (8.21) yield $h^+ = h^-$ and $\phi_i = \pm 45°$. Then, assuming that $\phi_i = +45°$ for the layers with $h_i = h_i^+$, whereas $\phi_i = -45°$ for the layers with $h_i = h_i^-$, we get from Eq. (8.22)

$$h = h^+ + h^- = \frac{2N_{xy}}{\overline{\sigma}_1}$$

The optimal laminate, as follows from the foregoing derivation, corresponds to a $\pm 45^\circ$ angle-ply structure as shown in Fig. 8.2b.

8.2. Composite laminates of uniform strength

Consider again the panel in Fig. 8.1 and suppose that unidirectional plies or fabric layers, that form the panel are orthotropic, i.e., in contrast to the previous section, we do not now neglect stresses σ_2 and τ_{12} in comparison with σ_1 (see Fig. 3.29). Then, the constitutive equations for the panel in a plane stress state are specified by the first three equations in Eqs. (5.35), i.e.,

$$\begin{aligned} N_x &= B_{11}\varepsilon_x + B_{12}\varepsilon_y + B_{14}\gamma_{xy} \\ N_y &= B_{21}\varepsilon_x + B_{22}\varepsilon_y + B_{24}\gamma_{xy} \\ N_{xy} &= B_{41}\varepsilon_x + B_{42}\varepsilon_y + B_{44}\gamma_{xy} \end{aligned} \tag{8.23}$$

where, in accordance with Eqs. (4.72), (5.28), and (5.42)

$$\begin{aligned} B_{11} &= \sum_{i=1}^{k} h_i\left(\overline{E}_1^{(i)}\cos^4\phi_i + \overline{E}_2^{(i)}\sin^4\phi_i + 2E_{12}^{(i)}\sin^2\phi_i\cos^2\phi_i\right) \\ B_{12} = B_{21} &= \sum_{i=1}^{k} h_i\left[\overline{E}_1^{(i)}\nu_{12}^{(i)} + \left(\overline{E}_1^{(i)} + \overline{E}_2^{(i)} - 2E_{12}^{(i)}\right)\sin^2\phi_i\cos^2\phi_i\right] \\ B_{22} &= \sum_{i=1}^{k} h_i\left(\overline{E}_1^{(i)}\sin^4\phi_i + \overline{E}_2^{(i)}\cos^4\phi_i + 2E_{12}^{(i)}\sin^2\phi_i\cos^2\phi_i\right) \\ B_{14} = B_{41} &= \sum_{i=1}^{k} h_i\left(\overline{E}_1^{(i)}\cos^2\phi_i - \overline{E}_2^{(i)}\sin^2\phi_i - E_{12}^{(i)}\cos 2\phi_i\right)\sin\phi_i\cos\phi_i \\ B_{24} = B_{42} &= \sum_{i=1}^{k} h_i\left(\overline{E}_1^{(i)}\sin^2\phi_i - \overline{E}_2^{(i)}\cos^2\phi_i + E_{12}^{(i)}\cos 2\phi_i\right)\sin\phi_i\cos\phi_i \\ B_{44} &= \sum_{i=1}^{k} h_i\left[\left(\overline{E}_1^{(i)} + \overline{E}_2^{(i)} - 2\overline{E}_1^{(i)}\nu_{12}^{(i)}\right)\sin^2\phi_i\cos^2\phi_i + G_{12}^{(i)}\cos^2 2\phi_i\right] \end{aligned} \tag{8.24}$$

and $\overline{E}_{1,2}^{(i)} = \dfrac{E_{1,2}^{(i)}}{1 - \nu_{12}^{(i)}\nu_{21}^{(i)}}$, $E_{12}^{(i)} = \overline{E}_1^{(i)}\nu_{12}^{(i)} + 2G_{12}^{(i)}$.

In the general case, the panel can consist of layers made of different composite materials. Using the optimality criterion developed in the previous section for fibrous structures, we suppose that the fibers in each layer are directed along the lines of principal strains, or principal stresses because $\tau_{12}^{(i)} = G_{12}\gamma_{12}^{(i)}$ for an orthotropic layer and the condition $\gamma_{12}^{(i)} = 0$ is equivalent to the condition $\tau_{12}^{(i)} = 0$ (see Section 2.4). Using the third equation in Eqs. (4.69), we can write these conditions as

$$2(\varepsilon_y - \varepsilon_x)\sin\phi_i\cos\phi_i + \gamma_{xy}\cos 2\phi_i = 0 \tag{8.25}$$

This equation can be satisfied for all the layers if we take

$$\varepsilon_x = \varepsilon_y = \varepsilon, \quad \gamma_{xy} = 0 \tag{8.26}$$

Then, Eqs. (8.23) yield

$$N_x = (B_{11} + B_{12})\varepsilon, \quad N_y = (B_{21} + B_{22})\varepsilon, \quad N_{xy} = (B_{41} + B_{42})\varepsilon$$

These equations allow us to find the strain, i.e.,

$$\varepsilon = \frac{N_x + N_y}{B_{11} + 2B_{12} + B_{22}} \tag{8.27}$$

and to write two relationships specifying the optimal structural parameters of the laminate

$$(B_{11} + B_{12})N_y - (B_{21} + B_{22})N_x = 0$$
$$(B_{41} + B_{42})(N_x + N_y) - (B_{11} + 2B_{12} + B_{22})N_{xy} = 0$$

Substitution of B_{mn} from Eqs. (8.24) results in the following explicit form of these conditions

$$\begin{aligned}
&\sum_{i=1}^{k} h_i\Big[\overline{E}_1^{(i)}\Big(1+\nu_{12}^{(i)}\Big)\Big(N_x\sin^2\phi_i - N_y\cos^2\phi_i\Big)\\
&\qquad + \overline{E}_2^{(i)}\Big(1+\nu_{21}^{(i)}\Big)\Big(N_x\cos^2\phi_i - N_y\sin^2\phi_i\Big)\Big] = 0\\
&\sum_{i=1}^{k} h_i\Big\{(N_x+N_y)\Big(\overline{E}_1^{(i)} - \overline{E}_2^{(i)}\Big)\sin\phi_i\cos\phi_i\\
&\qquad - N_{xy}\Big[\overline{E}_1^{(i)}\Big(1+\nu_{12}^{(i)}\Big) + \overline{E}_2^{(i)}\Big(1+\nu_{21}^{(i)}\Big)\Big]\Big\} = 0
\end{aligned} \tag{8.28}$$

To determine the stresses that act in the optimal laminate, we use Eqs. (4.69) and (8.26) that specify the strains in the principal material coordinates of the layers as $\varepsilon_1 = \varepsilon_2 = \varepsilon$,

$\gamma_{12} = 0$. Applying constitutive equations, Eqs. (4.56), substituting ε from Eq. (8.27) and writing the result in explicit form with the aid of Eqs. (8.24), we arrive at

$$
\begin{aligned}
\sigma_1^{(i)} &= \frac{\overline{E}_1^{(i)}}{S_i}\left(1+\nu_{12}^{(i)}\right)(N_x+N_y) \\
\sigma_2^{(i)} &= \frac{\overline{E}_2^{(i)}}{S_i}\left(1+\nu_{21}^{(i)}\right)(N_x+N_y) \\
\tau_{12}^{(i)} &= 0
\end{aligned}
\tag{8.29}
$$

where

$$S_i = \sum_{i=1}^{k} h_i\left[\overline{E}_1^{(i)}\left(1+\nu_{12}^{(i)}\right) + \overline{E}_2^{(i)}\left(1+\nu_{21}^{(i)}\right)\right]$$

is the laminate stiffness coefficient.

If all the layers are made from the same material, Eqs. (8.28) and (8.29) are simplified as

$$
\begin{aligned}
&\sum_{i=1}^{k} h_i\left[N_x \sin^2\phi_i - N_y \cos^2\phi_i + n\left(N_x \cos^2\phi_i - N_y \sin^2\phi_i\right)\right] = 0 \\
&\sum_{i=1}^{k} h_i\left[m(N_x+N_y)\sin\phi_i \cos\phi_i - (1+n)\,N_{xy}\right] = 0
\end{aligned}
\tag{8.30}
$$

$$\sigma_1^{(i)} = \sigma_1 = \frac{N_x+N_y}{h(1+n)}, \quad \sigma_2^{(i)} = \sigma_2 = \frac{n(N_x+N_y)}{h(1+n)}, \quad \tau_{12}^{(i)} = 0 \tag{8.31}$$

in which

$$n = \frac{E_2(1+\nu_{21})}{E_1(1+\nu_{12})}, \quad m = \frac{E_1 - E_2}{E_1(1+\nu_{12})}, \quad h = \sum_{i=1}^{k} h_i$$

Laminates of uniform strength exist under the following restrictions

$$\frac{n}{1+n} \leq \frac{N_x}{N_x+N_y} \leq \frac{1}{1+n}, \quad \left|\frac{N_{xy}}{N_x+N_y}\right| \leq \frac{1-n}{2(1+n)}$$

For the monotropic model of a unidirectional ply considered in the previous section, $n = 0$, $m = 1$, and Eqs. (8.30) reduce to Eqs. (8.9) and (8.10).

To determine the thickness of the optimal laminate, we should use Eqs. (8.31) in conjunction with one of the strength criteria discussed in Chapter 6. For the simplest case, using the maximum stress criterion in Eqs. (6.2), the thickness of the laminate can be found from the following conditions $\sigma_1 = \overline{\sigma}_1$ or $\sigma_2 = \overline{\sigma}_2$, so that

$$h_1 = \frac{N_x + N_y}{(1+n)\overline{\sigma}_1}, \quad h_2 = \frac{n(N_x + N_y)}{(1+n)\overline{\sigma}_2} \tag{8.32}$$

Obviously, for the optimal structure, we would like to have $h_1 = h_2$. However, this can happen only if material characteristics meet the following condition

$$\frac{\overline{\sigma}_2}{\overline{\sigma}_1} = n = \frac{E_2(1+\nu_{21})}{E_1(1+\nu_{12})} \tag{8.33}$$

The results of calculations for typical materials whose properties are listed in Tables 3.5 and 4.4 are presented in Table 8.2. As can be seen, Eq. (8.33) is approximately valid for fabric composites whose stiffness and strength in the warp and fill directions (see Section 4.6) are controlled by fibers of the same type. However, for unidirectional polymeric and metal matrix composites, whose longitudinal stiffness and strength are governed by the fibers and transverse characteristics are determined by the matrix properties, $\overline{\sigma}_2/\overline{\sigma}_1 \ll n$. In accordance with Eqs. (8.32), this means that $h_1 \ll h_2$, and the ratio h_2/h_1 varies from 12.7 for glass–epoxy to 2.04 for boron–epoxy composites. Now, return to the discussion presented in Section 4.4.2 from which it follows that in laminated composites, transverse stresses σ_2 reaching their ultimate value, $\overline{\sigma}_2$, cause cracks in the matrix, which do not result in failure of the laminate whose strength is controlled by the fibers. To describe the laminate with cracks in the matrix (naturally, if cracks are allowable for the structure under design), we can use the monotropic model of the ply and, hence, the results of optimization are presented in Section 8.1.

Consider again the optimality condition Eq. (8.25). As can be seen, this equation can be satisfied not only by strains in Eqs. (8.26), but also if we take

$$\tan 2\phi_i = \frac{\gamma_{xy}}{\varepsilon_x - \varepsilon_y} \tag{8.34}$$

Since the left-hand side of this equation is a periodic function with period π, Eq. (8.34) determines two angles, i.e.,

$$\phi_1 = \phi = \frac{1}{2}\tan^{-1}\frac{\gamma_{xy}}{\varepsilon_x - \varepsilon_y}, \quad \phi_2 = \frac{\pi}{2} + \phi \tag{8.35}$$

Table 8.2
Parameters of typical advanced composites.

Parameter	Fabric–epoxy composites			Unidirectional-epoxy composites				Boron–Al
	Glass	Carbon	Aramid	Glass	Carbon	Aramid	Boron	
$\overline{\sigma}_2/\overline{\sigma}_1$	0.99	0.99	0.83	0.022	0.025	0.012	0.054	0.108
n	0.85	1.0	1.0	0.28	0.1	0.072	0.11	0.7

Thus, the optimal laminate consists of two layers, and the fibers in both layers are directed along the lines of principal stresses. Suppose that the layers are made of the same composite material and have the same thickness, i.e., $h_1 = h_2 = h/2$, where h is the thickness of the laminate. Then, using Eqs. (8.24) and (8.35), we can show that $B_{11} = B_{22}$ and $B_{24} = -B_{14}$ for this laminate. After some transformation involving elimination of γ_{xy}^0 from the first two equations of Eqs. (8.23) with the aid of Eq. (8.34) and similar transformation of the third equation from which ε_x^0 and ε_y^0 are eliminated using again Eq. (8.34), we get

$$N_x = (B_{11} + B_{14}\tan 2\phi)\varepsilon_x^0 + (B_{12} - B_{14}\tan 2\phi)\varepsilon_y^0$$

$$N_y = (B_{12} - B_{14}\tan 2\phi)\varepsilon_x^0 + (B_{11} + B_{14}\tan 2\phi)\varepsilon_y^0$$

$$N_{xy} = (B_{44} + B_{14}\cot 2\phi)\gamma_{xy}^0$$

Upon substitution of coefficients B_{mn} from Eqs. (8.24) we arrive at

$$N_x = \frac{h}{2}\left[(\overline{E}_1 + \overline{E}_2)\varepsilon_x^0 + (\overline{E}_1\nu_{12} + \overline{E}_2\nu_{21})\varepsilon_y^0\right]$$

$$N_y = \frac{h}{2}\left[(\overline{E}_1\nu_{12} + \overline{E}_2\nu_{21})\varepsilon_x^0 + (\overline{E}_1 + \overline{E}_2)\varepsilon_y^0\right]$$

$$N_{xy} = \frac{h}{4}\left[\overline{E}_1(1-\nu_{12}) + \overline{E}_2(1-\nu_{21})\right]\gamma_{xy}^0$$

Introducing average stresses $\sigma_x = N_x/h$, $\sigma_y = N_y/h$, and $\tau_{xy} = N_{xy}/h$ and solving these equations for strains, we have

$$\varepsilon_x^0 = \frac{1}{E}(\sigma_x - \nu\sigma_y), \quad \varepsilon_y^0 = \frac{1}{E}(\sigma_y - \nu\sigma_x), \quad \gamma_{xy}^0 = \frac{\tau_{xy}}{G} \tag{8.36}$$

where

$$E = \frac{1}{2(E_1 + E_2)}\left[2E_1E_2 + \frac{E_1^2(1-\nu_{12}^2) + E_2^2(1-\nu_{21}^2)}{1-\nu_{12}\nu_{21}}\right]$$
$$\nu = \frac{E_1\nu_{12} + E_2\nu_{21}}{E_1 + E_2}, \quad G = \frac{E}{2(1+\nu)} \tag{8.37}$$

Changing strains for stresses in Eqs. (8.35), we can write the expression for the optimal orientation angle as

$$\phi = \frac{1}{2}\tan^{-1}\frac{2\tau_{xy}}{\sigma_x - \sigma_y} \tag{8.38}$$

It follows from Eqs. (8.36) that a laminate consisting of two layers reinforced along the directions of principal stresses behaves like an isotropic layer, and Eqs. (8.37) specify the

Table 8.3
Effective elastic constants of an optimal laminate.

Property	Glass–epoxy	Carbon–epoxy	Aramid–epoxy	Boron–epoxy	Boron–Al	Carbon–carbon	Al_2O_3–Al
Elastic modulus, E (GPa)	36.9	75.9	50.3	114.8	201.1	95.2	205.4
Poisson's ratio, ν	0.053	0.039	0.035	0.035	0.21	0.06	0.176

elastic constants of the corresponding isotropic material. For typical advanced composites, these constants are listed in Table 8.3 (the properties of unidirectional plies are taken from Table 3.5). Comparing the elastic moduli of the optimal laminates with those for quasi-isotropic materials (see Table 5.1), we can see that for polymeric composites the characteristics of the first group of materials are about 40% higher than those for the second group. However, it should be emphasized that whereas the properties of quasi-isotropic laminates are universal material constants, the optimal laminates demonstrate characteristics shown in Table 8.3 only if the orientation angles of the fibers are found from Eqs. (8.35) or (8.38) and correspond to a particular distribution of stresses σ_x, σ_y, and τ_{xy}.

As follows from Table 8.3, the modulus of a carbon–epoxy laminate is close to the modulus of aluminum, whereas the density of the composite material is lower by a factor of 1.7. This is the theoretical weight saving factor that can be expected if we change from aluminum to carbon–epoxy composite in a thin-walled structure. Since the stiffness of both materials is approximately the same, to find the optimal orientation angles of the structure elements, we can substitute in Eq. (8.38) the stresses acting in the aluminum prototype structure. A composite structure designed in this way will have approximately the same stiffness as the prototype structure and, as a rule, higher strength because carbon composites are stronger than aluminum alloys.

To evaluate the strength of the optimal laminate, we should substitute strains from Eqs. (8.36) into Eqs. (4.69) and thence these strains in the principal material coordinates of the layers – into constitutive equations, Eqs. (4.56), that specify the stresses σ_1 and $\sigma_2(\tau_{12} = 0)$ acting in the layers. Applying the appropriate failure criterion (see Chapter 6), we can evaluate the laminate strength.

Comparing Tables 1.1 and 8.3, we can see that boron–epoxy optimal laminates have approximately the same stiffness as titanium (but is lighter by a factor of about 2). Boron–aluminum can be used to replace steel with a weight saving factor of about 3.

For preliminary evaluation, we can use a monotropic model of unidirectional plies neglecting the stiffness and load-carrying capacity of the matrix. Then, Eqs. (8.37) take the following simple form

$$E = \frac{E_1}{2}, \quad \nu = 0, \quad G = \frac{E_1}{4} \tag{8.39}$$

As an example, consider an aluminum shear web with thickness $h = 2\,\text{mm}$, elastic constants $E_a = 72\,\text{GPa}$, $\nu_a = 0.3$ and density $\rho_a = 2.7\,\text{g/cm}^3$. This panel is loaded with shear stress τ. Its shear stiffness is $B_{44}^{\mathrm{a}} = 57.6\,\mathrm{GPa}\cdot\mathrm{mm}$ and the mass of a unit surface is $m_{\mathrm{a}} = 5.4\,\mathrm{kg/m^2}$. For the composite panel, taking $\sigma_x = \sigma_y = 0$ in Eq. (8.38) we have $\phi = 45°$. Thus, the composite panel consists of $+45°$ and $-45°$ unidirectional layers of the same thickness. The total thickness of the laminate is $h = 2$ mm, i.e., the same as for an aluminum panel. Substituting $E_1 = 140$ GPa and taking into account that $\rho = 1.55\,\mathrm{g/cm^3}$ for a carbon–epoxy composite which is chosen to substitute for aluminum we get $B_{44}^{\mathrm{c}} = 70\,\mathrm{GPa}\cdot\mathrm{mm}$ and $m_{\mathrm{c}} = 3.1\,\mathrm{kg/m^3}$. The stresses acting in the fiber directions of the composite plies are $\sigma_1^{\mathrm{c}} = \pm 2\tau$. Thus, the composite panel has a 21.5% higher stiffness and its mass is only 57.4% of the mass of a metal panel. The composite panel also has higher strength because the longitudinal strength of unidirectional carbon–epoxy composite under tension and compression is more than twice the shear strength of aluminum.

The potential performance of the composite structure under discussion can be enhanced if we use different materials in the layers with angles ϕ_1 and ϕ_2 specified by Eqs. (8.35). According to the derivation of Obraztsov and Vasiliev (1989), the ratio of the layers' thicknesses is

$$\frac{h_2}{h_1} = \frac{\overline{E}_1^{(1)} - \overline{E}_2^{(1)}}{\overline{E}_1^{(2)} - \overline{E}_2^{(2)}}$$

and the elastic constants in Eqs. (8.37) are generalized as

$$\overline{E} = \frac{E}{1-\nu^2} = \frac{\overline{E}_1^{(1)}\overline{E}_1^{(2)} - \overline{E}_2^{(1)}\overline{E}_2^{(2)}}{\overline{E}_1^{(1)} + \overline{E}_1^{(2)} - \overline{E}_2^{(1)} - \overline{E}_2^{(2)}}$$

$$\nu = \frac{\overline{E}_1^{(1)}\overline{E}_1^{(2)}\left(\nu_{12}^{(1)} + \nu_{12}^{(2)}\right) - \overline{E}_2^{(1)}\overline{E}_2^{(2)}\left(\nu_{21}^{(1)} + \nu_{21}^{(2)}\right)}{\overline{E}_1^{(1)}\overline{E}_1^{(2)} - \overline{E}_2^{(1)}\overline{E}_2^{(2)}}$$

Superscripts 1 and 2 correspond to layers with orientation angles ϕ_1 and ϕ_2, respectively.

8.3. Application to optimal composite structures

As stated in the introduction to this chapter, there exists special composite structures for which the combination of the specific properties of modern composites with the appropriate design concepts and potential of composite technology provide a major improvement of these structures in comparison with the corresponding metal prototypes. Three such special structures, i.e., geodesic filament-wound pressure vessels, composite flywheels, and an anisogrid lattice structure are described in this section.

8.3.1. Composite pressure vessels

As the first example of the application of the foregoing results, consider filament-wound membrane shells of revolution, that are widely used as pressure vessels, solid propellant

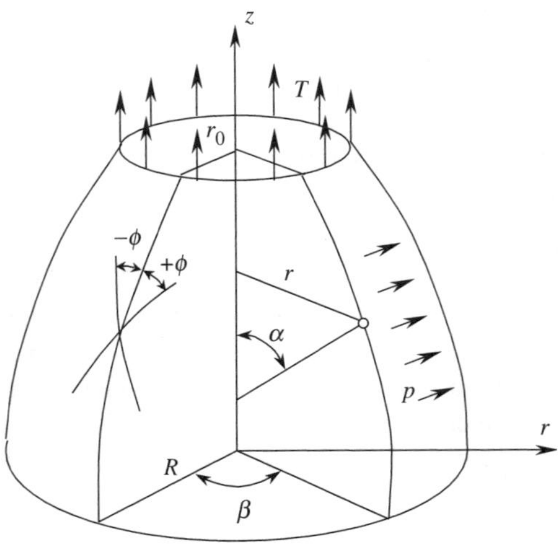

Fig. 8.4. Axisymmetrically loaded membrane shell of revolution.

rocket motor cases, tanks for gases and liquids, etc. (see Figs. 4.14 and 7.51). The shell is loaded with uniform internal pressure p and axial forces T uniformly distributed along the contour of the shell cross section $r = r_0$ as in Fig. 8.4. Meridional, N_α, and circumferential, N_β, stress resultants acting in the shell follow from the corresponding free body diagrams of the shell element and can be written as (see e.g., Vasiliev, 1993)

$$N_\alpha = -Q\frac{\left[1+(z')^2\right]^{1/2}}{rz'}$$
$$N_\beta = -\frac{1}{z'}\left[1+(z')^2\right]^{1/2}\left\{pr - \frac{Qz''}{z'\left[1+(z')^2\right]}\right\} \tag{8.40}$$

where $z(r)$specifies the form of the shell meridian, $z' = \mathrm{d}z/\mathrm{d}r$ and

$$Q = Tr_0 + \frac{p}{2}\left(r^2 - r_0^2\right) \tag{8.41}$$

Let the shell be made by winding an orthotropic tape at angles $+\phi$ and $-\phi$ with respect to the shell meridian as in Fig. 8.4. Then, N_α and N_β can be expressed in terms of stresses σ_1, σ_2, and τ_{12}, referred to the principal material coordinates of the tape with the aid of Eqs. (4.68), i.e.,

$$N_\alpha = h\left(\sigma_1\cos^2\phi + \sigma_2\sin^2\phi - \tau_{12}\sin 2\phi\right)$$
$$N_\beta = h\left(\sigma_1\sin^2\phi + \sigma_2\cos^2\phi + \tau_{12}\sin 2\phi\right) \tag{8.42}$$

in which h is the shell thickness. Stresses σ_1, σ_2, and τ_{12} are related to the corresponding strains by Hooke's law, Eqs. (4.55), as

$$\varepsilon_1 = \frac{1}{E_1}(\sigma_1 - \nu_{21}\sigma_2), \quad \varepsilon_2 = \frac{1}{E_2}(\sigma_2 - \nu_{12}\sigma_1), \quad \gamma_{12} = \frac{\tau_{12}}{G_{12}} \tag{8.43}$$

whereas strains $\varepsilon_1, \varepsilon_2$, and γ_{12} can be expressed in terms of the meridional, ε_α, and circumferential, ε_β, strains of the shell using Eqs. (4.69), i.e.,

$$\begin{aligned} \varepsilon_1 &= \varepsilon_\alpha \cos^2\phi + \varepsilon_\beta \sin^2\phi \\ \varepsilon_2 &= \varepsilon_\alpha \sin^2\phi + \varepsilon_\beta \cos^2\phi \\ \gamma_{12} &= (\varepsilon_\beta - \varepsilon_\alpha)\sin 2\phi \end{aligned} \tag{8.44}$$

Since the right-hand parts of these three equations include only two strains, ε_α and ε_β, there exists a compatibility equation linking $\varepsilon_1, \varepsilon_2$, and γ_{12}. This equation is

$$(\varepsilon_1 - \varepsilon_2)\sin 2\phi + \gamma_{12}\cos 2\phi = 0$$

Writing this equation in terms of stresses with the aid of Eqs. (8.43), we have

$$\left[\frac{\sigma_1}{E_1}(1+\nu_{21}) - \frac{\sigma_2}{E_2}(1+\nu_{12})\right]\sin 2\phi + \frac{\tau_{12}}{G_{12}}\cos 2\phi = 0$$

In conjunction with Eqs. (8.42), this equation allows us to determine stresses as

$$\begin{aligned} \sigma_1 &= \frac{1}{2hC}\left\{(N_\alpha + N_\beta)\left[1 + \frac{2G_{12}}{E_2}(1+\nu_{12})\tan^2 2\phi\right] + \frac{N_\alpha - N_\beta}{\cos 2\phi}\right\} \\ \sigma_2 &= \frac{1}{2hC}\left\{(N_\alpha + N_\beta)\left[1 + \frac{2G_{12}}{E_1}(1+\nu_{21})\tan^2 2\phi\right] - \frac{N_\alpha - N_\beta}{\cos 2\phi}\right\} \\ \tau_{12} &= \frac{G_{12}\tan 2\phi}{hC\cos 2\phi}\left[N_\beta\left(\frac{1+\nu_{21}}{E_1}\sin^2\phi + \frac{1+\nu_{12}}{E_2}\cos^2\phi\right)\right. \\ &\qquad \left. - N_\alpha\left(\frac{1+\nu_{21}}{E_1}\cos^2\phi + \frac{1+\nu_{12}}{E_2}\sin^2\phi\right)\right] \end{aligned} \tag{8.45}$$

where

$$C = 1 + G_{12}\left(\frac{1+\nu_{21}}{E_1} + \frac{1+\nu_{12}}{E_2}\right)\tan^2 2\phi$$

Now, assume that in accordance with the results presented in the previous section the optimal shell is reinforced along the lines of principal stresses, i.e., in such a way that

$\tau_{12} = 0$. In accordance with the last equation of Eqs. (8.43), for such a shell $\gamma_{12} = 0$ and, as follows from Eqs. (8.44), $\varepsilon_\alpha = \varepsilon_\beta = \varepsilon_1 = \varepsilon_2$.

Putting $\tau_{12} = 0$ in the last equation of Eqs. (8.45), we can conclude that for the optimal shell

$$\frac{N_\beta}{N_\alpha} = \frac{1 - (1-n)\cos^2\phi}{n + (1-n)\cos^2 2\phi} \tag{8.46}$$

where as above

$$n = \frac{E_2(1+\nu_{21})}{E_1(1+\nu_{12})} \tag{8.47}$$

Substituting N_α and N_β from Eqs. (8.40) into Eq. (8.46), we arrive at the following equation for the meridian of the optimal shell

$$\frac{rz''}{z'\left[1+(z')^2\right]} = \frac{pr^2}{Q} - \frac{1-(1-n)\cos^2\phi}{n+(1-n)\cos^2\phi} \tag{8.48}$$

The first two equations of Eqs. (8.45) yield the following expressions for stresses acting in the tape of the optimal shell

$$\sigma_1 = \frac{\sigma_2}{n} = \frac{N_\alpha}{h\left[n+(1-n)\cos^2\phi\right]}, \qquad \tau_{12} = 0 \tag{8.49}$$

Taking into account that in accordance with Eqs. (8.45)

$$\sigma_1 + \sigma_2 = \frac{1}{h}(N_\alpha + N_\beta)$$

we arrive at the following relationships

$$\sigma_1 = \frac{N_\alpha + N_\beta}{h(1+n)}, \qquad \sigma_2 = \frac{n(N_\alpha + N_\beta)}{h(1+n)}, \qquad \tau_{12} = 0$$

which coincide with Eqs. (8.31).

Substituting N_α from the first equation of Eqs. (8.40) into Eq. (8.49), we have

$$\sigma_1 h = -\frac{Q\left[1+(z')^2\right]^{1/2}}{rz'\left[n+(1-n)\cos^2\phi\right]} \tag{8.50}$$

Assume that the optimal shell is a structure of uniform stress. Differentiating Eq. (8.50) with respect to r and taking into account that according to the foregoing assumption

$\sigma_1 = constant$, we arrive at the following equation in which z'' is eliminated with the aid of Eq. (8.48)

$$\frac{d}{dr}\left\{rh\left[n + (1-n)\cos^2\phi\right]\right\} - h\left[1 - (1-n)\cos^2\phi\right] = 0 \tag{8.51}$$

This equation specifies either the thickness or the orientation angle of the optimal shell.

Consider two particular cases. First, consider a fabric tape of variable width $w(r)$ being laid up on the surface of the mandrel along the meridians of the shell of revolution to be fabricated. Then, $\phi = 0$, and Eq. (8.51) takes the form

$$\frac{d}{dr}(rh) - nh = 0$$

The solution of this equation is

$$h = h_R\left(\frac{r}{R}\right)^{n-1} \tag{8.52}$$

where $h_R = h(r = R)$ is the shell thickness at the equator $r = R$ (see Fig. 8.4). Assuming that there is no polar opening in the shell ($r_0 = 0$) or that it is closed ($T = pr_0/2$), we have from Eq. (8.41), $Q = pr^2/2$. Substituting this result into Eqs. (8.48) and (8.50), we obtain

$$\frac{rz''}{z'\left[1 + (z')^2\right]} = 2 - n \tag{8.53}$$

$$\sigma_1 = -\frac{pr}{2z'h}\sqrt{1 + (z')^2} \tag{8.54}$$

Integrating Eq. (8.53) with the condition $1/z' = 0$ for $r = R$ which means that the tangent line to the shell meridian is parallel to the axis z at $r = R$ (see Fig. 8.4), we arrive at

$$z' = -\frac{r^{2-n}}{\sqrt{R^{2(2-n)} - r^{2(2-n)}}} \tag{8.55}$$

Further integration results in the following parametric equation for the shell meridian

$$\frac{r}{R} = (1-t)^{\lambda}$$

$$\frac{z}{R} = \lambda\int_0^t t^{-\frac{1}{2}}(1-t)^{-\lambda}dt = \lambda B_x\left(\frac{1}{2}, 1-\lambda\right)$$

$$\lambda = \frac{1}{2(2-n)}$$

Here, B_x is the β-function (or the Euler integral of the first type). The constant of integration is found from the condition $z(r = R) = 0$. Meridians corresponding to various n-numbers are presented in Fig. 8.5. For $n = 1$ the optimal shell is a sphere, whereas for $n = 2$ it is a cylinder. As follows from Eq. (8.52), the thickness of the spherical ($n = 1$) and cylindrical ($n = 2, r = R$) shells is a constant. Substituting Eqs. (8.52) and (8.55) in Eq. (8.54) and taking into account Eq. (8.49), we have

$$\sigma_1 = \frac{\sigma_2}{n} = \frac{pR}{2h_R}$$

This equation allows us to determine the shell thickness at the equator $(r = R), h_R$, matching σ_1 or σ_2 with material strength characteristics.

As has been noted already, the shells under study can be made by laying up fabric tapes of variable width, $w(r)$, along the shell meridians. The tape width can be related to the shell thickness, $h(r)$, as

$$kw(r)\delta = 2\pi rh(r) \tag{8.56}$$

where k is the number of tapes in the shell cross section (evidently, k is the same for all the cross sections) and δ is the tape thickness. Substituting $h(r)$ from Eq. (8.52), we get

$$w(r) = \frac{2\pi h_R r^n}{\kappa\delta\, R^{n-1}}$$

Consider the second special case – a shell made by winding unidirectional composite tapes at angles $\pm\phi$ with respect to the shell meridian (see Fig. 8.4). The tape width, w_0, does not depend on r, and its thickness is δ. Then, the relevant equation similar to Eq. (8.56) can be written as

$$\frac{kw_0\delta}{\cos\phi(r)} = 2\pi rh(r)$$

where k is the number of tapes with angles $+\phi$ and $-\phi$. Thus, the shell thickness is

$$h(r) = \frac{kw_0\delta}{2\pi r\cos\phi(r)} \tag{8.57}$$

It can be expressed in terms of the thickness value at the shell equator $h_R = h\,(r = R)$ as

$$h(r) = h_R\frac{R\cos\phi_R}{r\cos\phi(r)} \tag{8.58}$$

where $\phi_R = \phi\,(r = R)$. It should be noted that this equation is not valid for the shell part in which the tapes are completely overlapped close to the polar opening.

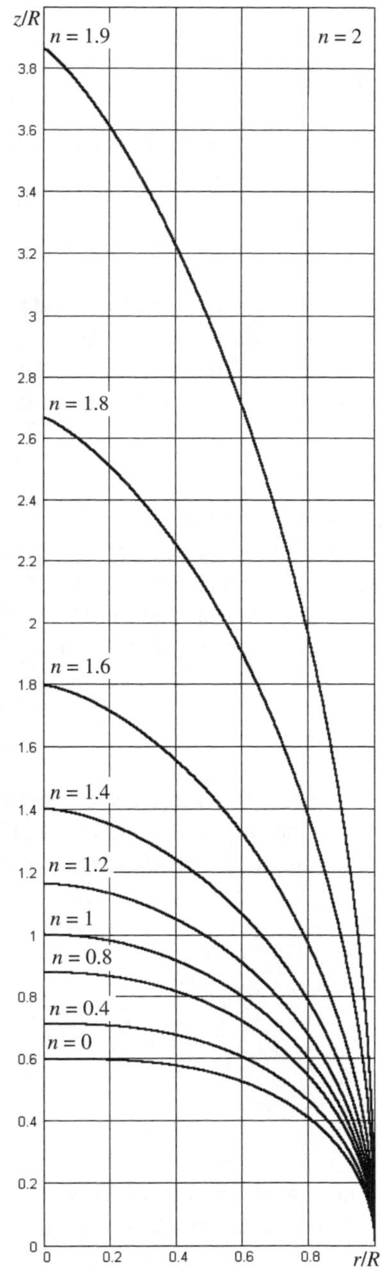

Fig. 8.5. Meridians of optimal composite shells.

Substituting $h(r)$ from Eq. (8.58) in Eq. (8.51), we arrive at the following equation for the tape orientation angle

$$r\frac{\mathrm{d}\phi}{\mathrm{d}r}\cdot\frac{\sin\phi[n-(1-n)\cos^2\phi]}{\cos\phi[1-(1-n)\cos^2\phi]}=1$$

The solution of this equation that satisfies the boundary condition $\phi(r=R)=\phi_R$ is presented as follows

$$r\big[1-(1-n)\cos^2\phi(r)\big]^{\frac{1-n}{2}}\cos^n\phi(r)=R\big[1-(1-n)\cos^2\phi_R\big]^{\frac{1-n}{2}}\cos^n\phi_R \quad (8.59)$$

Consider monotropic filament-wound shells. As noted in the previous section, the simplest and sufficiently adequate model of unidirectional fibrous composites for design problems is the monotropic model, which ignores the stiffness of the matrix. For this model, we should take $n=0$ in the foregoing equations. Particularly, Eq. (8.59) yields in this case

$$r\sin\phi(r)=R\sin\phi_R \quad (8.60)$$

This is the equation of a geodesic line on the surface of revolution. Thus, in the optimal filament-wound shells the fibers are directed along the geodesic trajectories. This substantially simplifies the winding process because the tape placed on the surface under tension automatically takes the form of the geodesic line, provided there is no friction between the tape and the surface. As follows from Eq. (8.60), for $\phi=90°$, the tape touches the shell parallel to radius

$$r_0=R\sin\phi_R \quad (8.61)$$

and a polar opening of this radius is formed in the shell (see Fig. 8.4).

Transforming Eq. (8.48) with the aid of Eqs. (8.60) and (8.61) and taking $n=0$, we arrive at the following equation which specifies the meridian of the optimal filament wound shell

$$\frac{z''}{z'\big[1+(z')^2\big]}=\frac{2r}{r^2-\eta^2}-\frac{r_0^2}{r\big(r^2-r_0^2\big)} \quad (8.62)$$

where

$$\eta^2=r_0^2-\frac{2T}{p}r_0$$

Integrating Eq. (8.62) with due regard for the condition $1/z'(R)=0$ which, as above, requires that for $r=R$ the tangent to the meridian be parallel to z-axis, we have

$$z'=-\frac{r\big(r^2-\eta^2\big)\sqrt{R^2-r_0^2}}{\sqrt{R^2\big(r^2-r_0^2\big)\big(R^2-\eta^2\big)^2-r^2\big(R^2-r_0^2\big)\big(r^2-\eta^2\big)^2}} \quad (8.63)$$

Using this equation to transform Eq. (8.50) in which we take $n = 0$ and substituting h from Eq. (8.58), we obtain the following equation for the longitudinal stress in the tape

$$\sigma_1 = \frac{p(R^2 - r_0^2) + 2r_0 T}{2Rh_R \cos^2 \phi_R} \tag{8.64}$$

As can be seen, σ_1 does not depend on r, and the optimal shell is a structure reinforced with uniformly stressed fibers.

Such fibrous structures are referred to as isotensoids. To study the types of isotensoids corresponding to the loading shown in Fig. 8.4, factor the expression in the denominator of Eq. (8.63). The result can be presented as

$$z' = -\frac{r(r^2 - \eta^2)}{\sqrt{(R^2 - r^2)(r^2 - r_1^2)(r^2 + r_2^2)}}, \tag{8.65}$$

where

$$r_{1,2}^2 = \left(\frac{R^2}{2} - \eta^2\right)\left\{\sqrt{\frac{R^2}{R^2 - r_0^2}\left[1 + \frac{(3R^2 - 4\eta^2)r_0^2}{(R^2 - 2\eta^2)^2}\right]} \pm 1\right\} \tag{8.66}$$

It follows from Eq. (8.65) that the parameters R and r_1 are the maximum and minimum distances from the meridian to the rotation axis. Meridians of isotensoids corresponding to various loading conditions are shown in Fig. 8.6. For $p = 0$, i.e., under axial tension, a hyperbolic shell is obtained with the meridian determined as

$$r^2 - z^2 \tan^2 \phi_R = R^2$$

This meridian corresponds to line 1 in Fig. 8.6. For $\phi_R = 0$, the hyperbolic shell degenerates into a cylinder (line 2). Curve 3 corresponds to $T = pr_0/2$, i.e., to a shell for which the polar opening of radius r_0 is closed. For the special angle $\phi_R = \phi_0 = 54°44'$, the shell degenerates into a circular cylindrical shell (line 2) as discussed in Section 8.1. For $T = 0$, i.e., in the case of an open polar hole, the meridian has the form corresponding to curve 4. The change in the direction of axial forces T yields a toroidal shell (line 5). Performing integration of Eq. (8.65) and introducing dimensionless parameters

$$\bar{r} = \frac{r}{R}, \quad \bar{z} = \frac{z}{R}, \quad \bar{r}_0 = \frac{r_0}{R}, \quad \bar{\eta} = \frac{\eta}{R}$$

we finally arrive at

$$\bar{z} = \frac{k_2 - \bar{\eta}^2}{\sqrt{1 - k_2}} F(k, \theta) + \sqrt{1 - k_2} E(k, \theta) \tag{8.67}$$

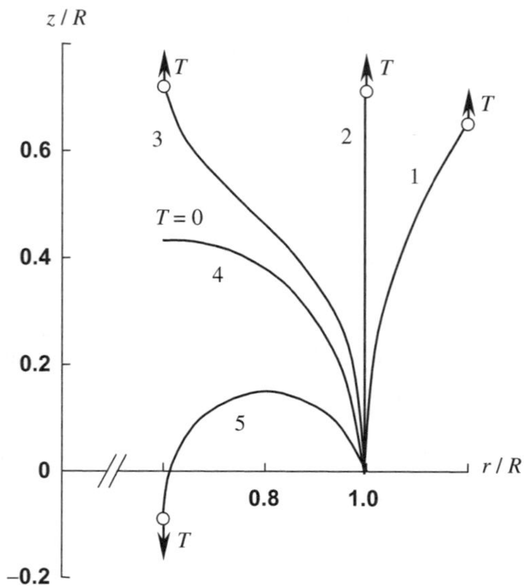

Fig. 8.6. Isotensoids corresponding to various loading conditions.

where

$$F(k,\,\theta)=\int_0^\theta \frac{\mathrm{d}\theta}{\sqrt{1-k^2\sin^2\theta}},\qquad E(k,\,\theta)=\int_0^\theta \sqrt{1-k^2\sin^2\theta}\mathrm{d}\theta$$

are the first-kind and the second-kind elliptic integrals and

$$k_{1,\,2}=\left(\frac{1}{2}-\overline{\eta}^2\right)\left\{\pm\sqrt{\frac{1}{1-\overline{r}_0^2}\left[1+\frac{3-4\overline{\eta}^2}{\left(1-2\overline{\eta}^2\right)^2}\right]}-1\right\}$$

$$k^2=\frac{1-k_1}{1-k_2},\qquad \sin\theta=\sqrt{\frac{1-\overline{r}^2}{1-k_1}}$$

As an application of the foregoing equations consider the optimal structure of the end closure of the pressure vessel shown in Fig. 4.14. The cylindrical part of the vessel consists of $\pm\phi_R$ angle-ply layer with thickness h_R that can be found from Eq. (8.64) in which we should take $T = pr_0/2$, and a circumferential ($\phi = 90°$) layer whose thickness is specified by Eq. (8.18), i.e.,

$$h_{90}=h_R(3\cos^2\phi_R-1)$$

The polar opening of the dome (see Fig. 4.14) is closed. So $T = pr_0/2$, $\overline{\eta} = 0$, and the dome meridian corresponds to curve 3 in Fig. 8.6. As has already been noted, upon

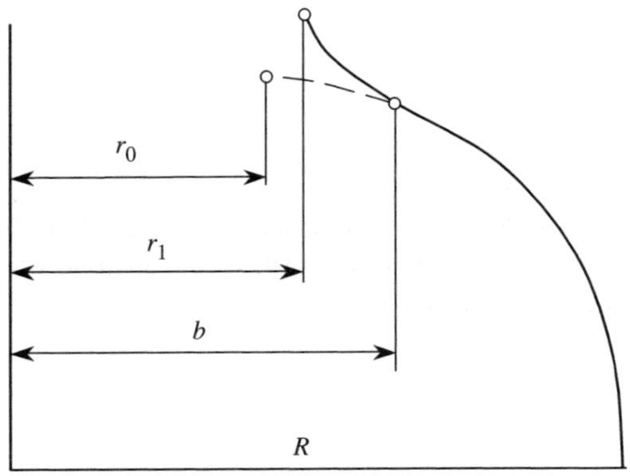

Fig. 8.7. Combined meridian of the pressure vessel dome.

winding, an opening of radius r_0 is formed at the shell apex. However, the analysis of Eq. (8.65) for r_1 that determines the minimum distance from the meridian to the z-axis (see Fig. 8.7) shows that r_1 is equal to r_0 only if a shell has an open polar hole (curve 4 in Fig. 8.6). For a pressure vessel whose polar hole is closed, $r_1 \geq r_0$ and the equality takes place only for $\phi = 0$, i.e., for $r_1 = r_0 = 0$. In real vessels, polar holes are closed with rigid polar bosses as shown in Fig. 8.8. The meridian of the shell under consideration can be divided into two segments. For $R \geq r \geq b$, the meridian corresponds to curve 3 in Fig. 8.6 for which $T = pr_0/2$ and $\eta = 0$. In Fig. 8.7 this segment of the meridian is shown with a solid line. The meridian segment $b \geq r \geq r_0$, where the shell touches the polar boss, corresponds to curve 4 in Fig. 8.6 for which $T = 0$. In Fig. 8.7, this segment of the meridian is indicated with the dashed line. The radius b in Figs. 8.7 and 8.8 can be set as the coordinate of an inflection point of this curve determined by the condition $z''(r = b) = 0$. Differentiating Eq. (8.65) and taking $\eta = 0$ for the closed

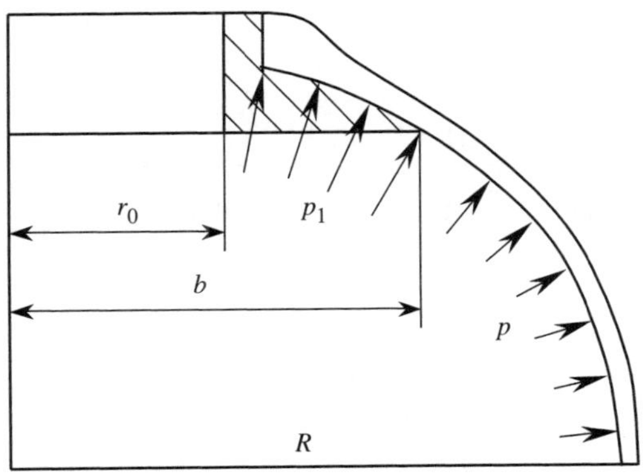

Fig. 8.8. Isotensoid dome with a polar boss.

polar opening, we get

$$b=\sqrt{\frac{3}{2}}\cdot r_0=1.225r_0$$

Since the segment $(b-r_0)$ is relatively small, we can assume that the contact pressure p_1 between the shell and the boss is uniform. Then, from the condition of boss equilibrium (the hole in the boss is closed), we have

$$p_1=\frac{pb^2}{b^2-r_0^2} \tag{8.68}$$

Constructing the combined meridian, we should take into account that functions $z(r)$ and $z'(r)$ must be continuous for $r=b$. Finally, using Eqs. (8.65) and (8.67), we obtain:
For $R\geq r\geq b$ $(T=pr_0/2, \eta=0)$

$$z'=-\frac{r^3\sqrt{R^2-r_0^2}}{\sqrt{R^6\left(r^2-r_0^2\right)-r^6\left(R^2-r_0^2\right)}} \tag{8.69}$$

and

$$\bar{z}=\frac{k_2}{\sqrt{1-k_2}}F(k,\theta_1)+\sqrt{1-k_2}E(k,\theta_1)$$

where

$$k_{1,2}=\frac{1}{2}\left(\pm\sqrt{\frac{1+3\bar{r}_0^2}{1-\bar{r}_0^2}}-1\right),\quad \sin\theta_1=\sqrt{\frac{1-\bar{r}^2}{1-k_1}},\quad k^2=\frac{1-k_1}{1-k_2}$$

For $b\geq r\geq r_0$ $(T=0,\ \eta=r_0)$

$$z'=-\frac{rb^2\sqrt{\left(r^2-r_0^2\right)\left(R^2-r_0^2\right)}}{\sqrt{R^6\left(b^2-r_0^2\right)^2-r^2b^4\left(r^2-r_0^2\right)\left(R^2-r_0^2\right)}} \tag{8.70}$$

and

$$\bar{z}=-\frac{m_1}{\sqrt{m_1+m_2}}[F(m,\theta_2)-F(m,\theta_2^*)]+\sqrt{m_1+m_2}[E(m,\theta_2)-E(m,\theta_2^*)]$$
$$+\frac{k_2}{\sqrt{1-k_2}}F(k,\theta_1^*)+\sqrt{1-k_2}E(k,\theta_1^*)$$

where

$$m_{1,2} = \frac{\bar{r}_0^2}{2}\left[\sqrt{1 - \frac{4(\bar{b}^2 - \bar{r}_0^2)^2}{\bar{b}^4\bar{r}_0^4(1-\bar{r}_0^2)}} \pm 1\right], \quad m^2 = \frac{m_2}{m_1 + m_2},$$

$$\cos\theta_2 = \sqrt{\frac{\bar{r}^2 - \bar{r}_0^2}{m_2}}, \quad \sin\theta_1^* = \sqrt{\frac{1-\bar{b}^2}{1-k_1}}$$

$$\cos\theta_2^* = \sqrt{\frac{\bar{b}^2 - \bar{r}_0^2}{m_2}}, \quad \bar{b} = \frac{b}{R}$$

Meridians plotted in accordance with Eqs. (8.69) and (8.70) and corresponding to various values of parameter $\bar{r}_0$ specifying the radius of the polar opening (which is closed) are presented in Fig. 8.9. The curve $\bar{r}_0 = 0$ corresponds to a shell reinforced along the meridians and is the same as that for the curve $n = 0$ in Fig. 8.5. This isotensoid shape can be readily obtained experimentally if we load a balloon reinforced along the meridians with internal pressure as in Fig. 8.10.

Stresses acting along the fibers of the shells whose meridians are presented in Fig. 8.9 are determined by Eq. (8.50) in which we should take $n = 0$. Substituting h from Eq. (8.58),

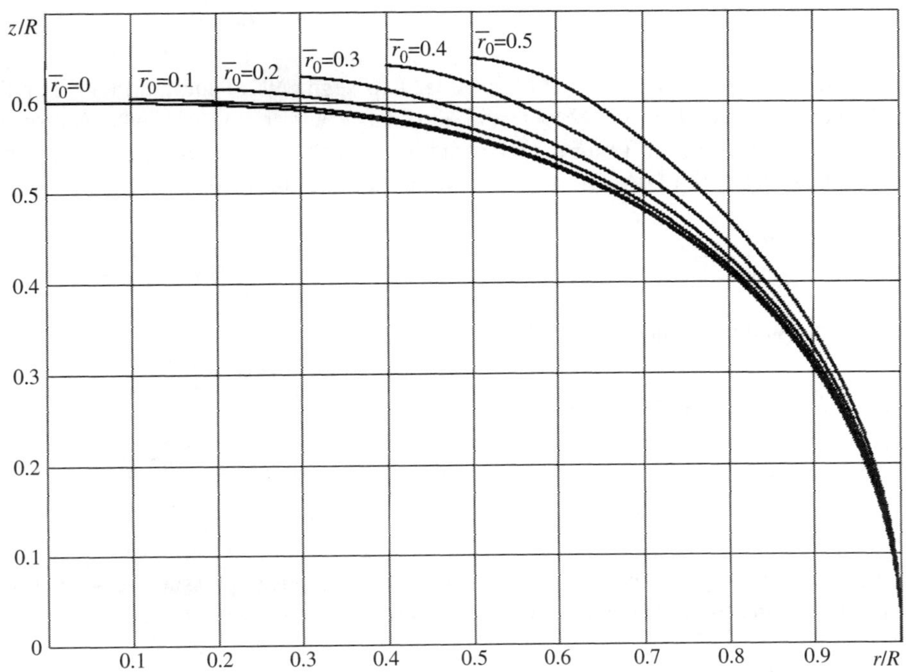

Fig. 8.9. Meridians of isotensoids corresponding to various normalized radii of the polar openings $\bar{r}_0 = r_0/R$.

Fig. 8.10. A model isotensoid reinforced along the meridians.

ϕ from Eqs. (8.60), (8.61), we should consider two segments of the meridian. For the first segment, we take $T = pr_0/2$ and z' in accordance with Eq. (8.69), whereas for the second one we substitute z' from Eq. (8.70) and put $T = 0$, $p = p_1$, where p_1 is specified by Eq. (8.68). For both segments, we arrive at the same result, i.e.,

$$\sigma_1 = \frac{pR}{2h_R \cos^2 \phi_R}$$

The shell mass and internal volume can be found as

$$M = 2\pi\rho h_R \cos\phi_R \int_{r_0}^{R} \sqrt{1 + (z')^2} \frac{\mathrm{d}r}{\cos\phi}$$

$$V = \pi \int_{r_0}^{R} z' r^2 \mathrm{d}r$$

where ρ is the density of the material. The mass of a composite pressure vessel is often evaluated by using the parameter μ in the equation

$$M = \mu \frac{p_\mathrm{u} V}{\overline{\sigma}_1/\rho}$$

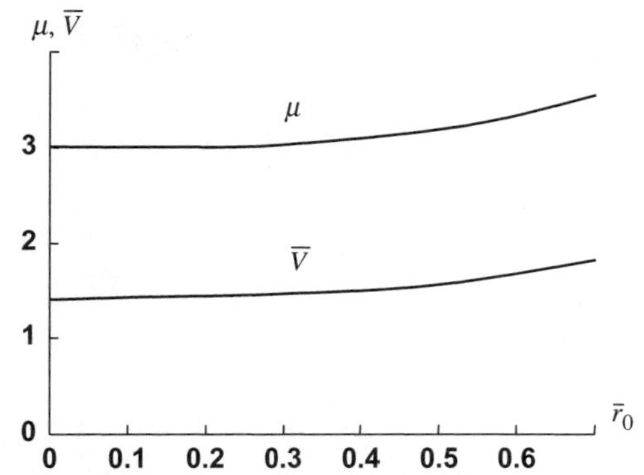

Fig. 8.11. Mass efficiency parameter μ and the normalized internal volume $\overline{V} = V/R^3$ of the isotensoid pressure vessel as functions of the polar opening radius.

Here, p_u is the ultimate pressure, and $\overline{\sigma}_1/\rho$ is the specific strength of the material. The variation of the parameter μ and the normalized internal volume $\overline{V} = V/R^3$ as a function of the radius of the polar opening are shown in Fig. 8.11.

8.3.2. Spinning composite disks

As the second example of optimal composite structure, consider a disk rotating around its axis with an angular velocity ω. Let the disk be reinforced with fibers making angles $+\phi$ and $-\phi$ with the radius as in Fig. 8.12 and find the optimal trajectories of the fibers (Kyser, 1965; Obraztsov and Vasiliev, 1989). The radial, N_r, and circumferential, N_β, stress resultants are related to the stresses σ_1 acting in the composite material along the fibers by Eqs. (8.42). Using the monotropic material model and putting $\sigma_2 = 0$ and $\tau_{12} = 0$

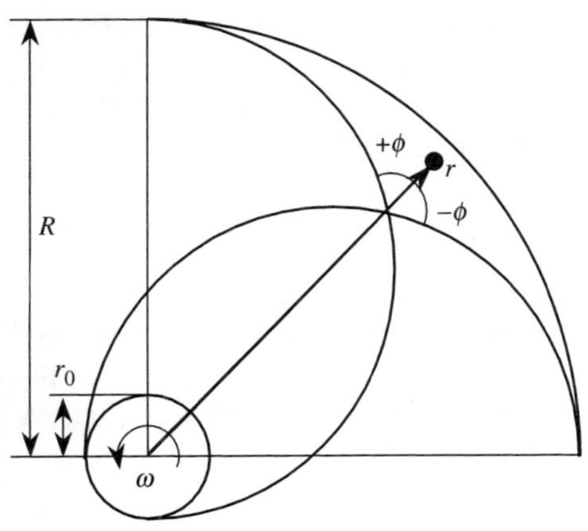

Fig. 8.12. Fibers' trajectories in the spinning disk.

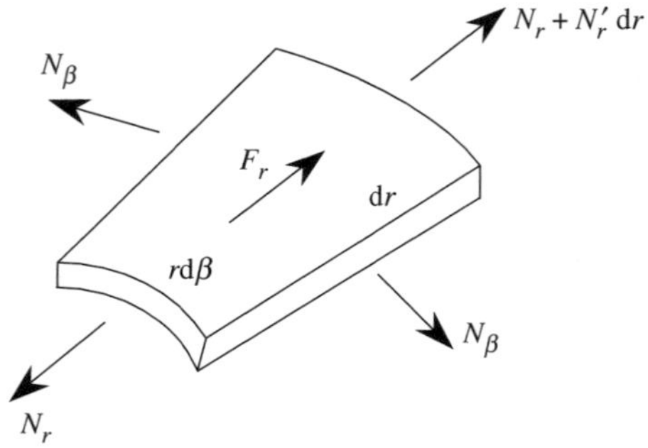

Fig. 8.13. Forces acting on the disk element.

in these equations, we get

$$N_r = h\sigma_1 \cos^2\phi, \quad N_\beta = h\sigma_1 \sin^2\phi \tag{8.71}$$

Consider a disk element shown in Fig. 8.13. The equilibrium condition yields

$$(rN_r)' - N_\beta + F_r = 0 \tag{8.72}$$

where $(\ )' = d(\ )/dr$ and $F_r = \rho h\omega^2 r^2$ with ρ being the material density. The disk thickness is specified by Eq. (8.57), i.e.,

$$h(r) = \frac{kw_0\delta}{2\pi r \cos\phi} \tag{8.73}$$

where k is the number of fibrous tapes passing through the circumference $r =$ constant, and w_0 and δ are the tape width and thickness.

For a disk of uniform strength, we take $\sigma_1 = \overline{\sigma}_1$ in which $\overline{\sigma}_1$ is the ultimate stress for the unidirectional composite under tension along the fibers. Correspondingly, we take $\omega = \overline{\omega}$, where $\overline{\omega}$ is the ultimate angular velocity of the disk. Then, substituting Eq. (8.73) into Eqs. (8.71) and the obtained stress resultants N_r and N_β into Eq. (8.72), we arrive at the following equation for the fibers' angle

$$r\phi' \sin\phi\cos\phi + \sin^2\phi = \frac{1}{\overline{\sigma}_1}\rho\overline{\omega}^2 r^2 \tag{8.74}$$

The solution of this equation must satisfy the boundary conditions. For a disk with radius R and with a central opening of radius r_0 as in Fig. 8.12, we must have $N_r = 0$ at $r = r_0$ and $r = R$. Taking into account the first expression in Eqs. (8.71), we arrive at the following boundary conditions for Eq. (8.74)

$$\phi(r = r_0) = \frac{\pi}{2}, \quad \phi(r = R) = \frac{\pi}{2} \tag{8.75}$$

Since Eq. (8.75) is of the first order, its solution can, in general, satisfy only one of these conditions. Using the second condition in Eqs. (8.75), we get

$$\sin\phi = \frac{R}{r}\sqrt{1 - \frac{\lambda}{2}\left(1 - \frac{r^4}{R^4}\right)} \tag{8.76}$$

where

$$\lambda = \frac{\rho\overline{\omega}^2 R^2}{\overline{\sigma}_1} \tag{8.77}$$

Applying the first condition in Eqs. (8.75), we arrive at the following equation specifying the parameter λ

$$\lambda = \frac{2}{1 + \left(\frac{r_0}{R}\right)^4} \tag{8.78}$$

In conjunction with Eq. (8.77), this result enables us to determine the maximum value of the disk angular velocity, i.e.,

$$\overline{\omega}^2 = \frac{2\overline{\sigma}_1}{\rho R^2\left(1 + \overline{r}_0^4\right)} \tag{8.79}$$

where $\overline{r}_0 = r_0/R$. It follows from Eq. (8.79) that the maximum value of the ultimate angular velocity corresponds to $\overline{r}_0 = 0$, i.e., to a disk without a central opening, for which Eq. (8.79) reduces to

$$\overline{\omega}_m^2 = \frac{2\overline{\sigma}_1}{\rho R^2} \tag{8.80}$$

Note that relatively small central openings have practically no effect on the ultimate angular velocity. For example, for $\overline{r}_0 = 0.1$, Eq. (8.79) gives $\overline{\omega}$ which is only 0.005% less than the maximum value $\overline{\omega}_m$ following from Eq. (8.80).

For further analysis, we take $\overline{r}_0 = 0$ and consider disks without a central opening. Then, Eq. (8.78) yields $\lambda = 2$, and Eq. (8.76) for the fiber angle becomes

$$\sin\phi = \frac{r}{R} \tag{8.81}$$

To find the fiber trajectory, consider Fig. 8.14 showing the tape element in Cartesian x, y and polar r, β coordinate frames. As follows from the figure,

$$x = r\sin\beta, \quad y = r\cos\beta, \quad \tan\phi = \frac{r\,\mathrm{d}\beta}{\mathrm{d}r} \tag{8.82}$$

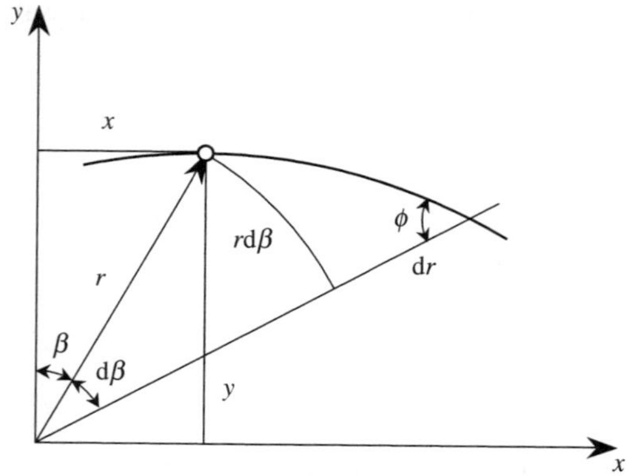

Fig. 8.14. A tape element in Cartesian and polar coordinate frames.

Applying the last equation of Eqs. (8.82) and using Eq. (8.81) for ϕ, we arrive at the following differential equation

$$\frac{\mathrm{d}\beta}{\mathrm{d}r} = \frac{1}{\sqrt{R^2 - r^2}}$$

whose general solution is

$$\sin(\beta - \beta_0) = \frac{r}{R} \tag{8.83}$$

in which β_0 is the constant of integration. Changing β and r to x and y with the aid of Eqs. (8.82), we can write Eq. (8.83) in Cartesian coordinates as

$$\left(x - \frac{R}{2}\cos\beta_0\right)^2 + \left(y - \frac{R}{2}\sin\beta_0\right)^2 = \frac{R^2}{4}$$

For each value of β_0, this equation specifies a circle with radius $R/2$ and center located on the circumference $r = R/2$. Changing β_0, we get the system of circles shown in Fig. 8.15.

Composite disks can be efficiently used as inertial accumulators of mechanical energy – flywheels such as that shown in Fig. 8.16. Note that the disk in Fig. 8.16 can be made using the technology described in Section 4.5.2. The disk composite structure is made by winding onto an inflated elastic mandrel similar to that shown in Fig. 8.10. After the shell with the appropriate winding patterns is fabricated, the pressure is continuously reduced and the shell is compressed in the axial direction between two plates. Once the shell is transformed into a disk, the resin in the composite material is cured.

The maximum kinetic energy that can be stored and the mass of the disk are

$$E = \pi\bar{\omega}^2\rho\int_0^R hr^3\mathrm{d}r, \quad M = 2\pi\rho\int_0^R hr\mathrm{d}r$$

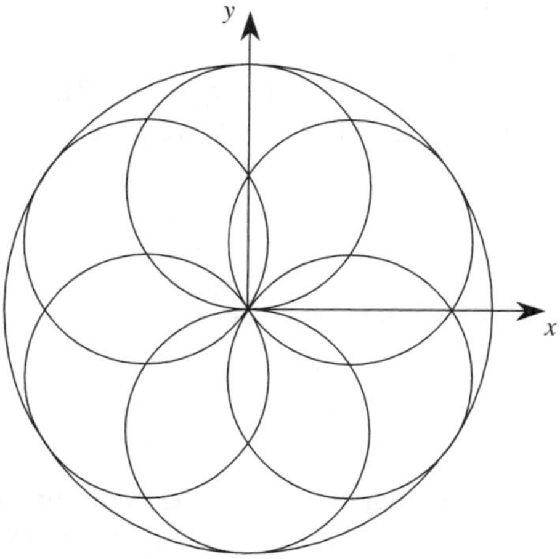

Fig. 8.15. Fiber patterns in the spinning optimal composite disk.

Fig. 8.16. Carbon–epoxy flywheel.

Substituting h in accordance with Eq. (8.73) and using Eq. (8.81) for the winding patterns, we get

$$E = \frac{\pi}{8} R^3 \rho \overline{\omega}^2 k w_0 \delta, \quad M = \frac{\pi}{2} R\rho\, k w_0 \delta \tag{8.84}$$

where k, w_0, and δ are specified in notations to Eq. (8.73). Transforming Eq. (8.84) for E with the aid of the corresponding equation for M, we have

$$E = \frac{1}{4} M R^2 \overline{\omega}^2$$

Substituting $\overline{\omega}$ from Eq. (8.80), we finally arrive at

$$E = \frac{M\overline{\sigma}_1}{2\rho}$$

Introducing the linear circumferential velocity at the outer circumference $r = R$ of the disk as $\overline{v}_R = \overline{\omega} R$ and using Eq. (8.80) for $\overline{\omega}$, we obtain the following result

$$\overline{v}_R = \sqrt{\frac{2\overline{\sigma}_1}{\rho}} \tag{8.85}$$

which means that $\overline{v}_R$ depends only on the material longitudinal specific strength $(\overline{\sigma}_1/\rho)$. The results of calculations for the composites whose properties are listed in Table 3.5 are presented in Table 8.4. Note that in order to use Eq. (8.85) for $\overline{v}_R$ we must substitute $\overline{\sigma}_1$ in N/m^2, ρ in kg/m^3 and take into account that $1\,N = 1\,kg \cdot m/s^2$.

8.3.3. Anisogrid composite lattice structures

Anisogrid (anisotropic grid) composite lattice structures (Vasiliev et al., 2001; Vasiliev and Razin, 2006) are usually made in the form of a cylindrical shell consisting of helical and circumferential (hoop) unidirectional composite ribs formed by continuous winding shown in Fig. 8.17 (see also Section 4.7). In the process of winding, glass, carbon, or aramid tows impregnated with epoxy resin are placed into the grooves formed in the elastic coating that covers the surface of the mandrel (see Fig. 8.17).

Table 8.4
Maximum values of the circumferential velocities for fibrous composite disks of uniform strength.

Composite material	Glass–epoxy	Carbon–epoxy	Aramid–epoxy
$\overline{v}_R$ (m/s)	1309	1606	1946

Fig. 8.17. Winding of a composite lattice structure.

Fig. 8.18. Removal of elastic coating.

After curing, elastic coating is pulled out of the structure as shown in Fig. 8.18.

Cylindrical anisogrid lattice structures with given diameter D and length L are characterized with six design variables, i.e., (Fig. 8.19)

- the shell thickness (the height of the rib cross section), h,
- the angle of helical ribs with respect to the shell meridian, ϕ,

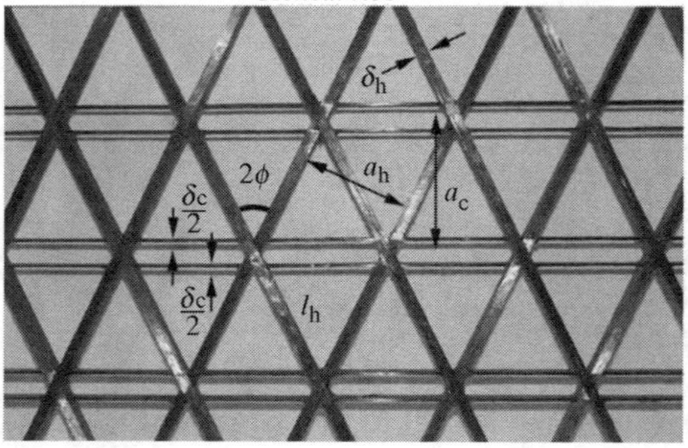

Fig. 8.19. Lattice structure.

- the widths of the helical and the circumferential (hoop) ribs cross sections, δ_h and δ_c (for the structure in Fig. 8.19, δ_c is the total width of the adjacent hoop ribs),
- the spacings of the helical and the hoop ribs, a_h and a_c, taken along the normal elements to the axes.

The ribs are the principal load-bearing elements of the structure, whereas the skin, the presence of which can be justified by design requirements, is not considered as a load-bearing element in the design of lattice structures. Moreover, the skin thickness, being treated as a design variable, degenerates in the process of optimization because the skin contribution to the mass of the structure is higher than that to the structural strength and stiffness. Thus, the optimal lattice structure design does not require a skin. If the actual structure needs a skin, its thickness and composition are pre-assigned to meet the operational and manufacturing requirements.

High performance and weight efficiency of composite lattice structures are provided by unidirectionally reinforced ribs that have a high strength and stiffness. In comparison with the known isogrid structures (Rehfield et al., 1980), consisting of helical and circumferential ribs forming equilateral triangles and having the same cross-sectional dimensions, the anisogrid structures under consideration provide additional mass savings because the thicknesses of the helical and circumferential ribs are different and are found, as well as the angle of the helical ribs, in the process of optimal design.

Anisogrid carbon–epoxy lattice structures are used as interstage sections of space launchers (Bakhvalov et al., 2005) and are normally designed for axial compression as the main loading case.

Consider the design of a cylindrical lattice shell with given diameter, D, and length, L. The shell is loaded with axial compressive force, P (see Fig. 5.33). For the shell referred to axial coordinate x and circumferential coordinate y, the constitutive equations, (Eqs. (5.5)), can be presented as

$$
\begin{aligned}
N_x &= B_{11}\varepsilon_x^0 + B_{12}\varepsilon_y^0, \quad N_y = B_{21}\varepsilon_x^0 + B_{22}\varepsilon_y^0, \quad N_{xy} = B_{44}\gamma_{xy}^0 \\
M_x &= D_{11}\kappa_x + D_{12}\kappa_y, \quad M_y = D_{21}\kappa_x + D_{22}\kappa_y, \quad M_{xy} = D_{44}\kappa_{xy}
\end{aligned}
\tag{8.86}
$$

in which, in accordance with Eqs. (4.189), and (5.36) for the stiffness coefficients,

$$B_{11} = 2E_h h\overline{\delta}_h c^4, \quad B_{12} = B_{21} = B_{44} = 2E_h h\overline{\delta}_h c^2 s^2$$
$$B_{22} = 2E_h h\overline{\delta}_h s^4 + E_c h\overline{\delta}_c, \quad D_{mn} = B_{mn} h^2/12 \tag{8.87}$$

Here, subscripts 'h' and 'c' correspond to the helical and circumferential ribs and

$$\overline{\delta}_h = \frac{\delta_h}{a_h}, \quad \overline{\delta}_c = \frac{\delta_c}{a_c}, \quad c = \cos\phi, \quad s = \sin\phi \tag{8.88}$$

The mass of the structure consists of the mass of helical and circumferential ribs, i.e.,

$$M = M_h + M_c, \quad M_h = n_h L_h h\delta_h \rho_h, \quad M_c = n_c L_c h\delta_c \rho_c$$

Here (see Fig. 8.19),

$$n_h = \frac{2\pi}{a_h} Dc, \quad n_c = \frac{L}{a_c} \tag{8.89}$$

are the numbers of helical and circumferential ribs in the shell with diameter D and length L,

$$L_h = \frac{L}{c}, \quad L_c = \pi D$$

are the lengths of the ribs. Finally, we get

$$M = \pi DLh\rho_h(2\overline{\delta}_h + \overline{\rho}\overline{\delta}_c) \tag{8.90}$$

where $\overline{\rho} = \rho_c/\rho_h$ and ρ_c, ρ_h are the densities of the circumferential and helical ribs.

To design the lattice structure, we must find the structural parameters, i.e., h, ϕ, δ_h, δ_c, a_h, a_c delivering the minimum value of the mass of the structure in Eq. (8.90) and satisfying the set of constraints providing

- sufficient strength in the helical ribs under compression,
- local stability of the segments of helical ribs between the points of intersection (nodal points),
- global stability of the lattice shell under axial compression.

Consider the foregoing constraints. The stresses σ_h acting in helical ribs which take an axial compressive force P can be found using the free-body diagram, i.e.,

$$P = n_h \sigma_h h\delta_h \cos\phi$$

Using Eq. (8.89) for n_h and notations in Eqs. (8.88), we get

$$\sigma_h = \frac{P}{2\pi Dh\overline{\delta}_h c^2} \tag{8.91}$$

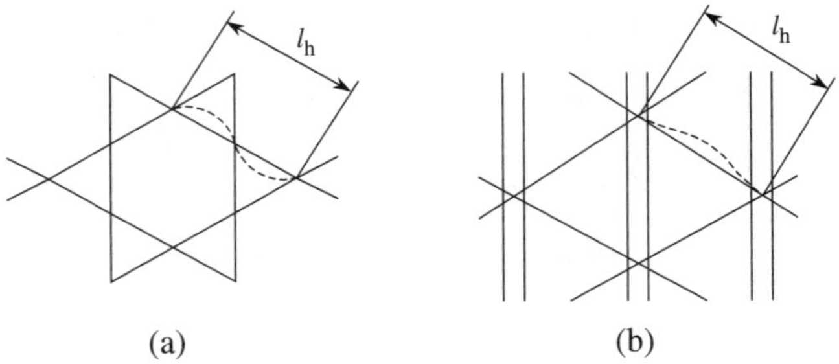

Fig. 8.20. Local buckling of helical ribs.

Thus, the strength constraint $\sigma_h \leq \overline{\sigma}$, in which $\overline{\sigma}$ is the strength of helical rib under compression, can be written as

$$P \leq 2\pi D h \overline{\delta}_c c^2 \overline{\sigma} \tag{8.92}$$

Helical ribs can experience local buckling under compression which shows itself as a local bending of the rib segments between the nodal points. For typical lattice structures, local buckling is presented in Fig. 8.20. The critical stress causing local buckling is determined by the Euler formula (see e.g., Vasiliev, 1993), i.e.,

$$\sigma_{cr} = k \frac{\pi^2 E_h I_h}{l_h^2 A_h} \tag{8.93}$$

The thickness of lattice structures h is usually greater than the width of the helical ribs δ_c, and local buckling occurs in the plane of the lattice structure. In this case, the moment of inertia and the area of the rib cross section are

$$I_h = \frac{h}{12}\delta_h^3, \quad A_h = h\delta_h \tag{8.94}$$

The length of the rib segment l_h shown in Figs. 8.19 and 8.20 can be expressed in terms of the design variables as

$$l_h = \frac{a_h}{\sin 2\phi} \tag{8.95}$$

and the parameter k in Eq. (8.93) is a coefficient depending on the boundary conditions. In general, k depends on the mutual location and bending stiffnesses of the helical and circumferential ribs. For both the structures in Fig. 8.20, design can be performed taking approximately $k = 4$. The local buckling constraint has the form $\sigma_h \leq \sigma_{cr}$ in which σ_h is the stress in the helical ribs specified by Eq. (8.91). Using Eqs. (8.94) and (8.95),

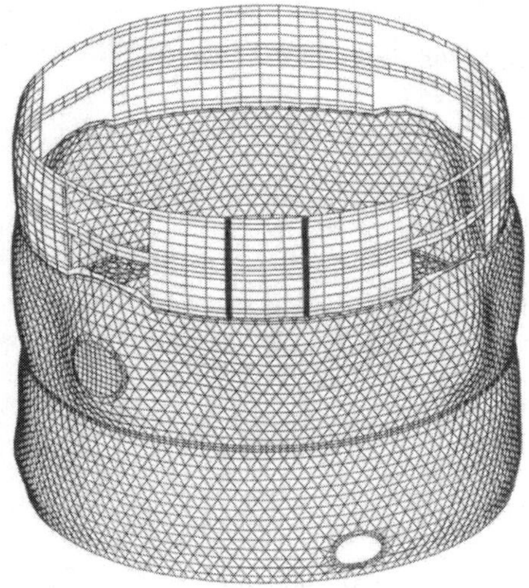

Fig. 8.21. Nonsymmetrical buckling of a lattice interstage structure.

we finally obtain

$$P \leq \frac{2}{3}\pi^3 k E_{\mathrm{h}} D h \bar{\delta}_{\mathrm{h}}^3 s^2 c^4 \tag{8.96}$$

Consider the global (shell-type) buckling constraint. Under axial compression, lattice cylindrical shells experience two modes of buckling: axisymmetric and nonsymmetric. The axisymmetric buckling results in shell bending accompanied by longitudinal waves. Computer simulation of nonsymmetric buckling of the Proton-M launcher's anisogrid interstage section whose winding is shown in Fig. 8.17 (Bakhvalov et al., 2005) is presented in Fig. 8.21 (the upper part of the structure is a dummy metal Proton-M section).

Taking into account that the bending stiffness for lattice structures D_{mn} can be expressed in terms of the membrane stiffnesses B_{mn} with the aid of the last expression in Eqs. (8.87), we can arrive at the following equations for the critical axial forces corresponding to the axisymmetric, $P_{\mathrm{cr}}^{\mathrm{a}}$, and nonsymmetric, $P_{\mathrm{cr}}^{\mathrm{n}}$, modes of buckling (Vasiliev and Razin, 2001)

$$P_{\mathrm{cr}}^{\mathrm{a}} = \frac{2\pi}{\sqrt{3}} h \sqrt{B_{11}B_{22} - B_{12}^2}, \quad P_{\mathrm{cr}}^{\mathrm{n}} = \frac{2\pi\sqrt{2}}{\sqrt{3}} h \sqrt{B_{44}\left(B_{12} + \sqrt{B_{11}B_{22}}\right)}$$

Substituting stiffnesses B_{mn} in accordance with Eqs. (8.87), we get

$$P_{\mathrm{cr}}^{\mathrm{a}} = \frac{2\pi\sqrt{2}}{\sqrt{3}} h^2 c^2 \sqrt{E_{\mathrm{h}} E_{\mathrm{c}} \bar{\delta}_{\mathrm{h}} \bar{\delta}_{\mathrm{c}}} \tag{8.97}$$

$$P_{\mathrm{cr}}^{\mathrm{n}} = \frac{4\pi\sqrt{2}}{\sqrt{3}} E_{\mathrm{h}} h^2 \bar{\delta}_{\mathrm{h}} c^2 s^2 \sqrt{1 + \sqrt{1 + \frac{E_{\mathrm{c}}\bar{\delta}_{\mathrm{c}}}{2E_{\mathrm{h}}\bar{\delta}_{\mathrm{h}} s^4}}} \tag{8.98}$$

Apparently, the lowest value of P_{cr} specified by Eqs. (8.97) and (8.98) should be selected for the design. Using Eqs. (8.97) and (8.98), we arrive at the following condition:

$$P_{cr}^{a} \leq P_{cr}^{n} \quad \text{if} \quad s \leq \frac{1}{2}\sqrt[4]{\frac{E_c\overline{\delta}_c}{E_h\overline{\delta}_h}} \tag{8.99}$$

As shown below, buckling of typical lattice structures corresponds to the axisymmetric mode because P_{cr}^{a} is usually less than P_{cr}^{n}. So, the global buckling constraint $P \leq P_{cr}$ can be taken in the form

$$P \leq \frac{2\pi\sqrt{2}}{\sqrt{3}}h^2c^2\sqrt{E_hE_c\overline{\delta}_h\overline{\delta}_c} \tag{8.100}$$

Thus, we should minimize the mass of the structure M in Eq. (8.90) subject to the constraints in Eqs. (8.92), (8.96), and (8.100). As can be seen, the structure under consideration is specified with four design variables: h, ϕ, $\overline{\delta}_h$, and $\overline{\delta}_c$. The main problem we face is associated with the fact that the constraints in Eqs. (8.92), (8.96), and (8.100) are written in the form of inequalities. To convert them to equalities, let us introduce safety factors $n \geq 1$ for all the modes of failure, i.e., for fracture of the helical ribs under compression, n_s, global buckling, n_b, and local buckling, n_l. As a result the constraints can be re-written in the following form

$$\frac{2\pi}{Pn_s}D\overline{\sigma}\,h\overline{\delta}_hc^2 = 1 \tag{8.101}$$

$$\frac{2\pi\sqrt{2}}{\sqrt{3}Pn_b}h^2c^2\sqrt{E_hE_c\overline{\delta}_h\overline{\delta}_c} = 1 \tag{8.102}$$

$$\frac{2\pi^3k}{3Pn_l}E_hDh\overline{\delta}_h^3c^4s^2 = 1 \tag{8.103}$$

We should also take into account that

$$c^2 + s^2 = 1 \tag{8.104}$$

The idea of the method of minimization of safety factors (Vasiliev and Razin, 2001) is implemented as follows. We apply Eqs. (8.101)–(8.104) to express the design variables in terms of the safety factors and minimize the mass of the structure with respect to both sets of unknown variables – the design parameters and the safety factors. If one of the parameters n is equal to unity, the corresponding constraint is active. If $n = n_0 > 1$, the constraint is satisfied with some additional safety factor n_0. For active constraints, function $M(n)$ has positive gradients, so to optimize the structure, we should take $n = 1$, and no further optimization is required. So, following this procedure, we should express h, ϕ, $\overline{\delta}_h$, and $\overline{\delta}_c$ in terms of the safety factors n_s, n_b, and n_l. Using Eq. (8.101), we get

$$c^2 = \frac{Pn_s}{2\pi D\overline{\sigma}\, h\overline{\delta}_h} \tag{8.105}$$

Substitution of this result into Eqs. (8.102) and (8.103) yields

$$\frac{\overline{\delta}_c}{\overline{\delta}_h} = \frac{3D^2 n_b^2 \overline{\sigma}^2}{2n_s^2 E_h E_c h^2} \tag{8.106}$$

$$s^2 = \frac{6n_l D\overline{\sigma}^2 h}{\pi k P n_s^2 E_h \delta_h} \tag{8.107}$$

Substituting further Eqs. (8.105) and (8.107) into Eq. (8.104), we obtain

$$\overline{\delta}_h = \frac{6n_l \overline{\sigma}^2 Dh}{\pi k P n_s^2 E_h} + \frac{Pn_s}{2\pi D\overline{\sigma} h} \tag{8.108}$$

Now, Eqs. (8.106) and (8.108) enable us to express the mass of the structure, Eq. (8.90) in terms of only one design variable – the shell thickness h, i.e.,

$$M = L\rho_c \left(\frac{12D^2 n_l \overline{\sigma}^2 h^2}{Pn_s^2 k E_h} + \frac{3Pn_b^2 D^2 \overline{\sigma}\,\overline{\rho}}{4n_s E_h E_c h^2} + \frac{9D^4 \overline{\sigma}^4 n_l n_b \overline{\rho}}{Pn_s^4 k E_h^2 E_c} + \frac{Pn_s}{\overline{\sigma}} \right) \tag{8.109}$$

Applying the condition $\partial M/\partial h = 0$, we have

$$h^4 = \frac{P^2 n_b^2 n_s \overline{\rho}}{16 n_l E_c \overline{\sigma}} \tag{8.110}$$

Substituting this result into Eq. (8.109), we arrive at

$$M = L\rho_c \left(\frac{9D^4 \overline{\sigma}^4 n_l n_b^2 \overline{\rho}}{Pn_s^4 k E_h} + \frac{6D^2 \overline{\sigma}\, n_b}{n_s E_h} \sqrt{\frac{n_l \overline{\sigma}\,\overline{\rho}}{k n_s E_c}} + \frac{Pn_s}{\overline{\sigma}} \right) \tag{8.111}$$

It follows from this equation that the mass of the structure M increases with an increase in the buckling safety factors n_b and n_l, and to minimize the mass, we must take the minimum allowable values of these factors, i.e., $n_b = 1$ and $n_l = 1$. This means that the buckling constraints in Eqs. (8.102) and (8.103) are active. To find the strength safety factor n_s, we need to put $\partial M/\partial n_s = 0$, where M is defined by Eq. (8.111). As a result, we have

$$n_s = \overline{\sigma}\left(\frac{144D^4\overline{\rho}}{P^2kE_h^2E_c}\right)^{1/5} \tag{8.112}$$

Taking into account that $n_s \geq 1$, then equation Eq. (8.112) yields

$$P \leq P_s = \frac{12D^2\overline{\sigma}^2}{E_h}\sqrt{\frac{\overline{\sigma}\,\overline{\rho}}{kE_c}} \tag{8.113}$$

So, we have two design cases. For $P < P_s$, we have $n_s > 1$, and the strength constraint, Eq. (8.101) is not active. There exists some safety factor for this mode of failure specified by Eq. (8.112). For $P > P_s$, we have $n_s = 1$, and the strength constraint becomes active, so all three constraints are active in this case.

To study these two cases, introduce the following mass and force parameters

$$m = \frac{4M}{\pi D^2L}, \quad p = \frac{4P}{\pi D^2} \tag{8.114}$$

Then, Eq. 8.113 gives

$$p_s = \frac{4P_s}{\pi D^2} = \frac{48\overline{\sigma}^2}{\pi E_h}\sqrt{\frac{\overline{\sigma}\,\overline{\rho}}{kE_c}} \tag{8.115}$$

Consider the case $p \leq p_s$. Substituting n_s specified by Eq. (8.112) into Eqs. (8.105), (8.110), (8.111) and using Eq. (8.114), we arrive at the following equations for the parameters of the optimal structure

$$\begin{aligned}
\overline{h} &= \frac{h}{D} = \frac{1}{4}\left(\frac{48\pi^4k^2\overline{\rho}^3}{E_hE_c^3}p^4\right)^{1/10} \\
\tan\phi &= \frac{1}{2}, \quad \phi = 26.565^\circ \\
\overline{\delta}_h &= \frac{5}{4\pi}\left(\frac{108\pi^2E_c}{k^4E_h^3\overline{\rho}}p^2\right)^{1/10} \\
\overline{\delta}_c &= \frac{\overline{\delta}_c}{2\overline{\rho}} \\
m &= \frac{25\rho_h}{8}\left(\frac{72\overline{\rho}\,p^3}{\pi^2kE_h^2E_c}\right)^{1/5}
\end{aligned} \tag{8.116}$$

Consider the case $p \geq p_s$, repeating the derivation of Eqs. (8.116) and taking $n_s = 1$, we have

$$
\begin{aligned}
\overline{h} &= \frac{h}{D} = \left(\frac{\pi^2 k\overline{\rho}}{E_c\overline{\sigma}}p^2\right)^{1/4} \\
\tan^2\phi &= \frac{p_s}{4p} \\
\overline{\delta}_h &= \frac{2}{\pi\sin 2\phi}\sqrt{\frac{3\overline{\sigma}}{kE_h}} \\
\overline{\delta}_c &= \frac{p_s\overline{\delta}_h}{2\overline{\rho}\,p} \\
m &= \frac{p\rho_h}{\overline{\sigma}}\left(1+\frac{p_s}{4p}\right)^2
\end{aligned}
\tag{8.117}
$$

For $p = p_s$ Eqs. (8.116) and (8.117) yield the same results. Note that these equations are universal ones, i.e., they do not include the structural dimensions.

The Eqs. (8.116) and (8.117) are valid subject to the conditions in Eqs. (8.99). Substituting the parameters following from Eqs. (8.117) in the second of these conditions, we can conclude that the axisymmetric mode of shell buckling exists if

$$
p \leq p_0 = p_s\sqrt{\frac{1}{2}\left(\sqrt{\frac{2E_h\overline{\rho}}{E_c}} + \sqrt{\frac{2E_h\overline{\rho}}{E_c}} - 1\right)} \tag{8.118}
$$

Analysis of this result confirms that the calculated value of p_0 corresponds to an axial force that is much higher than the typical loads for existing aerospace structures. So, the nonsymmetric mode of buckling does not occur for typical lattice structures.

As an example, consider an interstage section of a space launcher with $D = 4\,\text{m}$ designed to withstand an axial force $P = 15\,\text{MN}$. The ribs are made from carbon–epoxy composite with the following properties: $E_h = E_c = 90\,\text{GPa}$, $\overline{\sigma} = 450\,\text{MPa}$, $\rho_h = \rho_c = 1450\,\text{kg/m}^3$. Taking $k = 4$ and calculating p, p_s, and p_0 using Eqs. (8.114), (8.115), and (8.118), we get $p = 1.2\,\text{MPa}$, $p_s = 1.45\,\text{MPa}$, $p_0 = 1.6\,\text{MPa}$. As can be seen, $p < p_s$ and the optimal parameters of the structure are specified by Eqs. (8.116) which give the following results

$$
\begin{aligned}
&\overline{h} = 0.009, \quad \phi = 26.565^\circ, \quad \overline{\delta}_h = 0.05 \\
&\overline{\delta}_c = 0.025, \quad m = 6.52\,\text{kg/m}^3
\end{aligned}
$$

Consider a design in which there are 120 helical ribs in the shell cross section and that the lattice structure corresponds to that in Fig. 8.20b. In this case, the calculation yields

$a_h = 188\,\text{mm}$ and $a_c = 210\,\text{mm}$. For a structure with $D = 4\,\text{m}$, we have $h = 36$ mm, $\delta_h = 9.4\,\text{mm}$, $\delta_c = 2.35\,\text{mm}$. The mass of the unit surface is $6.52\,\text{kg/m}^2$. To confirm the high weight efficiency of this lattice structure, note that the composite section with this mass corresponds to a smooth or stringer stiffened aluminum shell with the efficient thickness $h = 2.4\,\text{mm}$. The axial stress induced in this shell by an axial force $P = 15\,\text{MN}$ is about 500 MPa, which is higher than the yield stress of typical aluminum alloys.

8.4. References

Bakhvalov, Yu.O., Molochev, V.P., Petrovskii, S.A., Barynin, V.A., Vasiliev, V.V. and Razin, A.F. (2005). Proton-M composite interstage structures: design, manufacturing and performance. In *Proc. European Conf. Aerospace Sci.*, July 4–7, 2005, Moscow, CD-ROM.

Kyser, A.C. (1965). Uniform-stress spinning filamentary disk. *AIAA Journal*. July, 1313–1316.

Obraztsov, I.F. and Vasiliev, V.V. (1989). Optimal design of composite structures. In *Handbook of Composites: Vol. 2, Structure and Design* (C.T. Herakovich and Yu.M. Tarnopol'skii eds.). Elsevier, Amsterdam, pp. 3–84.

Rehfield, L.W., Deo, R.B. and Renieri, G.D. (1980). Continuous filament advanced composite isogrid: a promising design concept. In *Fibrous Composites in Structural Design* (E.M. Lenoe, D.W. Oplinger and J.L. Burke, eds.). Plenum Publishing Corp., New York, pp. 215–239.

Vasiliev, V.V. (1993). *Mechanics of Composite Structures*. Taylor & Francis, Washington.

Vasiliev, V.V., Barynin, V.A. and Razin, A.F. (2001). Anisogrid lattice structures – survey of development and application. *Composite Struct.* **54**, 361–370.

Vasiliev, V.V. and Razin, A.F. (2001). Optimal design of filament-wound anisogrid composite lattice structures. In *Proc. 16th Annual Tech. Conf. American Society for Composites*, September 9–12, 2001, Blacksburg, VA, USA. (CD-ROM).

Vasiliev, V.V. and Razin, A.F. (2006). Anisogrid composite lattice structures for spacecraft and aircraft applications. *Composite Struct.* **76**, 182–189.

AUTHOR INDEX

[Plain numbers refer to text pages on which the author (or his/her work) is cited. Boldface numbers refer to the pages where bibliographic references are cited.]

SUBJECT INDEX